# BIOGEOGRAPHY, TIME, AND PLACE: DISTRIBUTIONS, BARRIERS, AND ISLANDS

## TOPICS IN GEOBIOLOGY

### Series Editors:

**Neil H. Landman,** American Museum of Natural History, New York, landman@amnh.org
**Douglas S. Jones,** University of Florida, Gainesville, Florida, dsjones@flmnh.ufl.edu

---

*Current volumes in this series*

Volume 29: **Biogeography, Time, and Place: Distributions, Barriers, and Islands**
Willem Renema
Hardbound, ISBN 978-1-4020-6373-2, 2007

Volume 28: **Paleopalynology: Second Edition**
Alfred Traverse
Hardbound, ISBN 978-1-4020-5609-5, 2007

Volume 27: **Neoproterozoic Geobiology and Paleobiology**
Shuhai Xiao and Alan J. Kaufman
Hardbound, ISBN 978-1-4020-5201-9, October 2006

Volume 26: **First Floridians and Last Mastodons: The Page-Ladson Site in the Aucilla River**
S. David Webb
Hardbound, ISBN 978-1-4020-4325-3, October 2006

Volume 25: **Carbon in the Geobiosphere – Earth's Outer Shell –**
Fred T. Mackenzie and Abraham Lerman
Hardbound, ISBN 978-1-4020-4044-3, September 2006

Volume 24: **Studies on Mexican Paleontology**
Francisco J. Vega, Torrey G. Nyborg, María del Carmen Perrilliat, Marison Montellano-Ballesteros, Sergio R. S. Cevallos-Ferriz and Sara A. Quiroz-Barroso
Hardbound, ISBN 978-1-4020-3882-2, February 2006

Volume 23: **Applied Stratigraphy**
Eduardo A. M. Koutsoukos
Hardbound, ISBN 978-1-4020-2632-4, January 2005

Volume 22: **The Geobiology and Ecology of *Metasequoia***
Ben A. LePage, Christopher J. Williams and Hong Yang
Hardbound, ISBN 978-1-4020-2631-7, March 2005

Volume 21: **High-Resolution Approaches in Stratigraphic Paleontology**
Peter J. Harries
Hardbound, ISBN 978-1-4020-1443-7, September 2003

Volume 20: **Predator–Prey Interactions in the Fossil Record**
Patricia H. Kelley, Michal Kowalewski, Thor A. Hansen
Hardbound, ISBN 978-0-306-47489-7, January 2003

Volume 19: **Fossils, Phylogeny, and Form**
Jonathan M. Adrain, Gregory D. Edgecombe, Bruce S. Lieberman
Hardbound, ISBN 978-0-306-46721-9, January 2002

Volume 18: **Eocene Biodiversity**
Gregg F. Gunnell
Hardbound, ISBN 978-0-306-46528-4, September 2001

For detailed information on our books and series please visit: www.springer.com

# Biogeography, Time, and Place: Distributions, Barriers, and Islands

Edited by

WILLEM RENEMA
*Nationaal Natuurhistorisch Museum Naturalis*
*Leiden, The Netherlands*

Springer

A C.I.P. Catalogue record for this book is available from the Library of Congress

ISBN 978-1-4020-6373-2
ISBN 978-1-4020-6374-9

Published by Springer
P.O. Box 17, 3300 AA Dordrecht. The Netherlands

*www.springer.com*

*Cover illustrations:*
Central photo: Position of the Australian continent during the Oligocene (34-24 Ma). Drawing by Willem Renema after Hall, 1998

Top row photos:
Opisthostoma species shell shapes that are now known from Borneo. Drawings by Jaap J. Vermeulen, 1994

*Printed on acid-free paper.*

# Aims & Scope Topics in Geobiology Book Series

**Topics in Geobiology** series treats geobiology – the broad discipline that covers the history of life on Earth. The series aims for high-quality, scholarly volumes of original research as well as broad reviews. Recent volumes have showcased a variety of organisms including cephalopods, corals, and rodents. They discuss the biology of these organisms – their ecology, phylogeny, and mode of life – and in addition, their fossil record, i.e., their distribution in time and space.

Other volumes are more theme-based such as predator–prey relationships, skeletal mineralization, paleobiogeography, and approaches to high-resolution stratigraphy, and cover a broad range of organisms. One theme that is at the heart of the series is the interplay between the history of life and the changing environment. This is treated in skeletal mineralization and how such skeletons record environmental signals and animal–sediment relationships in the marine environment.

The series editors also welcome any comments or suggestions for future volumes.

Series Editors:
Douglas S. Jones dsjones@flmnh.ufl.edu
Neil H. Landman landman@amnh.org

# Contents

# Contributors

**Cees van Achterberg**
Nationaal Natuurhistorisch Museum
Naturalis
P.O. Box 9517 2300
RA Leiden, The Netherlands
achterberg@naturalis.nnm.nl

**J.W. Arntzen**
Nationaal Natuurhistorisch Museum
Naturalis
P.O. Box 9517
2300 RA Leiden, The Netherlands
arntzen@nnm.nl

**Gert D. van den Bergh**
Nationaal Natuurhistorisch Museum
Naturalis
P.O. Box 9517
2300 RA Leiden, The Netherlands
Vos@naturalis.nl

**René W.R.J. Dekker**
Nationaal Natuurhistorisch Museum
Naturalis
P.O. Box 9517
2300 RA Leiden, The Netherlands
dekker@naturalis.nl

**Stephen K. Donovan**
Nationaal Natuurhistorisch Museum
Naturalis
P.O. Box 9517
2300 RA Leiden, The Netherlands
Donovan@naturalis.nl

**Daryl P. Domning**
Department of Anatomy, Howard
University
520 W Street NW, Washington, DC
20059
ddomning@howard.edu

**Charles H.J.M. Fransen**
Nationaal Natuurhistorisch Museum
Naturalis
P.O. Box 9517
2300 RA Leiden, The Netherlands
Fransen@naturalis.nl

**Dirk Gassmann**
Institute of Biology, University of Leiden
c/o Nationaal Natuurhistorisch Museum
Naturalis
P.O. Box 9517
2300 RA Leiden, The Netherlands
gassmann@rulsfb.leidenuniv.nl

**E. Gittenberger**
Nationaal Natuurhistorisch Museum.
Naturalis
P.O. Box 9517
2300 RA Leiden, The Netherlands
Gittenberger@naturalis.nnm.nl

**Lars W. van den Hoek Ostende**
Nationaal Natuurhistorisch Museum
Naturalis
P.O. Box 9517
2300 RA Leiden, The Netherlands
hoek@naturalis.nl

**Bert W. Hoeksema**
Nationaal Natuurhistorisch Museum
Naturalis
P.O. Box 9517
2300 RA Leiden, The Netherlands
Hoeksema@naturalis.nl

**Rienk de Jong**
Nationaal Natuurhistorisch Museum
Naturalis
P.O. Box 9517
2300 RA Leiden, The Netherlands
jong@naturalis.nnm.nl

**Roger W. Portell**
Florida Museum of Natural History
P.O. Box 117800
University of Florida
Gainesville, FL 32611
portell@flmnh.ufl.edu

**Willem Renema**
Nationaal Natuurhistorisch Museum
Naturalis
P.O. Box 9517
2300 RA Leiden
The Netherlands
Renema@naturalis.nl

**P. Sá-Sousa**
Conservation Biology Unit,
Department of Biology
University of Evora, Pólo da Mitra
7002-554 Évora, Portugal
psasousa@uevora.pt

**Jan van Tol**
Department of Entomology
Nationaal Natuurhistorisch Museum
Naturalis
P.O. Box 9517
2300 RA Leiden, The Netherlands
Tol@naturalis.nl

**John de Vos**
Nationaal Natuurhistorisch Museum
Naturalis
P.O. Box 9517
2300 RA Leiden, The Netherlands
Vos@naturalis.nl

**Frank P. Wesselingh**
Nationaal Natuurhistorisch Museum
Naturalis
P.O. Box 9517
2300 RA Leiden, The Netherlands
and Biology Department
University of Turku
SF 20014 Turku, Finland
wesselingh@naturalis.nnm.nl

**Cor F. Winkler Prins**
Nationaal Natuurhistorisch Museum
Naturalis
P.O. Box 9517
2300 RA Leiden, The Netherlands
winkler@nnm.nl

# Introduction

Paleontology and zoology have much to offer one another, but exchanges between these fields have been limited. Prejudices that the fossil record is too scattered and poorly preserved, and that zoologists think on irrelevant evolutionary time scales, have often formed barriers in their cooperation. In this book we hope to show that even in poorly fossilising taxa, the inclusion of geological data in the analysis can improve the outcome.

Patterns of biodiversity have long attracted the attention of both biologists and palaeontologists. Although many taxonomic groups have their origins in the Palaeozoic, their current distribution patterns are usually dominated by Cenozoic overprints, making even the breakup of Gondwana (Late Jurassic) relevant to understanding present distributional patterns. The inclusion of geological data in any analysis of present-day distribution patterns leads to a better understanding of them.

The most pervasive impact of changing palaeogeographic configurations on biogeography is the (dis)appearance of migration routes. For example, the closure of the Isthmus of Panama prevented, first, the migration of deep marine faunal elements, and progressively shallower marine members until it was completely closed. From that moment onwards, it was a total barrier for the dispersal of the marine flora and fauna, but formed a corridor for terrestrial fauna.

Other examples of similar processes are, perhaps, less obvious. Insects associated with freshwater need terrestrial environments. Corridors can be provided by island arcs resulting from subduction zones, enabling island hopping by terrestrial insects and plants (Chapters 1, 2, and 3). Continuing plate movements have led to the destruction of these pathways and to the origination of others. The reconstruction of the presence of islands along arcs is often very difficult due to a poor fossilisation potential, as well as the fact that old arc systems have either been accreted or subducted, and needs interaction between zoologists and paleogeographers.

At larger time scales (millions to tens of millions of years), rafting of taxa on (micro)continents might also affect the distribution of taxa. As an example, marsupials formerly occurred on the Gondwana-connected continents of Australia and South America. In tectonically complex regions such as the Indo-Malaysian Archipelago and Southern Europe, rafting has resulted in the tectonic mixing of faunas due to the changing position of continental fragments.



*W. Renema (ed.), Biogeography, Time, and Place: Distributions, Barriers, and Islands,* 1–4

Paleogeographical events, including rafting of microcontinents and formation and accretion of island arcs, cannot be summarised as a sequence of vicariance events like phylogenetic trees. Phylogenetic trees thus do not necessarily follow area relationships. Dispersal of mobile taxa further complicates these patterns, as shown in Chapter 1.

"Classical" island faunas occur in these tectonically active areas. By studying the fossil mammal faunas on some present-day islands in the Aegean Sea it can be shown that, although these are currently surrounded by deep water, during the Miocene the faunas were still balanced and did not show signs of isolation from the mainland. The opposite holds for the island faunas retrieved from the fissures on Gargano, now part of mainland Italy (Chapter 10).

Arntzen and Sá Sousa (Chapter 12) studied the pattern in genetic isolation and morphological differentiation in briefly (14 ka) differentiated lizard populations. They found that genetic drift is the main force driving in population divergence.

Again there are other, less obvious, examples of the development of island faunas. Long-lived lakes can be seen as freshwater islands surrounded by otherwise hostile terrestrial environments. The Miocene (23–8 Ma) Pebas long-lived lake allowed the study of the rise and fall of species flocks in bivalve molluscs (Chapter 9). Comparison with other (predominantly Recent Miocene) long-lived lakes showed that morphological change and the development of species flocks are facilitated by the availability of empty biotopes.

Speciation due to the isolation of populations on "islands", using gastropods as an example, is much more common than generally thought (Chapter 11). Recognising the hostile environment surrounding the habitable biotopes just requires another perspective.

A second important subject in (paleo)biogeography is recognising biodiversity patterns and the processes responsible for their generation. As Wilson and Rosen (1998) have shown, especially in this kind of study, knowledge of the fossil record is needed to distinguish between several of the current models explaining high diversity. Most of these models can be classified into three basic concepts: (1) "centre of origin" or "cradle", in which it is assumed that taxonomic richness results from a concentration of evolutionary appearances; (2) "centre of accumulation", in which species originate in other regions, but over time, their geographical ranges change to converge in the high-diversity focus; and (3) "centre of survival" or "refugium", which assumes that high-diversity results from the survival of formerly much more widespread highly diverse fauna (Wilson and Rosen, 1998, and references therein). In Chapter 7 an example of a centre of survival is described. After a period of widespread shallow water reef development that culminated in the Visean (Early Carboniferous), the number and extent of tropical reefs declined during the Pennsylvanian (Webb, 2002), after which they were more diverse and occurred in a broader geographical range during the Permian (Wahlman, 2002). The Late Carboniferous reef-associated brachiopods survived mainly in the area of the present-day Cantabrian Mountains and Arctic Canada. These faunas seeded the widespread and diverse Permian reef-associated brachiopod faunas. Early representatives of specialised groups were first found in these refugia.

Both Fransen (Chapter 4) and Hoeksema (Chapter 5) show that the current maximum of marine biodiversity cannot be explained by one of the abovementioned models alone. By studying the modern distribution and phylogeny of host-associated shrimps, it proved to be impossible to distinguish between overlapping species ranges in the central part of the area and inwardly directed, convergent dispersal of peripherally originated species (Chapter 4). The "Coral Triangle" or the centre of maximum marine biodiversity is located in Southeastern Asia. Although this name is often used, it has never been defined. In Chapter 5 a review is given of the history of this name, and the diversity patterns of many marine biota. A method using target species is proposed. It is shown that large-scale environmental variation during the ice ages has strongly affected the position of the marine centre of maximum diversity. The Sunda and Sahul shelves were aerially exposed, providing a migration route for mammals towards Java, Sumatra, and Borneo (Chapter 10), and the coral reefs could not survive there. On the other hand, it is suggested that in the geologically complex area in eastern Indonesia many (partly) isolated subbasins led to high speciation rates during the Plio-Pleistocene (Chapter 5).

Most coral genera have a long fossil record extending back to the Eocene or older. The centre of maximum marine diversity is a relatively young (Late Oligocene/Early Miocene) phenomenon where reef-associated corals are concerned (Wilson and Rosen, 1998). In Chapter 6 this pattern is placed in a geological perspective, in which the geological history of the diversity of reef-associated larger benthic foraminifera is discussed. The faunas show a longitudinal shift in diversity from the western Tethys in the Eocene to the central Indo-West Pacific in the Miocene and younger. Major radiation occurred in the Eocene, Oligocene, and Plio-Pleistocene. Contrary to reef-associated corals, the Oligocene and earliest Miocene were as diverse as the Pleistocene. Whether these contrasting patterns are real, or due to a poor record of Oligocene corals needs further attention.

A takeaway lesson from this book is that patterns are not always what they seem if you look at them without a spatial or temporal reference. This cannot be generated without documenting fossil or recent occurrences of taxa.

## Acknowledgements

The following persons all critically reviewed contributions to this book: W.E. Boles, R.P. Brown, C.H.C. Brunton, Xx Carter, S. Caranza, C.J. Cleal, A. Currant, D. Dudgeon, R.A. Gastaldo, S. de Grave, M. Harzhauser, Holloway, R.W. Jones, P. Lunt, M.L. Martínez Chacón, C. Meyer, D. McFarlane (two chapters), G. Paulay, D.A Polhemius, R. Preece, D.L.J. Quicke, D.G. Reid, E. Robinson, G. Rosenberg, G.J. Vermeij, and N. Wahlberg.

## References

Wahlman, G.P., 2002, Upper Carboniferous-Lower Permian (Bashkirian-Kungurian) mounds and reefs, in: Kiessling, W., Flügel, E., and Golonka, J. (eds), *Phanerozoic Reef Patterns. SEPM Special Publication*, **72**: 271–338.

Webb, G.E., 2002, Latest Devonian and Early Carboniferous reefs: depressed reef building after the Middle Paleozoic collapse, in: Kiessling, W., Flügel, E., and Golonka, J. (eds), *Phanerozoic Reef Patterns, SEPM Special Publication* **72**: 239–269.

Wilson, M.E.J. and Rosen, B.R., 1998, Implications of paucity of corals in the Paleogene of SE Asia: plate tectonics or Centre of Origin, in: Hall, R. and Holloway, J.D. (eds), *Biogeography and Geological Evolution of Southeast Asia*, Backhuys, Leiden, pp. 165–195.

Chapter 1

# Global Disjunctions and Flying Insects

RIENK DE JONG AND CEES VAN ACHTERBERG
Nationaal Natuurhistorisch Museum Naturalis, P.O. Box 9517, 2300 RA, Leiden, The Netherlands, jong@naturalis.nnm.nl, achterberg@naturalis.nnm.nl



**Abstract** Total evidence scenarios for the origin of wide and disjunct distributions in two groups of flying insects, butterflies and parasitic wasps, are discussed, with emphasis on the possible role of the break-up of Gondwana. All six families of butterflies (Hesperiidae, Papilionidae, Pieridae, Lycaenidae, Riodinidae, and Nymphalidae) and two families of wasps (Stephanidae and Braconidae) have been examined for obvious disjunct distributions. The evidence for the impact of the fragmentation of Pangea and Gondwana on the global distribution of butterflies and both wasp families is considered to be weak. For the basal lineages of Braconidae, occupation of the niche of galls and "pseudogalls" (e.g., aphids) has been a more important driving force than vicariance. Obviously, dispersal and extinction played an important role before and after the break-up of the continents. They are integral parts of the evolution of life, ongoing processes, now and then punctuated by vicariance events, of which the traces will become obliterated with time. Dominance of vicariance events for the evolution of the families studied in this paper is considered unlikely.



*W. Renema (ed.), Biogeography, Time, and Place: Distributions, Barriers, and Islands,* 5–44

## 1. Introduction

From the fossil record we know that the distribution of extant taxa has not been a reliable indicator of the distribution of these taxa or their ancestors in the past. The family Camelidae (camels), of North American origin but now as native animals confined to parts of Asia and South America (e.g., Honey et al., 1998), and the order Proboscidea (elephants and relatives), of African origin but once distributed on all continents except Australia and Antarctica (Shoshani and Tassy, 1996), are good examples. But there are many more, including several groups of insects (e.g., Coope, 1973; Eskov, 2002), where the disjunction or apparent shift in distribution is due to extinction. Likewise, the fossil record documents not only different past distributions and extinctions, but also large-scale movements into and out of landmasses according to opportunities provided, as well as fragmentation of populations and biota by the formation of barriers.

Disjunct distributions, particularly across southern continents, are widely considered the result of vicariance events (e.g. continental drift; for butterflies, see, e.g., Shields, 1979; Miller and Brown, 1979; Shields and Dvorak, 1979; Brown, 1987; Miller and Miller, 1989, 1997; Grehan, 1991; at a more general level, Humphries and Parenti, 1999). This would be the inevitable conclusion, if vicariance were the only process by which disjunctions could originate. Since this is obviously not the case, one would expect to find justification for such a claim in the form of evidence (e.g., fossils, age determination). Without such evidence, the claims are no more than possible scenarios, storytelling.

Implicit in the idea of the occurrence of taxa in Gondwana before it broke up is the notion of an age pre-dating the break-up and in case of worldwide distributions even going back to before the break-up of Pangea (*c.* 200 Ma) (e.g., Brown, 1987, with respect to butterflies).

In this paper we examine possible scenarios for the origin of wide and disjunct distributions in two groups of flying insects, butterflies, and parasitic wasps of the families Stephanidae and Braconidae, with emphasis on the possible role of the break-up of Gondwana. The two groups are very different in size and age. The Lepidoptera probably arose early in the Jurassic. The distribution of the basal-most family, Micropterygidae, clearly shows traces of the occurrence in Pangea (Gibbs et al., 2004). The butterflies (approximately 17,000 species) are a crown group of the more than 40 superfamilies, and possibly arose in the Late Cretaceous. In contrast, the Stephanidae are a much smaller group, about 300 species, known as fossils from the middle Cenozoic onwards, but probably very much older, while the Braconidae are a much larger group, with probably more than 100,000 species, and as old as the Late Jurassic. Therefore, we can expect to find very different distributional patterns. The break-up of Pangea and Gondwana is taken for granted, but there are divergent views about the details. If in the future the ideas on the geological evolution change, we might need to change our conclusions as well. Croizat's (1984: 55) axiom that "what geological theories say is of no account the moment these theories contradict the data from biogeography as such" is rejected, as biogeographic data as such cannot contradict geological theories. It is the *a priori* interpretation of the data (e.g., as being the result of vicariance) that can contradict geological theories. We do not take an *a priori* position here. We are not interested in posing novel geological hypotheses based on a presupposed historical process of faunal development; we are interested in which historical processes have played what role. We have chosen a total evidence approach here: all possible pieces of evidence are examined separately and in conjunction to see if and where their lines intersect and form a picture of the early evolution of the groups under study.

## 2. Conjecture and Evidence

History means motion in time. Historical biogeography means motion in time and space. Motion is only observable between at least two points in time or space. If we find a taxon, be it recent or fossilized, in a particular place, it is evidence of just one point in time and space, nothing more. It must have had a history, but it does not tell us anything about it. Any story about the history of such a taxon is conjecture. To learn more about its history we need at least one more point in time (e.g., fossil of the same taxon or its ancestor), or in space (same taxon: disjunction; sister taxon: complete disjunction not necessary), or time and space (fossil of same taxon or ancestor in other place). In other words, we need evidence.

If a taxon occurs in two separate places (disjunction) there are three possible historical explanations:

(a) Vicariance
(b) Dispersal from one area into the other
(c) Dispersal from a third area into the two others, followed by extinction in the third area

From the distribution alone it is not possible to choose between these options without additional information that in part must come from another discipline, i.e., geology. Vicariance is the fragmentation of the distribution of a biota by the formation of a barrier. This is a geological or ecological event, and if we want to explain the disjunction by such an event, we must be confident that the event has indeed taken place. Even if this has been the case, we can only correlate the disjunction with the geological or ecological event, if the age of the divergence of the taxa constituting the disrupted biota correlates with the age of the geological or ecological event. As long as this information is lacking, any statement about the cause of the disjunction is conjecture.

Evidently when dealing with terrestrial organisms, those organisms that cannot cross saltwater barriers on their own are the best indicators of vicariance rather than dispersal. But even in such cases one must be careful and not too easily jump to conclusions. Data presented by Trewick (2000), on wingless insects belonging to different orders dispersing between New Zealand and Chatham Islands, Winkworth et al. (2002) on plant lineages in New Zealand thought to be of Gondwanan age but actually being relatively recent arrivals, and Gittenberger (personal communication, 2005) on gastropods belonging to the northern genus *Balea* occurring on the Tristan da Cunha archipelago in the South Atlantic Ocean, indicate that dispersal can occur well beyond our imagination, and may lead to a pattern similar to vicariance. In other words, our power of imagination is not a good guide in drawing conclusions.

In biogeography two approaches are widespread, a taxon approach and a pattern (or area) approach. The taxon approach aims at an explanation of the distribution of individual taxa. It is often part of a taxonomic revision. As far as it explains an individual distribution (e.g., an Australian–South American disjunction) by a general event (e.g., the break-up of Gondwana), it is a form of inductive reasoning that remains conjecture as long as it has not been tested. An example of this form of reasoning is given by an explanation of the distribution of the butterfly tribe Brephidiini (Lycaenidae). The tribe consists of one monotypic genus, *Oraidium*, in South Africa, and one genus, *Brephidium*, with one species in southern Africa and two from the southern USA to Venezuela. The disjunction in the genus *Brephidium* is supposed to be the result of the separation of Africa and the Americas, or even of the break-up of Pangea (Brown, 1987; Miller and Miller, 1997). If this would be true, why then is the same disjunction not found in the sister group (which is of the same age) or in any of the seven lycaenid clades (if we consider the Riodinidae a separate family) that are older (in the phylogenetic tree of Eliot, 1973)? A general explanation for a special case only should make one suspicious, to say the least. Clearly a claim of Pangean age should be accompanied by a test of the age of the divergence in *Brephidium*. Without such a test, the claim remains a conjecture. As matters stand in this case, even without a test of age, a Pangean age is highly unlikely, since it would predate the origin of the angiosperms (Qiu et al., 1999; Wikström et al., 2004) on which almost all Lepidoptera (of which *Brephidium* is a crown clade) are dependent. We shall return to this line of reasoning below.

The pattern (or area) approach makes use of general area cladograms or generalized tracks as the resultant of a number of individual area cladograms or tracks (see, e.g., Forey et al., 1992; Craw et al., 1999; Humphries and Parenti, 1999). It does not aim at explaining the distribution of taxa, but the history of areas or biota. Recurrent vicariant distribution patterns are explained as the result of a geological vicariance event (even to the point that if such a geological event is unknown, it is considered a shortcoming of the geologists; see the remark by Croizat in the Introduction; see also Humphries and Parenti, 1999: 145: "Biological data represented by area cladograms … present a considerable challenge to geologists"). The idea behind this is that it is unlikely that a number of independent, random dispersal events will lead to the same pattern. The conclusion may be correct, at least for part of the individual disjunctions, but how do we know for which part? It is as with phylogenetic relationship: similarity may be due to common descent, but how do we know to which extent? In phylogenetic analysis, similarity is commonly rejected as a measure of relationship. For biogeographic inference it is rejected here as well. Any vicariance explanation of a disjunction, be it an individual case or a recurrent pattern, is considered conjecture here, if it is not supported by evidence that the divergence of the taxa agrees in age with the vicariance event. Similarly, a dispersal explanation is considered conjecture if it is not supported by evidence that the divergence of the taxa is younger than the geological vicariance event (or if a vicariance event is not known altogether). It means that for every individual case of disjunction, evidence relating to its origin (age determination, fossils) must be sought for and analysed.

With the acceptance of the idea that we need evidence of age, the problems are not solved, but start. There are three possible sources of evidence, fossils, molecular clock, and dependence on other organisms whose age is known. Fossil insects are rare, at least for the groups we are examining in this paper (as will be explained below). Evidently, fossils can only give a minimum age and their explanatory power is, consequently, weak. The application of a molecular clock has been much debated (see, e.g., Strauss, 1999; Donoghue and Smith, 2004) and is often thought to be controversial. Apart from intrinsic problems with saturation and differential mutation rate in codon position, genes, and lineages, calibration points may be difficult to find and extrapolation should not be overextended (e.g., Graur and Martin, 2004). Still we agree with Bromham et al. (1999) that a sloppy clock is better than no clock. The third source of evidence, dependence on other organisms of which the age is known, is fossil or molecular evidence in disguise, since the age of the latter organisms must have been determined by fossils or molecules.

With all reservations that have to be made, the present authors are of the opinion that all available evidence must be examined and that weak evidence is better than no evidence at all. If later the evidence used turns out to be false, we may have to adjust or reject the original hypothesis, but this is how science proceeds. Conjecture may be a good start for a line of research, but without evidence it remains what it is. Alfred Wegener's idea of drifting continents was a conjecture to begin with, but physical evidence was needed to get it widely accepted. Note that it was not biogeographic evidence that turned the balance. If continental drift had been found a physical impossibility, no biogeographic evidence could have saved the concept.

In the present paper we are concerned with disjunctions, particularly across southern continents. A disjunction is considered to be of potential Gondwanan age if it concerns sister taxa endemic to Gondwanan fragments. A number of such cases will be dealt with. If the sister group of an endemic taxon in a Gondwanan fragment is unknown (making the application of a molecular clock impossible) and there are no fossils known of the right age in the same continental fragment, we do not imply that the taxon is not of Gondwanan age, we can only hope to draw conclusions when the phylogeny has been clarified.

## 3. Evolution of Continents

Continental drift is a process that started when continents were formed for the first time, and will continue until the driving force, the energy inside the Earth, is exhausted. In this context it is an oversimplification to speak of the break-up of Pangea or of Gondwana, or even the final separation of Australia from Antarctica, as an "event". Not only is it an ongoing process in terms of barrier formation (vicariance), but what constitutes a barrier for one kind of organism, may not even be noticed by another kind of organism. Thus, for some organisms the barrier between Australia and Antarctica may have become insurmountable as soon as the land contact was severed, for others the new strait did not act as a barrier until it was much wider, maybe millions of years later. Lawver and Gahagan (1998) describe how Australia may have been separated from Antarctica by a shallow seaway since the early Cenozoic (65 Ma), while the complete separation of the two continents by the formation of oceanic crust took place much later, by the end of the Eocene (*c.* 34 Ma). Similar scenarios probably apply to the separation of other fragments, the areas across which exchange could take place becoming increasingly narrow and acting as a filter route for a longer or shorter time, before complete separation. For terrestrial organisms the depth of the sea would seem to be of minor importance: whether it is 1 cm or 1 km deep, a butterfly will drown in it. So the formation of epicontinental seas separating the subaerial part of continental plates must have been the beginning of interruption in gene flow, long before the continents became separated by the forming of ocean crust between them. On the other hand, organisms differ in their ability to cross such barriers. For some organisms the seaway must have started to act as a barrier earlier than for others. However, since evidence is lacking it is useless to speculate on a date for each kind of organism. For the present purpose, the ages for the various vicariance events are supposed to be as follows (taken from the paleomaps of Smith et al. (1994), and the data provided by McLoughlin (2001), unless otherwise stated):

(a) Early Jurassic Period (200 Ma). Pangea: all southern continents plus North America form a single supercontinent. There is a narrow seaway between North America, Greenland and Eurasia (Europe mainly consisted of islands); Eurasia itself is widely separated from the southern continental block.

(b) Middle Jurassic Period (175 Ma). North America starts to rift from South America in the west and Africa in the east.

(c) Middle–Late Jurassic Period (160 Ma). The separation of North America from the southern block (Gondwana) is complete; an epicontinental sea (Turgai Strait) separates Europe from Asia. Rifting starts between West Gondwana (South America and Africa) and East Gondwana (the remaining part).

(d) Early Cretaceous (140 Ma). Separation of East and West Gondwana is more or less complete, but South America remains connected to Antarctica either by a land bridge or by a chain of islands; according to Hallam (1994) western Gondwana was completely separated from eastern Gondwana in the Tithonian (last stage of Late Jurassic, 150–145 Ma; according to fossils of various groups of marine organisms in East Africa, Madagascar and the southern Andes); southern Atlantic Ocean starts to open up; North America and Europe are more or less connected across the northern Atlantic.

(e) Early Cretaceous (130 Ma). Rifting of India and Madagascar from Antarctica is almost complete; South America and Africa still broadly connected in their central parts; rifting of Australia starts.

(f) Early Cretaceous (105 Ma). Narrow connection between South America and West Africa is severed; narrow land connection (or chain of islands) between South America and Antarctica (the movements of the many blocks involved not yet entirely understood); an epicontinental sea divides North America into a western part, connected to Asia across Beringia, and an isolated eastern part.

(g) Late Cretaceous (90 Ma). Madagascar is separated from India; Australia is still connected to Antarctica; isolation of South America stronger (but probably still a chain of islands between its southern tip and Antarctica), all connection with Africa being severed between 106 and 84 Ma (Pitman et al., 1993) (Hallam, 1994: between 95 and 80 Ma).

(h) Late Cretaceous (80 Ma). New Zealand rifts from Antarctica, and the rift between Australia and Antarctica proceeds eastwards (see also Metcalfe, 1998: fig. 8c).

(i) Late Cretaceous (70 Ma). A North–South epicontinental sea separates western Africa from the rest of Africa; eastern North America and Europe are divided into a number of larger islands. Fossil evidence indicates limited migration of vertebrates between North and South America near the Cretaceous–Cenozoic boundary (Hallam, 1994), South America becoming increasingly isolated thereafter, until the Panama Isthmus was closed in the Pliocene.

(j) Paleocene (60 Ma). The northern tip of the Antarctic Peninsula is adjacent to the southern tip of South America, but the South Georgia and South Orkney blocks lying in between become detached and gradually migrate eastwards making a terrestrial crossing increasingly difficult and practically impossible (except for long-distance dispersal) from the Oligocene onwards (Hallam, 1994); epicontinental sea across Africa dries up; India reaches the Equator, but according to Metcalfe (1999) there was an initial contact between India and Eurasia at this time (not emergent); North America and Europe become more coherent and are more or less connected at northern latitudes; Asia and North America are broadly connected through Beringia through most of the Cenozoic (Hopkins, 1967).

(k) Middle Eocene (45 Ma). Southern continents are well separated, and India and Australia drift further northwards; according to Hallam (1994), a narrow land bridge may have linked Australia to Antarctica up to 38 Ma, but the evidence for this is not clear, and in the reconstructions by Hall (1998, 2002), Australia is completely detached by 50 Ma, while according to Lawver and Gahagan (1998) a shallow seaway separated Australia from Antarctica as early as the early Cenozoic (65–60 Ma); according to Metcalfe (1999) there was major indentation between India and Eurasia at this time, but in his fig. 8d (i.e., Metcalfe, 2001: Fig. 13d) there was still a wide shallow sea between the two; because of the appearance of large terrestrial mammals in India in the Eocene, Hallam (1994) noted that a major land corridor between India and Asia had become established by the Middle Eocene.

(l) Oligocene (30 Ma). India has almost completely crossed the Equator, but see above; with Australia completely detached from Antarctica, cold ocean currents can encircle Antarctica and glaciation starts. According to Kemp (1981) the land ice had reached the sea by 30 Ma, but Hallam (1994) recorded a much later start of significant growth of the ice cap, in the Middle Miocene; whatever was the case, from the Late Eocene–Early Oligocene (*c.* 38 Ma) onwards, faunal exchange between Australia and South America across Antarctica was not possible for terrestrial organisms, and in view of the narrow connections and/or stepping stones between South America and Australia on the one hand, and Antarctica on the other, it must have been a sweepstake passage, maybe as far back as the beginning of the Cenozoic.

(m) Early Miocene (20 Ma). India is broadly connected to Asia; connection between North America and Europe across the northern Atlantic is still more or less intact in the form of stepping stones.

(n) Middle Miocene (15 Ma). Collision of Australian and Asian plates; according to Audley-Charles (1993), there is no evidence of land between Australia, New Guinea, and Sulawesi before 6 Ma.

## 4. Butterflies (Hesperioidea and Papilionoidea)

### 4.1. General

Analysis of morphological data (de Jong et al., 1996) and molecular data (Wahlberg et al., 2005) supports a phylogeny of the families of the butterflies as in Fig. 1. There is very little evidence on the age of the butterflies. Fossils are extremely rare, and less than 90 specimens are known to date. The oldest known fossil, from the Upper Paleocene of Denmark, is supposed to represent a hesperioid taxon (Kristensen and Skalski, 1999), i.e., belonging to the basal-most family of the butterflies. The next oldest fossil can be attributed to the Papilionidae (*Praepapilio*) from the Eocene Green River Shale of Colorado (Durden and Rose, 1978). Remarkably, the Papilionidae are also the next family to appear in the phylogeny. In view of the rarity of fossil butterflies we cannot attach much importance to these ages. Remarks on the "near-modern" appearance of the fossils, suggesting that the butterfly ancestor must be much older (e.g., Brown, 1987; Miller and Brown, 1989; Zeuner, 1961) do not give us an estimated age either. Moreover, what is "near-modern"? And appearance (similarity) is not the

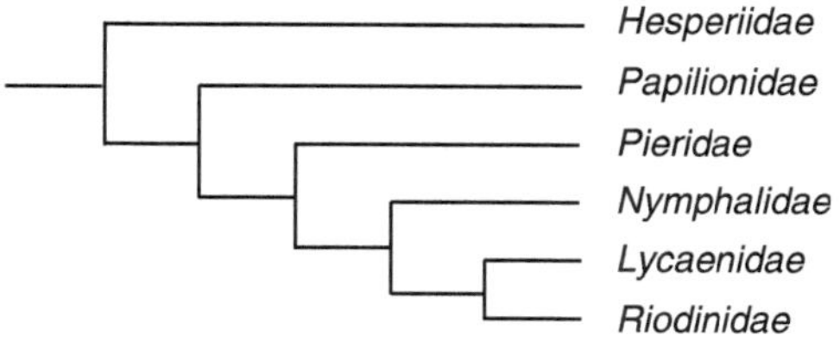


Fig. 1 Cladogram of the families of butterflies.

same as apomorphy. Preliminary attempts to apply a molecular clock to the phylogeny of the butterflies (de Jong, unpublished) resulted in an estimated age of 50–60 Ma for the early divergence, but this is considered an underestimation due to saturation, and much more work needs to be done here. So all we can say at the moment is that the butterflies are at least *c.* 55 My old, and apparently older.

We can also, be it very roughly, estimate the maximum age of the butterflies. All butterflies (except a few taxa that secondarily switched to carnivory), as well as all Lepidoptera except the basal-most clades, are dependent on angiosperms for larval food (Powell et al., 1999). Contrary to butterfly fossils, fossil remains of angiosperms are abundant. As far as palynological and other fossil evidence goes, angiosperms first radiated in pre-Barremian Cretaceous times, 130–145 Ma (e.g., Lovis, 1989; Willis and McElwain, 2002). Even if this is an underestimate by *c.* 40 My, it is still too young for a Pangean origin and diversification. Applying a molecular clock Wikström et al. (2001, 2004) estimate the earliest diversification of the angiosperms at *c.* 180 My. Considering the position of the butterflies in the crown of the Lepidoptera tree and that almost all Lepidoptera are dependent on angiosperms, the occurrence of butterflies at the time of Pangea seems out of the question. Hence, worldwide disjunctions cannot be attributed to the vicariance event of the break-up of Pangea. There is no reason to suppose that the butterflies are too young to have lived in Gondwana before it fragmented. However, this is not the same as stating that disjunctions across Gondwana fragments are due to the break-up of Gondwana. Observing that the butterflies as a taxon are old enough to have lived in Gondwana does not imply that subordinate taxa (species, genera, even subfamilies) are old enough. So each case of disjunction (vicarious distribution) must be examined separately.

Vicarious distributions do not occur at the family level. All families are worldwide or semi-worldwide. As stated above, the occurrence of butterflies in Pangea is out of the question, at least on the basis of present-day knowledge. Hence, worldwide distributions of families must have come about by dispersal.

Disjunctions in taxa of species rank have not been considered for two reasons. First, for an understanding of the relationships of populations of a single species on more than two fragments, morphological characters are usually inadequate, and molecular characters are lacking so far. Second, species with a distribution across two or more Gondwana fragments are either migratory (e.g., *Vanessa cardui*, Nymphalidae, worldwide except South America), or species of open habitats confined to the Old World tropics (e.g., *Eurema hecabe*, Pieridae, and *Zizula hylax*, Lycaenidae) and hardly disjunct, except for intervening seas and deserts. For these species a dispersal scenario is more likely, the more so while they are not basal species in their families: there are numerous taxa (also higher taxa) of the same or older age that are confined to one of the Gondwana fragments suggesting that they originated after fragmentation (unless we assume large-scale extinction that is not supported by any evidence). It is meaningful in this respect that there is no species restricted to South America and any or all other fragments of Gondwana. A special case of disjunction between the Old World and the New World is presented by the nymphalid species *Hypolimnas misippus*. It is found throughout the Oriental and Afrotropical regions, northern South America, Central America (sporadically), the Antilles, and sporadically in the eastern USA (Ackery et al., 1995; DeVries, 1987; Scott, 1986). Corbet and Pendlebury (1956: 212) "believed that the species reached the New World in the old slave ships which first sailed between the West Indies and Africa nearly 400 years ago", but Scott (1986) does not exclude the possibility that the species crossed the Atlantic under its own power. In this connection Stoneham's (1965: 12) record that "in February, 1917 there was a big migration of many thousands of this species six hundred miles off Mosamedes (Angola) in the South Atlantic Ocean. Many came aboard the troopship 'Omrah' " is relevant. Migrations are also known from Central America (DeVries, 1987). The reason for supposing an African origin of the American populations and not vice versa, is twofold. First, all other (18) species of the genus are restricted to the Old World. Second, the female of *H. misippus* mimics the model *Danaus chrysippus*, which is widespread in the Old World, but does not occur in the New World. In the following sections some relevant cases from each butterfly family will be examined.

## 4.2. Hesperiidae

Being the basal-most (and, thus, oldest) family of the butterflies one can expect to find vicarious distributions due to the break-up of Gondwana in this family in the first place. The family is currently divided into six subfamilies (Ackery et al., 1999): Coeliadinae (Palaeotropics), Pyrrhopyginae (Neotropics), Pyrginae (cosmopolitan), Heteropterinae (semi-cosmopolitan, absent from Australian region), Trapezitinae (Australian), and Hesperiinae (cosmopolitan). The family does not occur in New Zealand. The interrelationships of the subfamilies are still not entirely clear. Pyrginae may not be monophyletic, and Hesperiinae may be paraphyletic in terms of Trapezitinae. Evans (1951) believed "the Pyrrhopyginae to be co-ancestral with the Old-World Coeliadinae and to represent the evolution of the branch that accompanied America in its Wegener journey from Europe and West Africa to where we now find it". Morphological (de Jong et al., 1996) and molecular information (Wahlberg et al., 2005), however, indicate that the subfamily Coeliadinae is sister to the rest of the Hesperiidae. It does not preclude the possibility that the earliest divergence was the result of the break-up of Gondwana, as later dispersals may have completely wiped out the traces of a southern origin of the sister of the Coeliadinae, but to arrive at this conclusion we either need fossil evidence, or evidence that the split is old enough. Such evidence is lacking so far.

The Coeliadinae themselves have a vicarious distribution, and are found in the Afrotropical, Oriental, and Australian regions (with an extension into the Eastern Palaearctic) (Fig. 2). The pattern is very similar to that of the bird genus *Aviceda* (Accipitridae) as depicted by Craw et al. (1999: 70). These authors call such patterns "classic Indian Ocean (Gondwanic) patterns", a typical example of inductive reasoning where the pattern is, at the same time, its own explanation. As for the Coeliadinae, there is no vicariant pattern between sister taxa on Gondwana fragments. The tree in Fig. 3 is based on a parsimony analysis of all (nine) genera and 22 morphological characters (de Jong, unpublished). The tree suggests an early radiation

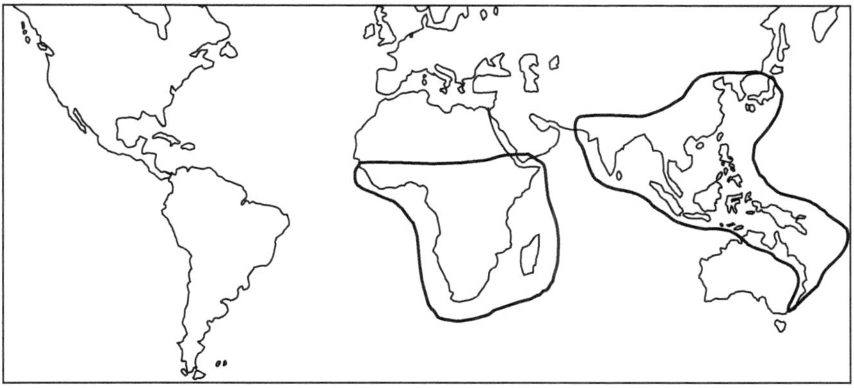

Fig. 2 Distribution of Coeliadinae (Lepidoptera: Hesperiidae).

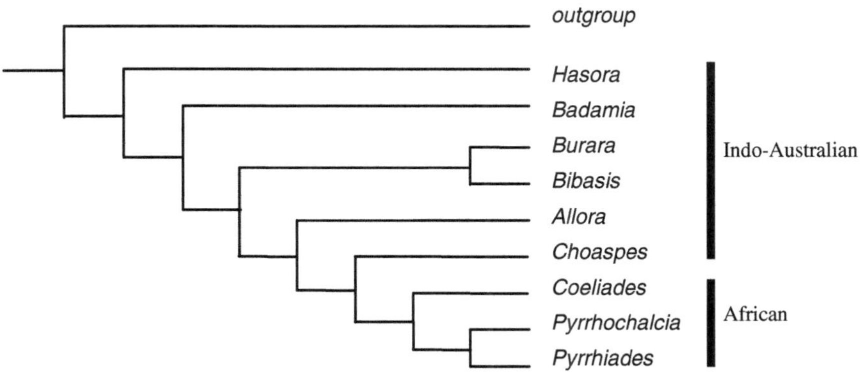


Fig. 3 Cladogram of genera of Coeliadinae (Lepidoptera: Hesperiidae) based on morphological characters (de Jong, in preparation).

in the Oriental region with extension into the Papuan region, before reaching Africa, not exactly a vicarious Gondwana pattern. So far the age of the Coeliadinae remains unknown. There are no fossils known. Molecular studies are on the way (Warren, personal communication, 2006). The foodplants, with more than 20 plant families recorded, are too varied to give a clue. But even without an indication of age (the split between the African taxa and *Choaspes* should be more than 100 My old) a Gondwana pattern appears to be ill founded. If the subfamily would really be that old we should also expect to find members in South America.

The genus *Celaenorrhinus* (Pyrginae) is one of the largest genera of the Hesperiidae, consisting of some 90–95 species. It is the only pantropical genus in the family (Fig. 4). Among other characters the genus is characterized by peculiar scent-distributing organs at the base of the ventral side of the abdomen of the male (de Jong, 1982). The phylogeny of the genus is still unknown. In the three parts in which the distribution area is divided, South America, Africa (including Madagascar), and south as well as southeast Asia, there are yellow-spotted and white-spotted species, and per area the species may not be monophyletic, i.e., the interrelationship between the areas may not be simple in terms of *Celaenorrhinus* species. Note that the genus hardly transcends Wallace's Line to the east: there is only one Oriental species reaching the Moluccas. So there is no indication that the genus has ever been represented on the Australian plate. The Old World distribution is not essentially different from that of the bird genus *Polihierax* (Falconidae) as depicted by Craw et al. (1999: 71). This distribution pattern is called a "classic Indian Ocean (Gondwanic)" pattern (see also the Coeliadinae, above), even though no Gondwana fragment in the eastern part is involved. The connection between Africa and South America was severed at least 80 Ma. Acanthaceae are the main foodplants for *Celaenorrhinus*, while also Verbenaceae, Balsaminaceae, Urticaceae, and Oleaceae have been recorded. The origin of all these families has been estimated at *c*. 40 Ma, except the last family that should have originated *c*. 60 Ma

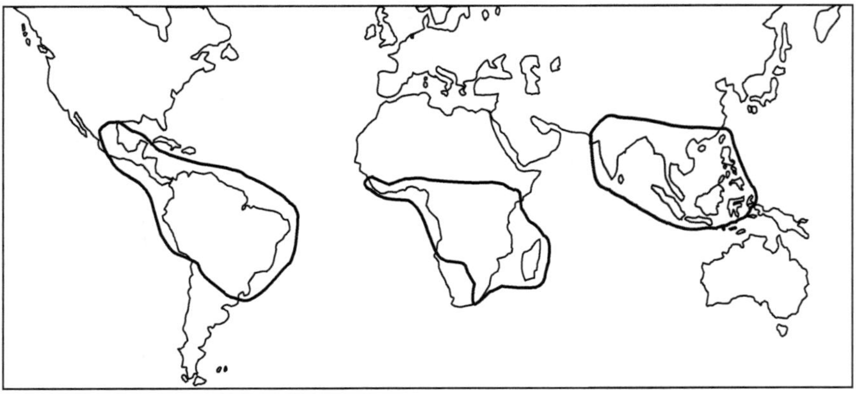

Fig. 4 Distribution of *Celaenorrhinus* (Lepidoptera: Hesperiidae).

(Wikström et al., 2004). These ages are much too recent to accept a Gondwana distribution for *Celaenorrhinus* in Africa and South America. Above we concluded that we should expect to find traces of the Coeliadinae in South America, if this subfamily is over 100 My old. We can expand this reasoning now: *Celaenorrhinus*, belonging to the sister group of the Coeliadinae, must have originated well after the Coeliadinae; if the distribution of *Celaenorrhinus* was a Gondwana relic, then we should certainly expect to find traces of Coeliadinae in South America. Since we do not find such traces, and since all available evidence is against a Gondwanan age for *Celaenorrhinus*, we must conclude that the vicarious distribution is not due to the break-up of Gondwana. It not only applies to the Old–New World disjunction, but also to a more regional scale: the two apparently monophyletic Malagasy species, *C. ambra* and *C. humbloti*, must have originated from a single dispersal event from East Africa, where the very similar *C. zanqua* occurs.

Although no vicariance is involved, the position of a relative of *Celaenorrhinus* should be mentioned here, namely *Euschemon*. This monotypic genus is restricted to areas along the eastern coast of Australia in Queensland and northern New South Wales (Braby, 2000). The species is exceptional, as the male has a wing coupling device (frenulum and retinaculum) that is not found in any other butterfly, but is widespread among other Lepidoptera. For this reason it has been placed in a family, or at least a subfamily, of its own in the past. The "primitive" wing coupling device and the restricted distribution have contributed to ideas of great antiquity, but so far there are no indications that this could be true. In a morphological analysis (de Jong et al., 1996) *Euschemon* appeared, as sister to *Celaenorrhinus*. This did not come as a surprise, since no other potentially closely related genera were involved. According to Scott and Wright (1990) *Euschemon* is most similar to Coeliadinae, but this claim is not substantiated and could be based on symplesiomorphy. Anyway, such a relationship contradicts the result of the parsimony analysis by de Jong et al. (1996). Parsons (1998) thinks *Euschemon* most closely related to the Papuan genus *Chaetocneme*. *Euschemon* feeds on Monimiaceae (Braby, 2000), a family that *c*. 80 Ma came into being (Wikström et al., 2004). It does not contradict the presence of *Euschemon* on Gondwanan Australia, but it is not an argument in favour of it either.

## 4.3. Papilionidae

The Papilionidae are by far the best-studied family of butterflies, but it does not mean that there is agreement about the relationships within the family. Generally three subfamilies are recognized (Ackery et al., 1999), Baroniinae, Parnasiinae, and Papilioninae. The classical hypothesis of their interrelationship, as for instance found by Miller (1987), is Baroniinae, Parnassiinae, and Papilioninae. However, this hierarchy was not recovered in the morphological study by de Jong et al. (1996). In the latter study most trees showed a sister group relationship between the Parnassiinae and the rest of the family. In a study based on mitochondrial and nuclear genes, Caterino et al. (2001) did find a sister group relationship between the Baroniinae and the remainder of the family, but the Parnassiinae turned out paraphyletic. In a study combining morphological and molecular data, Wahlberg et al. (2005) found a hierarchy as follows: (Papilioninae (Baroniinae Parnassiinae)). Baroniinae (comprising a single species restricted to the Sierra Madre del Sur in southwest Mexico) and Parnassiinae (with a number of genera in the Palaearctic, one genus, *Parnassius*, extending into the Nearctic) are strictly northern hemisphere groups; Papilioninae on the other hand, is cosmopolitan (but absent from New Zealand). The two described European fossil papilionids (Early Oligocene: *Thaitis ruminiana* Scudder; and Late Miocene: *Doritites bosniaski* Rebel) fall well within Parnassiinae; the two North American fossil papilionids (Middle Eocene: *Praepapilio colorado* Durden and Rose; and *P. gracilis* Durden and Rose), however, lack apomorphies that can link them to any of the extant subfamilies (de Jong, personal observation). Vicarious distributions across southern continents are only found in Papilioninae. In this subfamily three or four tribes are distinguished. We shall discuss vicarious distributions in two of the tribes, Troidini and Papilionini.

The Troidini are remarkable for their oligophagy, as they use only a limited number of species of the family Aristolochiaceae (mainly of the genus *Aristolochia*) for larval food. It is an old plant family, its roots going back to *c.* 130 Ma, with the genus *Aristolochia* having originated *c.* 80 Ma (Wikström, 2004). While the Aristolochiaceae are almost cosmopolitan, the Troidini are restricted to the warmer parts of the world (Fig. 5). Surprisingly, they are completely absent from the mainland of Africa, though there is a monotypic genus, *Pharmacophagus*, in Madagascar. De Jong (2003) dealt with the supposed sister taxa *Cressida* (Australia, Timor, New Guinea) and *Euryades* (Argentine). Applying a molecular clock he came to the conclusion that the split was too young to have been caused by the break-up of Gondwana. The position of *Pharmacophagus* is uncertain. Parsons (1996a, b), using mainly larval characters, came to the conclusion that it was sister to the Papuan *Ornithoptera*, and he considered the split to be the result

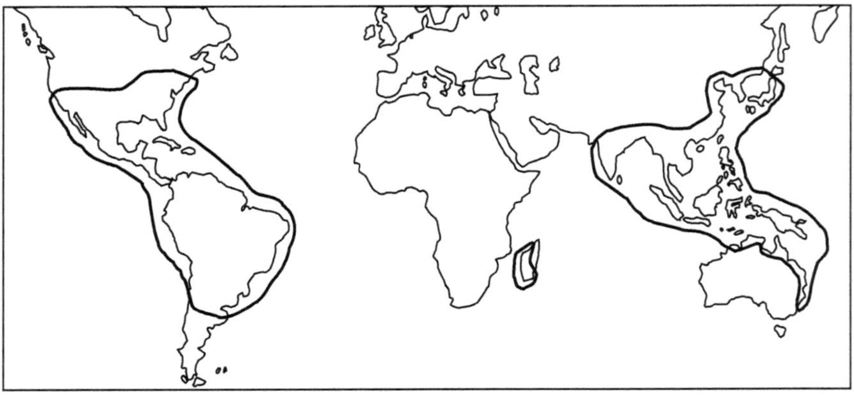

Fig. 5 Distribution of Troidini (Lepidoptera: Papilionidae).

of the fragmentation of Gondwana. Such a relationship seems highly unlikely geographically, one should rather expect an India–Madagascar split, i.e., a sister group relationship between.

*Troides* or another genus represented in India and *Pharmacophagus*. But analyses of more extensive morphological data (Miller, 1987, Tyler et al., 1994) and of molecular data (Morinaka et al., 1999) indicate a still different relationship that is, moreover, not unambiguous, although the three analyses agree in placing *Pharmacophagus* rather basally (Fig. 6). With so many uncertainties it does not seem sensible to speculate on the geographic history of *Pharmacophagus*. Yet an origin by vicariance is highly unlikely. *Pharmacophagus* lives on *Aristolochia* (Ackery et al., 1995), like most of the other Troidini. The estimated age of *Aristolochia* is (*c*. 80 Ma) is much younger than the estimated age of the separation of the India–Madagascar block from Antarctica (*c*. 130 Ma).

Zakharov et al. (2004) analysed sequences of mitochondrial and nuclear DNA for 51 of the about 205 species of the worldwide genus *Papilio* (Papilionini). One of the interesting results is the common ancestry of the Australian endemic *P. anactus* with the African–Malagasy species groups of *P. phorcas*, *P. rex*, and *P. delalandei*. They estimated divergence times, calibrating the clock at several points using geological vicariance events. As stated above, starting from vicariance prevents investigation of the process, since the process has been chosen *a priori*. Moreover, as

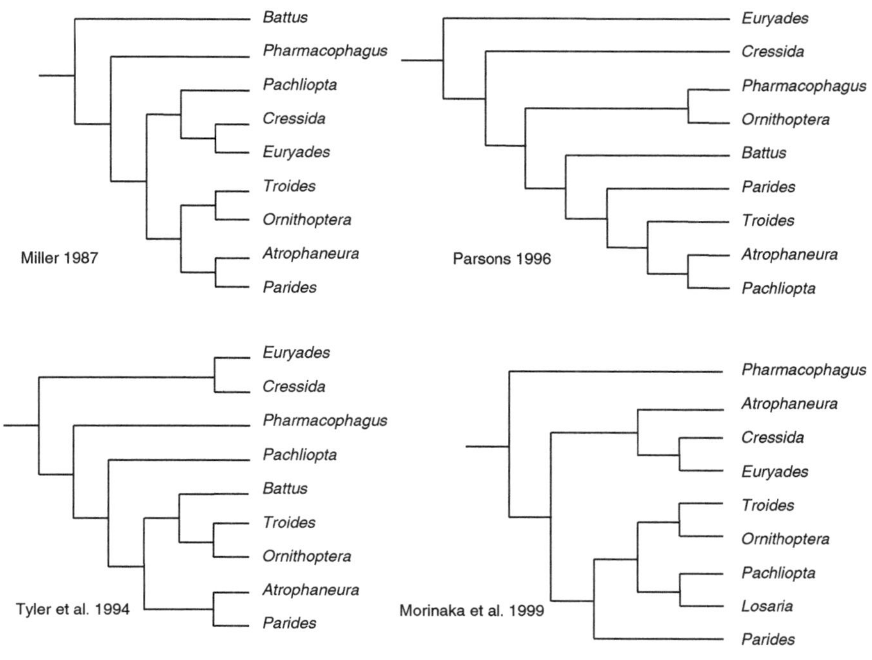


Fig. 6 Four cladograms of genera of Troidini (Lepidoptera: Papilionidae), based on different sets of characters.

vicariance event for the group just mentioned they used the separation of Australia from Antarctica (35 Ma), whereas they should have taken the much older separation of Africa (*c.* 130 Ma) from Antarctica, since that was the moment the African and Australian taxa became separated. The larvae of all these species live on Rutaceae. The estimated age of this family, based on a molecular clock, is 47 Ma, but putative fossils have been recorded from 67 Ma (Wikström et al., 2004). In both cases this is much younger than the separation of Africa from Antarctica. Hence, a vicariance explanation for this disjunction is unlikely. As Zakharov (personal communication, 2004) explained to the senior author, the occurrence in Africa, although not said so explicitly, was indeed considered the result of long-distance dispersal, but then, again, the *a priori* idea is that the contact between Africa and Australia was through Antarctica, and that is exactly what we should investigate and compare to other possibilities. Zakharov et al. (2004) estimated the early diversification of *Papilio* at 55–65 Ma. It may be correct, but it should be based on calibration points that are independent on the processes we want to investigate.

## 4.4. Pieridae

The oldest fossil with a pierid apomorphy (*Coliates proserpina* Scudder, 1875) is from the Early Oligocene. It is too poor to be assigned to a lower taxonomic rank. Four subfamilies are currently recognized, namely, Pierinae, Coliadinae, Dismorphiinae, and Pseudopontiinae (Ackery et al., 1999). Klots (1931–1932) considered Coliadinae subordinate to Pierinae, and Pseudopontiinae and Dismorphiinae sister groups, but his scheme was intuitive and in morphological cladistic analyses, Dismorphiinae appear as sister to the rest of the family, with the relationships of the three other subfamilies unresolved (de Jong et al., 1996). A recent molecular study will remedy this unsatisfactory situation (Braby, personal communication, 2005), but we shall not anticipate on that.

Pierinae and Coliadinae are both cosmopolitan, but absent from New Zealand (except for accidental introduction). Pseudopontiinae contains a single monotypic genus, *Pseudopontia*, in the rain forests of Africa (Sierra Leone to Uganda, Angola and Zambia). Dismorphiinae show a wide disjunction (Fig. 7). The subfamily is predominantly Neotropical (about 100 species in 6 genera, one species rarely straying into southern Texas), but there is a small genus, *Leptidea* (6–7 species) in the Palaearctic. The *Leptidea* species are all very similar externally and in genitalia, suggesting a relatively recent diversification. The interrelationships of the seven genera of the Dismorphiinae are still unknown. All members of the subfamily feed on Fabaceae (as far as known), a family that is estimated to be *c.* 80 My old (Wikström et al., 2004). Supposing the Dismorphiinae are not older than the Fabaceae, the wide distribution cannot be attributed to the break-up of Pangea, the only time South America and the Palaearctic were (indirectly) connected. Hence, dispersal as well as extinction must have been the underlying processes. The New World Dismorphiinae are all forest dwellers, from lowland rain forest to montane cloud forest (DeVries, 1987). The Palaearctic *Leptidea* are equally forest dwellers,

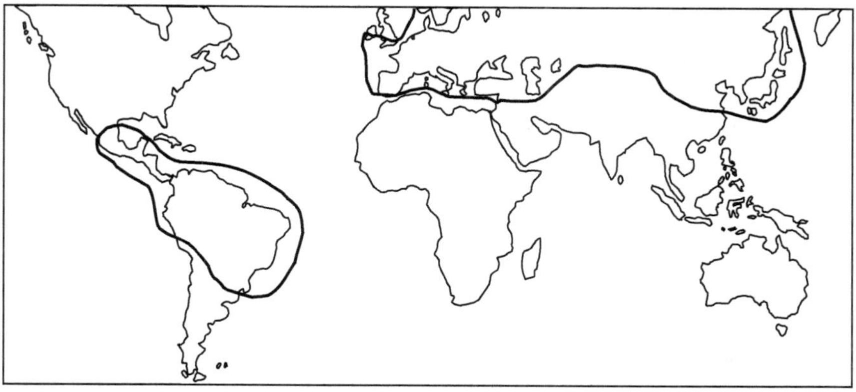

Fig. 7 Distribution of Dismorphiinae (Lepidoptera: Pieridae).

but obviously not in tropical forest. Either *Leptidea* adapted to a cooler climate, or the South American Dismorphiinae to tropical conditions. In this context it may be of interest that particularly during the Early Eocene full tropical multistratal rain forest occurred as far as 30° N, and subtropical conditions may have extended to 60° N (Behrensmeyer et al., 1992).

## 4.5. Lycaenidae

With about 5,000 species the Lycaenidae form the second largest family of the butterflies. Although cosmopolitan in distribution, it is mainly an Old World group in diversity (de Jong, 2003). The number of subfamilies varies with the author. Eliot (1973) recognized eight subfamilies, but later (1990) he reduced this number to five, one of which was Riodininae, considered a separate family before, and nowadays again. The phylogeny of the family as given by Eliot (1973), although based on many morphological characters, is mainly intuitive. All 33 tribes recognized occur in the Old World, where 28 are endemic. Five tribes, belonging to three subfamilies in both Eliot's 1973 and 1990 schemes, extend to the New World, and only two of these, belonging to a single subfamily (three tribes if we include Central America, but still a single subfamily) are found in South America as well. The interrelationship of the tribes is still uncertain. There is no cladistic analysis of all tribes based on morphological characters available. So far, published molecular analyses of lycaenid taxa are based on a too limited set of taxa to be very helpful. Recent results (Wahlberg et al., 2005) of a combined analysis of morphological and molecular characters, however, indicate that the first split in the lineage was between Curetinae and the rest of the Lycaenidae. Curetinae consists of a single genus, *Curetis*, with 18 species (Eliot, 1990). Almost all species are restricted to the Oriental region, but one species slightly extends into the Palaearctic, and there is one species in the Papuan region. The tribes represented in South American belong to Lycaeninae sensu Eliot (1990), of which most species, genera and tribes occur in the Old World. There are no endemic tribes in South America. The information so far suggests an early differentiation of the Lycaenidae in the Old World in relative isolation, with much later (and only limited) extensions into the Nearctic and finally, into the Neotropics. There is no evidence of vicariance between Gondwana fragments at the tribal level.

A remarkable disjunction is shown by the little tribe Spalgini, consisting of two genera only (Eliot, 1973): *Spalgis* (about 10 species, Palaeotropics) and *Feniseca* (monotypic, Nearctic) (Fig. 8). The tribe is remarkable for the pupae resembling a monkey's head in miniature. Later Eliot (1990) added *Taraka* (two species, Oriental region) to the tribe, but the reason for this is not clear and we follow Eliot (1973) here, as did Vane-Wright and de Jong (2003). The larvae are carnivorous, feeding on aphids and coccids. *Feniseca* is not only the sole representative of the tribe in the New World, but of the whole subfamily Miletinae to which the tribe belongs. Since North America separated from Gondwana in the Middle Jurassic, long before the butterflies are supposed to have originated, and since the early diversification of the Lycaenidae apparently took place in the Old World, the disjunction cannot be attributed to the break-up of Gondwana (unless we want to speculate on a South American origin of *Feniseca*, followed by dispersal to the North and extinction in South America). Rather the disjunction must be understood as the result of large-scale extinction in the Holarctic.

An interesting case of disjunction occurs within the subfamily Lycaeninae. The subfamily has an almost worldwide distribution, but is absent from Australia and South America (there is one species in Guatemala) (Fig. 9). Two tribes (Bozano and Weidenhoffer, 2001; Eliot, 1973: sections) are recognized, Lycaenini and Heliophorini. The former is (with a variable number of genera) Holarctic, but one widespread species extends through the mountains of East Africa as far south as Malawi, two species are found in South Africa, and four species in

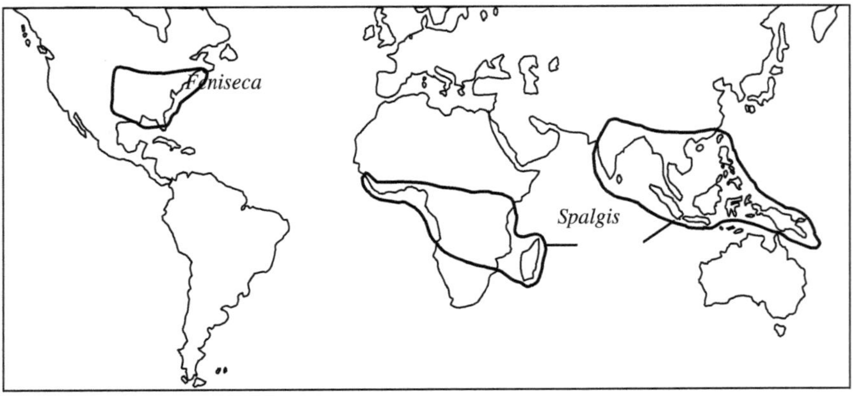


Fig. 8 Distribution of Spalgini (Lepidoptera: Lycaenidae).

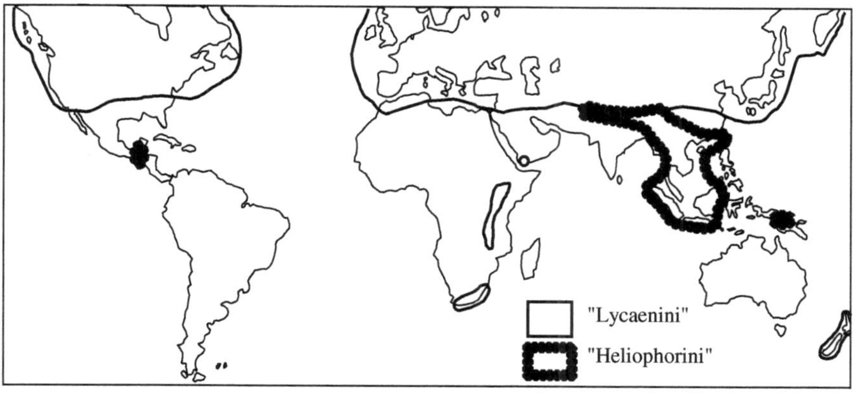


Fig. 9 Distribution of Lycaeninae (Lepidoptera: Lycaenidae).

New Zealand. The Heliophorini are found (with three genera) in the Oriental region, the mountains of Papua New Guinea and a mountain valley in Guatemala. Such a distribution is not explicable by any hypothesis of tectonic events. Either the distribution is due to dispersal, or the tribes are not monophyletic, or both. Preliminary results of a molecular study (based on mitochondrial as well as nuclear genes) indicate that, indeed, the Heliophorini are not monophyletic the genus *Iophanus* from Guatemala was not available for DNA analysis), but constitute the successive basal offshoots of the lineage, apparently followed by the ancestor of the New Zealand species, which became the sister of the rest of the Lycaeninae (van Dorp and de Jong, 2004). In addition, the two South African species proved to be sister to a strictly Palaearctic species group and not to be closely related to the Holarctic species (*Lycaena phlaeas* Linnaeus) that descends from the Middle East to Malawi. Figure 10 is a simplified tree based on the mitochondrial gene CO1. Miller and Brown (1979) hypothesized a Pangean origin of the Lycaeninae. The long isolation of New Zealand as well as the worldwide distribution would ask for that. However, the subfamily is restricted to Polygonaceae for larval food. This family is supposed to have originated *c.* 40 Ma (Wikström et al., 2004), long after New Zealand and Africa became separated from other fragments of Gondwana. Hence, the remarkable distribution of the Lycaeninae must be due to dispersal.

The apparent absence of a Gondwanan radiation in Lycaenidae, the restriction of almost all tribes to the Old World and the absence of a South American radiation except in two crown groups (Eumaeini and Polyommatini, both of the Polyommatinae) are all indicative of a relatively recent evolution of the family, in spite of the large number of species. In this respect it is interesting to note that two recent molecular studies suggest a relatively recent radiation in two speciose genera. Applying a molecular clock, Megens et al. (2004) estimated the earliest radiation in the Oriental-Papuan genus *Arhopala* (with about 200 species, making it the largest genus in Lycaenidae) at *c.* 11 Ma, while Kandul et al. (2004) estimated the origin on the Palaearctic genus *Agrodiaetus* (about 90 species) at 2.51–3.85 Ma.

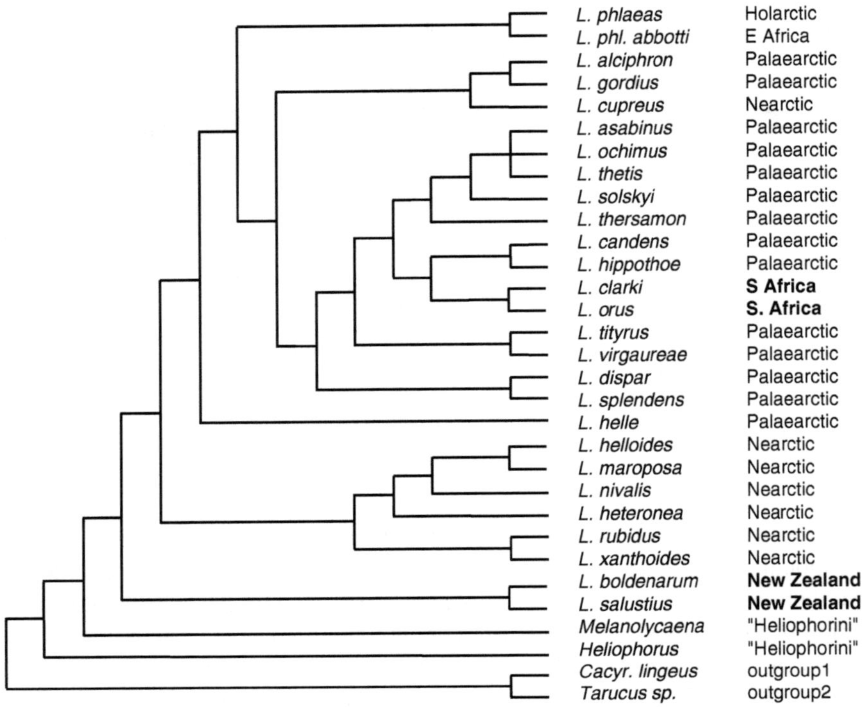


Fig. 10 Simplified cladogram of Lycaeninae (Lepidoptera: Lycaenidae) based on sequences of the mtDNA gene COI (details will be published elsewhere).

## 4.6. Riodinidae

Riodinidae have often been considered to form a subfamily of the Lycaenidae (e.g., Ackery et al., 1999), but nowadays they are considered to form a family in its own right. Recent analyses indicate a sister group relationship (Campbell et al., 2000; Wahlberg et al., 2005). With about 1,500 species worldwide, it is a modestly large family. The distribution of diversity is very different from the Lycaenidae. Of the approximately 140 genera only 13 occur in the Old World, the remaining genera occurring overwhelmingly in the Neotropics. All Old World taxa have been placed by Hall and Harvey (2002) in the subfamily Nemeobiinae. The family is not represented in Australia, except for one Papuan species just reaching Cape York Peninsula (Braby, 2000). Africa has a single genus (*Abisara*) that also occurs in the Oriental region, but Madagascar has an endemic genus with three species (Ackery et al., 1995). There is only a single species (*Hamearis lucina*) in West Palaearctic. Monophyly of the Nemeobiinae is supported by their restriction to plants of the related families Primulaceae and Myrsinaceae for larval food, while the New World Riodinidae feed on 35 or more plant families (e.g., DeVries, 1997) but not on Primulaceae or Myrsinaceae although both families occur there. In the molecular analysis by Campbell et al. (2000), based on sequences in a single nuclear gene, *wingless*, including 12 genera among which two from the Old World, the latter did not come out as sister to the New World taxa. By far most of the New World taxa fall within the subfamily Riodininae, but a few genera are placed in the Euselasiinae, in which Hall and Harvey (2002) also include the genera *Corrachia* and *Styx*, both sometimes considered subfamilies on their own. In the analysis by Cambell et al. (2000) the Riodininae are sister to the Old World taxa, while Euselasiinae (with *Euselasia* as exemplar taxon) is sister to the Riodininae (with nine examplar taxa) and Nemeobiinae (with two exemplar taxa) combined. This result does not agree with the relationships found by de Jong et al. (1996) on the basis of morphological characters, nor with the scheme proposed by Hall and Harvey (2002): Riodininae (Euselasiinae Nemeobiinae). Whatever the correct hierarchy, the molecular and morphological data so far exclude a sister group relationship between the Old World taxa and the New World taxa combined, in either scheme the Nemeobiinae are sister to only a part of the New World Riodinidae. The monophyly of the Nemeobiinae, implying that there was a single ancestor, while there was already a New World radiation, points to dispersal rather than vicariance. The absence of an Australian development precludes an Australian–South American link. According to Wikström et al. (2004), the Primulaceae and Myrsinaceae combined originated *c.* 75 Ma. This is too young for an African–South American link except by dispersal. However, if Nemeobiinae originated from a dispersing ancestor, a North American–East Asian link could also come into consideration.

As to the age of the Riodinidae we have very little information. Hall et al. (2004) reported on a fossil in Dominican amber, 15–25 Ma (at the same time demonstrating that records of older riodinid fossils are unreliable due to lack of relevant details). They hypothesized that the divergence of the fossil species, *Voltinia dramba*, from the extant *Voltinia danforthi* could be dated at 40–50 Ma, based on the availability of a dispersal route over a more or less continuous landmass, the proto-Greater Antilles. This should be tested by applying a molecular clock to the divergence of the genus *Voltinia* from its sister (which, then, should be older than 50 Ma). However, if dispersal is involved, any event older than 15–25 Ma would explain the presence of the fossil species on Hispaniola.

## 4.7. Nymphalidae

The Nymphalidae are the most speciose (about 6,000 species) and most diverse family of butterflies. In the most recent analysis, based on the sequences of one mitochondrial and two nuclear genes (Wahlberg et al., 2003), 12 subfamilies and 39 tribes are recognized. The trees obtained were rooted with *Libythea*, since rooting with a species of the families Pieridae or Lycaenidae proved to be difficult, probably because of long-branch attraction.

Previous morphological as well as molecular studies had indicated that *Libythea* (or rather the subfamily Libytheinae) has a sister group relationship with the rest of the family. However, in a recent study combining morphological and molecular data (Wahlberg et al., 2005) the genus is not the basal offshoot of the nymphalid lineage, but the result of a later speciation event, making it sister to the Danainae. According to Wahlberg (personal communication, 2005) this less basal position is due to poor taxon sampling. The position of Libytheinae is of crucial importance in dating the evolution of the Nymphalidae. If it is the basal offshoot, dating the origin of Libytheinae will set a maximum age for all other Nymphalidae, if it is sister to the Danainae, it will only set a maximum age for that subfamily. Apart from *Libythea* (Old World, 7–9 species) the subfamily comprises the genus *Libytheana* (New World, 1–4 species) (Fig. 11). The number of species recognized depends on the treatment of island populations (Madagascar, Mauritius, Caribbean islands). According to Shields (1979) the worldwide distribution is due to the presence of the genus in Pangea, and the occurrence of an endemic species on the Marquesas (*Libythea collenetti*) is evidence of the initial opening of the Pacific Ocean in the Jurassic on an expanding Earth. The hypothesis of an expanding Earth does not seem to have many adherents today. Apart from that, the idea that a species high up in the phylogenetic tree of the butterflies came into being in the Jurassic (146–200 Ma), does not seem to be very realistic. The larvae of Libytheinae are only known to feed on *Celtis* (Ulmaceae) (Ackery et al., 1995; Braby, 2000; Igarashi and Fukuda, 2000; Scott, 1986; and others). According to Wikström et al. (2001) the genus *Celtis* originated *c.* 25 Ma. A molecular clock estimate for the subfamily Celtidoideae gives *c.* 67 Ma and fossils attributed to this group have been dated 69 Ma (Wikström et al., 2004). Two butterfly fossils have been recorded as belonging to Libytheinae, *Prolibythea vagabunda* Scudder, 1889, and *Barbarothea florissanti* Scudder, 1892, both from the Early Oligocene of Florissant, Colorado, but the allocation of the fossils is based on general similarity, and no apomorphy of the Libytheinae is visible (de Jong, personal observation), making the fossils useless for the determination of a minimum age. Several species of Libytheinae are known to be strong migrants, which may account for their often very wide ranges. A phylogeny of the subfamily was published by

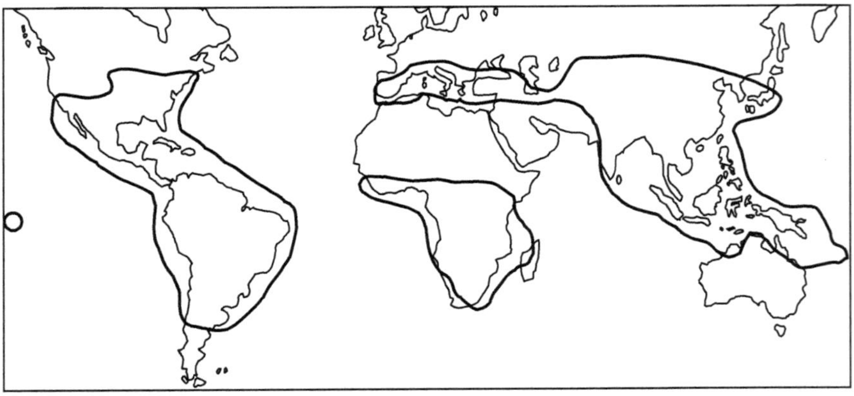

Fig. 11 Distribution of Libytheinae (Lepidoptera: Nymphalidae).

Kawahara (2003), but without underlying character distribution. It shows *Libythea* and *Libytheana* as sister groups. As a consequence, there is an Old World–New World vicariance. However, none of the groups are restricted to a Gondwana fragment, so a sister group relationship between Gondwana fragments is not apparent. In view of the age of the foodplants only a vicariance Australia–South America would come into consideration, but the South American *L. carinenta* is not found in Chili (except as a migrant (Peña and Ugarte, 1996) ) and the southern half of Argentina where we would expect to find it in case of such a vicariance, and in Australia only the widespread *L. geoffroy* (Burma to Loyalty Islands and Rennell Island) occurs, restricted to patches in the far north (Braby, 2000). This, together with the migratory behaviour suggests that the worldwide distribution is the result of dispersal rather than vicariance, and this must certainly be the case for the origin of the Marquesas endemic as there is no geological evidence for a dry land route to the Marquesas. If the phylogeny given by Kawahara (2003) is correct, then the occurrence on the Marquesas must be of long-standing dispersal, since *L. collenetti* is sister to the rest of the *Libythea* species combined.

The genus *Vanessa* offers a good example of the untenable conclusions about vicariance drawn from disjunctions alone as in panbiogeography. This popular genus (red admirals and painted ladies) was subdivided by Field (1971) into three genera, *Vanessa*, *Bassaris*, and *Cynthia*. This distinction is rarely followed today, but for the sake of convenience we shall deal with them as subgenera here. Craw (1990), accepting the subdivision into three genera, did a cladistic analysis. Figure 12 is based on Craw's tree, but has been slightly modified to accomodate some recent additions and insights. The genus is absent from the Afrotropical and Neotropical regions (Fig. 13). All speciation events are interpreted by Craw as "an underlying pattern of geographical vicariance ... upon which is superimposed 'noise' caused by secondary, across-barrier, dispersal events" (Craw et al., 1999: 18: Figs. 1–5). The interpretation is in accordance with panbiogeographic principles, but not necessarily with history. Thus, the Hawaiian endemic *V. tameamea* and its Oriental sister group originated

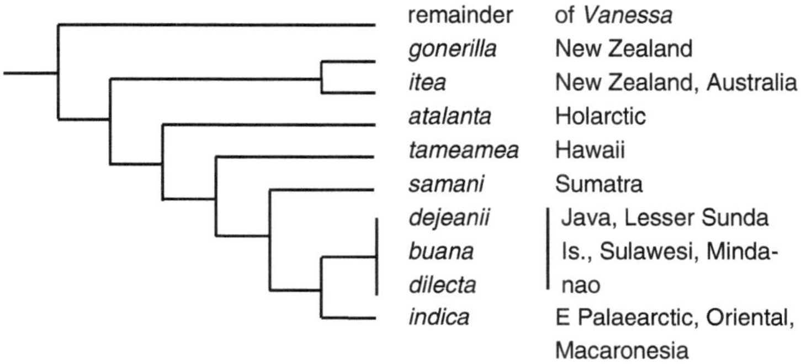


Fig. 12 Cladogram for part of species of Vanessa (Lepidoptera: Nymphalidae) (mainly after Craw, 1990, with some recent information added).

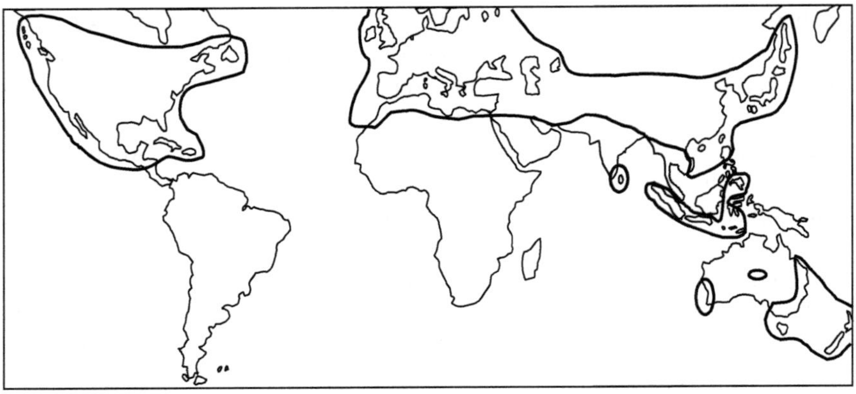

Fig. 13 Distribution of *Vanessa* (*Vanessa*) (Lepidoptera: Nymphalidae).

by a vicariance event that is unknown in geology. Similarly, the group from Australia and New Zealand (*Bassaris*) and its Holarctic-Oriental sister group (*Vanessa* s.s.) must have originated by an impossible vicariance event, since the two areas never formed a single area to the exclusion of other areas. One could think of a Pangean origin with subsequent extinction in Africa and South America to save one's presupposition of vicariance, at the same time rejecting the accumulated knowledge on the origin and evolution of the angiosperms (the genus uses a number of plant families for larval food, the oldest of which, Asteraceae, is estimated to have originated *c.* 70 Ma; Wikström et al., 2004), but in this way everything, and thus nothing, can be explained. It cannot be without meaning that some of the strongest mirgatory butterfly species in the world, *V.* (*Cynthia*) *cardui* and *V. atalanta*, belong to this genus, while in recent years the North American *V.* (*Cynthia*) *virginiensis* crossed the Atlantic Ocean and became established in Portugal (Maravalhas, 2003). The only logical conclusion must be that the representation in Australia and New Zealand originated from dispersal from the north, and not from a geological vicariance event. Note that there is some similarity with the case of the subfamily Lycaeninae dealt with above. Also note that there is a continent-wide disjunction in *V. indica*: Canary Islands, Madeira, and East Palaearctic-Orientalis. The cause of this disjunction is unknown. Possibly the species was competed out by an invading *V. atalanta* over much of its original area, but so far this is speculation. Even so, the apparent mobility in this genus proves that the evolution of life is not necessarily dependent on the evolution of the Earth.

In both examples from the Nymphalidae described above endemic taxa from Pacific islands were involved, and the conclusion or at least the suggestion in the literature was that the occurrence in the Pacific was due to a former land connection and/or a closed Pacific Ocean. One more example is given here of a distribution that seemingly crosses the Pacific Ocean. Grehan (1991: 102, Fig. 14)

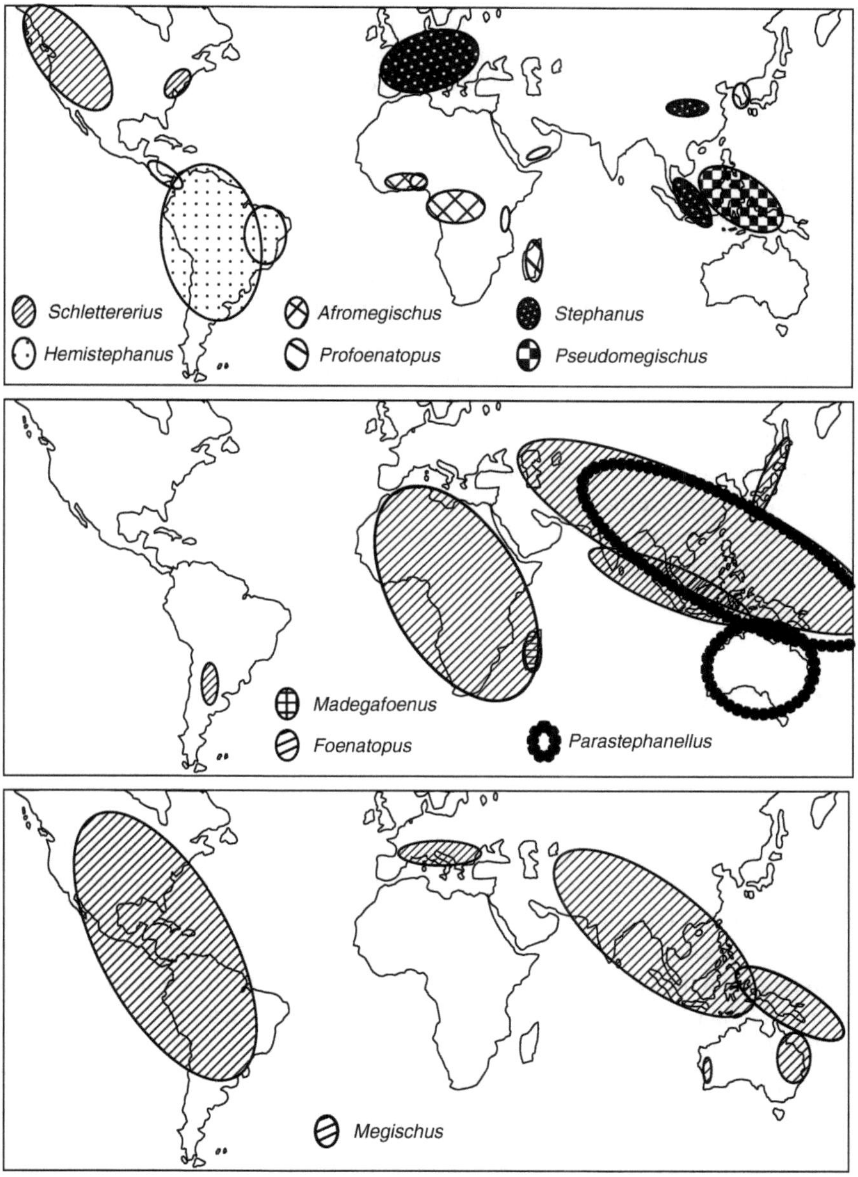


Fig. 14 Distribution of the genera of Stephanidae (Hymenoptera). Because of paucity of available material and, therefore, recorded localities, the ranges have only roughly been indicated.

illustrates the distribution of the Ithomiinae (Neotropics and Papuan region), basing himself on Shields (1976) who considered it an example of a distribution that dated back to pre-continental drift (have the continents ever been at rest?) and pre-seafloor spreading times. Shields (1989), in his turn, although not mentioning Ehrlich (1958), apparently based himself on this author who considered Ithomiinae to consist of the South American Ithomiini and the Papuan Tellervini. However, Ackery and Vane-Wright (1984) and Ackery et al. (1999) could not find support for such a close relationship to the exclusion of Danaini, and considered the three groups as tribes of the Danainae, with unresolved phylogeny. Whether the Libytheinae are sister to the rest of the Nymphalidae, or to Danainae only, divergence in the latter cannot be older than Libytheinae (see above). Hence, even if in future Tellervini and Ithomiini would turn out to be sisters, an origin due to a vicariance event is unlikely.

Understandably in such a large and diverse family there are many more instances of very wide and disjunct distributions, but space prevents us to go further into detail. It may suffice to mention here that a search for Gondwanan elements among the Australian Nymphalidae did not yield support for the existence of such elements (de Jong, 2003).

## 5. Hymenoptera

### 5.1. General

Hymenoptera are as old as or older than the Lepidoptera. They have a much better fossil record than the Lepidoptera, and very much better than the butterflies. The oldest fossils, belonging to the extant superfamily Xyeloidea, are known from Middle or Late Triassic (Rasnitsyn, 2002; Schlüter, 2000). The monophyletic grouping Apocrita, the most highly evolved Hymenoptera, start to appear in the fossil record in the early Jurassic Period (Megalyridae). Analyses using only morphological data (Vilhelmsen, 2001) and using morphological, as well as molecular data (Schulmeister et al., 2002) point to Orussoidea as sister to the Apocrita. This superfamily is known from Jurassic fossils. The higher-level phylogeny of the Hymenoptera, and of the Apocrita in particular, is still insufficiently known (for a discussion, see Ronquist et al., 1999). Here we deal with two families from the Apocrita, Stephanidae, and Braconidae. In the tree presented by Vilhelmsen (2001) for the basal lineages of the Hymenoptera, Stephanidae are the first offshoot of the Apocrita lineage, followed by Megalyridae, and then the remainder of the Apocrita. However, since only six exemplar genera of the Apocrita were involved in the analysis, we cannot attach too much importance to it. In the trees discussed by Ronquist et al. (1999), these families sometimes appear prior to the Aculeata (wasps, bees, ants) and Ichneumonoidea, sometimes later. The Ichneumonoidea (including the Braconidae) are considered to be sister to the Aculeata, although this position is not confirmed by all analyses.

## 5.2. Stephanidae

The Stephanidae are a small group with about 300 extant species, mainly restricted to tropical and subtropical forests, where they are idiobiont ectoparasitoids of larvae of holometabolous insects living in dead wood. The oldest known fossils are from the Eocene (Rasnitsyn, 2002), when both known subfamilies, Schlettereriinae and Stephaninae, were already present (van Achterberg, 2002). Both Shaw (1988) and Rasnitsyn (1988) considered Megalyridae and Stephanidae to be sister groups, but they did not give convincing synapomorphies. A reanalysis of Rasnitsyn's data not surprisingly show for the Stephanidae a position remote from the Megalyridae and nearer to the Ichneumonoidea (Ronquist et al., 1999). In the scheme of Vilhelmsen (2001) the Stephanidae are the oldest taxon of the Apocrita, with the Megalyridae splitted off later and possibly being the sister group of the Gasteruptiidae. In a number of trees found by Schulmeister et al. (2002) the Stephanidae were sister to the Orussoidea. The Stephanidae have a very different morphology to the Megalyridae, suggesting either a more distant relationship with the Megalyridae or a very long evolutionary separation. Since Orussoidea as well as Megalyridae are known from the Jurassic (Rasnitsyn, 2002), it would seem that the Stephanidae may also be of early Jurassic age, and the known Eocene fossils would give a very strong underestimate of the age of the group. In that case they would be much older than the butterflies, and hence, other scenarios could have played a role in the distribution of the group, since they would have been present before the break-up of Pangea. On the other hand, if they branched off after Ichneumonoidea, as in Figs. 7 and 9 of Ronquist et al. (1999), they may have originated in the Early Cretaceous, excluding their presence in Pangea. In both cases they originated before the Proctotrupidae, which family appears in the fossil record since Early Cretaceous.

Figures 14 and 15 show the relationships and distribution of the genera according to van Achterberg (2002). Although it is a very old and worldwide group, there are only approximately 300 extant species. Representation on Gondwana fragments is strikingly absent in the early diversification of the family. If not due to extinction (for which there is no evidence) it indicates that the group arose in the northern hemisphere (Laurasia) after the fragmentation of Pangea. This agrees well with an Early Cretaceous or Late Jurassic origin. At the time (most of) the Oriental region, except for India, was part of the Asian block. Only at a later phase the southern hemisphere was reached, possibly in conjunction with the Eocene warming of the northern hemisphere. As a result we do not find groups with primitive characters in the southern hemisphere, but we do find highly derived characters, such as the reduction and final disappearance of wing veins in *Foenatopus* and, less strongly expressed, *Profoenatopus*. There is no Australian–South American vicarious distribution of the genera, another indication that the family was not present in Gondwana before the break-up. As a consequence, the extant presence of the family in Gondwana fragments must be due to dispersal from the north, and the only vicarious distribution between genera, *Afromegischus* and *Madegafoenus*, must also be the result of dispersal.

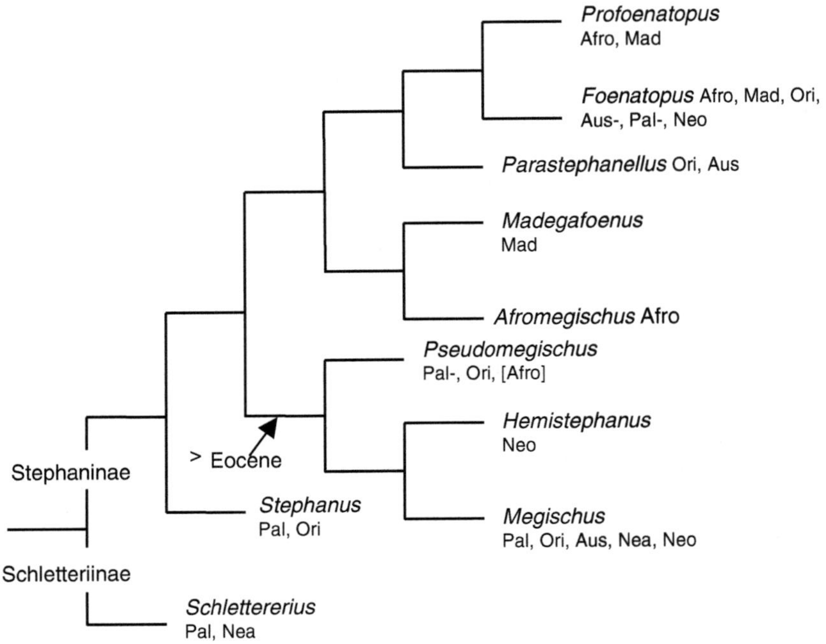


Fig. 15 Cladogram of Stephanidae (Hymenoptera) (after van Achterberg, 2002). Abbreviations for faunistic regions: Afro = Afrotropical (excluding Madagascar); Aus = Australian; Aus- = id., but excluding Australia and New Zealand; Mad = Madagascar; Nea = Nearctic; Neo = Neotrop-ical; Ori = Oriental (including Wallacea); Pal = Palaearctic; Pal- = id., but southern part.

The most basal species group of *Megischus*, the *M. anomalipes* group, has an aberrant disjunct distribution. Of the four species included, one has a western Palaearctic distribution, one a Nearctic distribution, and two are only known from Australia (one West, the other East). There is no geological vicariance event that can account for this disjunction. In view of the apparently northern origin of the family, we must conclude that an early member of *Megischus* managed to reach Australia from the north (as did so many butterflies; de Jong, 2003) and became extinct in southeast Asia, possibly outcompeted by recent species. *Megischus* underwent an extensive radiation in southeast Asia, probably in the last 20 My concurring with the development of evergreen tropical lowland rain forests since the Early Miocene (Morley, 1999).

The North American boreotropical elements could escape to Central and South America during the Pliocene (about 5 Ma) because of the formation of the Panama Isthmus. The Afrotropical region became severely affected by the late Neogene desiccation; in Madagascar the rainforests became probably restricted to tiny pockets at the eastern coast. In the African continent the climatic change was more severe because of the mid-Cenozoic uplift related to the formation of the East African rift system. It resulted in many more extinctions within the equatorial rainforests than in other equatorial regions (Morley, 1999). It accounts for the depauperate flora (many characteristic angiosperm groups are absent or rare) and fauna of the African rainforests. It is amply shown by the remaining fauna of Stephanidae: what is left are some old rainforest elements in Madagascar and on the continent only some derived groups, which are obviously (considering their colour pattern) inhabitants of a more or less open savanna habitat.

### 5.3. Braconidae

Mainly due to the relentless research on their DNA and anatomy by Dr. D.L.J. Quicke (Imperial College, London) we are getting a better understanding of the early history of this family. Fossils of Braconidae and their sister group, Ichneumonidae are known from the earliest Cretaceous. It would thus seem that in (Late?) Jurassic the Ichneumonoidea branched off the lineage, which would become the Aculeata (with the bees branching off in the early Cenozoic to coevolve with the angiosperms). We leave the intermediate Eoichneumonidae out of consideration; this group is only known from the Early Cretaceous and probably extinct now (Rasnitsyn, 2002). Starting from the aculeate lineage with the plesiomorphous telescopic condition of the second and third metasomal tergites, the segments became hinged (Ichneumonidae) to immovably (Braconidae) connected to each other. This development to strengthen the basal part of the metasoma to become a "pressure-plate" is connected with the type of substrate to be penetrated. The earliest parasitoids search for wood-boring larvae, a trait already present in the symphytan ancestors, the woodwasps (Siricoidea) and the parasitic woodwasps (Orussidae). Two other substrates contain living proteins: galls and eggs that are harder to penetrate. The tree in Fig. 16 is based on DNA sequences (28S and 18S; Belshaw and Quicke, 2002) of the major groups and agrees with their morphology. It shows the early development of the lineage was with some very aberrant basal taxa. According to Quicke et al. (1999) also the aberrant Hybrizontinae (listed as Paxylommatinae) might belong here. The Trachypetinae (only known from Australia; the biology is unknown) have an aberrant biochemistry (Dowton et al., 2002). The basal lineage of the Sigalphoids includes the Masoninae (mainly southern hemisphere, penetrating the Nearctic from the Neotropics; the biology is unknown), which have a highly aberrant morphology (van Achterberg, 1995). Remarkable for a group of probably Jurassic age with an early radiation in the southern hemisphere, there are no traces of Africa having been involved in this radiation.

Since Africa became isolated from other Gondwanan fragments *c.* 105 Ma, it would appear that the early radiation of the family took place after that date. However, the oldest fossils known of the Braconidae are from the Early Cretaceous in Asia (Rasnitsyn, 1983, 2002), well before Africa became isolated. Since there has never been a contact between Asia and Gondwana to the exclusion of Africa, either long-distance dispersal (between the northern hemisphere and Australia through South America) or large-scale extinction in a globally distributed group (in Africa), or both have been the underlying processes (supposed that the apparent

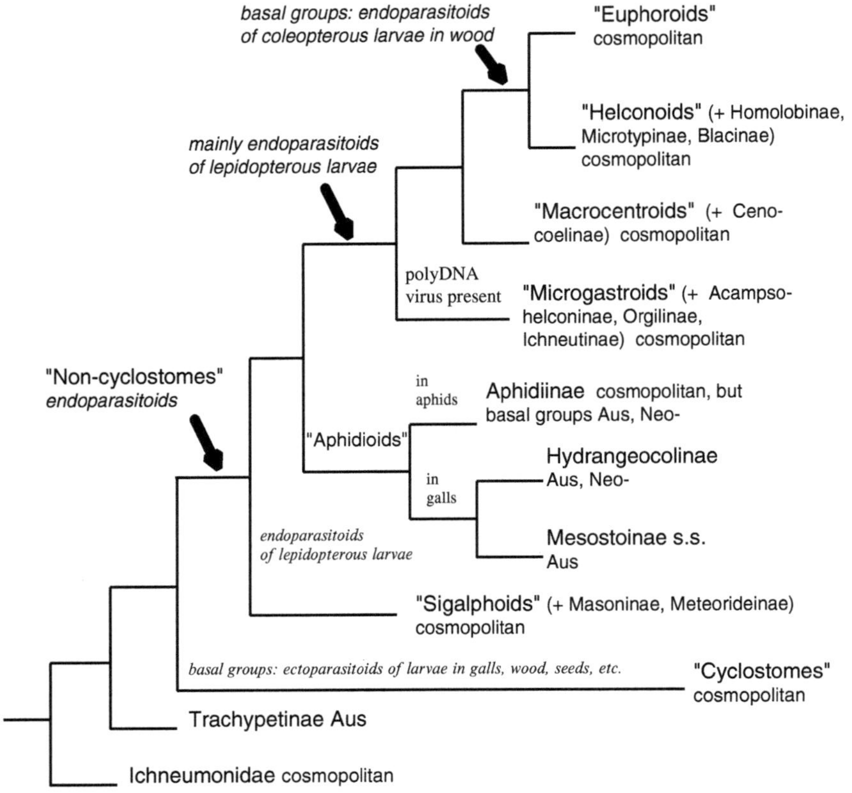


Fig. 16 Simplified cladogram of Braconidae (Hymenoptera) (mainly after Belshaw and Quicke, 2002). Abbreviations for regions: Aus = Australian; Neo- = Chile and part of Argentina.

absence from Africa is not due to undercollecting). We cannot conclude from the available data that the Braconidae are of either northern or southern origin.

The main division of the Braconidae is between the "cyclostomes" and relatives, and the remainder of the Braconidae (all koinobiont endoparasitoids, with pupation outside the host, except the small subfamily Meteorideinae, a basal member of the Sigalphoids). The cyclostomes are in general idiobiont ectoparasitoids or koinobiont endoparasitoids of holometabolous larvae with pupation inside the host. The non-cyclostomes form the sister group of the cyclostomes and their basal lineages are of much interest biogeographically. The Mesostoinae *s.s.* are restricted to Australia but Belshaw and Quicke (2002) include the Hydrangeocolinae and some genera of the tribe Austrohormiini, which extend the range to New Zealand and the Neotropical region. The Hydrangeocolinae are found in Australia and South America (Chile and part of Argentina), and the early development of the Aphidiinae also seems to have been restricted to Australia and the *Nothofagus* forests of Chile

(*Pseudephedrus*, Chile, followed by the probable sister groups *Choreopraon* [manuscript name], New Zealand, and *Parephedrus*, Australia; Belshaw et al., 2000). The Mesostoinae are gall-formers on *Banksia* (Proteaceae) (Austin and Dangerfield, 1998). The Hydrangeocolinae are parasitoids of Cecidomyiidae larvae in galls on *Mikania* (Asteraceae) (Oda et al., 2001) and supposedly on *Hydrangea* (Hydrangeaceae) species. The Aphidiinae are endoparasitoids of adult aphids, which may be considered "pseudogalls" on leaves from a biological viewpoint. According to Morley (1999) the genus *Nothofagus* arose about 90 Ma in a southern belt centred on South America and Eastern Gondwana and only invaded the Malesian area during Miocene (but Wikström et al., 2004, estimated the age of the Fagaceae at *c.* 60 Ma). The vicarious distribution in Hydrangeocolinae and the early radiation in Aphidiinae agree well with an origin by the severance of the connection between Australia and South America (through Antarctica). The family is certainly old enough for that, but whether the genera involved are old enough is a matter to be checked in future.

The biological data indicate that the occupation of the niche of galls and "pseudogalls" has been an important driving force in the basal evolution of the group. In some groups of the cyclostomes galls are parasitized (especially in the Braconinae), eggs are penetrated (ovo–larval parasitism), seeds are used (some Doryctinae, first for the larvae in the seeds but some becoming completely phytophagous), or became gall-formers themselves (some Rhyssalinae). Surprisingly, also in the cyclostomes these specialized groups are almost exclusively southern hemisphere in distribution, and most diverse in the Neotropical region.

The cyclostomes in South America have become most diverse in the basal groups such as Rhyssalinae and Doryctinae. It may be the result of less severe effects of the terminal Eocene cooling event in South America compared to Africa and southeast Asia (Morley, 1999). The Afrotropical and Oriental (southeast Asia) regions became most diverse in the more derived Braconinae, possibly because many of the old elements were wiped out and were replaced by the more derived Braconinae. In the northern hemisphere derived Aphidiinae (because of the radiation of aphids in the temperate regions) and derived groups as Opiinae and Alysiinae became very diverse (following the radiation of the dipterous hosts, especially of the leafminers). For the Braconidae in general the strong radiation of the Lepidoptera in the Mesozoic and Cenozoic (itself triggered by the radiation of angiosperms) must have been very important, offering a wide variety of food and niches.

## 6. Discussion

Evidence for the impact of the fragmentation of Pangea and Gondwana on the global distribution of butterflies is, at best, weak and even if we find better evidence in future then still the present-day distribution of the butterflies appears to be overwhelmingly the result of dispersal. Also in Stephanidae, although probably much older than the butterflies and certainly old enough to have been present in Gondwana, dispersal seems to be the main process that has shaped present-day distribution. If Stephanidae ever lived in Gondwana before it became fragmented, then there are no traces left. In the Braconidae, at least as old as the Stephanidae, the earliest diversification bears traces of vicariance events, but at least 96% of the recent taxa of Braconidae belong to cosmopolitan groups, indicating that also in Braconidae it is mainly dispersal that has shaped the distribution across the globe.

In a recent study, Sanmartín and Ronquist (2004) combined and analyzed a large data set of 54 animal and 19 plant phylogenies in an attempt to establish the relative roles played by vicariance and dispersal in shaping southern hemisphere biotas. One of their interesting conclusions is that the impact of the Gondwanan break-up is greater on animal distribution than on plant distribution. Their study also emphasized that "the biogeographic history of the Southern Hemisphere cannot be entirely reduced to a simple sequence of vicariance events" (Sanmartin and Ronquist (2004: 216) and that dispersal has also played a part, as is becoming increasingly clear from recent studies. We think this is an understatement. For their study they selected published phylogenies meeting four criteria, among which (1) monophyletic groups, distributed almost exclusively in the southern hemisphere, and claimed to be of Gondwanan origin; and (2) represented in at least two (preferably three or more) of the areas studied. For the purpose of their study the selection is understandable, but it can hardly be considered a random or representative sample of southern hemisphere biota, being a very small fraction of southern biota, and a biased one too. Interchange of faunal elements between northern and southern continents after the break-up of Gondwana is well documented by fossils (see for instance the Introduction) and can be deduced from the occurrence of extant taxa on northern as well as on southern fragments that cannot be attributed to a former Pangean distribution because of age or because of restriction to only one southern and one northern fragment (e.g., de Jong, 2001, 2003).

In a most interesting review of the biogeography of insects based on a wealth of fossil data starting in the Devonian (360–415 Ma), Eskov (2002) convincingly argumented that there are an increasing number of "Gondwanan" insect taxa that have been encountered as fossils in Eurasia and North America (the same, but to a lesser degree, is also the case with several "Laurasian" taxa of which fossil representatives are found in the southern hemisphere), indicating that extant "Gondwanan" elements could well be relics of a former broader distribution, and that the links between the taxa on the Gondwana fragments may not be direct, but through the northern hemisphere. Eskov (2002: 434) concludes that "the value of the modern range appears to be of very limited significance" in drawing conclusions about a Gondwanan origin of a taxon. De Jong (2003) further argued that some extant "Gondwanan" taxa may not be relics of a once broader distribution, but later (post-break-up) invaders from the north that became extinct in the northern hemisphere when tropical conditions disappeared from there. In the absence of fossils the latter conclusion or dispersal between Gondwana fragments are inevitable if the divergence time between the sister groups on the Gondwana fragments is younger than the vicariance event (break-up).

In this paper we have examined global distribution patterns in (some) flying insects in an attempt to understand the underlying processes of dispersal, vicariance, and extinction. We do not pretend to give an exhaustive explanation of the distribution of the respective groups across the globe, not to mention the history of the biotas to which these insects belong. Distribution history is an integral part of the evolution of a species (and higher categories). In his thesis on the Lycaenidae of the Australasiatic region, Toxopeus (1930) aptly called the species a function of space and time. Individuals, populations, and species are not objects moving at random wherever their legs, wings, or the wind bring them. They are constrained by internal and external factors. They do not evolve in geographic isolation only; they evolve all the time, also while on the move. They originate and become extinct. The constraints lead to a pattern that forms the basis of a hypothesis of evolution that has been called *taxon pulse* by Erwin (1981, 1985). Starting from this hypothesis and a well-supported phylogeny of the subgenus *Inseliellum* of *Simulium* (Diptera), Spironello and Brooks (2003) successfully tested aspects of the MacArthur–Wilson theory of island biogeography (MacArthur and Wilson, 1967), while explaining the distribution and species richness of these flies across Pacific islands. Although Spironello and Brooks (2003) dealt with relatively recent events, similar processes must have been going on since there was terrestrial life and there were islands. It would be most challenging to examine how far these principles can also be applied at a larger, maybe even continental scale, not only explaining distributions of individual taxa, but also unevenly distributed radiations. A question that comes to mind is why in the Palaearctic the satyrine Nymphalidae are the dominant grass-feeding butterflies, while in the Nearctic the hesperiine Hesperiidae are the principal grass-feeders. Such questions are, however, far beyond the scope of the present paper. The message we like to convey is that dispersal and extinction are not processes that started after the continents had become separated. They do not act in isolation but are integral parts of the evolution of life, ongoing processes, now and then punctuated by vicariance events, the traces of which will become obliterated with time. It is most challenging to try finding these traces, but any suggestion of the distribution of diversity across the globe being due to vicariance events in the first place is, as a general statement, not likely to be true. Possibly the classic Greek had this in mind when they asserted παντα ρεί, ουδεν μενεί (everything flows, nothing remains).

## Acknowledgements

The senior author is grateful to Michael F. Braby, Andrew V. Z. Brower, Ming-Min Lee, Sören Nylin, Naomi E. Pierce, Felix A. H. Sperling, Roger U. Vila, Andrew D. Warren, Niklas Wahlberg, and Evgueni Zakharov, the co-authors of a study on the fylogeny of the butterflies based on morphological and molecular data (Wahlberg et al., 2005), for permission to make use of some, at the time still unpublished results. He wants to thank Niklas Wahlberg in particular for suggestions and comments on the manuscript. The junior author expresses his gratitude to Donald Quicke for long-standing cooperation and discussions, and for useful comments on the manuscript.

## References

Achterberg, C. van, 1995, Generic revision of the subfamily Betylobraconinae (Hymenoptera: Braconidae) and other groups with modified fore tarsus, *Zoologische Verhandelingen* **298**: 1–242.

Achterberg, C. van, 2002, A revision of the Old World species of *Megischus* Brullé, *Stephanus* Jurine and *Pseudomegischus* gen. nov., with a key to the genera of the family Stephanidae (Hymenoptera: Stephanoidea), *Zoologische Verhandelingen* **339**: 3–206.

Ackery, P.R., Jong, R. de, and Vane-Wright, R.I., 1999, The Butterflies: Hedyloidea, Hesperioidea and Papilionoidea, in: Kristensen (ed.), *Lepidoptera, Moths and Butterflies, Volume 1: Evolution, Systematics and Biogeography*, Walter de Gruyter, Berlin and New York, pp. 263–300.

Ackery, P.R., Smith, C.R., and Vane-Wright, R.I., 1995. *Carcasson's African Butterflies*, CSIRO, East Melbourne.

Ackery, P.R., and Vane-Wright, R.I., 1984, *Milkweed Butterflies, Their Cladistics and Biology*, British Museum (Natural History), London.

Audley-Charles, M.G., 1993, Geological evidence bearing upon the Pliocene emergence of Seram, an island colonizable by land plants and animals, in: Edwards, I.D., Macdonald, A.A., and Proctor, J. (eds), *Natural History of Seram*, Intercept, Andover, UK, pp. 13–18.

Austin, A.D., and Dangerfield, P.C., 1998, Biology of *Mesostoa kerri* Austin and Wharton (Insecta: Hymenoptera: Braconidae: Mesostoinae), an endemic Australian wasp that causes stem galls on *Banksia marginata* Cav, *Australian Journal of Botany* **46**: 559–569.

Behrensmeyer, A.K., Damuth, J.D., DiMichele, W.A., Potts, R., Sues, H.-D., and Wing, S.L., 1992, *Terrestrial Ecosystems Through Time*, The University of Chicago Press, Chicago and London.

Belshaw, R., Dowton, M., Quicke, D.L.J., and Austin, A.D., 2000, Estimating ancestral geographical distributions: a Gondwanan origin for aphid parasitoids? *Proceedings Royal Society of London* **267**: 491–496.

Belshaw, R., and Quicke, D.L.J., 2002, Robustness of ancestral state estimates: evolution of life history strategy in Ichneumonoid parasitoids, *Systematic Biology* **51**: 450–477.

Bozano, G.C. and Weidenhoffer, Z., 2001, *Guide to the Butterflies of the Palearctic Region. Lycaenidae, Part 1*, Omnes Artes, Milano.

Braby, M.F., 2000, *Butterflies of Australia* (2 volumes), CSIRO, Collingwood, VIC.

Bromham, L., Phillips, M.J., and Penny, D., 1999, Growing up with dinosaurs: molecular dates and the mammalian radiation, *Trends in Ecology and Evolution* **14**: 113–118.

Brown Jr., K.S., 1987, Biogeography and evolution of Neotropical butterflies, in: Whitmore, T.C. and Prance, G.T. (eds), *Biogeography and Quaternary History in Tropical America*, Clarendon Press, Oxford, UK, pp. 66–104.

Campbell, D.L., Brower, A.V.Z., and Pierce, N.E., 2000, Molecular evolution of the *wingless* gene and its implications for the phylogenetic placement of the butterfly family Riodinidae (Lepidoptera: Papilionoidea), *Molecular Biology and Evolution* **17**: 684–696.

Caterino, M.S., Reed, R.D., Kuo, M.M., and Sperling, F.A.H., 2001, A partitioned likelihood analysis of swallowtail butterfly phylogeny (Lepidoptera: Papilionidae), *Systematic Biology* **50**: 106–127.

Coope, G.R., 1973, Tibetan species of dung beetle from Late Pleistocene deposits in England, *Nature* **245**: 335–336.

Corbet, A.S., and H.M. Pendlebury, 1956, *The Butterflies of the Malay Peninsula*, 2nd Edition, Oliver and Boyd, Edinburgh and London.

Craw, R., 1990, New Zealand biogeography: a panbiogeographic approach, *New Zealand Journal of Zoology* **16**: 527–547.

Craw, R.C., Grehan, J.H., and Heads, M.J., 1999, *Panbiogeography. Tracking the History of Life*, Oxford University Press, New York and Oxford.

Croizat, L., 1984, Mayr vs. Croizat: Croizat vs. Mayr – an enquiry, *Tuatara* **27**: 49–66.

DeVries, P.J., 1987, *The Butterflies of Costa Rica and Their Natural History*, Princeton University Press, Princeton, NJ.

DeVries, P.J., 1997, *The Butterflies of Costa Rica and Their Natural History, Volume II: Riodinidae*, Princeton University Press, Princeton, NJ.

Donoghue, P.C.J., and Smith, M.P. (eds), 2004, *Telling the Evolutionary Time; Molecular Clocks and the Fossil Record*, CRC Press, Boca Raton, London, New York, Washington, DC.

Dorp, K. van, and Jong, R. de, 2004, Origin and relationship of the Copper butterflies (*Lycaena*) of South Africa and New Zealand, *Fourth Southern Connections Conference 2004, Programme and Abstracts*: 29.

Dowton, M., Belshaw, R., Austin, A.D., and Quicke., D.L.J., 2002, Simultaneous molecular and morphological analysis of Braconid relationships (Insecta: Hymenoptera: Braconidae) indicates independent mt-tRNA gene inversions within a single wasp family, *Journal of Molecular Evolution* **54**: 210–226.

Durden, C.J., and Rose, H., 1978, Butterflies from the middle Eocene: the earliest occurrence of fossil Papilionoidea (Lepidoptera), *Pearce-Sellards Series: Texas Memorial Museum* **29**: 1–25.

Ehrlich, P., 1958, The comparative morphology, phylogeny and higher classification of the butterflies (Lepidoptera: Papilionoidea), *Kansas University Science Bulletin* **39**: 305–370.

Eliot, J.N., 1973, The higher classification of the Lycaenidae (Lepidoptera): a tentative arrangement, *Bulletin British Museum (Natural History), Entomology* **28**: 373–505.

Eliot, J.N., 1990, Notes on the genus *Curetis* Hübner (Lepidoptera, Lycaenidae), *Transactions Lepidopterological Society of Japan* **41**: 201–225.

Erwin, T.L., 1981, Taxon pulses, vicariance, and dispersal: an evolutionary synthesis illustrated by carabid beetles, in: Nelson, G. and Rosen, D.E. (eds), *Vicariance Biogeography, a Critique*, Columbia University Press, NY, pp. 159–183.

Erwin, T.L., 1985, The taxon pulse: a general pattern of lineage radiation and extinction among carabid beetles, in: Ball, G.E. (ed.) *Taxonomy, Phylogeny and Zoogeography of Beetles and Ants*, Dr W. Junk Publishers, Dordrecht, The Netherlands, pp. 437–472.

Eskov, K. Yu., 2002. Geographical history of insects, in: Rasnitsyn, A.P. and Quicke, D.L.J. (eds), *History of Insects*, Kluwer Academic, Dordrecht, Boston, and London: 427–435.

Evans, W.H., 1951, *A Catalogue of the American Hesperiidae in the British Museum (Natural History). Part 1, Introduction and Group A. Pyrrhopyginae*, Trustees of the British Museum, London.

Field, W.D., 1971, Butterflies of the genus *Vanessa* and of the resurrected genera *Bassaris* and *Cynthia* (Lepidoptera: Nymphalidae), *Smithsonian Contributions to Zoology* **84**: 1–105.

Forey, P.L., Humphries, C.J., Kitching, I.J., Scotland, R.W., Siebert, D.J., and Williams, D.M., 1992, *Cladistics. A Practical Course in Systematics*, Clarendon Press, Oxford, UK.

Gibbs, G., Kobayashi, Y., Suzuki, H., and Hashimoto, S., 2004, Micro-moths trace the history of mega-continents, *Programme and Abstracts Southern Connection 2004*, Cape Town, p. 56.

Graur, D., and Martin, W., 2004, Reading the entrails of chickens: molecular timescales of evolution and the illusion of precision, *Trends in Genetics* **20**: 80–86.

Grehan, J.R., 1991, A panbiogeographic perspective for pre-Cretaceous Angiosperm-Lepidoptera coevolution, in: P.Y. Ladiges, P.Y., Humphries C.J., and Martinelli, L.W. (eds), *Austral Biogeography*, CSIRO, Australia, pp. 91–110.

Hall, J.P.W., and Harvey, D.J., 2002, A survey of androconial organs in thre Riodinidae (Lepidoptera), *Zoological Journal of the Linnean Society* **136**: 171–197.

Hall, J.P., Robbins, R.K., and Harvey, D.J., 2004, Extinction and biogeography in the Caribbean: new evidence from a fossil riodinid butterfly in Dominican amber, *Proceedings of the Royal Society of London B* **271**: 797–801.

Hall, R., 1998, The plate tectonics of Cenozoic SE Asia and the distribution of land and sea, in: Hall, R., and Holloway, J.D. (eds), *Biogeography and Geological Evolution of SE Asia*, Backhuys Publishers, Leiden, pp. 99–131.

Hall, R., 2002, Cenozoic geological and plate tectonic evolution of SE Asia and the SW Pacific: computer-based reconstructions, model and animations, *Journal of Asian Earth Sciences* **20**: 353–434.

Hallam, A., 1994, *An Outline of Phanerozoic Biogeography*, Oxford University Press, Oxford.

Honey, J.G., Harrison, J.A., Prothero, D.R., and Stevens, M.S., 1998, Camelidae, in: Janis, C.M., Scott, K.M., and Jacobs, L.L. (eds), *Evolution of Tertiary Mammals of North America, Volume 1: Terrestrial Carnivores, Ungulates, and Ungulatelike Mammals*, Cambridge University Press, Cambridge, NY, pp. 439–462.

Hopkins, D.M., 1967, The Cenozoic history of Beringia – a synthesis, in: Hopkins, D.M. (ed.), *The Bering Land Bridge*, Stanford University Press, Stanford, CA, pp. 451–484.

Humphries, C.J., and Parenti, L.R., 1999, *Cladistic Biogeography*, 2nd Edition, Oxford University Press, Oxford.

Igarashi, S., and Fukuda, H., 2000, *The Life Histories of Asian Butterflies*, vol. 2, Tokyo University Press, Tokyo.

Jong, R. de, 1982, Secondary sexual characters in *Celaenorrhinus* and the delimitation of the genus (Lepidoptera, Hesperiidae), *Journal of Natural History* **16**: 695–705.

Jong, R. de, 2001, Faunal exchange between Asia and Australia in the Tertiary as evidenced by recent butterflies, in: Metcalfe, I., Smith, J.M.B., Morwood, M., and Davidson, I. (eds), *Faunal and Floral Migrations and Evolution in SE Asia-Australasia*, A.A. Balkema Publishers, Lisse, pp. 133–146.

Jong, R. de, 2003, Are there butterflies with Gondwanan ancestry in the Australian region? *Invertebrate Systematics* **17**: 143–156.

Jong, R. de, Vane-Wright, R.I., and Ackery, P.R., 1996, The higher classification of butterflies (Lepidoptera): problems and prospects, *Entomologica Scandinavica* **27**: 65–101.

Kandul, N.P., Lukhtanov, V.A., Dantchenko, A.V., Coleman, J.W.S., Sekercioglu, C.H., Haig, D., and Pierce, N.E., 2004, Phylogeny of *Agrodiaetus* Hübner 1822 (Lepidoptera: Lycaenidae) inferred from mt DNA sequences of *COI* and *COII* and nuclear sequences of *EF1*-α: karyotype diversification and species radiation, *Systematic Biology* **53**: 278–298.

Kawahara, A.Y., 2003, Rediscovery of *Libythea collenettei* Poulton and Riley (Nymphalidae: Libytheinae) in the Marquesas, and a description of the male, *Journal of the Lepidopterists' Society* **57**: 81–85.

Kemp, E.M., 1981, Tertiary palaeogeography and the evolution of Australian climate, in: Keast, A. (ed.), *Ecological Biogeography of Australia*, Dr. W. Junk B.V. Publishers, The Hague, pp. 33–49.

Klots, A.B., 1931–1932, A generic revision of the Pieridae (Lepidoptera), together with a study of the male genitalia, *Entomologica Americana* **12**: 139–204 (1931), 205–242 (1932).

Kristensen, N.P. and Skalski, A.W., 1999, Phylogeny and Palaeontology, in: Kristensen, N.P. (ed.), *Lepidoptera, Moths and Butterflies, Volume. 1: Evolution, Systematics, and Biogeography, Handbook of Zoology*, IV (35), Walter de Gruyter, Berlin and New York, pp. 7–25.

Lawver, L.A., and Gahagan, L.M., 1998, Opening of the Drake Passage and its impact on Cenozoic ocean circulation, in: Crowley, T.J. and Burke, K.C. (eds), *Tectonic Boundary Conditions for Climate Reconstructions, Oxford Monographs on Geology and Geophysics* **39**: 212–223.

Lovis, J.D., 1989, Timing, exotic terranes, angiosperm diversification, and panbiogeography, *New Zealand Journal of Zoology* **16**: 713–729.

MacArthur, R.H. and Wilson, E.O., 1967, *The Theory of Island Biogeography*, Princeton University Press, Princeton, NJ.

Maravalhas, E., 2003, *As Borboletas de Portugall*, Apollo Books, Stenstrup, Denmark.

McLoughlin, S., 2001, The breakup history of Gondwana and its impact on pre-Cenozoic floristic provincialism, *Australian Journal of Botany* **49**: 271–300.

Megens, H.-J., van Moorsel, C.H.M., Piel, W.H., Pierce, N.E., and Jong, R. de, 2004, Tempo of speciation in a butterfly genus from the Southeast Asian tropics, inferred from mitochondrial and nuclear DNA sequence data, *Molecular Phylogenetics and Evolution* **31**: 1181–1196.

Metcalfe, I., 1998, Palaeozoic and Mesozoic geological evolution of the SE Asian region: multidisciplinary constrains and implications for biogeography, in: Hall, R. and Holloway, J.D. (eds), *Biogeography and Geological Evolution of SE Asia*, Backhuys Publishers, Leiden, pp. 25–41.

Metcalfe, I., 1999, Gondwana dispersion and Asian accretion: an overview, in: Metcalfe, I. (ed.), *Gondwana Dispersion and Asian Accretian*, Balkema, Rotterdam, pp. 9–28.

Metcalfe, I., 2001, Palaeozoic and Mesozoic tectonic evolution and biogeography of SE Asia-Australasia, in: Metcalfe, I., Smith, J.M.B, Morwood, M., and Davidson, I. (eds), *Faunal and Floral Migrations and Evolution in SE Asia-Australasia*, A.A. Balkema Publishers, Lisse, pp. 15–34.

Miller, J.S., 1987, Phylogenetic studies in the Papilioninae (Lepidoptera : Papilionidae), *Bulletin of the American Museum of Natural History* **186**: 365–512.

Miller, J.Y. and Brown, F.M. 1989, A new Oligocene fossil butterfly, *Vanessa amerindica* (Lepidoptera: Nymphalidae), from the Florissant formation, Colorado, *Bulletin Allyn Museum* **126**: 1–9.

Miller, L.D. and Brown, F.M., 1979, Studies in the Lycaeninae (Lycaenidae). 4. The higher classification of the American Coppers, *Bulletin Allyn Museum* **51**: 1–30.

Miller, L.D. and Miller, J.Y., 1989, The biogeography of West Indian butterflies (Lepidoptera: Papilionoidea, Hesperioidea): a vicariance model, in: Woods, C.A. (ed.), *Biogeography of the West Indies, Past, Present, and Future*, Sandhill Crane Press, Gainesville, FL, pp. 229–262.

Miller, L.D. and Miller, J.Y., 1997, Gondwanan butterflies: the Africa-South America connection, *Metamorphosis* **3** (Suppl.): 42–51.

Morinaka, S., Maeyama, T., Maekawa, K., Erniwati, D., Prijono, S.N., Ginarsa, I.K., Nakazawa, T., and Hidaka, T., 1999, Molecular phylogeny of birdwing butterflies based on the representatives in most genera of the tribe Troidini (Lepidoptera: Papilionidae), *Entomological Science* **2**: 347–358.

Morley, R.J., 1999, *Origin and Evolution of Tropical Rain Forests*, Wiley, Chichester, UK.

Oda, R.A.M., Valverde de Macédo, M., and Quicke, D.L.J., 2001, First biological data for *Aspilodemon* Fischer (Hymenoptera: Braconidae: Hydrangeocolinae): parasitoids of Cecidomyiid fly galls on Asteraceae in Brazil, *Journal of Hymenoptera Research* **10**: 126–130.

Parsons, M.J., 1996a, A phylogenetic reappraisal of the birdwing genus Ornithoptera (Lepidoptera: Papilionidae: Troidini) and a new theory of its evolution in relation to Gondwanan vicariance biogeography, *Journal of Natural History* **30**: 1707–1736.

Parsons, M., 1996b, Gondwana evolution of the troidine swallowtails (Lepidoptera: Papilionidae): cladistic reappraisals using mainly immature stage characters, with focus on the birdwings *Ornithoptera* Boisduval, *Bulletin of the Kitakyushu Museum of Natural History* **15**: 43–118.

Parsons, M.J., 1998, *The Butterflies of Papua New Guinea. Their Systematics and Biology*, Academic Press, San Diego, CA.

Peña, L.E. and Ugarte, A.J., 1996, *Las Mariposas de Chile*, Editorial Universitaria, Santiago de Chile.

Pitman III, W.C., Cande, S., LaBrecque, J., and Pindell, J., 1993, Fragmentation of Gondwana: the separation of Africa from South America, in: Goldblatt, P. (ed.), *Biological Relationships Between Africa and South America*, Yale University Press, New Haven and London, pp. 15–34.

Powell, J.A., Mitter, C., and Farrell, B., 1999, Evolution of Larval Food Preferences in Lepidoptera, in: Kristensen, N.P. (ed.), *Lepidoptera, Moths and Butterflies, Volume. 1: Evolution, Systematics and Biogeography*, Walter de Gruyter, Berlin and New York, pp. 403–422.

Qiu, Y.-L., Lee, J., Bernasconi-Quadroni, F., Soltis, D.E., Soltis, P.S., Zanis, M., Zimmer, E.A., Chen, Z., Savolainen, V., and Chase, M.W., 1999, The earliest angiosperms: evidence from mitochondrial, plastid and nuclear genomes, *Nature* **402**: 404–407.

Quicke, D.L.J., Lopez-Vaamonde, C., and Belshaw, R., 1999, The basal Ichneumonidae (Insecta, Hymenoptera): 28S D2 rDNA considerations of the Brachycyrtinae, Labeninae, Paxylommatinae and Xoridinae, *Zoologica Scripta* **28**: 203–210.

Rasnitsyn, A.P., 1983, Ichneumonoidea (Hymenoptera) from the lower Cretaceous of Mongolia. *Contributions of the American Entomological Institute* **20**: 259–265.

Rasnitsyn, A.P., 1988, An outline of evolution of the hymenopterous insects (Order Vespida), *Oriental Insects* **22**: 115–145.

Rasnitsyn, A.P., 2002, Superorder Vespidea Laicharting, 1781. Order Hymenoptera Linné, 1758 (=Vespida Laicharting, 1781), in: Rasnitsyn, A.P. and Quicke, D.L.J. (eds), *History of Insects*, Kluwer Academic, Dordrecht, Boston and London, pp. 242–254.

Ronquist, F., Rasnitsyn, A.P., Roy, A., Eriksson, K., and Lindgren, M., 1999, Phylogeny of the Hymenoptera: a cladistic reanalysis of Rasnitsyn's (1988) data, *Zoologica Scripta* **28**: 13–50.

Sanmartín, I. and Ronquist, F., 2004, Southern Hemisphere biogeography inferred by event-based models: plant versus animal patterns, *Systematic Biology* **53**: 216–243.

Schlüter, T., 2000. *Moltenia rieki* n. gen., n. sp. (Hymenoptera: Xyelidae?), a tentative sawfly from the Molteno Formation (Upper Triassic), South Africa, *Paläontologische Zeitschrift* **74**: 75–78.

Schulmeister, S., Wheeler, W.C., and Carpenter, J.M., 2002, Simultaneous analysis of the basal lineages of Hymenoptera (Insecta) using sensitivity analysis, *Cladistics* **18**: 455–484.

Scott, J.A., 1986, *The Butterflies of North America*, Stanford University Press, Stanford, CA.

Scott, J.A., and Wright, D.M., 1990, Butterfly fylogeny and fossils, in: Kudrna, O. (ed.), *Butterflies of Europe, Volume 2: Introduction to Lepidopterology*, Aula Verlag, Wiesbaden, pp. 152–208.

Shaw, S.R., 1988, *Carminator*, a new genus of Megalyridae (Hymenoptera) from the Oriental and Australian regions, with a commentary on the definition of the family, *Systematic Entomology* **13**: 101–113.

Shields, O., 1976, Fossil butterflies and the evolutionof Lepidoptera, *Journal of Research on the Lepidoptera* **15**: 132–143.

Shields, O., 1979, Evidence for initial opening of the Pacific Ocean in the Jurassic, *Palaeogeography, Palaeoclimatology, Palaeoecology* **26**: 181–220.

Shields, O., 1989, Systematic position of Libytheidae, Diphylogeny of Rhopalocera and Heteroceran Ancestry of Rhopalocera (Lepidoptera), *Tyô to Ga* **40**: 197–228.

Shields, O. and Dvorak, S.K., 1979, Butterfly distribution and continental drift between the Americas, the Caribbean and Africa, *Journal of Natural History* **13**: 221–250.

Shoshani, J., and Tassy, P. (eds), 1996, *The Proboscidea. Evolution and Palaeoecology of Elephants and Their Relatives*, Oxford University Press, Oxford, UK.

Smith, A.G., Smith, D.G., and Funnel, B.M., 1994, *Atlas of Mesozoic and Cenozoic Coastlines,* Cambridge University Press, Cambridge, UK.

Spironello, M. and Brooks, D.R., 2003, Dispersal and diversification: macroevolutionary implications of the MacArthur-Wilson model, illustrated by *Simulium* (*Inseliellum*) Rubstov (Diptera: Simuliidae), *Journal of Biogeography* **30**: 1563–1573.

Stoneham, H.F., 1965, *The Butterflies of Western Kenya. Fasc. 5*, Stoneham Museum Publications, Nairobi.

Strauss, E., 1999, Can mitochondrial clocks keep time? *Science* **283**: 1435–1438.

Toxopeus, L.J., 1930, *De soort als functie van plaats en tijd, getoetst aan de Lycaenidae van het Australaziatische gebied*, H.J. Paris, Amsterdam.

Trewick, A.S., 2000, Molecular evidence for dispersal rather than vicariance as the origin of flightless insect species on the Chatham Islands, New Zealand, *Journal of Biogeography* **27**: 1189–1200.

Tyler, H., Brown, K.S., and Wilson, K., 1994, *Swallowtail Butterflies of the Americas*, Scientific Publishers, Gainesville, FL.

Vane-Wright, R.I., and Jong, R. de, 2003, The butterflies of Sulawesi: annotated checklist for a critical island fauna, *Zoologische Verhandelingen* **343**: 3–267.

Vilhelmsen, L., 2001, Phylogeny and classification of the extant basal lineages of the Hymenoptera (Insecta), *Zoological Journal of the Linnean Society* **131**: 393–442.

Wahlberg, N., Weingartner, E., and Nylin, S., 2003, Towards a better understanding of the higher systematics of Nymphalidae (Lepidoptera: Papilionoidea), *Molecular Phylogenetics and Evolution* **28**: 473–484.

Wahlberg, N., Braby, M.F., Brower, A.V.Z., Jong, R. de, Lee, M.-M., Nylin, S., Pierce, N.E., Sperling, F.A.H., Vila, R.U., Warren, A.D., and Zakharov, E., 2005, Synergistic effects of combining morphological and molecular data in resolving the phylogeny of butterflies and skippers, *Proceedings of the Royal Society B* 272: 1577–1586.

Wikström, N., Savolainen, V., and Chase, M.W., 2001, Evolution of the angiosperms: calibrating the family tree, *Proceedings Royal Society London B* **268**: 2211–2220.

Wikström, N., Savolainen, V., and Chase, M.W., 2004, Angiosperm divergence times: congruence and incongruence between fossils and sequence divergence estimates, in: Donoghue, P.C.J. and Smith, M.P. (eds), *Telling the Evolutionary Time; Molecular Clocks and the Fossil Record*, CRC Press, Boca Raton, London, New York, Washington DC, pp. 142–165.

Willis, K.J., and McElwain, J.C., 2002, *The Evolution of Plants*, Oxford University Press, Oxford, UK.

Winkworth, R.C., Wagstaff, S.J., Glenny, D., and Lockhart, P.J., 2002, Plant dispersal N.E.W.S. from New Zealand, *Trends in Ecology and Evolution* **17**: 514–520.

Zakharov, E.V., Caterino, M.S., and Sperling, F.A.H., 2004, Molecular phylogeny, historical biogeography, and divergence time estrimates for Swallowtail Butterflies of the genus *Papilio* (Lepidoptera: Papilionidae), *Systematic Biology* **53**: 193–215.

Zeuner, F.E., 1961, Notes on the evolution of the Rhopalocera (Lep.), *Verhandlungen des XI Internationalen Kongresses für Entomologie 1* (1960) [1961]: 310–313.

Chapter 2

# Zoogeography of Freshwater Invertebrates of Southeast Asia, with Special Reference to Odonata

JAN VAN TOL[1] AND DIRK GASSMANN[2]
[1]Department of Entomology, Nationaal Natuurhistorisch Museum Naturalis, P.O. Box 9517, 2300 RA Leiden, The Netherlands, Tol@naturalis.nl
[2]Institute of Biology, University of Leiden c/o Nationaal Natuurhistorisch Museum Naturalis, P.O. Box 9517, 2300 RA Leiden, The Netherlands, gassmann@rulsfb.leidenuniv.nl



**Abstract** The present knowledge of the historical biogeography of aquatic invertebrate groups is reviewed. Most orders of aquatic insects have a fossil record starting in the Early Permian, or Middle Carboniferous (Odonata), making even the break-up of Gondwana (Late Jurassic) relevant to understanding present distributional patterns. The complex geological history of Southeast Asia is summarized, and geological area cladograms presented. Biogeographical studies are seriously hampered by the limited information on subaerial history of the various islands and terranes. The historical biogeography of the Platycnemididae (Odonata), with special reference to the subfamily Calicnemiinae, is presented as one of the first examples of such a study of a widespread group. The species of



*W. Renema (ed.), Biogeography, Time, and Place: Distributions, Barriers, and Islands,* 45–91

southeast Asia derived from African Platycnemididae. Malesian Calicnemiinae derived from ancestors on the mainland of Asia, and may have dispersed along the Izu-Bonin Arc (40–50 Ma), or along the Late Cretaceous "Inner Melanesian Arc" sensus Polhemus. A clade of the genera *Lieftinckia* and *Risiocnemis* (Solomon Islands and the Philippines) represents a more recent westward dispersal of the Calicnemiinae, via the Caroline and Philippine Arcs during the Oligocene.

Various other more limited phylogenetic reconstructions and biogeographical analyses of other freshwater invertebrates, particularly Odonata and Hemiptera, are discussed. Areas of endemism on New Guinea are generally congruent with geological entities recognized, e.g., the microterranes along the northern margin of New Guinea. Special attention is paid to the fauna of Sulawesi. Area cladistic reconstructions based on distribution patterns and phylogenetic reconstructions of, e.g., *Protosticta* Selys (Odonata, Platystictidae) and genera and species of Chlorocyphidae (Odonata), show a pattern of (northern arm (southwest arm – central and southeastern arm)), which is a reflection of the geological history of the island.

Biogeographical patterns recognized in freshwater invertebrates of Malesia do not principally differ from those found in strictly terrestrial taxa. The distribution of land and water seems to be the driving force in speciation during the Cenozoicum. It is unresolved whether rafting of biotas on the various island arcs, or congruent patterns in dispersal, are to be considered the underlying principle. The extreme habitat requirements and poor dispersal power of many species involved seem to make a dispersal scenario unlikely. However, recent studies show that such habitat specialization may develop rapidly.

> *Facts such as these can only be explained by a bold acceptance of vast changes in the surface of the earth.* (Wallace, 1860: 177)

## 1. Introduction

Recently, de Bruyn et al. (2004) found extensive genetic divergence between wild populations of the giant freshwater prawn *Macrobrachium rosenbergii* (De Man) in southeast Asia. This species of prawn occurs in the wild from Pakistan to Australia and on some Pacific islands, and it is cultured widely around the world in more than 40 countries (Mather and de Bruyn, 2003). It is of high economic importance for some regions in southeast Asia, with harvesting of wild populations alone exceeding a value of US$800 million in 1998. In the 1990s, harvest of several stocks in culture experienced a decline, presumably due to inbreeding. Consequently, wild populations are important sources of genetic diversity to overcome inbreeding problems, but *M. rosenbergii* is rapidly declining in the wild due to overharvesting and habitat loss. Mating between specimens of different parts of the species range resulted in reduced larval survival, although heterosis (hybrid vigour) was found for other populations from the same region. Obviously, a better understanding of the genetic diversity is needed to contribute to improved aquaculture of this species. Study of the variation in 16S ribosomal DNA (de Bruyn et al., 2004) proved to be successful in describing the evolutionary relationships in this species, and supported previous allozyme and morphological work that had identified an eastern and a western form (Holthuis, 1995). The boundary between both "forms" proved to be Huxley's (1868) line (Fig. 1), the biogeographically based division of the Oriental and Australian regions running between Palawan and the rest of the Philippines in the north, then southward between Borneo and Sulawesi (Celebes), and between Bali and Lombok. It only differs from Wallace's (1863) line in the position of the Philippines. Recent data, based on morphological studies, showed that the distribution of these species differs in details from a division as by Huxley's line (Wowor, 2004, cf. Fig. 1).

While it may be true that not all lack of knowledge on the zoogeography and phylogenetic relationships of species has similar economical implications as the example of *M. rosenbergii*, it may serve as an example that the historical relationships of aquatic invertebrates and their distributions are still poorly examined even for better known species. It may also demonstrate that phylogenetic and biogeographical understanding is not only a scientific problem, but may also have practical, e.g., economic, consequences.

The example of *Macrobrachium* also raises another issue, namely the mechanism or mechanisms by which present patterns have evolved. Based on current knowledge of the palaeogeography of the region, island-hopping along terranes or island arcs during the Cenozoicum has been hypothesized to explain the present patterns in some groups. Such a mechanism may seem likely for groups such as birds, butterflies, cicadas, and can even be defended for mostly aquatic groups with

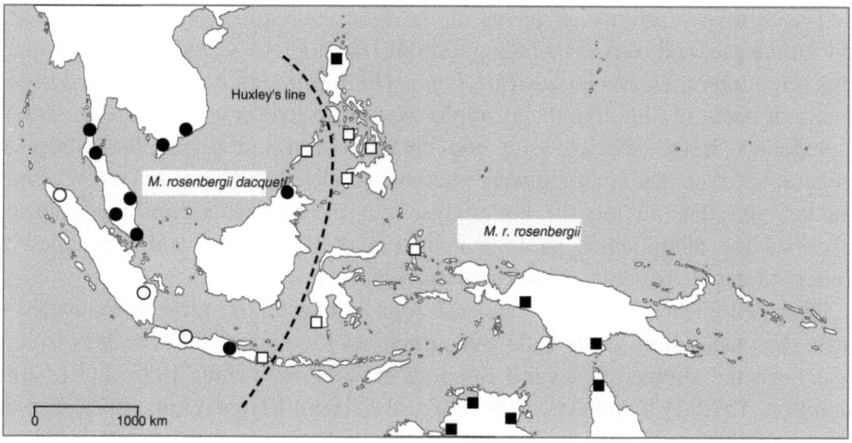


Fig. 1 Distribution of *Macrobrachium rosenbergii* (de Man) and *M. dacqueti* (Sunier) in southeast Asia (black symbols, after de Bruyn et al. (2004); some essential additional records from Wowor (2004) with open symbols).

a terrestrial adult stage such as the caddisflies. However, such a mechanism seems less likely for prawns such as *Macrobrachium*, although they are also known from some Pacific islands. *M. rosenbergii* has a tolerance for salt, which potentially increases the dispersal power, but distribution patterns of marine organisms in the Indo-West Pacific are typically related to patterns of ocean currents (Briggs, 1974; Hoeksema, this volume). A study at the molecular level could possibly unravel whether "human-mediated dispersal" may have played a role (Diamond, 1988). A similar unlikely distribution pattern was studied by Austin (1999) in the lizard *Lipinia noctua* (Lesson), which does occur in human settlements, and was probably transported in canoes by the Polynesians as far as the Marquesas Islands, Tuamotu and Hawaii.

In this paper we will examine the distribution patterns of aquatic invertebrates in southeast Asia, especially in relation to the geological history of the region. The historical relationships and the present distributions of most groups of aquatic invertebrates are insufficiently known to follow the example of Turner et al. (2001) in reconstructing generalized area cladistic relationships based on aquatic invertebrates. We here present a summary of the much-scattered knowledge of various taxonomic groups, and also demonstrate the congruences of various area cladograms based on reconstructions of phylogenies as compared to palaeogeographical reconstructions. Special attention will be paid to Odonata, the dragonflies and damselflies. New data are available for this order of insects, especially those obtained by the junior author for the calicnemiine Platycnemididae. We will also summarize results of some more limited studies of the senior author and others. Finally, the data from aquatic invertebrates will be compared with present knowledge of the area relationships obtained from other groups.

Schuh and Stonedal (1986), and more recently Turner et al. (2001), tried to reconstruct the historical biogeography of the southeast Asian region. Turner et al. used such diverse groups as plants of the families Sapindaceae, Euphorbiaceae, and Rubiaceae, and insects including cicadas (Homoptera, Cicadidae), semiaquatic bugs (Hemiptera, *Haloveloides*, *Halobates*, *Halovelia*, and *Xenobates*) and several genera of beetles. Although the examples were selected for carefully reconstructed phylogenies at the species level and detailed information on distribution, the "general patterns that emerged were weakly supported and [did] not allow general conclusions". The authors did not analyse why the reconstruction failed, but they described the complicated geological history of the region, and mentioned the process of active dispersal of biotas along island arcs.

The geological history of southeast Asia is one of the most complicated on earth. Reconstructions of the palaeogeography of the region since the Mesozoicum have been the subject of several research groups (e.g., Hall, 1998, 2001, 2002; Hamilton, 1979; Hill and Hall, 2003; Kroenke, 1996; Rangin et al., 1990; Yan and Kroenke, 1993) and have thus greatly improved in details, but information on the scale necessary for biogeographical studies of terrestrial organisms is still scarce. It is, for instance, still poorly known which areas were submerged for a shorter or longer period of time during their history. And, although it is known that some islands in the region have moved along the Pacific or Philippine plates over a long distance during the last 10–15 My, their positions during this period differ significantly between the various studies. The analysis is further complicated by island arcs of the past that have been (nearly) fully absorbed by plate movements later in the geological history. We will describe present knowledge of land masses, microcontinents and island arcs as far as relevant for the present study. Since many extant families of some insect orders, e.g., the Odonata, are already known since the Jurassic, even details of the break-up of Pangaea are relevant. The palaeogeography of the Cenozoicum of the Malay archipelago and the West Pacific is described in more detail to enable comparison of the area cladograms at the generic or species group level in selected families.

## 2. History of Aquatic Invertebrates

It may be questioned how far a geological history may be traced back in patterns of extant taxa. Is it reasonable to reconstruct distributional histories of groups from as long ago as the break-up of Pangaea? Apart from knowledge of palaeogeography, it is necessary to know how long families, or even genera and species, have existed. Since estimates based on molecular data are sparse and their reliability under discussion, we are dependent on data of the fossil record up to now.

Several observations indicate that even species may persist for many millions of years. Kathirithamby and Grimaldi (1993) mention a record of *Bohartilla megalognata* Kinzelbach, an extant species of Strepsiptera, from the Miocene Dominican amber (20 Ma), while Rasnitsyn (2002) mentions that such examples are even available for Baltic amber (*c*. 40 Ma).

It is generally known that the fossil record is incomplete and biased. Carle (1995), for instance, discussed the overwhelming abundance of dragonflies of lentic habitats in the fossil record, while most species of extant anisopteran families are obligate inhabitants of streams and seepage areas. Such species are, however, rarely preserved as fossils, since they have small population sizes and their habitats are less suitable for preservation of fossil specimens. Since small stream habitats have permanently existed at least since the Jurassic, the inhabitants of this habitat have been able to survive up to today without significant morphological change, while faunas of lentic habitats became extinct when ponds and lakes dried up. Consequently, when a new lentic habitat developed, the settlement of other biotic lineages provided new opportunities for local evolution.

Here we will examine the data of the age of various groups of invertebrates, especially insects, based on the fossil record. The affinities of the orders of the Insecta have recently been extensively discussed by Wheeler et al. (2001), while Rasnitsyn and Quicke (2002) provide a thorough summary of the knowledge of the geological history of the insect orders. Sinitshenkova (2002b) provides a summary for the aquatic insect orders in chronological order, including an interpretation in ecological context.

Many orders of aquatic insects, or at least those with aquatic larval stages, are known in the fossil record as early as the Early Permian, namely in the terminology of Rasnitsyn and Quicke (2002), Ephemerida, Hemiptera, Coleoptera, Neuroptera, and Trichoptera. The earliest fossils are those of Libellulida (i.e., Odonata) from the middle Carboniferous (*c.* 325 Ma). Somewhat later in the fossil record appear the Corydalida and Perlida (i.e., Plecoptera), from middle Permian (c. 299–250 Ma), while the Diptera are not known from the Palaeozoicum, but only from the Middle Triassic (*c.* 228–245 Ma) onwards. We do not discuss extinct orders with aquatic stages in the present context.

Not only many of the present orders, but also many extant superfamilies have a long geological record, and are known from the Mesozoicum. Most superfamilies of the Odonata are recorded from the Late Jurassic or Early Cretaceous (150–135 Ma) (Rasnitsyn and Pritykina, 2002). New studies of fossils show that all superfamilies of the Odonata had developed before the Cretaceous (135 Ma). Various extant families of the suborder Calopterygina are known from the supercontinents Gondwana (Brazil) as well as from Laurasia (England) from that period.

For the other groups of aquatic insects that will be discussed below, the following data are available.

Ephemerida (i.e., Ephemeroptera). Several superfamilies (Oligoneuroidea, Ephemeroidea, Leptophlebioidea) are known from the Early Cretaceous (Kluge and Sinitshenkova, 2002).

Perlida (i.e., Plecoptera). A group with many plesiomorphic characters. Fossils are uncommon in most deposits, since virtually all species are rheophilic and such species hardly enter the fossil record (see above). The oldest fossils known are Permian (c. 299–250 Ma) (Sinitshenkova, 2002a). Recent families seem to be much younger. Nemouridae are only known from the Early Cretaceous. Permian stoneflies were widely distributed and are known from both the northern and southern hemisphere, including Australia, South Africa, and Antarctica. Stoneflies were common during the Jurassic.

The superfamilies of the aquatic hemipteran infraorder Nepomorpha all appear in the fossil record in the Late Triassic (*c.* 210 Ma), while the earliest Gerromorpha (semiaquatic water bugs of the superfamily Hydrometroidea) are known from the Early Cretaceous (Shcherbakov and Popov, 2002). Fossils from the Santana formation of Brazil indicate that all modern families of Heteroptera had evolved by at least the Cretaceous (D.A. Polhemus, personal communication, 2005).

Although the order Trichoptera is known from the early Permian onwards, extant families appear later in the fossil record, e.g., Rhyacophilidae from Middle Jurassic, and most other groups even much later (e.g., Hydropsychidae from the Eocene, *c.* 50 Ma, only) (Ivanov and Sukatsheva, 2002).

In summary, the fossil record indicates that most groups of insects had developed as early as 150–200 Ma. During the break-up of Gondwana, that started in the Late Jurassic (*c.* 152 Ma) (cf. McLoughlin, 2001), but continued in more extensive form during the Cretaceous, most families here under discussion were represented.

## 3. Geological History of Southeast Asia

### 3.1. Geological History of the Southeast Asian Continent, the Malay Archipelago, and the West Pacific

Since most groups had developed as early as the end of the Palaeozoicum, much of the present knowledge on the geological history of the region is relevant for the reconstruction of the history of present distributions.

The study of the geological and tectonic evolution of southeast Asia has been intensified during the last decades. Various summarizing papers, also by biogeographers, are available, and an intriguing picture is emerging of the historical relationships of the presently existing land masses. The summary below will focus on the general patterns and on some details relevant for the distribution patterns of groups discussed. Few regions of the earth have changed so dramatically as southeast Asia during the last 100 My. Besides, this process of rapid change continues up to today. Not more than 10 Ma the position of the island of Halmahera (Moluccas) was northeast of the Bird's Head Peninsula of New Guinea, and Halmahera approximately had the position of Manus Island today by the end of the Oligocene (25 Ma) (Hall 2002: 407, see also Fig. 4). These data add a new dimension to our understanding of the evolution of the present distribution patterns of biotas. After our summary of the palaeogeography, the historical relationships of the various "areas of endemism" based on the geological reconstructions are discussed. These relationships are described in a so-called geological area cladogram.

### 3.2. Mesozoicum

By the end of the Permian (250 Ma) the continents were still connected as Pangaea. Several slivers of continent rifted northward towards Eurasia forming the Palaeotethys (between North China and the Cimmerian continent) and the Mesotethys (between the Cimmerian continent and the still connected continental parts of, e.g., Birma at southern latitudes) (Metcalfe, 2001). The Cimmerian continent included the so-called Sibumasu terrane, now forming parts of Thailand, the Malay Peninsula, and northern Sumatra (Fig. 2). The Sibumasu terrane amalgated with Indochina and South China during the Triassic (200–250 Ma). Based on the fossil record, it is presumed that this terrane has had a history above sea level since the Triassic. Another sliver of continent or arc of terranes, including Lhasa, West Burma, and Western Sulawesi, was separated from Gondwana during the late Triassic, opening the Ceno-Tethys. These terranes accreted to the Sibumasu terrane during the Cretaceous. Southwestern Borneo had a position at the southeastern margin of the Eurasian continent at least since the Jurassic.

India (with Madagascar and the Seychelles) became isolated from Africa at *c.* 130 Ma, wherafter India and the Seychelles separated from Madagascar 88 Ma (cf. Bossuyt and Milinkovitch, 2001). Eruption of the Deccan flood basalts resulted,

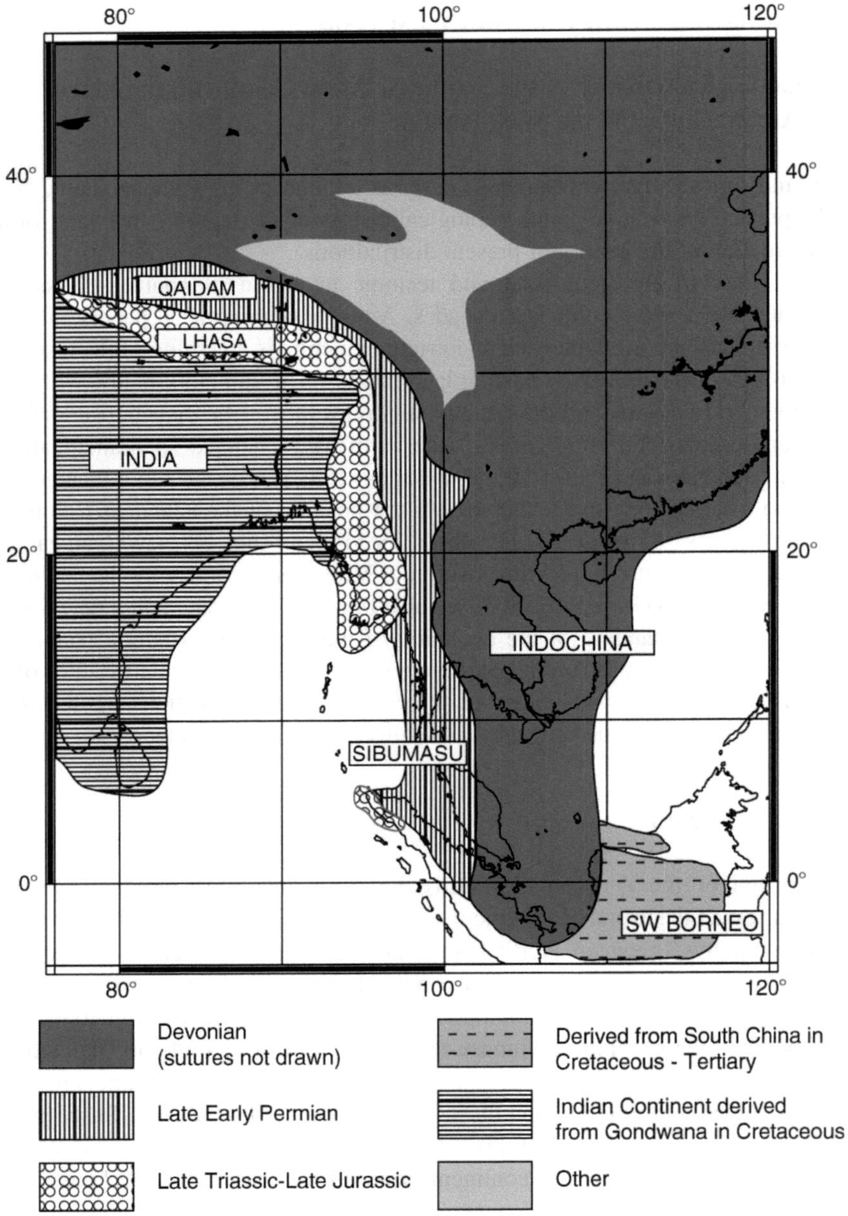


Fig. 2 Distribution of principal continental terranes of East and southeast Asia. Sutures, especially of Devonian age, not indicated (simplified after Metcalfe, 2001).

among other things, in the separation of India and the Seychelles block at *c*. 65 Ma (Braithwaite, 1984; McLoughlin, 2001). The Seychelles block became fixed to the African continent from that time. Only during the collision of India with the southern Asia continent (between 65 and 56 Ma, but according to McLoughlin (2001) *c*. 43 Ma), the southeastern corner of Asia with Indochina and the former Sibumasu terrane turned clockwise to its present more north–south orientation.

Australia separated from Gondwana at *c*. 85 Ma and rifted northward (Metcalfe, 2001; Hill and Hall, 2003). The northern margin of the Australian plate included at least the southeastern parts of present-day Sulawesi, Buton, Buru, Seram, as well as parts of New Guinea. During the process of rifting during the Cretaceous, a series of continental slivers became isolated along the passive northern margin. Some of these fragments are now part of the Central Highlands of New Guinea. This series of fragments is known as the "Inner Melanesian Arc" in biogeographical studies, and further discussed below under Cenozoicum, since it was presumably absorbed with the northern margin of the Australian craton during the Eocene.

## 3.3. Cenozoicum

The geological evolution of southeast Asia during the Cenozoicum has been extensively studied and discussed by Hall and collaborators (e.g., Hall, 2001, 2002) and with special attention to the northern margin of the Australian continent by Hill and Hall (2003). Hall's reconstructions, and particularly the terminology, are not fully congruent with those of Yan and Kroenke (1993) and Kroenke (1996) for the West Pacific region. Quarles van Ufford and Cloos (2005: Fig. 2) summarize the different models for the Cenozoic plate-tectonic history of New Guinea, while also providing a new summary of the tectonic evolution. The Cenozoic palaeogeography of the region in relation to biogeographical problems has also been discussed various times (Beuk, 2002b; Soulier-Perkins, 2000). The geological history of smaller parts of this region in relation to biogeography has been analysed, e.g., southeast Asia, Borneo and Sulawesi (Moss and Wilson, 1998), the Philippines (de Jong, 1996), the West Pacific (Keast and Miller, 1996; de Boer, 1995, de Boer and Duffels, 1996, 1997), or with special emphasis on New Guinea (e.g., Polhemus and Polhemus, 1998, 2002).

The general pattern arising from recent reconstructions can be described as follows. The collision of India with the southern margin of the Asian continent significantly changed the structure of that area between 65 and 56 Ma. The collision resulted in the orogeny of the Himalayas. It may have increased the land surface as well, but the amount of crustal shortening is unknown. Recent data (Krause et al., 1999; Bossuyt and Milinkovitch, 2001) suggest that the fauna that developed in India during that time has spread over the Oriental region since *c*. 60 Ma. The northward movement of Australia towards the Pacific plate that started 85 Ma has continued with relatively slow speed up to today, although the separation of Australia from Antarctica at *c*. 55 Ma increased the rate of convergence. From *c*. 43 Ma (Quarles van Ufford and Cloos, 2005), a southwest directed subduction of the Pacific plate started two subduction systems, one at the Papuan–Rennell–New Caledonian trench system, and a more northerly subduction zone at the New Guinea–Manus–Kilinailau–Solomon trench system. Several arc systems were formed during subduction and rotation of the plates (see discussion below for more detailed geology of the New Guinea region). A subduction at the western margin of the Pacific plate, north of the equator, formed the Izu-Bonin–Mariana Arc system, while at the same time the Philippine plate became a separate entity between the Australian and Pacific plate. The Philippine plate itself has a "complex rotation history" (Hall, 2002: 378), with a rotation of 50° between 50 and 40 Ma (Fig. 3), whereafter a period without rotation continued up to 25 Ma.

The most important reorganization of the plate boundaries occurred at *c.* 25 Ma (Hall, 2002). The New Guinea passive margin collided with the East Philippines–Halmahera–South Caroline Arc system, and the northwestern corner of the Australian plate collided with southeast Asia in the Sulawesi area. From that

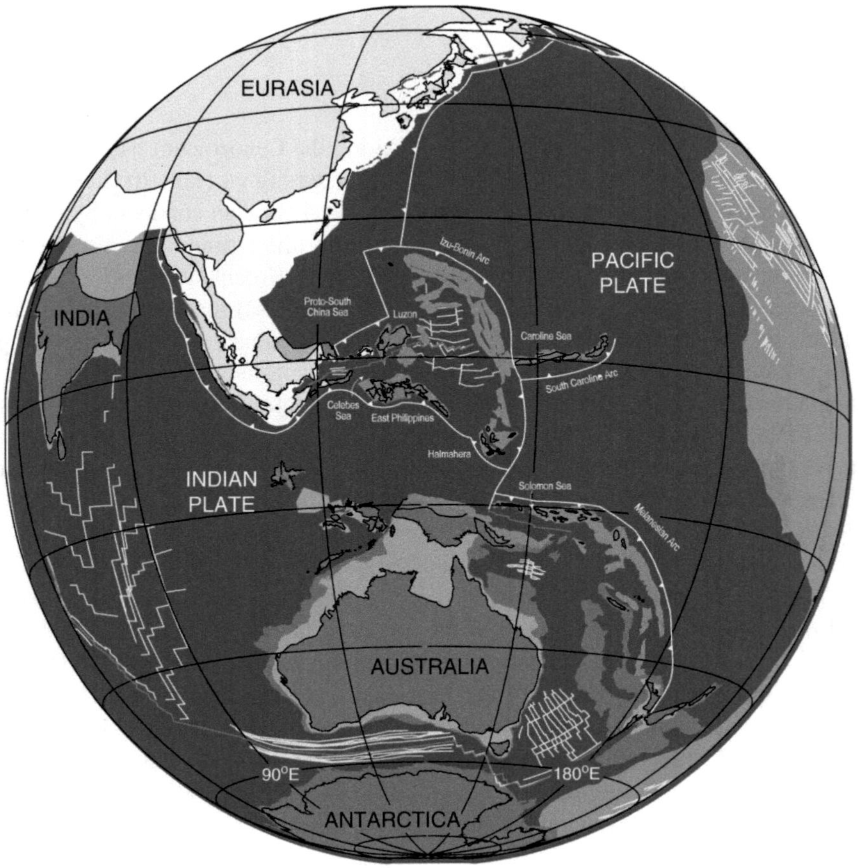


Fig. 3 Palaeogeographic reconstruction of Southeast Asia at 45 Ma (Middle Eocene) (from Hall, 2002). Note the position of southwestern Sulawesi approximately at its present position, of East Sulawesi at the northwestern corner of the Australian plate, and of northern Sulawesi at the margin of the Australian and the Philippine plates in an island arc with the east Philippines and Halmahera. The collision of the parts of Sulawesi only occurred during the Middle to Late Miocene (15–10 Ma).

time on, the Pacific plate became the driving force of the regional tectonic events. The northward movement of Australia caused the accretion of microcontinents north of New Guinea. The final large change in the tectonics of the region, possibly due to motion change of the Pacific plate, occurred at *c.* 5 Ma, with significant impact in the Taiwan–Philippine region, and uplift in southern Indonesia (Java to the Lesser Sunda islands).

Whether the events described above are relevant to the present distribution of freshwater invertebrates mainly depends on whether the area was subaerial for all the time. Although much new information has become available during the last 20 years, there is still much controversy. Examples are the palaeogeographic evolution of Sulawesi (Wilson and Moss, 1999) and of the Melanesian Arc. Sulawesi consists of a complex of fragments that only merged into its present position during the last 5 My. The southwestern arm is considered a part of the Asian continent, with the same position in relation to Borneo for at least *c.* 45 My. East Sulawesi originated in the northwestern corner of continental Australia, probably as early as the Early Eocene (56–49 Ma) (Hall, 2002, see also Fig. 3). The northern Sulawesi arm was formed much further north at the northern margin of the northward moving Australian plate in an island arc with the eastern Philippine islands at *c.* 45 Ma (Philippine Arc). It possibly docked with the southwestern arm in the Early Oligocene (34–29 Ma; Wilson and Moss, 1999), but alternatively as late as the Middle Miocene (*c.* 15 Ma) (Hill and Hall, 2003). With the opening of the Celebes Sea (Early Oligocene, 34–29 Ma) the western part of the Philippines shifted to a more northern position, while on the clockwise rotating Philippine plate parts of the eastern Philippines moved more towards their present positions. The northwest movement of the Australian plate slowly pushed the central and southeastern parts of Sulawesi towards their present positions. The relatively fast rotation of the Philippine plate caused a rapid change of positions of the islands along its margin (eastern Philippines, Halmahera) during the Miocene. According to Wilson and Moss (1999) the eastern arms of Sulawesi collided with central Sulawesi in the Early Miocene (23–16 Ma), but Hill and Hall (2003) reconstructed a Pliocene collision of these island fragments (Fig. 4).

Along the subduction zones and partly induced by turning of the plates, several island arcs were formed, displaced, and (partly) accreted or subducted again. During the Eocene (56–34 Ma), the area converged due to northward movement of the Australian plate. While the Australian plate subducted under the Philippine plate, the Philippine Arc was formed. At least during the Oligocene (34–23 Ma), while this zone was running more or less west to east, this island arc included from west to east peninsular northwestern Sulawesi, Mindanao, and Halmahera, including other parts of the Moluccas. Also during the Eocene, along the eastern margin of the Philippine Plate the north–south oriented Caroline Arc was formed at the collision zone with the Greater Pacific Plate. Due to backarc spreading during the Oligocene creating the Caroline Sea Plate, this island arc, consisting of fragments now part of northern New Guinea, started a nearly 90° clockwise movement. A third island arc, the Melanesian Arc, was created at the southern margin of the Pacific Plate at the subduction zone with the Australian plate. This process intensified during the Oligocene due to backarc spreading, creating the Solomon Sea Plate.

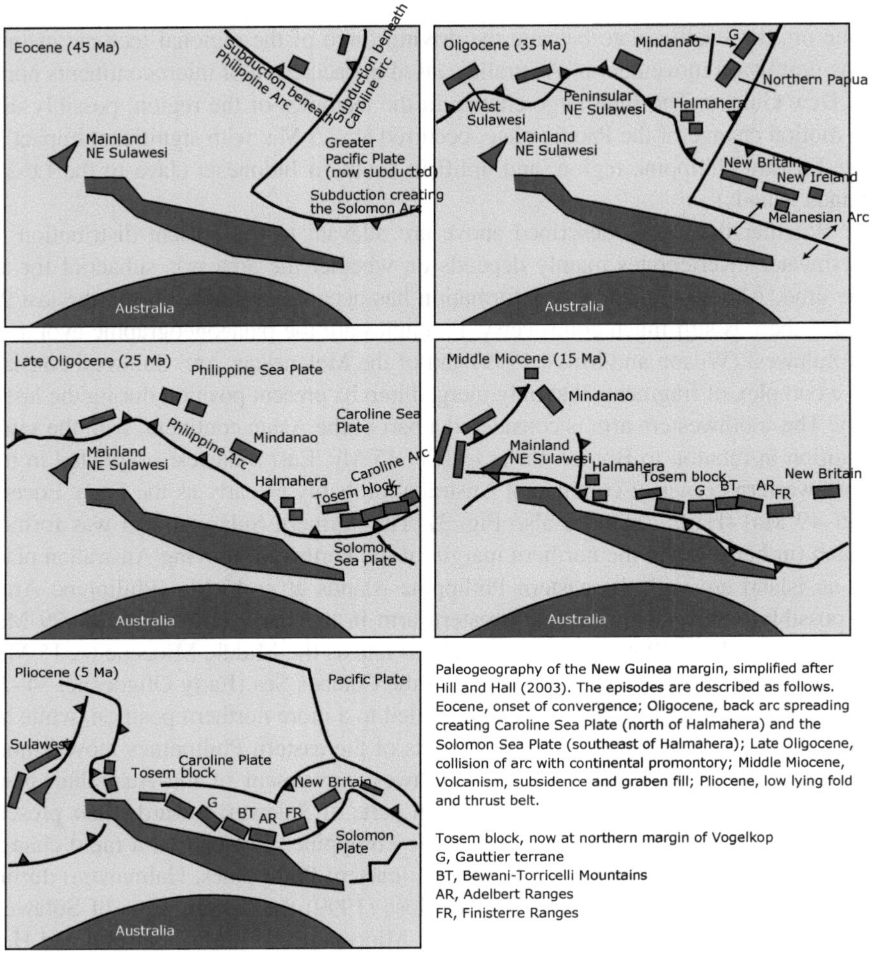


Fig. 4 Tectonic evolution of the New Guinea region (after Hill and Hall, 2003).

Around 25 Ma, the Philippine Arc and the Caroline Arc were more or less in line at the northern margin of the Australian and Solomon Sea Plates, while still rapidly rotating clockwise. From the Miocene onwards, the Melanesian Arc formed a continuation of the Caroline Arc in eastern direction.

These island arcs or island groups are considered relevant in biogeographical analysis. At *c.* 30 Ma the South Caroline Arc consisted of (from west to east) the Tosem Block (now northern Vogelkop), northern Papua (Irian Jaya), the Gauttier terrane, the Bewani–Torricelli Mountains, the Adelbert ranges, and the Finisterre ranges (Hill and Hall, 2003), and was situated northeast of the Australian continent (Fig. 4). The Melanesian Arc consisted of New Britain, New Ireland, and then to the south, the Solomon Islands, Vanuatu, Fiji, and Tonga (Hall, 2002; Hill and Hall, 2003). In a previous reconstruction, based on Hall (2002), Beuk (2002a) included central New Guinea, the Papuan Peninsula, northern New Guinea, Finisterre, and Bismarck/New Britain in the Caroline Arc, while the Melanesian Arc started south of New Britain with New Ireland.

Polhemus (1995) and Polhemus (1998) mentioned additional hypotheses on island arc systems in analysing the distribution patterns of aquatic insects with sister-group relationships between the Philippines and New Guinea, while not occurring in the Moluccas and Sulawesi. The reconstruction of this "Inner Melanesian Arc system" is partly visible in Yan and Kroenke (1993). This must have been a pre-Eocene, presumably Cretaceous, "arc" extending from Mindanao, a section of northern Australia that later became New Guinea, the Solomon Islands, and New Caledonia to New Zealand. Parts of this arc now may have a position in the highlands of New Guinea. The "arc" collided with the northern Australian continental plate during the Mesozoicum. Technically, the Inner Melanesian Arc cannot be considered an arc system, but a series of slivers of continental crust that became isolated during the process of rifting along the northern Australian margin (Polhemus, 1998). Recently, the name "Inner Melanesian Arc" was used by Quarles van Ufford and Cloos (2005) for an Eocene–Oligocene Arc including the Bewani–Torricelli Arc, the Papuan ophiolite belt, and (much further to the south) New Caledonia. In a more recent publication, Polhemus and Polhemus (2002) relied more on Hall (1998) for their palaeogeographic interpretations, but their terminology is different from Beuk (2002a).

## 3.4. Geological Area Cladogram

A geological area cladogram was first presented for the West Pacific by de Boer (1995), and further elaborated by de Boer and Duffels (1997) and Beuk (2002a, b) (Fig. 5). The area cladograms were based on the geological reconstructions of southeast Asia by Daly et al. (1991), Pigram and Davies (1987), and Rangin et al. (1990), and several papers describing the history of smaller parts of the area.

Three island arcs were distinguished, namely the West Pacific Arc (from west to east consisting of central Philippines, northern/eastern Sulawesi, central New Guinea, Papuan Peninsula, northern New Guinea, Finisterrre, Bismarck archipelago, northeastern Solomons), the eastern Philippine–Halmahera Arc (from north to south consisting of eastern Philippines and the Halmahera Arc, and possibly also the Marianas and Yap), and the Southwest Pacific Arc (from north to south consisting of Solomon Islands, Vanuatu, Fiji, and Tonga). Most areas in the island arcs coincide with areas of endemism for cicadas (Homoptera, Cicadidae) (Beuk, 2002a: 248). Particularly the West Pacific Arc is believed to be relevant for the dispersal of many groups of animals. It should be realized that several parts of the area did not belong to any of the arcs, but were a group of microcontinents with a history more connected with Australia. Beuk (2002a) presented an update of this view. He considered the eastern Philippines and Halmahera not related to an arc system. His South Caroline (as Carolina) Arc system (at *c.* 30 Ma, late Oligocene) consisted from west to east of central New Guinea, Papuan

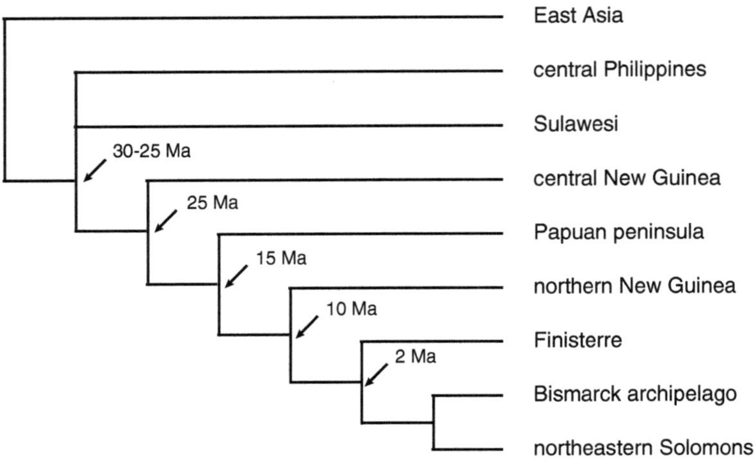


Fig. 5 Geological area cladogram of southeast Asia (from Beuk, 2002).

Peninsula, northern New Guinea, Finisterre, Bismarck, and New Britain, while the north–south oriented Melanesian Arc consisted of Bismarck/New Ireland, Solomon Islands, Vanuatu, Fiji, and Tonga. The timing of the fragmentation sequence is also given in Fig. 5.

According to the reconstructions by Hall (1998, 2001, 2002) and Hill and Hall (2003) the geological history of the Philippines and Sulawesi is more complex than that presented by Daly et al. (1991). Especially the position of Luzon is distinctly different, since it was formed at the northern margin of the Philippine plate by southward subduction of the Pacific plate at 45 Ma. The southwestern peninsula of Sulawesi is supposed to have the same position in relation to Borneo since at least the Middle Eocene (45 Ma). The northern peninsula was part of an island arc at the southern margin of the West Philippine Basin, while the eastern peninsula had a position on the westernmost part of the Australian plate (see Fig. 3).

The present reconstructions (Hill and Hall, 2003) differ in various ways relevant to biogeographical analysis. First, the Philippine Arc (Mindanao, Halmahera) continued to the west with peninsular northeastern Sulawesi at least during the Oligocene (35 Ma). This island arc continued to the east in the Melanesian Arc, where the Moluccas and New Britain seem to have had a position rather close to each other during the Oligocene. The Caroline Arc formed the continuation of the Philippine Arc to the north at the subduction zone of the Great Pacific Plate. Due to backarc spreading creating the Caroline Sea Plate, mentioned above, the Philippine Arc and the Caroline Arc more or less formed one arc system during the Late Oligocene (25 Ma). The Melanesian Arc began to form one line with the Caroline Arc during the Miocene. The counterclockwise rotation of New Britain and New Ireland was induced by the spreading of the Solomon Plate during the Pliocene only.

# 4. Distribution Patterns

## 4.1. Introduction

Very few revisions with an extensive cladistic reconstruction of the phylogeny of aquatic groups are available for southeast Asian taxa, and such examples are uncommon even if all terrestrial biotas are considered (see Turner et al. 2001 for an overview). It is, therefore, not feasible to construct a generalized area cladogram based on aquatic taxa. We even doubt whether the construction of a generalized area cladogram as presently used is methodologically sound for an area as southeast Asia with reticulate relationships of areas of endemism. It is necessary to estimate the timing of splitting events in the original cladograms based on independent data. Geological evidence of minimum ages of areas of endemism may reveal molecular clock data for splitting events in various taxonomic groups. Such data are needed, since effects of random dispersal, local extinctions, vicariance events without splitting of lineages, apart from the usual incertainties in phylogenetic trees based on misinterpretations of homologies, will disturb the process of construction of a generalized area cladogram. As has been noticed before in other words, a taxon can only belong to one historical entity, but an area may be part of more than one entity. This may be due to amalgation, splitting, or displacement of the area under study as compared to another area.

For an area for which so few cladograms are available, not all of them are equally useful. To resolve area relationships, it is minimally needed to study the taxonomy and phylogeny of a group of predominantly parapatric taxa. So, even when well-founded phylogenies have been published, some studies are hardly useful in the reconstruction of area relationships. Up to now, more extensive phylogenies have been published for several groups of aquatic Hemiptera of southeast Asia (Andersen 1991, 1998, Damgaard et al. 2000, Damgaard and Zettel 2003, Polhemus, 1994, 1996, Polhemus and Polhemus, 1987, 1988, 1990, 1994, 2002). For some insect orders, e.g., the Plecoptera, the phylogenetic relationships of the families seem to be intimately connected with the break-up of Pangaea, and various examples have been included below.

The methodology of direct comparison between palaeogeography and phylogenetic relationships is not uncontroversial. Eskov (2002) discussed the "Gondwanan" ranges of recent taxa. He mentioned several examples of presumably Gondwanan groups, which appeared to have representatives in the fossil record of Eurasia or North America. Consequently, such present-day "Gondwanan" groups are only relics of a wider, possibly even global distribution, which may or may not have included Gondwana during its break-up. In conclusion, reconstructions in zoogeography have to be based on all available evidence (total evidence tree).

The main order of this chapter is taxonomical, but papers with special attention for the Sulawesi fauna will be discussed in section 4.4.

## 4.2. Odonata

### *4.2.1. Odonata: Ancient Families as "Gondwanan" Elements in Australia and South America*

As the oldest extant group of pterygote insects, it may not be surprising that Gondwanan distributions are still recognizable in Odonata at the family level. According to Watson (1982), possibly up to 40% of the Australian fauna should be considered of Gondwana origin, i.e., that there are sister-group relationships between the fauna of Australia and South America. This problem was extensively discussed by Carle (1995), when he re-analysed the phylogeny of the ancient Anisoptera. The extant dragonfly superfamilies were all well established before the break-up of Pangaea, and dispersal of the groups was made possible by the so-called trans-pangaeian mountain system. Carle (1995: 394–395) concluded that "repeated north-south congruences with early anisopteran phylogeny indicate that the trans-pangaeian montane dispersal route was persistent yet tenuous". Such a route during the Mesozoicum is probably the cause of the occurrence of several primitive genera of anisopteran superfamilies in the eastern USA.

Carle (1995) presented a new phylogenetic hypothesis based on morphological characters of ancient families of Odonata. Several of these families were redefined based on his new analysis of characters, and the distributions of the new groups further discussed. The Gomphoidea has been mentioned several times as an example of a Gondwanan element in the Australian fauna. It is an ancient group indeed and has a fossil record extending as early as the Jurassic. The Petaluridae are represented in the southern hemisphere with the subfamily Petalurinae in Australia (*Petalura* Leach), New Zealand (*Uropetala* Selys), and Chile (*Phenes* Rambur). One fossil petalurid species is known from the Jurassic of Europe.

The next monophyletic group is formed by the Aeshnoidea and Libelluloidea, of which the Austropetaliidae are the most plesiomorphic. The Austropetaliidae are known from Tasmania and eastern Australia (*Austropetalia* Tillyard), and two genera in Chile, another example of Gondwanan distribution. All species of this family are confined to seepages or small streams; the larvae of most, if not all, species are semiterrestrial.

Carle (1995) also re-evaluated the status of the genus *Neopetalia* Cowley (one species, confined to Chile), and concluded that it represents a family on its own, and forms the sister-group of the non-cordulegastrid Libelluloidea. All other genera formerly included in the Neopetaliidae were placed in the Austropetaliidae (see above). The non-cordulegastrid Libelluloidea are the most speciose group of all extant dragonflies. According to Carle, this adaptive radiation started *c.* 140 Ma in Antarctica.

### *4.2.2. Odonata: Calicnemiinae*

Recently, taxonomy and phylogeny of the calicnemiine Platycnemididae of southeast Asia have been studied extensively (Gassmann, 1999, 2000; Gassmann and Hämäläinen, 2002; Dijkstra, unpublished). A reconstruction of the phylogeny of this subfamily was published by Gassmann (2005). The characters used in the analysis, and details how the results were obtained will not further be discussed here.

Both subfamilies of the Platycnemididae, namely the Platycnemidinae and Calicnemiinae, are found in the Afrotropical, Palaearctic, Oriental, and Papuan regions. The family is absent from Australia and the New World, and, remarkably, from Sulawesi. The subfamily Calicnemiinae is widespread in India and Indochina, especially in the mountainous regions around the Himalayas. Its distribution in Malesia is illustrated in Fig. 6. On some islands, several genera show significant radiation at the species level. For example, the rather widespread genus *Coeliccia* Kirby of southeast Asia is very speciose in Borneo. Many other well-defined genera have restricted ranges within Malesia, e.g., *Idiocnemis* Selys is confined to New Guinea and the adjacent islands, *Risiocnemis* Cowley is restricted to the Philippines, and *Lieftinckia* Kimmins is confined to the Solomon Islands. Several smaller, but distinctly different, genera have even smaller ranges, e.g., *Asthenocnemis* Lieftinck (Palawan), *Arrhenocnemis* Lieftinck, *Lochmaeocnemis* Lieftinck, *Cyanocnemis* Lieftinck, and *Torrenticnemis* Lieftinck (all New Guinea).

The simplified version of the cladogram (Fig. 7) will be discussed here in relation to the present distributions of the taxa, mainly genera. The substitution of taxa for areas of endemism will also present a basis for a hypothesis on the history of the distributional patterns. Two genera of Platycnemidinae, *Copera* Kirby, and *Platycnemis* Burmeister, were used as outgroup.

At the base of the cladogram we find various Afrotropical genera (*Arabicnemis* Waterston, *Allocnemis* Selys, *Stenocnemis* Karsch, *Mesocnemis* Karsch). The sister-group of all species found in southeast Asia is *Leptocnemis cyanops* Selys, a species confined to the Seychelles. According to the present analysis, partly based on selected species of various genera, the ancestor of the genera *Calicnemia*, and *Indocnemis* Laidlaw plus *Coeliccia* is sister to all other Calicnemiinae. All taxa of this group are represented in the mainland of southeast Asia, but *Coeliccia* is

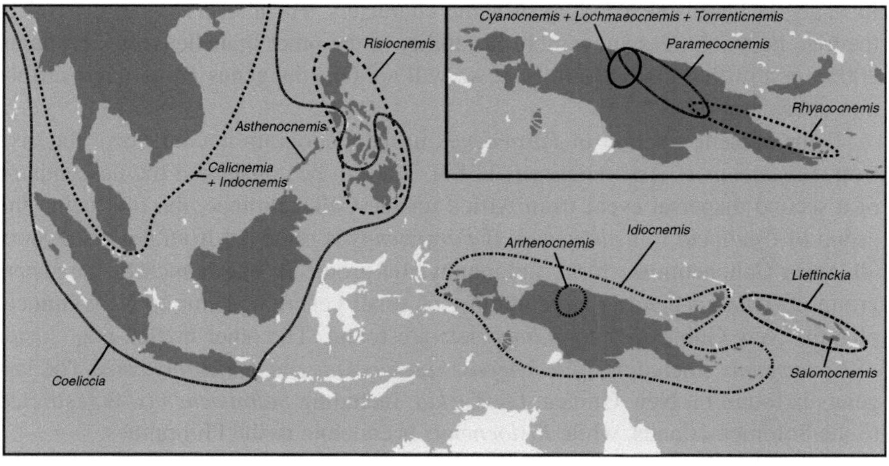


Fig. 6 Distribution of the Calicnemiinae (Odonata, Platycnemididae).

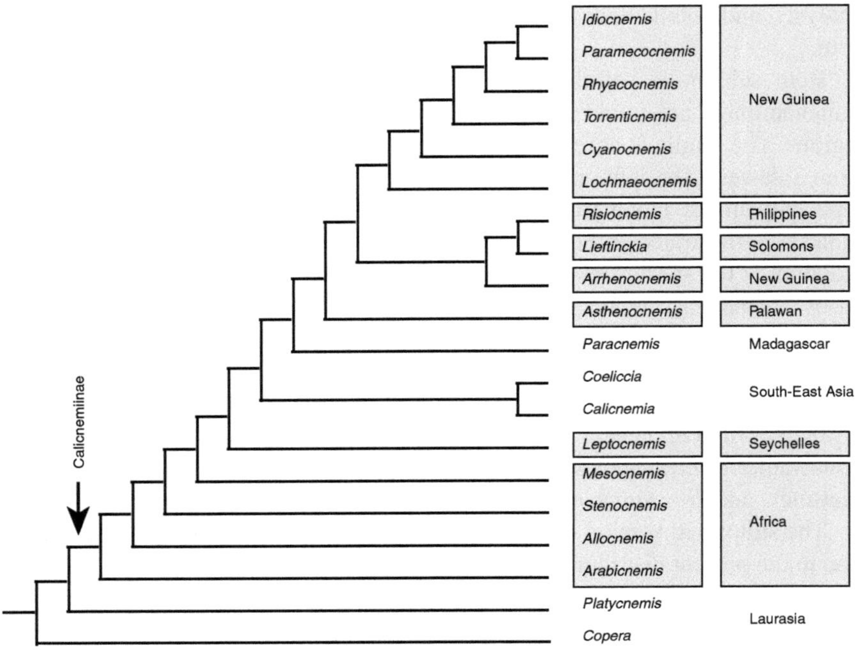


Fig. 7 Simplified cladogram of the Calicnemiinae (Odonata: Platycnemididae). The distribution of the clades (area cladogram) is given as shaded areas to the right.

also widespread in Sundaland and parts of the Philippines. Remarkably, the sistergroup of this clade consists of the genus *Paracnemis* Martin, which is restricted to Madagascar, plus, as a sister to *Paracnemis*, all other taxa of southeast Asia. However, the position of *Paracnemis* in the cladogram is still somewhat uncertain. In analyses based on recoding of some characters, *Paracnemis* is more basal in the tree, but such trees show more instability in the other branches (cf. Gassmann, 2005, for a further discussion). Here we will not take the genus *Paracnemis* further into consideration.

If the present position of *Paracnemis* in the cladogram is confirmed, the taxa from *Idiocnemis* to *Asthenocnemis* in Fig. 7 are to be considered the descendants of a second dispersal event from Africa for the Calicnemiinae, the first being the group of *Coeliccia* to *Calicnemia*. If *Paracnemis* is removed from the discussion, all Asian Calicnemiinae form a monophyletic group. In one branch of the sistergroup of *Asthenocnemis* a large number of small genera endemic to New Guinea, plus the New Guinean genus *Idiocnemis* are found. The other branch is a cluster of *Risiocnemis* (including *Igneocnemis*), *Lieftinckia*, and *Arrhenocnemis*. The last genus is found on New Guinea, *Lieftinckia*, including *Salomocnemis*, is restricted to the Solomon Islands, while *Risiocnemis* is endemic to the Philippines.

The following biogeographical scenario arises from the cladogram. The Calicnemiinae of southeast Asia are derived from African Platycnemididae. Two distinct lineages can be recognized. One clade, with *Calicnemia* and *Coeliccia*, is widespread and speciose in the Oriental region including the Philippines, but does not occur east of Borneo in the Malay Archipelago. At the base of the sister-clade, we find *Asthenocnemis stephanodera* Lieftinck, a species confined to Palawan. The sister-group of *Asthenocnemis* are all remaining Calicnemiinae. In this group, two monophyletic clades can be distinguished. One, including the genus *Idiocnemis*, is completely confined to the Papuan region, while the other represents a Papuan and Philippine clade. In this clade, the genus *Arrhenocnemis* Lieftinck from New Guinea is the sister-group of an eastern Papuan and Philippine group, namely the eastern Papuan genus *Lieftinckia* Kimmins (including *Salomocnemis* Lieftinck, all from Solomon Islands) as one monophyletic group and the Philippine genus *Risiocnemis* Cowley (including *Igneocnemis* Hämäläinen).

In this scenario Malesian Calicnemiinae derived from ancestors on the mainland of Asia, including Palawan. This group may have dispersed along the eastern margin of the Philippine plate, along an arc that was formed by subduction of the Pacific plate. This so-called Izu-Bonin Arc, which was formed 40–50 Ma, is the basis of the "northern dispersal scenario" of Beuk (2002a: 279). The Izu-Bonin Arc must be considered the northern continuation of the Caroline Arc during the Eocene. Alternatively, Polhemus (1995) and Polhemus and Polhemus (1998) hypothesized a Late Cretaceous island arc (Inner Melanesian Arc) (Mindanao to New Zealand) as a means for the dispersal route of Papuan groups of aquatic Heteroptera with distinct Asian mainland affinities. As described above, this island arc collided with the northern magin of the Australian terrane during or even before the Eocene. Taxa that reached this corner of the Pacific, later may have used the (South) Caroline Arc while still situated far north from its present position, and much later its continuation to the south, the Melanesian Arc. The taxa that evolved during that time all have remarkable autapomorphies and are presently recognized as separate genera. They have evolved on the terranes or microcontinents of the "Inner Melanesian Arc" at the northern margin of the Australian plate; some of these terranes have a subaerial history since the Late Cretaceous.

The clade of *Lieftinckia* and *Risiocnemis* then presumably represents a westward dispersal of *Risiocnemis* from the Solomon Islands into the Philippines. As already mentioned above, the subfamily is absent from Sulawesi and the Moluccas. This may contribute to our understanding of the dating the dispersal of this group. It may be hypothesized that the Philippine Arc has played an important role in the evolution of this group. This arc collided with the Australian plate at *c.* 25 Ma (Late Oligocene). The spreading of the Philippine genus *Risiocnemis* can then be dated at *c.* 15–20 Ma (Early Miocene). The clade with the small genera distinctly represents a reflection of the tectonic history of the Caroline Arc at the subduction zone of the (Great) Pacific Plate.

The mechanism of dispersal via the Caroline Arc has already been discussed various times. Beuk (2002b) showed that the (South) Caroline Arc had a westward extension (here named Philippine Arc) via Halmahera and the eastern Philippines to southeast Asia, a southern route via northern Sulawesi, or a northern route via the northwestern Philippines at *c.* 30 Ma. This scenario is based on reconstructions by Hall (2002: 405). In that case, the absence of Platycnemididae in Halmahera can only be understood in this model if this group became locally extinct, or if we have to presume that no freshwater was available at a certain period of time. A similar pattern of distribution has been found in several groups of aquatic Heteroptera, including sagorine Naucoridae (Nepomorpha), and the *Rhagovelia novacaledonica* group (Fig. 12), gerromorph heteropterans with poor dispersal power (Polhemus, 1995). According to Polhemus, however, such patterns resulted from the long, pre-Eocene northwest-southeast trending arc system. Such a system extended from New Zealand through the Solomons to Mindanao, but not including Halmahera nor Sangihe. Unfortunately, very little is known from the history of this arc system. It is also not clear from the description in Polhemus whether an arc movement in western or eastern direction is hypothesized, also since the text includes at least one evident mistake "and has apparently transported continental fragments from the Vogelkop Peninsula *eastward* [recte: westward] to near Celebes". Based on Hill and Hall (2003), we suppose that such sister-group relationships could also have evolved during the Oligocene, when parts of the Philippine and Caroline arc systems were relatively close to each other. More information on the tectonic history of the Moluccas seems to be crucial for a further understanding of the dispersal opportunities.

The colonization of the mainland of southeast Asia should be linked with the presence of *Leptocnemis* of the Seychelles at the basis of all southeast Asian species. As described above, the non-African lineages then split off *c.* 88 Ma (early Late Cretaceous), while the separation of India from the Seychelles is dated *c.* 65 Ma. In this scenario, the absence of the Platycnemididae from Australia asks for a local extinction in that continent, a not uncommon phenomenon for tropical groups. It seems that Gondwanan (sub)tropical groups have more rarely survived in Australia than groups confined to temperate habitats.

#### *4.2.3. Odonata: Platystictidae*

The Platystictidae, or forest damselflies, represents a distinct group of the suborder Zygoptera of the Odonata. The presumed monophyly of the group is based on the presence of the so-called post-cubital vein, a character not present in any other recent species of dragonfly (e.g., Bechly, 1996). Presently, three subfamilies are recognized, the speciose and widely distributed Platystictinae of southeast Asia, the recently established Sinostictinae of southern China (Wilson, 1997), and the Palaemnematinae of Middle and South America (e.g., Calvert, 1931, 1934; Kennedy, 1938) (Fig. 8). Four genera are recognized in the Platystictinae, *Platysticta* Selys, *Protosticta* Selys, *Drepanosticta* Laidlaw, and *Sulcosticta* Van Tol (see Van Tol, 2005), one in the Sinostictinae, *Sinosticta* Wilson, and one in the Palaemnematinae, *Palaemnema* Selys. The total number of species described per subfamily is presently (Van Tol, unpublished) 124 in Platystictinae, two in Sinostictinae and 42 in Palaemnematinae.

The present global distribution seems to go back to at least the Cretaceous (Van Tol and Müller, 2003). The family was presumably distributed across Laurasia. The climate of that time was tropical, and Europe and America were still connected.

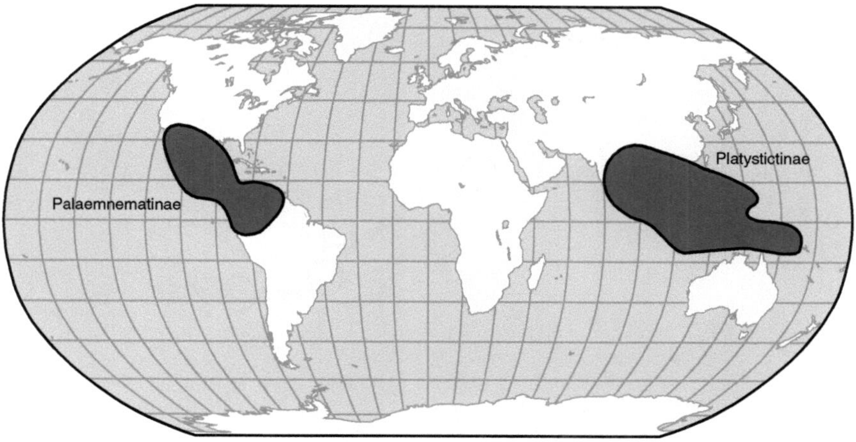


Fig. 8 Global distribution of Platystictidae (after van Tol and Müller, 2003).

After their separation, the climate became less favourable for tropical biotas, and the ancestors of the present Platystictidae were forced to move southward in both America and Eurasia. The presence of *Palaemnema* in South America possibly dates back only 3 My, following the emergence of the Panama Isthmus (Coates, 1999). Comparable distribution patterns of southeast Asia and Central America have been found in some other groups as well, e.g., the plant genus *Spathiphyllum* (Araceae). If further, e.g., molecular, studies will confirm such an early separation of both subfamilies, the morphology of both groups has remained remarkably stable over the last 60 My. The structure of the male appendages, for instance, hardly differs between species of *Palaemnema* and of *Drepanosticta*.

The phylogeny of the southeast Asian Platystictinae is poorly understood. The generic characters of wing venation seem to be rather useless. Since presumed sister-species are presently assigned to two different genera, the generic diagnoses ask for rigorous redefinition. Nevertheless, some distinct groups characterized by one or more unique autapomorphies can be distinguished, providing a first base for zoogeographic analysis. Such a group is the *Drepanosticta lymetta* group, which is characterized by the unique structure of the hind margin of the posterior lobe of the pronotum (Fig. 9). The group is distributed from Luzon to eastern New Guinea, with (partly undescribed) species known from Luzon, Siquijor, Mindanao, Halmahera, and New Guinea, and one species on Java. This pattern shows a largely congruent relationship with the Philippine–Caroline Arc and its continuation to the northwest.

The subfamily Platystictinae shows its highest structural diversity in the mainland of southeast Asia, in some sense extending over the Greater Sunda islands, but a few groups show extreme radiation on various islands, such as the genus *Protosticta* Selys on Sulawesi (van Tol, 2000) and the genus *Drepanosticta* Laidlaw on the Philippine islands (van Tol, 2005). Structural differentiation seems to decrease in eastern direction towards New Guinea.

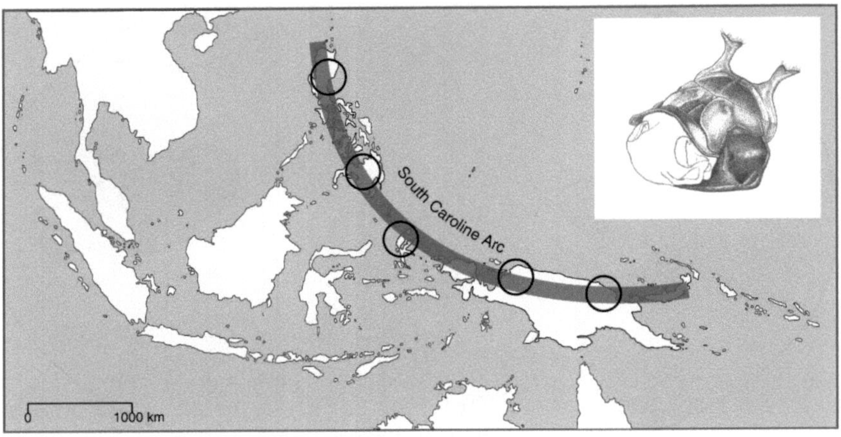


Fig. 9 Distribution of the *Drepanosticta lymetta* group.

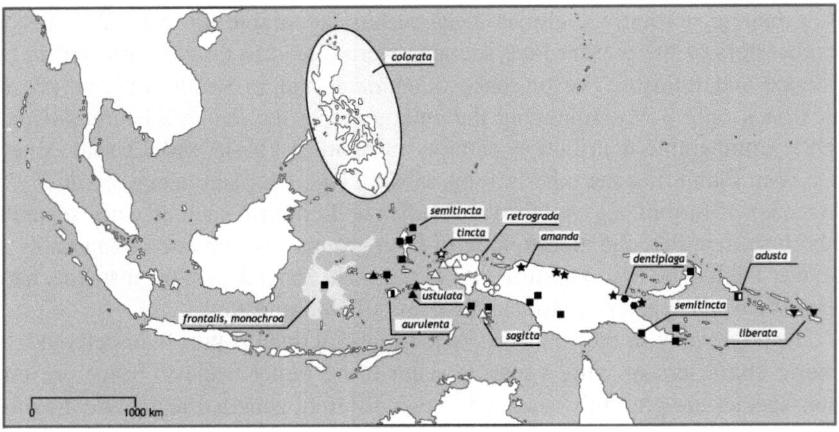


Fig. 10 Distribution of the *Rhinocypha tincta* group.

*4.2.4. Odonata:* Rhinocypha tincta *Complex (Chlorocyphidae)*

"The geographical distribution of this subspecies [i.e., *Rhinocypha tincta semitincta* Selys] is puzzling, but I am still unable to differentiate between the various populations from remote localities. Some of the specimens from the Solomon Islands seem absolutely inseparable from topotypical *semitincta* of Halmahera, with which I have actually compared them" (Lieftinck, 1949a: 27). The distribution of this subspecies of chlorocyphid damselfly (Fig. 10) also includes the easternmost part of New Guinea (Papuan Peninsula), the Baliem valley (central New Guinea), the Kai and Aru islands, the Sula islands and a very restricted part of central Sulawesi (Lieftinck, 1938, 1949a, own observations). Records from Cape York have not been confirmed in the 20th century (Watson, Theischinger and Abbey, 1991: 173).

What most puzzled Lieftinck was, of course, the distance between populations of this taxon that were morphologically inseparable. Apart from *R. t. semitincta*, many more taxa in this complex are distinguished, of which several inhabit the areas between the populations assigned to *R. tincta semitincta*. The *R. tincta*-group is distributed (Fig. 10) from the Philippines to New Britain and the Solomon Islands with the following taxa: *Rhinocypha colorata* (Hagen) widespread in the Philippines and considered the sister-species of *R. tincta* or a subspecies of *R. tincta*, *R. frontalis* Selys and *R. monochroa* Selys, and possibly also *R. phantasma* Lieftinck, from Sulawesi, the typical *R. tincta*, which is only known from Waigeo, subspecies *R. tincta sagitta* Lieftinck occurring on Salawati and in the southern part of the Bird's Head of New Guinea. Further east, in the northern parts of the Berau Peninsula, and in the isthmus of western New Guinea, we find *R. tincta retrograda* Lieftinck, along the north coast of New Guinea occurs *R. tincta amanda* Lieftinck, except for the Finisterre range and adjacent areas, where *R. tincta dentiplaga* Lieftinck is found. Further eastward, specimens from Bougainville and the Shortland islands are assigned to *R. tincta adusta* Lieftinck. Finally *R. liberata* Lieftinck inhabits Ugi and Guadalcanal. According to Lieftinck (1949b), *R. liberata* is the sister-group to the Moluccan *R. ustulata*.

Although the phylogenetic relationships of these taxa are poorly understood, their distributions are congruent with a series of tectonic events also found in patterns of other taxa. The series of subspecies (at least *retrograda*, *amanda*, *dentiplaga* and *adusta*) along the northern coastal margin of New Guinea reflects the pattern of the Caroline island arc north of New Guinea that partly accreted with New Guinea during the Late Miocene and Pliocene. If tectonic events and present distributions have to be related, the Caroline Arc is the most likely pathway for this complex to reach the area. The distribution of many taxa particularly reflects the palaeogeography during the Oligocene. The distribution of the widespread *R. tincta semitincta,* occurring on the Moluccas and the Solomon Islands, but absent from the area in between, seem to indicate an evolution since the Oligocene (35 Ma). The occurrence of this taxon in a very limited area in central Sulawesi may be an indication that a fragment of this area also formed part this island arc, but no palaeogeographical reconstruction confirms this observation. It could, however, explain the occurrence of Papuan elements in the Sulawesi fauna, and should be subject to further studies. The series of related species, such as those from the Philippines and Sulawesi, may have evolved on the Philippine island arc during the late Oligocene (25 Ma).

### *4.2.5. Odonata: the Genus* Macromia *Rambur (Corduliidae)*

*Macromia* Rambur is a virtually cosmopolitan genus of rheophilic dragonflies. With more than 120 species, *Macromia* is one of the largest genera of the Anisoptera. The Sulawesi species of this genus were studied by van Tol (1994), who also provided a reconstruction of phylogenetic relationships between species in southeast Asia.

The Papuasian representatives of this genus share at least four characters, including a small discoidal triangle in the hind wing and a minute pterostigma (Lieftinck, 1952, 1971). Lieftinck (1971) distinguished three groups among the

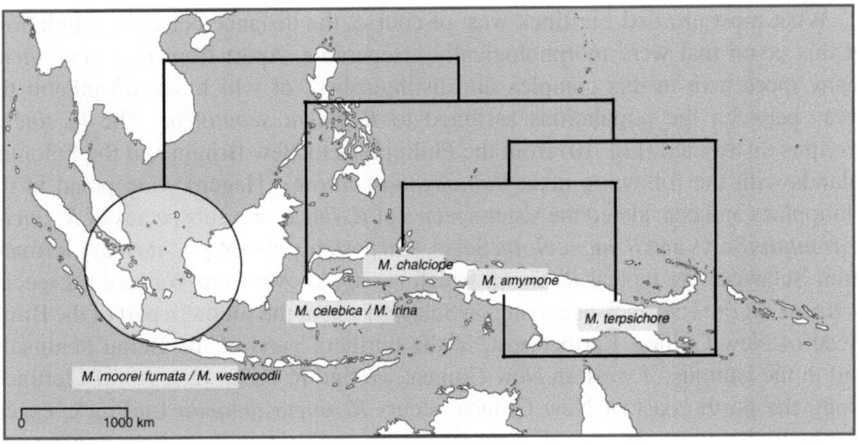


Fig. 11 Relationships of *Macromia* species in the Malay Archipelago, the distributions plotted on a map (after van Tol, 1994).

Papuan species, which are all but one confined to New Guinea, the Bismarck archipelago, Waigeu, and Misool, while one species, *M. chalciope* Lieftinck, is known from Schouten Island, Halmahera and Bacan. The genus *Macromia* is not further known from the Moluccas. About 15 species are known from the Malay Peninsula and the Greater Sunda islands. A preliminary grouping by Lieftinck was apparently not based on natural affinities. The Philippines are inhabited by three species, including at least one endemic.

Van Tol (1994) presented a phylogenetic tree of the Indo-Australian species groups of *Macromia* (also Fig. 11). It appeared that the groups as defined by Lieftinck were not corroborated by the analysis. The tree, rooted with the species of the mainland of southeast Asia, showed that the species of Sulawesi are the sister-group of the Papuan species. Secondly, *M. chalciope* Lieftinck from Halmahera appeared to be the sister-species of *M. terpsichore* Förster from northeast New Guinea, while these two species together formed the sister-group of *M. melpomene* Ris.

When the distributions of the species are used to define areas of endemism and are plotted on a map, the area cladogram (Fig. 11) is congruent with the geological area cladogram of Beuk (2002b).

## 4.3. Other Groups

### *4.3.1. Mollusca*

Although several species of Malesian freshwater molluscs were described as early as the late 19th century, they have remained poorly known up to now. Molluscs are rarely considered in biogeographical studies (Davis, 1982; Glaubrecht et al. 2003).

Glaubrecht et al. (2003) analysed the *Corbicula* freshwater bivalves (Corbiculidae) of southeast Asia, especially Sulawesi. The genus *Corbicula*

Megerle von Mühlfeld is a monophyletic taxon, in which all Old World species are the sister-group of the Japanese *C. japonica* Prime. Two species are widely distributed in Asia and introduced in Europe and North America, *C. fluminalis* O.F. Müller and *C. fluminea* O.F. Müller; some authors have lumped all described taxa under these two names. The first species is salt-tolerant and occurs in estuaries and similar habitats; it releases a veliger larva. The second species is more restricted to the lacustrine environment, and incubates embryos in the gills. More careful studies, e.g., in Japan, have revealed that the taxonomy is more complicated. For instance, some forms reproduce by androgenesis (using only the genome of spermatozoa). The genetic variation in the *Corbicula* species of Sulawesi has proved to be much more complicated, and these taxa cannot be assigned to only one or two species. Based on an analysis of morphological and molecular characters, at least five additional species were recognized, one on Sumatra and four on Sulawesi, all endemic to one of the large lakes of Sulawesi (Matana, Poso, Lindu). These lakes are presumably not older than 2 My. All these Sulawesi taxa reproduce sexually. This pattern of endemic taxa has also been found in the pachychilid gastropod genus *Tylomelania* Sarasin and Sarasin, and in ancylid molluscs (von Rintelen and Glaubrecht, 2003). The data on the timing of the evolution of these taxa was discussed by Glaubrecht et al. Molecular studies revealed a shallow polytomy based on mitochondrial COI sequences for *Corbicula* species from Japan to Australia, which suggests a relatively recent origin of these taxa, and such data were confirmed when the Indonesian taxa were included. Even a Pleistocene age is not unlikely, which would indicate that dispersal over larger distances should have played a major role in the evolution of the present distribution of the extant species. Long range dispersal by birds may have played a role. On the other hand, a phylogenetic analysis of the pachychilid gastropods of the genus *Brotia* H. Adams of southeast Asia showed that they reflected palaeogeographical events of the Cretaceous rather than of recent geological periods. In this case vicariance events rather than dispersal seem to have played the dominant role in distributional evolution.

### *4.3.2. Plecoptera*

Stoneflies, or Plecoptera, are most diverse in the temperate regions, and not common in tropical areas. Their poor flying capacities and specialized ecology make dispersal unlikely. The zoogeography of the groups was discussed by Zwick (2000), who did not follow previous conclusions by Illies (1965).

The suborder Antarctoperlaria is presumably an old group occurring in South America, New Zealand, and Australia, but absent from Africa, Madagascar, and India. The Permian *Euxenoperla* Riek is assigned to the Gripopterygidae, a family of the Antarctoperlaria, but the assignment is considered doubtful. Nevertheless, if the group is Cretaceous or older, an extinction in Africa and India is needed to explain the pattern, or dispersal is more likely in this group than presently presumed. The genera that are known from the Greater Sunda islands and the Philippines are all representatived of the suborder Arctoperlaria, which is typically holarctic. The Neoperlini originated in the Oriental region, and have even reached New Guinea to the east (one species of *Neoperla* Needham), and North America (also *Neoperla*) via the northwest. The genus *Neoperla* is also widespread in the Afrotropical region.

According to Zwick (2000) it is still uncertain whether Plecoptera patterns are old (pre-Cretaceous). In the case of old patterns, massive extinctions in the Arctoperlaria have to be postulated in Gondwana. However, if the patterns are more recent, extensive dispersal across seas must be hypothesized. A further study of the phylogeny of at least the *Neoperla* of southeast Asia would provide a valuable contribution to zoogeography of aquatic biotas.

### *4.3.3. Ephemeroptera*

Mayflies represent probably one of the best groups of aquatic insects for biogeographic studies. The larval stages are usually stenotopic, and imaginal and subimaginal stages are short-lived and weak flyers, which make them poor dispersers. Unfortunately, little information of the phylogeny of this group of insects is available. Soldán (2001) recently summarized systematic knowledge of the Oriental (including Malesian) Ephemeroptera fauna. Not more than 535 species of mayflies have been described from the Oriental region and its transition zones (Afghanistan, Himalayas, China and Papua New Guinea). About 35% of the nominal taxa are only known from the larval stage, and 39% only from the adult stage. The total number of species will probably be three to four times higher. For the species of southeast Asia, only very few taxonomic revisions are available, and Ulmer's (1939, 1940) treatment of the fauna of the Sunda islands needs both extension and updating. Zoogeography has been discussed by Edmunds (1979) for the Oriental fauna, and Edmunds and Polhemus (1990) for the Malay Archipelago, with special attention to Sulawesi.

The fauna of the Malay Archipelago, as far east as New Guinea, is nearly fully of Oriental origin, while Australian elements are rare. Many Oriental elements known from the Sunda islands (33 genera) do not reach Sulawesi (Edmunds and Polhemus, 1990). The fauna of New Guinea is relatively poor, and based on radiations of only seven clades, of which six are Oriental. Only the genus *Tasmanocaenis* is shared between Australia and New Guinea, and no other species of Australian mayfly occurs in New Guinea. The genus *Sulawesia* Peters and Edmunds is considered the only Australian element in the fauna of Sulawesi.

### *4.3.4. Hemiptera*

Systematics, phylogeny and zoogeography of the aquatic and semiaquatic Heteroptera of southeast Asia have intensively been studied by N.M. Andersen, P. Chen, N. Nieser, D.A. Polhemus, J.T. Polhemus, and H. Zettel. The results of the study of the Sulawesi fauna will be discussed below.

Andersen (1991, 1998) described the cladistic biogeography of the marine water striders of the families Hermatobatidae, Velidae, and Gerridae of the Indo-Pacific region. *Hermatobates* Carpenter is the only genus of Hermatobatidae, and species occur along continental coasts and islands in the Indo-Pacific, including the Red Sea, East Africa, Seychelles, Maldives, Ryukyu islands, Philippines, Indonesia, Australia, and Hawaii. One species is known from the Caribbean. Species of other families with similar ecological preferences usually have smaller distributional ranges (see Andersen, 1998, for details). For an analysis of the distributions, the Indo-Pacific region was divided in eight areas of endemism: Australia, East Asia, Indian Ocean, Malayan, Papuasia, Philippines, Sulawesi, and West Pacific. The analysis for the various groups revealed cladograms that were not necessarily congruent. The combination of all groups studied yielded two completely resolved area cladograms, which only differed in the position of the Indian Ocean, but a strict consensus of both is completely unresolved. One of the two trees is (Aus (W Pac (Papu (Sula (Phil (Ind-O (Mala and East Asia))))))). This area relationship distinctly indicates a trend in a western direction from Australia. Andersen (1998) remarks that, for instance, the *Halovelia* Bergroth species of East Africa, Madagascar and the islands of the Indian Ocean belong to the southeast Asian *H. malaya* group, and possibly have to be considered a late dispersal. The archipelago of volcanic islands and microcontinents between Australia and Asia during the Cenozoic must have been a perfect place for allopatric speciation. Fossils, however, show that *Halovelia* and *Halobates* have occupied a much wider area than at present, indicating that (local) extinction must have played a major role in shaping present distribution patterns.

D.A. Polhemus, frequently with J.T. Polhemus (e.g., Polhemus and Polhemus, 1998) discussed regional distribution patterns in more detail (Fig. 12). They surveyed and analysed the Nepomorpha and Gerromorpha fauna of New Guinea in detail, and have linked that to the palaeogeography of the region, although many early papers lack a cladistic analysis. Polhemus and Polhemus (2002) includes an analysis of the biogeography of the small waterstriders of the subfamily Trepobatinae based on a careful phylogeny of this taxon. This group is very speciose in New Guinea and the ranges show strong correlations with past tectonic events, such as the accretion of island arcs, although the various tribes appeared to have all differing biogeographic histories. All Metrobatini of New Guinea and Australia are the sister-group of the

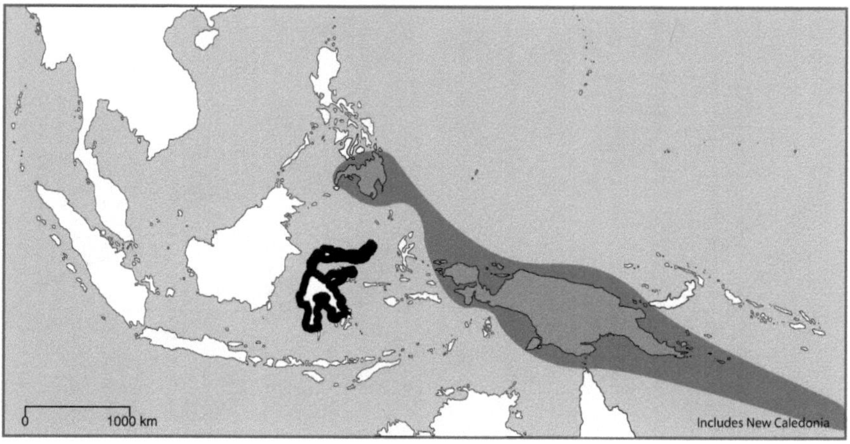


Fig. 12 Distribution of the *Rhagovelia novacaledonica* group (from Polhemus, 1995).

Neotropical *Metrobates* Uhler. This pattern indicates to a Gondwanan origin, the vicariance presumably due to separation of Australia and Antarctica at *c.* 85 Ma. Many taxa of New Guinea have distributions congruent with the geological entities as defined by Pigram and Davies (1987). The genus *Metrobatopis* Esaki (Fig. 13) is confined to the Solomons, the Bismarck archipelago and along the northern coast of New Guinea, portions of the Miocene-Pliocene Solomons Island Arc (cf. Kroenke, 1996). The genus *Andersenella* J.T. Polhemus and D.A. Polhemus only occurs on the Papuan Peninsula, which is, according to Polhemus and Polhemus (2002), an indication of the relatively recent accretion of this complex terrane with the rest of New Guinea. The genus *Metrobatoides* Polhemus and Polhemus is only known from the Torricelli mountains of northern New Guinea and the northern part of the central mountain ranges in the Mamberamo River Basin, which are both portions of accreted terranes. Three other genera of Metrobatini are extending over those parts of New Guinea that once formed part of the Australian continental craton. Such groups may have evolved on the Vogelkop microcontinent, which fused with the rest of New Guinea during the Late Miocene or Early Pliocene, or on the Halmahera Arc (considered part of the Philippine Arc in the present paper), which lay north of New Guinea from the Oligocene through the Miocene.

The genus *Calyptobates* Polhemus and Polhemus of the tribe Naboandelini is distributed over Australia, New Guinea, the Moluccas, Borneo, the Andaman Islands, and Sri Lanka, but not east of New Guinea. All other species occur in Africa. Since all three species of New Guinea occur south of the central mountain ranges, Polhemus and Polhemus presumed that the genus reached New Guinea via Australia, being a Gondwanan element. The Stenobatini, the third tribe of this group of waterstriders, includes one genus, *Thetibates* Polhemus and Polhemus, with a distribution pattern congruent with the Solomons Arc terranes.

It is hypothesized that most groups that evolved in Asia reached the New Guinea region via the so-called Papuan Arc. This arc collided with the northern

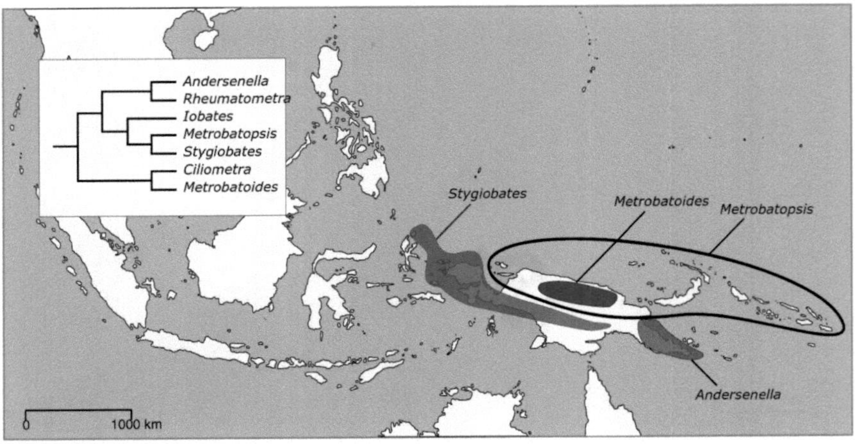


Fig. 13 Distribution of genera of Metrobatini (after Polhemus and Polhemus, 2002).

margin of Australia in the Late Eocene to Early Oligocene. Species that evolved in this arc can now be expected in central New Guinea and mountains of the Vogelkop Peninsula, and have outgroup relationships with groups to the west. Polhemus and Polhemus were able to identify various taxa with distribution patterns that reflect the Papuan Arc distribution, including Microveliinae, Gerridae (genus *Ptilomera* Amyot and Serville), and Naucoridae (genera *Nesocricos* and *Tanycricos*). Finally, some groups evolved in the Caroline Arc during the Oligocene. This island arc accreted to northeast New Guinea until the Early Pliocene. The distribution of *Rhagovelia biroi* Lundblad is considered to be an example of a Solomon Arc distribution of Miocene age.

Polhemus and Polhemus (1994) analysed the phylogenetic position of the Timorese endemic waterstrider *Aquarius lili* Polhemus and Polhemus (Gerridae). The genus *Aquarius* Schellenberg is widespread in the Palaearctic, Afrotropical, Oriental and Australian regions, but it is absent from Sulawesi, the Moluccas or the Lesser Sunda islands (see Andersen 1990). In a cladistic analysis, confirmed by Damgaard et al. (2000) and Damgaard and Zettel (2003), *Aquarius lili* appeared to be the sister-species of *A. adelaidis* (Dohrn), a species widespread in the mainland southeast Asia, the Philippines and the Greater Sunda islands, rather than of the Australian species. This result supports the geological fore-arc ridge hypothesis of Hamilton (1979), in which Timor is part of the Banda Arc fore-crest, rather than the deformed northwestern edge of the Australian continental plate. If Timor is part of the Banda arc system, rafting of Asian elements, such as *A. lili*, during the eastward migration of this island during the Miocene from a position near Celebes and Borneo to its present position is hypothesized. As stressed by Polhemus and Polhemus, such a scenario seems to be supported by records of a full-size Eocene anthracothere with southeast Asian affinities on Timor. A further zoogeographical analysis of Timor has to await further detailed cladistic analyses of groups including Timorese endemics. Polhemus and Polhemus (1994) mention a sister-group relationship of *Ptilomera timorensis* Hungerford and Matsuda (Heteroptera: Gerridae) with species occurring on the mainland of southeast Asia.

### *4.3.5. Trichoptera*

Distribution patterns of Australasian caddisflies Trichoptera were reviewed by Mey (2001), mainly at the family level. The origin of the families of the Trichoptera precedes the break-up of Pangaea for the Spicipalpia and Annulipalpia, but (most) families of the Integripalpia evolved after the break-up. The Asian and Australian faunas made contact when the Australian plate collided with island arcs in the "midwestern Pacific" from the Late Oligocene. Some of the arcs or island chains provided dispersal routes for some elements of the Trichoptera fauna. The dispersal capacity of various groups of caddisflies may, however, differ considerably.

The Gondwanan influence on the Oriental fauna after the docking of India has not been studied yet, but the distribution of the genus *Apsilochorema* Ulmer has been attributed to the accretion of terranes in southeast Asia before India reached the southern margin of Laurasia (Mey, 1998, 1999). *Apsilochorema* is a genus of the Hydrobiosidae (Spicipalpia), which contains 20 genera in South America, of

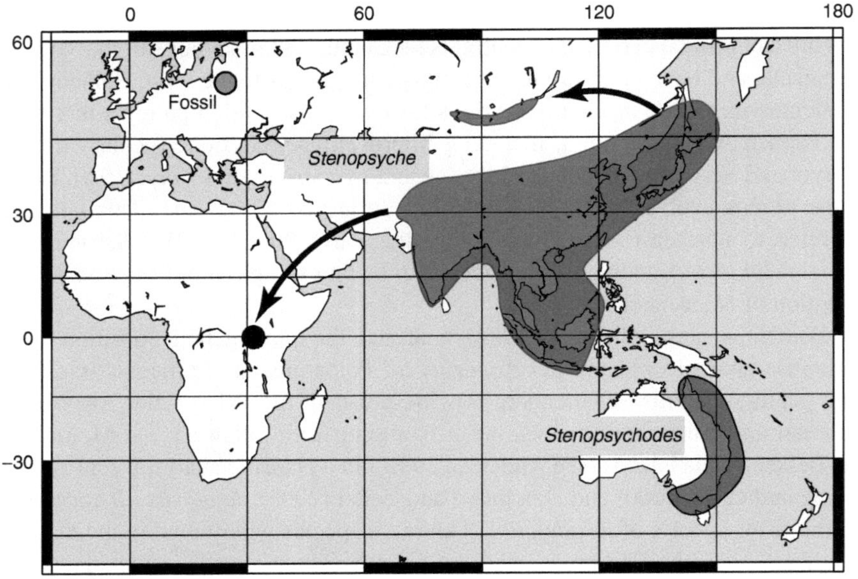


Fig. 14 Example of the distribution of Stenopsychidae (Trichoptera) (after Mey, 2001).

which only one has dispersed northward up to southern North America. Twenty-five genera of Hydrobiosidae are known from Australia. The genus *Apsilochorema* itself is widespread (with two subgenera) in Australia, Indonesia, the southeastern part of Asia, along the Pacific coast northward up to Japan, and westward through the Himalayas. There are disjunct populations in southern India and Sri Lanka, and the southern part of the Caspian Sea. Families like Goeridae, Lepidostomatidae, and Dipsodopsidae extend into the northern part of the Australian region, but are "clearly of Asian origin" (Mey, 2001: 263).

The distribution of the Stenopsychidae (Annulipalpia) (Fig. 14) shows distinct disjunctions, presumably due to poor dispersal capacities. Three genera are known of this family, *Pseudostenopsyche* Döhler in the southern Andes, *Stenopsychodes* in eastern Australia, and *Stenopsyche* MacLachlan in the mainland of southeast Asia including India and northward to Japan, and in the Greater Sunda islands in the Malay archipelago. The genus occurs disjunctly in southern Siberia, and central Africa, while it is known from Baltic amber in Europe. Mey (2001) suggests that this genus has had a much larger distribution in the past.

## 4.4. Sulawesi

The biogeography of Sulawesi has been the subject of studies since the 19th century. The remarkable mixture of Oriental and Australian elements, with endemics of isolated phylogenetic origin, has attracted attention since Wallace (e.g., 1860, 1863) travelled and studied the Malay archipelago. The study and analysis of the flora and fauna continues up to today (e.g., Knight and Holloway, 1990; Whitmore, 1981; Whitten et al., 1987; Holloway, 1997; van Oosterzee, 1997), while much new information on the geological history of the region has provided a basis for the understanding of the processes that presumably gave rise to the evolution of so many species. The details of the geological history have been summarized above. In summary, the southwestern arm may have a subaerial history for the last 60 My, while the other arms derive from the margin of the Australian plate (eastern arms) or the southern margin of the Philippine plate. The northern arm may have had a position not far from the SW arm since the Eocene (Wilson and Moss, 1999), or much later (Hall), the microcontinents of Banggai-Sula and Buton-Tukang Besi only accreted during the Pliocene. It is still uncertain to what extend they have a history above sea level. Some geological reconstructions include conclusions of floral or faunal analyses, so that the use of such data for biogeographical analyses requires caution. It is worthwhile to mention that western Sulawesi remained in or near tropical latitudes throughout the Cenozoicum (Wilson and Moss, 1999).

As in many other groups, the composition of the aquatic invertebrates of Sulawesi is characterized by a high percentage of island endemism, and the absence of groups that are common and speciose elsewhere in the Malay archipelago. All these phenomena are clearly related to the long isolation of Sulawesi. Also, many species are restricted to parts of Sulawesi, and the species distributions within the island frequently show distinct congruence with the geological history of the island. The present knowledge of the aquatic biotas is summarized here.

*Absence of widespread taxa.* Although absence is methodologically difficult in zoogeographical analyses, for several groups data are sufficiently abundant and reliable as a basis for discussion.

In Odonata there is a remarkable absence of some families that occur widely west of Sulawesi in the Greater Sunda islands, namely Amphipterygidae, Euphaeidae, Platycnemididae, and Cordulegastridae, while other families are poorly represented, e.g., Calopterygidae, Protoneuridae, and Gomphidae. The ranges of many Australian/Papuan taxa above the species level reach their western limits in Sulawesi (see below), but the families Isostictidae and Synthemistidae are not known west of Halmahera (Moluccas). As compared to the Philippines, the same families of odonates are absent as compared to the Greater Sunda islands.

*Affinities.* The affinities of the fauna of Sulawesi are predominantly Oriental. The analysis of the Ephemeroptera (Edmunds and Polhemus, 1990) showed that 20 genera of mayflies are known from Sulawesi, of which at least 3 genera are endemic. The other genera mainly have an Oriental distribution (14 genera shared with the Greater Sunda islands, 8 with the Philippines). Only six genera occurring on Sulawesi are shared with the Moluccas. In addition, Peters and Edmunds (1990) noted the depauperate nature of the Sulawesi Ephemeroptera fauna, since many genera found in Borneo, Java, Sumatra, and the southeast Asian mainland are lacking. The interpretation of all these figures is, however, severely hampered by incomplete data sets.

Among the exceptions of Oriental affinity are the genus *Echinobaetis* Mol of the Baetidae (Mol, 1989) and genus *Sulawesia* Peters and Edmunds of the atalophlebiine Leptophlebiidae (Peters and Edmunds, 1990). *Echinobaetis* is considered most closely related to *Jubabaetis* Müller–Liebenau from Luzon (Philippines). According to Edmunds and Polhemus (1990), *Sulawesia* is a member of an ancient Gondwanan lineage and the only genus of mayflies in the region that has southern (Australian) affinities, with its relatives known from Australia, New Zealand, southern South America, Madagascar, Sri Lanka, and southern India, but not from the Papuan region. *Sulawesia* seems to belong to a group of genera restricted to South America, Australia, and New Zealand, rather than to a group distributed in Madagascar, Sri Lanka, and India.

The Odonata of Sulawesi are also primarily of Oriental origin (van Tol, 1987a). Several elements of the Papuan fauna reach the western limit of their ranges in Sulawesi, including the genus *Nososticta* Hagen in Selys (Protoneuridae, 1 species in Sulawesi and 28 in the Papuan region), *Nannophlebia* Selys (Libellulidae, 1 species in Sulawesi, at least 8 in the Papuan region), and *Diplacina* Brauer (Libellulidae, 3 species in Sulawesi, at least 7 in the Papuan region) (cf. Lieftinck, 1949a). Although the sister-groups of some Sulawesi taxa are found in the Philippines, e.g., in *Drepanosticta* Laidlaw (Platystictidae) and *Diplacina,* both faunas further share only common and widespread species (cf. Hämäläinen and Müller, 1997). According to Vane-Wright (1990: 19) "Sulawesi shares by far the greatest portion of its genera [of butterflies] with the Philippines (…), but at the species level its strongest unique link is with the Moluccas." This observation cannot be confirmed for the Odonata, nor for other aquatic insects. Among the aquatic Heteroptera, Sulawesi has 41 out of 42 genera in common with the Greater Sunda islands or continental Asia. Polhemus and Polhemus (1990) hypothesized that most elements have crossed the Makassar Strait from Borneo.

*Island endemism.* Island endemism at the species level is extremely high for many groups in Sulawesi, e.g., 98% of the mammals (excluding bats) (Whitten et al., 1987), 95% in cicadas (Cicadidae) (Duffels, 1990), 76% in amphibians, and 29% in swallowtail butterflies (Whitten et al., 1987). Not many data are available for aquatic invertebrates, but Polhemus and Polhemus (1990) estimated 65% endemicity at the species level; no endemic genera are known from Sulawesi. Edmunds and Polhemus (1990) provide some data for Ephemeroptera, including two out of six genera of Baetidae endemic to Sulawesi. In the Odonata, nearly all species in various families with poor dispersal power are island endemics. Examples are the Platystictidae and Chlorocyphidae, both with 100% island endemics (although recent data confirm the occurrence of the Papuan *R. tincta semitincta* of the Moluccas in Central Sulawesi). Van Tol (1987a, b) noticed that Odonata species confined to streams in primary forest are nearly all endemics, while the secondary habitats are inhabited by widely distributed, dispersive species. Similarly, the percentage of endemicity at the species level in the aquatic Heteroptera is nearly 100% (instead of 65%) if species confined to rice paddies and similar habitats are excluded from the analysis (Polhemus and Polhemus, 1990).

*Isolated phylogenetic position.* Quite some species endemic to Sulawesi are distinctly different from other known species and have been attributed to separate genera. The occurrence of such taxa with isolated taxonomic position can be related

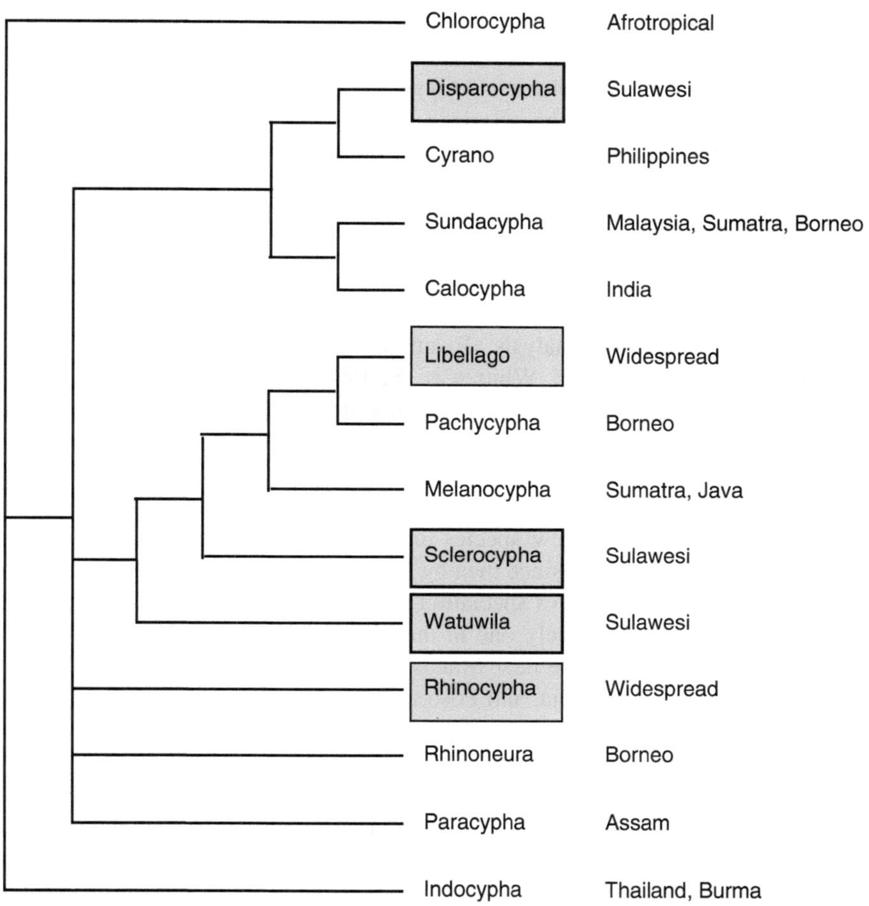


Fig. 15 Cladogram of genera of Chlorocyphidae (Odonata) (after van Tol, 1998). Genera occurring in Sulawesi in grey box, genera endemic to Sulawesi in bold border. Chlorocyphidae are very diverse on Sulawesi, even at the generic level.

to infrequent contact with the source area, usually a continental mainland, but in the case of Sulawesi presumably the Greater Sunda islands. Well-known examples on Sulawesi include mammal species such as anao, *Bubalus depressicornis* (H. Smith), and babirusa *Babyrousa babyrussa* (L.). Babirusa has no common ancestor since the Oligocene (30 Ma) (Whitten et al., 1987).

Among the Odonata, three endemic genera of Chlorocyphidae are recognized, namely *Disparocypha* Ris, *Sclerocypha* Fraser, and *Watuwila* van Tol. Apart from these genera, representatives of two more widespread genera also inhabit the island. An analysis of the genera of the Chlorocyphidae of southeast Asia (Fig. 15) (cf. van Tol, 1998 for details) shows that the Sulawesi genera belong to at least three monophyletic clades, which are all widespread in southeast Asia. The Sulawesi endemics *Sclerocypha* and *Watuwila* both split off early in a clade that also includes the widespread genus *Libellago* Selys. The aberrant genus *Disparocypha*, placed in a separate subfamily by Munz (1919), appears to be the sister-group of *Cyrano* Needham and Gyger, a genus confined to the Philippines. Although *Disparocypha* is characterized by many autapomorphies, it appears to be a relative recent clade in the Chlorocyphidae. Genera endemic to Sulawesi are also recognized in the Megapodagrionidae (*Celebargiolestes* Kennedy), and Libellulidae (*Celebophlebia* Lieftinck, *Celebothemis* Ris), but no phylogenetic analyses are available for these families.

*Areas of endemism within Sulawesi.* Species in many groups have small distributional ranges within Sulawesi, making the island extremely suitable for detailed biogeographical analysis. Examples include such diverse groups as macaques, carpenter bees (cf. Whitten et al., 1987), and cicadas (Duffels, 1990). Distribution data of freshwater invertebrates are available for some groups of Nepomorpha, Gerromorpha, Trichoptera, and Odonata. For aquatic and semi-aquatic Heteroptera, Polhemus and Polhemus (1990) provided a first overview and analysis, of which the distribution patterns of the veliid genus *Rhagovelia* Mayr are of particular interest. New species and distributional records of *Rhagovelia* species have been published since then by Nieser and Chen (1993, 1997) (Fig. 16). Based on congruent patterns of speciation, Polhemus and Polhemus identified five "centres of endemism", namely one in the Minahasa (northern arm), one around Kendari (Southeast arm), one near Makassar/Maros (southwestern arm), and two in central Sulawesi, the Toraja, and Poso centres. These areas are also recognized as separate entities in tectonic reconstructions.

Similar patterns were found when relationships of Odonata were used. Distribution maps or distributional data for Odonata taxa were published by Van Tol (1987b, 1994, 2000). An example is given in Fig. 17 for the red-coloured and superficially similar chlorocyphids *Watuwila*, *Libellago*, and *Sclerocypha*. *Watuwila* is only known from the southeastern peninsula, possibly an artefact. *Sclerocypha* is confined to the northern arm of Sulawesi, while the *Libellago*-complex inhabits the rest of Sulawesi. Within the last clade, the southwestern arm is sister to central and eastern Sulawesi.

As an example of a distribution pattern of two closely related species (or parapatric subspecies) one may use the distribution of *Diplacina m. militaris* Ris, and *D. militaris dumogae* Van Tol (Fig. 18). The genus *Diplacina* is of particular interest since it represents a distinct Philippine or Papuan relationship. The sister-species of *D. militaris s. l.* is *D. bolivarii* Selys, which is widespread in the Philippines (Luzon, Mindoro, Visayan regions). According to Hämäläinen and Müller (1997), the populations of *D. bolivarii* of Palawan and the Sulu islands are distinct and presumably represent another subspecies. Other species of *Diplacina* Brauer occur in the Moluccas and New Guinea, while three more species are known from the Philippines, including *D. braueri* Selys and *D. nana* Brauer, which are both closely related to *D. bolivarii*. Two more species of *Diplacina* are also known from Sulawesi, of which *D. torrenticola* van Tol belongs to the *D. militaris* species group. The affinities of *D. sanguinolenta* van Tol are uncertain, and cannot further be discussed. The common ancestor of

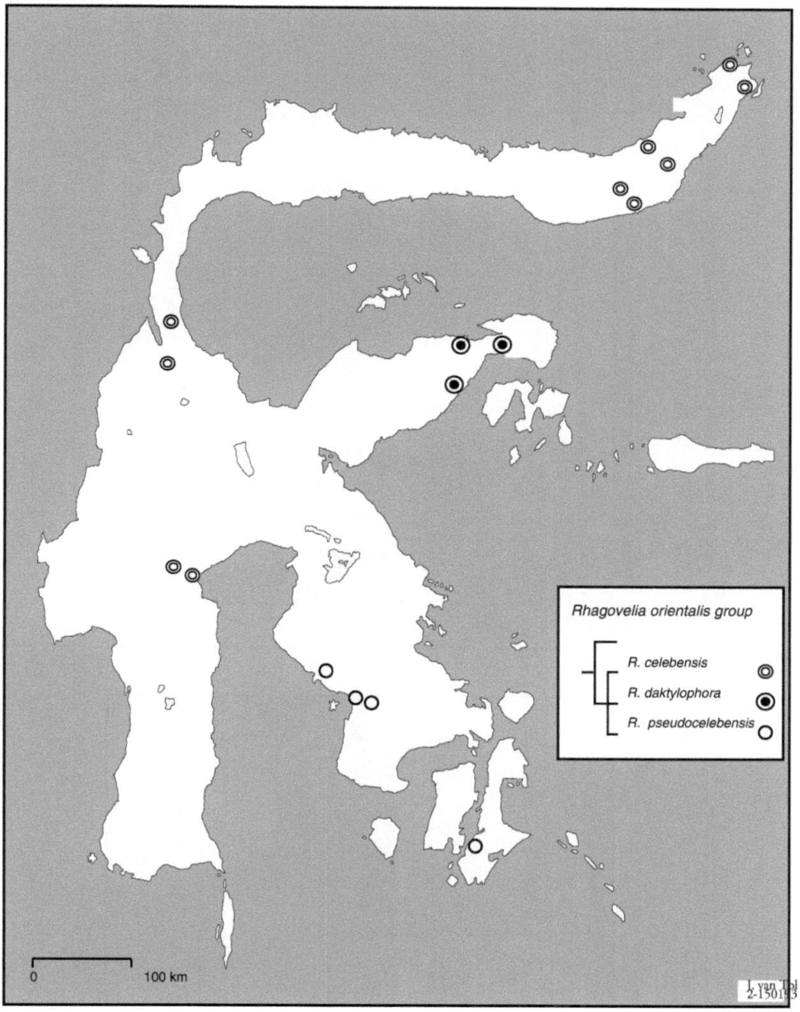


Fig. 16 Distribution of *Rhagovelia orientalis* group on Sulawesi.

the Philippine and Sulawesi *Diplacina* species presumably lived on one of the terranes in the Philippine Arc. Whether the occurrence on Sulawesi should be attributed to a dispersal event from a Philippine element of this arc to Sulawesi, or to a subaerial history of the northern Sulawesi Peninsula itself, is unknown. The dispersal power of *Diplacina* is presumably low. In particular, the absence of this genus from Borneo, even though it is known from Palawan, is remarkable. The subaerial history of the northern Sulawesi arm is still very uncertain, but Wilson and Moss (1999: 329) explicitly mention the existence of dispersal routes from the Philippines to Sulawesi "along vulcanic arcs, such as the long-lived arc along the North arm of Sulawesi".

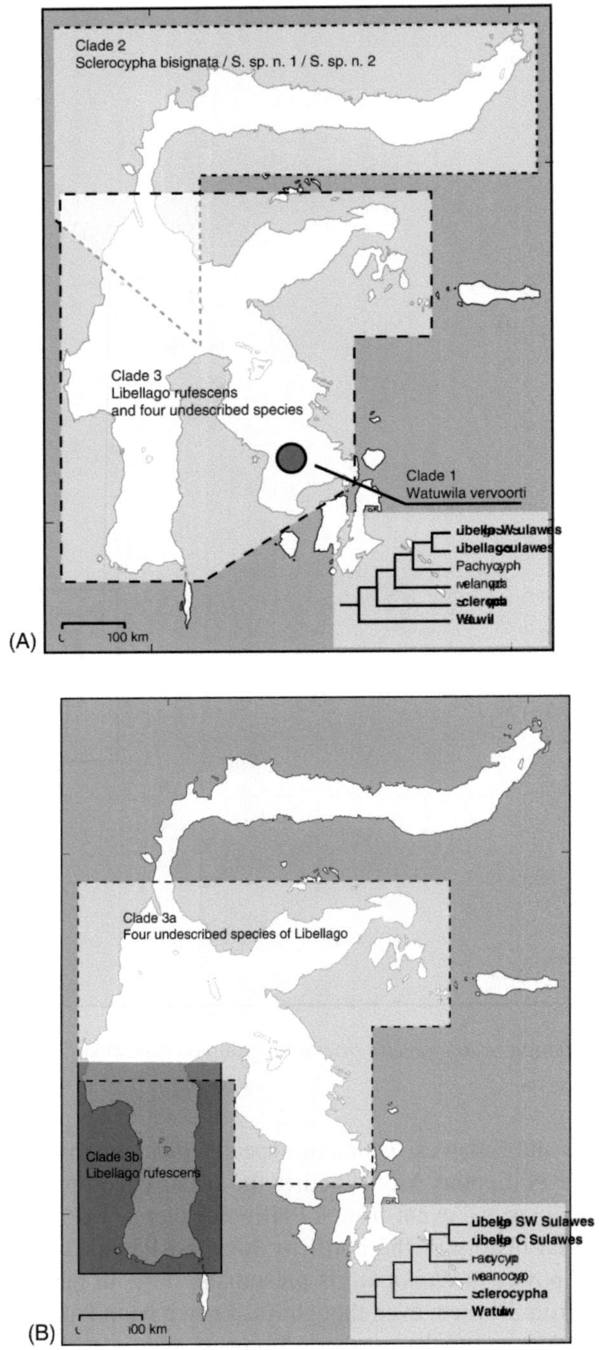


Fig. 17 Area relationships in Sulawesi based on phylogenetic relationships of *Libellago*, including *Sclerocypha*, in Sulawesi. (**A**) First division in cladogram. (**B**) Second division in cladogram.

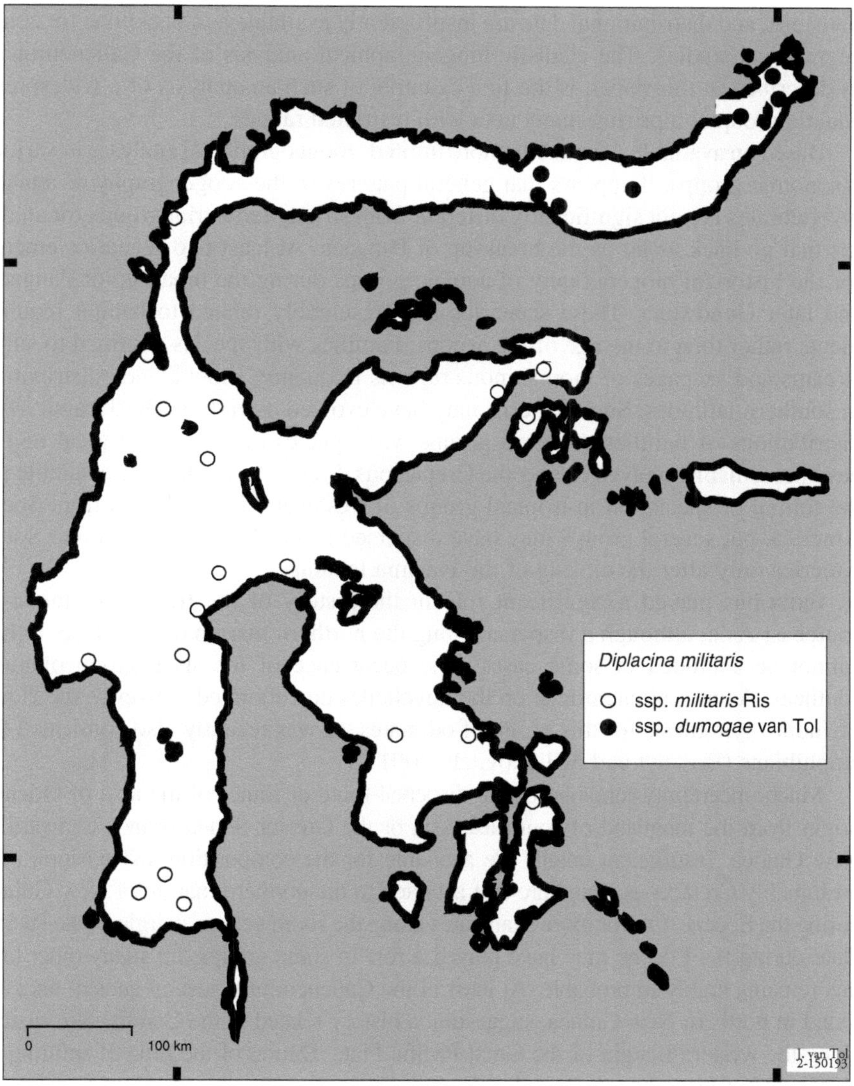


Fig. 18 Distribution of subspecies of *Diplacina militaris* Ris on Sulawesi.

# 5. Discussion

The number of groups for which a well-founded phylogenetic reconstruction of the southeast Asian taxa at the species level has been published is still low. The best studied groups are plants, and some groups of insects. Invertebrate taxa, including insects, of freshwaters remain particularly poorly studied. Actually, not only the phylogeny is poorly known, but even the basic data such as inventories, taxonomic revisions, and distributional data are insufficiently available as a backbone for zoogeographical studies. The cladistic biogeographical analysis of the Calicnemiinae, as described in this paper, is the first example of such an analysis of a widespread aquatic group comprising many taxa with restricted ranges.

Based on available, but mostly more limited, zoogeographical analyses in various taxonomic groups, it appears that general patterns in the zoogeography of aquatic invertebrates are not significantly different from strictly terrestrial groups for analyses that go back as far as the break-up of Pangaea. At least two scenarios emerge for the historical biogeography of aquatic groups during the break-up of Pangaea, and later Gondwana. These scenarios are presumably related to habitat requirements rather then to the age of the groups. Families with species confined to small streams and seepages of mountainous regions frequently show a wide distribution in southern latitudes. Such patterns may have evolved as early as the Jurassic. The distributions of families or genus-groups with species of the (sub)tropical region have presumably evolved during the Cretaceous. Presently, no data are available for the timing of speciation in tropical groups of freshwater invertebrates from South America, but several groups may have dispersed from Middle America into South America only after the closing of the Panama Isthmus.

India has played a significant role in the history of the freshwater fauna of southeast Asia, although a dispersal along the northern margin of the Meso-Tethys cannot be excluded in some cases. The occurrence of the sister-group of most southeast Asian Calicnemiinae on the Seychelles corroborated, however, the "Out-of-India" hypothesis for this group of odonates, as was recently also confirmed for amphibians (Bossuyt and Milinkovitch, 2001).

Much uncertainty remains for the dispersal route or routes of the taxa of Oriental origin from the mainland of southeast Asia or the Greater Sunda islands eastward to New Guinea. Insufficient details are available for the composition and position of a presumably Cretaceous island arc that accreted to the northern margin of New Guinea during the Eocene. One or more island arcs along the southwestern margin of the Pacific plate during the Eocene may have played a role in some groups, for many other taxa this remains highly improbable. At least in the Calicnemiinae several ancient taxa are found in northern New Guinea, suggesting a history related to the Caroline arc system along the western margin of the Great Pacific Plate. Dating of the time of splitting of clades, for instance based on molecular data, is not yet possible. In several orders of aquatic insects monophyletic groups show a pattern of distribution from the Philippines eastward to eastern New Guinea or even further into the Pacific, excluding Sulawesi and the Moluccas. Since such patterns cannot be explained by ecological or climatological conditions, the geological history of the area is presumed to be the causal factor. Although new information on the palaeogeography of southeast Asia has become available since the 1990s, the history of the island arcs and series of microcontinents along the southern and western margins of the Pacific plate (and partly also the Philippine plate) is still insufficiently known. A pre-Eocene, presumably Cretaceous, arc running from Mindanao to the Pacific, but excluding Sulawesi and the Moluccas, has been hypothesized (Polhemus, 1995), but not much geological information is available.

The special, "unbalanced" composition of the biota of Sulawesi is also confirmed for aquatic invertebrates. Most groups are of Oriental origin, but examples are available of monophyletic groups with Papuan affinities which do not occur west of Sulawesi. These patterns presumably go back as far as the early Mesozoicum, since when the southwestern arm of Sulawesi has remained in the same geographic position compared to Borneo as today. Endemicity at the species level is more than 50% for most groups, and is usually between 90% and 100% when only taxa confined to primary habitats are considered. Phylogenetically isolated taxa are also known among the Sulawesi freshwater invertebrates, e.g., in odonates, and such groups also have Oriental sister-group relationships. Although the geological origin of the other arms of Sulawesi is fairly well known now, the uncertainties of the subaerial history of these islands and their position in relation to New Guinea and parts of the Philippines are a serious handicap in understanding the distributional history of some remarkable taxa with Papuan affinities, e.g., in the genus *Rhinocypha* (Odonata). Area cladistic analyses based on taxa with restricted distributions within Sulawesi, defining areas of endemism, are congruent with the sequence of events in geological reconstructions.

The Malay archipelago is an area in which the distribution of land and sea has proved to be the primary driving force in the composition of the biotas. Many aquatic invertebrates, especially those confined to rain forests, are poor dispersers. Their distribution is the result of splitting, movement, and amalgation of the areas they inhabit, plus the more improbable oversea dispersal of the biotas themselves. The origin and disappearance, and also the significant movements, of terranes and island arcs in southeast Asia for at least the last 60 My, have provided the stage for the evolution of its highly diverse flora and fauna. A pattern caused by dispersal events by whole biotas may be difficult to distinguish from a pattern evolved by splitting of areas. Some authors simply state that dispersal should be considered the driving force, e.g., De Jong (2001: 135) "Whatever the land–sea distribution may have been since the mid Oligocene, animals and plants did disperse from Asia to Australia and *vice versa*." The question in zoogeography then will be how a certain monophyletic group has evolved, given the distribution of land and sea, and the biological characters of the species. While we are still searching for generalized patterns of area relationships, an analysis of the evolution of a particular group remains problematic.

The study of area relationships is hampered by phenomena as poorly corroborated phylogenetic reconstructions, no insight into the timing of speciation, insufficient data of land–water distributions in the past, and *random* dispersal of taxa. In the case of the Malay archipelago and the West Pacific, it is likely that at least the *effect* of dispersal has not been random, since survival after dispersal is strongly constrained by the seas surrounding the islands. It is methodologically still unexplored how patterns derived from rifting of terranes or island arcs, amalgation of such land masses, or active directional dispersal of the biotas should be distinguished. Also, not much effort has been made up to now to estimate the scale of dispersal. The distribution pattern of any species is, of course, generally due to small-step active or passive dispersal, depending on the life strategy of the species. Such dispersal can best be called *spreading* or *dispersion* over suitable habitat. Dispersal events over *unsuitable* habitat theoretically have another character. Specific characters, such as preferred habitat or ecological plasticity or adaptivity, rather than dispersal capacity per se, seem to determine the result of long-distance dispersal over geological times. For some groups of animals and plants, the drifting over sea of rafts of up to a few square kilometers might be one of the most likely methods of dispersal, as has especially been established for pieces of riverine forest. Based on these suppositions, species with very specific habitat requirements have little chance to establish populations on remote islands, since even successful active or passive transport to such a place does not guarantee survival of transported specimens. Even such an analysis meets many difficulties. Many species occurring on islands are extreme habitat specialists, but such a specialization may apparently evolve rather rapidly after a dispersal event, as has been shown for several groups of large and flightless insects of the Chatham islands (Trewick, 2000), but actually already for the Darwin finches of the Galapagos islands.

## Acknowledgements

This paper is mainly based on studies of geologists, systematic zoologists, and biogeographers. We herewith acknowledge their expertise, and apologize when their intentions have not fully been implemented in our synthesis. During the preparation we received helpful suggestions from our colleagues Dr. Rienk de Jong and Dr. Willem Renema. Dr. J.P. Duffels was our guide in the biogeography and palaeogeography of southeast Asia for many years. We gratefully acknowledge comments and suggestions on earlier drafts of this paper by Drs J.P. Duffels, R. de Jong, M. Hämäläinen, D.A. Polhemus, and an anonymous referee. The contribution by D. Gassmann was made possible by a grant of the Netherlands Organization for the Advancement of Science.

## References

Andersen, N.M., 1990, Phylogeny and taxonomy of water striders, genus *Aquarius* Schellenberg (Insecta, Hemiptera, Gerridae), with a new species from Australia, *Steenstrupia* **16**: 37–81.

Andersen, N.M., 1991, Cladistic biogeography of marine water striders (Insecta, Hemiptera) in the Indo-Pacific, *Australian Journal of Botany* **4**: 151–163.

Andersen, N.M., 1998, Marine water striders (Heteroptera, Gerromorpha) of the Indo-Pacific: cladistic biogeography and Cenozoic palaeogeography, in: R. Hall and J.D. Holloway (eds), *Biogeography and Geological Evolution of SE Asia*. Backhuys Publishers, Leiden, pp. 341–354.

Austin, C.C., 1999, Lizards took express train to Polynesia, *Nature* **397**: 113–114.

Bechly, G., 1996, Morphologische Untersuchungen am Flügelgeäder der rezenten Libellen und deren Stammgruppenvertreter (Insekten; Pterogota; Odonata) unter besonderer

Brücksichtigung der Phylogenetischen Sytsematik und des Grundplanes der *Odonata, *Petalura* **2**: 1–402.

Beuk, P.L.Th., 2002a, Cicadas spreading by island or spreading the wings? Historic biogeography of dundubiine cicadas of the Southeast Asian continent and archipelagos, Ph. D. thesis, University of Amsterdam.

de Boer, A.J., 1995, Islands and cicadas adrift in the West-Pacific, biogeographic patterns related to plate tectonics. *Tijdschrift voor Entomologie* **138**: 169–244.

de Boer, A.J. and Duffels, J.P., 1996, Historical biogeography of the cicadas of Wallacea, New Guinea and the West Pacific: a geotectonic explanation, *Palaeogeography, Palaeoclimatology, Palaeoecology* **124**: 153–177.

de Boer, A.J. and Duffels, J.P., 1997, Biogeography of Indo-Pacific cicadas east of Wallace's line, in: Keast, A. and Miller, S.E. (eds), *The Origin and Evolution of Pacific Island Biotas, New Guinea to Eastern Polynesia: Patterns and Processes*, SPB Academic, Amsterdam, pp. 297–330.

Bossuyt, F. and Milinkovitch, M.C., 2001, Amphibians as indicators of early tertiary "out-of-India" dispersal of vertebrates, *Science* **292**: 93–95.

Braithwaite, C.J.R., 1984, Geology of the Seychelles, in: Stoddart, D.R. (ed.), *Biogeography and Ecology of the Seychelles Islands*, Dr W. Junk, The Hague, pp. 17–38.

Briggs, J.C., 1974, *Marine Zoogeography*, McGraw-Hill, New York.

Calvert, P.P., 1931, The generic characters and the species of *Palaemnema* (Odonata: Agrionidae). *Transactions of the American Entomological Society* **57**: 1–111.

Calvert, P.P., 1934, Two Mexican species of *Palaemnema* (Odonata: Agrionidae). *Transactions of the American Entomological Society* **59**: 377–381.

Carle, F.L., 1995, Evolution, taxonomy, and biogeography of ancient Gondwanian libelluloides, with comments on anisopteroid evolution and phylogenetic systematics (Anisoptera: Libelluloidea), *Odonatologica* **24**: 383–424.

Coates, A.G., 1999, Lithostratigraphy of the Neogene strata of the Caribbean coast from Limon, Costa Rica to Colon, Panama, *Bulletins of American Paleontology* **357**: 17–40.

Daly, M.C., Cooper, M.A., Wilson, I., Smith, D.G., and Hooper, B.G.D., 1991, Cenozoic plate tectonics and basin evolution in Indonesia, *Marine and Petroleum Geology* **8**: 2–21.

Damgaard, J., Andersen, N.M., and Sperling, F.A.H., 2000, Phylogeny of the water strider genus *Aquarius* Schellenberg (Heteroptera: Gerridae) based on nuclear and mitochondrial DNA sequences and morphology, *Insect Systematics and Evolution* **31**: 71–90.

Damgaard, J. and Zettel, H., 2003, Genetic diversity, species phylogeny and historical biogeography of the *Aquarius paludum* group (Heteroptera, Gerridae), *Insect Systematics and Evolution* **34**: 313–328.

Davis, G.M., 1982, Historical and ecological factors in the evolution, adaptive radiation, and biogeography of freshwater mollusks, *American Zoologist* **22**: 375–395.

de Bruyn, M., Wilson, J.A., and Mather, P.B., 2004, Huxley's line demarcates extensive genetic divergence between eastern and western forms of the giant freshwater prawn, *Macrobrachium rosenbergii. Molecular Phylogenetics and Evolution* **30**: 251–257.

de Jong, R., 1996, The continental Asian element in the fauna of the Philippines as exemplified by Coladenia Moore, 1881 (Lepidoptera: Hesperiidae), *Cladistics* **12**: 323–348.

de Jong, R., 2001, Faunal exchange between Asia and Australia in the tertiary as evidenced by recent butterflies, in: Metcalfe, I., Smith, J.M.B., Morwood, M., and Davidson, I. (eds), *Faunal and Floral Migrations and Evolution in SE Asia–Australasia*, A.A. Balkema Publishers, Lisse, pp. 133–146.

Diamond, J.M., 1988, Express train to Polynesia, *Nature* **336**: 307–308.

Duffels, J.P., 1990, Biogeography of Sulawesi cicadas (Homoptera: Cicadoidea), in: Knight, W.D. and Holloway, J.D., *Insects and the Rain Forests of South East Asia* (*Wallacea*). Royal Entomological Society of London, London, pp. 63–72.

Edmunds, G.F., Jr., 1979, Biogeographical relationships of the Oriental and Ethiopian mayfly faunas, *Proceedings of the International Conference on Ephemeroptera* **2** (Krakow): 11–14.

Edmunds, G.F., Jr. and Polhemus, D.A., 1990, Zoogeographical patterns among mayflies (Ephemeroptera) in the Malay archipelago, with special reference to Celebes, in: Knight, W.D. and Holloway, J.D., Insects *and the Rain Forests of South East Asia* (*Wallacea*), Royal Entomological Society of London, London, pp. 49–56.

Eskov, K.Yu., 2002, Geographical history of insects, in: Rasnitsyn, A.P. and Quicke, D.L.J. (eds), *History of Insects*, Kluwer Academic, Dordrecht, pp. 427–435.

Gassmann, D., 1999, Taxonomy and distribution of the inornata species-group of the Papuan genus *Idiocnemis* Selys (Odonata: Zygoptera: Platycnemididae), *Invertebrate Taxonomy* **13**: 977–1005.

Gassmann, D., 2000, Revision of the Papuan *Idiocnemis bidentata*-group (Odonata: Platycnemididae), *Zoologische Mededelingen* **74**: 375–402.

Gassmann, D., 2005, The phylogeny of Southeast Asian and Indo-Pacific Calicnemiinae (Odonata, Platycnemididae), *Bonner Zoologische Beiträge* **53**: 37–80.

Gassmann, D. and Hämäläinen, M., 2002, A revision of the Philippine subgenus *Risiocnemis* (*Igneocnemis*) Hämäläinen (Odonata: Platycnemididae), *Tijdschrift voor Entomologie* **145**: 213–266.

Glaubrecht, M., von Rintelen, T., and Korniushin, A.V., 2003, Toward a systematic revision of brooding freshwater Corbiculidae in Southeast Asia (Bivalvia, Veneroida): on shell morphology, anatomy and molecular phylogenetics of endemic taxa from islands in Indonesia, *Malacologia* **45**: 1–40.

Hall, R., 1998, The plate tectonics of Cenozoic SE Asia and the distribution of land and sea, in: Hall, R. and Holloway, J.D. (eds), *Biogeography and Geological Evolution of SE Asia*. Backhuys Publishers, Leiden, pp. 99–131.

Hall, R., 2001, Cenozoic reconstructions of SE Asia and the southwest Pacific: changing patterns of land and sea, in: Metcalfe, I., Smith, J.M.B., Morwood, M., and Davidson, I. (eds), *Faunal and Floral Migrations and Evolution in SE Asia–Australalasia*, A.A. Balkema Publishers, Lisse, pp. 35–56.

Hall, R., 2002, Cenozoic geological and plate tectonic evolution of SE Asia and the SW Pacific: computer-based reconstructions, model and animations, *Journal of Asian Earth Sciences* **20**: 353–431.

Hämäläinen, M. and Müller, R.A., 1997, Synopsis of the Philippine Odonata, with lists of species recorded from forty islands, *Odonatologica* **26**: 249–315.

Hamilton, W., 1979, Tectonics of the Indonesian region, *United States Geological Survey Professional Paper* **1078**: 1–345.

Hill, K.C. and Hall, R., 2003, Mesozoic-Cenozoic evolution of Australia's New Guinea margin in a west Pacific context, in: Hillis, R.R. and Müller, R.D. (eds), *Evolution and Dynamics of the Australian Plate, Geological Society of America, Special Paper* **372**: 265–290.

Hoeksema, B.W., 2007, Delineation of the Indo-Malayan Centre of Maximum Marine Biodiversity: the Coral Triangle. *This volume*.

Holloway, J.D., 1997, Sundaland, Sulawesi and eastwards: a zoogeographic perspective, *Malayan Nature Journal* **50**: 207–227.

Holthuis, L.B., 1995, Notes on Indo-West Pacific Decapoda III to IX. VIII. The correct names of the subspecies of *Macrobrachium rosenbergii* (De Man, 1879), *Zoologische Mededelingen* **69**: 139–151.

Huxley, T.H., 1868, On the classification and distribution of the Alectoromorphae and Heteromorphae, *Proceedings of the Zoological Society of London* **1868**: 294–319.

Illies, J., 1965, Phylogeny and zoogeography of the Plecoptera, *Annual Review of Entomology* **10**: 117–140.

Ivanov, V.D. and Sukatsheva, I.D., 2002, Order Trichoptera Kirby, 1813, the caddisflies (= Phryganeida Latreille, 1810), in: Rasnitsyn, A.P. and Quicke, D.L.J. (eds), *History of Insects*, Kluwer Academic, Dordrecht, pp. 199–220.

Kathirithamby, J. and Grimaldi, D.A., 1993, Remarkable stasis in some Lower Tertiary parasitoids: descriptions, new records, and review of Strepsiptera in the Oligo-Miocene amber of the Dominican Republic, *Entomologica Scandinavica* **24**: 31–41.

Keast, A. and Miller, S.E., 1996, *The Origin and Evolution of Pacific Island Biotas, New Guinea to Eastern Polynesia: Patterns and Processes*, SPB Academic, Amsterdam.

Kennedy, C.H., 1938, *Palaemnema picicaudata, P. abbreviata* and *P. brucelli*, new dragonflies from Ecuador, with notes on other *Palaemnemas* (Odonata: Platystictidae), *Annals of the Entomological Society of America* **31**: 249–266.

Kluge, N.Ju. and Sinitshenkova, N.D., 2002, Order Ephemerida Latreille, 1810. The true mayflies (= Ephemeroptera Hyatt et Arms, 1891 (s.l.); = Euephemeroptera Kluge, 2000), in: Rasnitsyn, A.P. and Quicke, D.L.J. (eds), *History of Insects*, Kluwer Academic, Dordrecht, pp. 89–97.

Knight, W.J. and Holloway, J.D., 1990, *Insects and the Rain Forests of South East Asia* (*Wallacea*), Royal Entomological Society of London, London.

Krause, D.W., Rogers, R.R., Forster, C.A., Hartman, J.H., Buckley, G.A., and Sampson, S.D., 1999, The Late Cretaceous vertebrate fauna of Madagascar: implications for Gondwanan biogeography, *GSA Today* **9** (8): 1–7.

Kroenke, L.W., 1996, Plate tectonic development of the western and southwestern Pacific: Mesozoic to the present, in: Keast, A. and Miller, S.E. (eds), *The Origin and Evolution of Pacific Island Biotas, New Guinea to Easterm Polynesia: Patterns and Processes*, SPB Academic, Amsterdam, pp. 19–34.

Lieftinck, M.A., 1938, The dragonflies (Odonata) of New Guinea and neighbouring islands. Part V. Descriptions of new and little known species of the families Libellaginidae, Megapodagrionidae, Agrionidae (sens. lat.), and Libellulidae (genera *Rhinocypha, Argiolestes, Drepanosticta, Notoneura, Palaiargia, Papuargia, Papuagrion, Teinobasis, Nannophlebia, Synthemis*, and *Anacordulia*), *Nova Guinea* (n.s.) **2**: 47–128.

Lieftinck, M.A., 1949a, The dragonflies (Odonata) of New Guinea and neighbouring islands. Part VII. Results of the Third Archbold Expedition 1938–1939 and of the Le Roux Expedition 1939 to Netherlands New Guinea (II. Zygoptera), *Nova Guinea* (n.s.) **5**: 1–271.

Lieftinck, M.A., 1949b, Synopsis of the odonate fauna of the Bismarck Archipelago and the Solomon Islands, *Treubia* **20**: 319–374.

Lieftinck, M.A., 1952, On the Papuasian representatives of the genus *Macromia* Rambur, with descriptions of five new species and some larval forms (Odon.), *Treubia* **21**: 437–468.

Lieftinck, M.A., 1954, Handlist of Malaysian Odonata. A catalogue of the dragonflies of the Malay Peninsula, Sumatra, Java and Borneo, including the adjacent small islands, *Treubia* **22** (Suppl.): 202 pp.

Lieftinck, M.A., 1971, Studies in Oriental Corduliidae (Odonata), I, *Tijdschrift voor Entomologie* **114**: 1–63.

Mather, P.B. and de Bruyn, M., 2003, Genetic diversity in wild stocks of the giant freshwater prawn (*Macrobrachium rosenbergii*): implications for aquaculture and conservation, *NAGA, Worldfish Center Quarterly* **26** (4): 4–7.

McLoughlin, S., 2001, The breakup history of Gondwana and its impact on pre-Cenozoic floristic provincialism, *Australian Journal of Botany* **49**: 271–300.

Metcalfe, I., 2001, Palaeozoic and Mesozoic tectonic evolution and biogeography of SE Asia–Australasia, in: Metcalfe, I., Smith, J.M.B., Morwood, M., and Davidson, I. (eds), *Faunal and Floral Migrations and Evolution in SE Asia–Australasia.* A.A. Balkema Publishers, Lisse, pp. 15–34.

Metcalfe, I., Smith, J.M.B., Morwood, M., and Davidson, I., 2001, *Faunal and Floral Migrations and Evolution in SE Asia–Australasia.* A.A. Balkema Publishers, Lisse.

Mey, W., 1998, The distribution of *Apsilochorema* Ulmer, 1907: biogeographic evidence for the Mesozoic accretion of a Gondwana microcontinent to Laurasia, in: Hall, R. and Holloway, J.D. (eds), *Biogeography and Geological Evolution of SE Asia*, Backhuys Publishers, Leiden, pp. 91–98.

Mey, W., 2001, Australasian distributions in Trichoptera (Insecta) – a frequent pattern of a rare case? in: Metcalfe, I., Smith, J.M.B., Morwood, M., and Davidson, I. (eds), *Faunal and Floral Migrations and Evolution in SE Asia–Australasia*, A.A. Balkema Publishers, Lisse, pp. 255–268.

Mol, A.W.M., 1989, *Echinobaetis phagas* gen nov., spec. nov., a new mayfly from Sulawesi (Ephemeroptera: Baetidae), *Zoologische Mededelingen* **63**: 63–70.

Moss, S.J. and Wilson, M.E.J., 1998, Biogeographic implications of the tertiary palaeogeographic evolution of Sulawesi and Borneo, in: Hall, R. and Holloway, J.D. (eds.), *Biogeography and Geological Evolution of SE Asia*, Backhuys Publishers, Leiden, 133–163.

Munz, P.A., 1919, A venational study of the suborder Zygoptera (Odonata) with keys for the identification of genera, *Memoirs of the American Entomological Society* **3**: 1–78.

Nieser, N. and Chen, P., 1993, The *Rhagovelia* (Heteroptera: Veliidae) of Sulawesi (Indonesia). Notes on Malesian aquatic and semiaquatic bugs (Heteroptera), 4, *Tijdschrift voor Entomologie* **136**: 259–281.

Nieser, N. and Chen, P., 1997, Ten new species species of *Rhagovelia* (Heteroptera: Veliidae) from Sulawesi (Indonesia). Notes on Malesian aquatic and semiaquatic bugs (Heteroptera), 7, *Tijdschrift voor Entomologie* **140**: 17–41.

Peters, W.L. and Edmunds, G.F., Jr., 1990, A new genus and species of Leptophlebiidae: Atalophlebiinae from the Celebes (Sulawesi) (Ephemeroptera), in: Campbell, I.C. (ed.), *Mayflies and Stoneflies: Life Histories and Biology*, Kluwer Academic, Dordrecht, pp. 327–335.

Pigram, C.J. and Davies, H.L., 1987, Terranes and the accretion history of the New Guinean orogen, *BMR Journal of Australian Geology and Geophysics* **10**: 193–211.

Polhemus, D.A., 1994, Taxonomy, phylogeny and zoogeography of the genus *Cylindrostethus* Fieber in the Paleotropical region (Heteroptera: Gerridae), *Bishop Museum Occasional Papers* **38**: 1–34.

Polhemus, D.A., 1995, Two new species of *Rhagovelia* from the Philippines, with a discussion of zoogeographic relationships between the Philippines and New Guinea (Heteroptera: Veliidae), *Journal of the New York Entomological Society* **103**: 55–68.

Polhemus, D.A., 1996, Island arcs, and their influence on Indo-Pacific biogeography, in: Keast, A. and Miller, S.E. (eds), *The Origin and Evolution of Pacific Island Biotas, New Guinea to Easterm Polynesia: Patterns and Processes.* SPB Academic, Amsterdam, pp. 51–66.

Polhemus, D.A. and Polhemus, J.T., 1987, A new genus of Naucoridae (Hemiptera) from the Philippines, with comments on zoogeography, *Pan-Pacific Entomologist* **63**: 265–269.

Polhemus, D.A. and Polhemus, J.T., 1994, A new species of *Aquarius* (Heteroptera: Gerridae) from Timor, with notes on Timorese zoogeography, *Proceedings of the Entomological Society of Washington* **96**: 54–62.

Polhemus, D.A. and Polhemus, J.T., 1998, Assembling New Guinea: 40 million years of island arc accretion as indicated by the distributions of aquatic Heteroptera (Insecta), in: Hall, R. and Holloway, J.D. (eds), *Biogeography and Geological Evolution of SE Asia*, Backhuys Publishers, Leiden, pp. 327–340.

Polhemus, J.T. and Polhemus, D.A., 1988, Zoogeography, ecology and systematics of the genus *Rhagovelia* Mayr (Heteroptera: Veliidae) in Borneo, Celebes and the Moluccas, *Insecta Mundi* **2**: 161–230.

Polhemus, J.T. and Polhemus, D.A., 1990, Zoogeography of the aquatic aquatic Heteroptera of Celebes: regional relationships versus insular endemism, in: Knight, W.J. and Holloway, J.D. (eds), *Insects and the Rain Forests of Southeast Asia*, Royal Entomological Society of London, London, pp. 73–86.

Polhemus, J.T. and Polhemus, D.A., 2002, The Trepobatinae (Gerridae) of New Guinea and surrounding regions, with a review of the world fauna. Part 6. Phylogeny, biogeography, world checklist, bibliography and final taxonomic addenda, *Insect Systematics and Evolution* **33**: 253–290.

Quarles van Ufford, A. and Cloos, M., 2005, Cenozoic tectonics of New Guinea, *AAPG Bulletin* **89**: 119–140.

Rangin, C., Jolivet, L., Pubellier, M., and Tethys Pacific Working Group, 1990, A simple model for the tectonic evolution of southeast Asia and Indonesia region for the past 43 m.y., *Bulletin de la Société Géologique de France* (8) **VI** (6): 907–913.

Rasnitsyn, A.P., 2002, Cohors Libelluliformes Laicharting, 1781 (= Subulicornes Latreille, 1807, = Hydropalaeoptera Rohdendorf, 1968), in: Rasnitsyn, A.P. and Quicke, D.L.J. (eds), *History of Insects*, Kluwer Academic, Dordrecht, pp. 85–89.

Rasnitsyn, A.P. and Pritykina, L.N., 2002, Superorder Libellulidea Laicharting, 1781. Order Odonata Fabricius, 1792. The dragonflies, in: Rasnitsyn, A.P. and Quicke, D.L.J. (eds), *History of Insects*, Kluwer Academic, Dordrech, pp. 97–104.

Rasnitsyn, A.P. and Quicke, D.L.J., 2002, *History of Insects*, Kluwer Academic, Dordrecht.

Shcherbakov, D.E. and Popov, Yu.A., 2002, Superorder Cimicidea Laicharting, 1781. Order Hemiptera Linné, 1758. The bugs, Cicadas, Plantlice, Scale insects, etc. (= Cimicida Laicharting, 1781, = Homoptera Leach, 1815 + Heteroptera Latreille, 1810), in: Rasnitsyn, A.P. and Quicke, D.L.J. (eds), *History of Insects*, Kluwer Academic, Dordrecht, pp. 143–157.

Schuh, R.T. and Stonedahl, G.M., 1986, Historical biogeography of the Indo-Pacific: a cladistic approach, *Cladistics* **2**: 337–355.

Sinitshenkova, N.D., 2002a, Order Perlida Latreille, 1810. The stoneflies (=Plecoptera Burmeister, 1839), in: Rasnitsyn, A.P. and Quicke, D.L.J. (eds), *History of Insects*, Kluwer Academic, Dordrecht, pp. 281–287.

Sinitshenkova, N.D., 2002b, Ecological history of the aquatic insects, in: Rasnitsyn, A.P. and Quicke, D.L.J. (eds), *History of Insects*, Kluwer Academic, Dordrecht, pp. 388–426.

Soldán, T., 2001, Status of the systematic knowledge and priorities in Ephemeroptera studies: the Oriental region, in: Dominguez, E. (ed.), *Trends in Research in Ephemeroptera and Plecoptera*, Kluwer Academic, Dordrecht, pp. 53–65.

Soulier-Perkins, A., 2000, A phylogenetic and geotectonic scenario to explain the biogeography of the Lophopidae (Hemiptera, Fulgoromorpha), *Palaeogeography, Palaeoclimatology, Palaeoecology* **160**: 239–254.

van Tol, J., 1987a, The Odonata of Sulawesi (Celebes), Indonesia. An introduction. *Advances in Odonatology* **3**: 147–155.

van Tol, J., 1987b, The Odonata of Sulawesi and adjacent islands. Part 2. The genus *Diplacina* Brauer on Sulawesi, *Zoologische Mededelingen* **61**: 160–176.

van Tol, J., 1994, The Odonata of Sulawesi and adjacent islands. Part 3. The genus *Macromia* Rambur (Corduliidae), *Tijdschrift voor Entomologie* **137**: 87–94.

van Tol, J., 1998, The Odonata of Sulawesi and adjacent islands. Part 4. A new genus and species of Chlorocyphidae from South-East Sulawesi, *Zoölogische Verhandelingen* **323**: 441–448.

van Tol, J., 2000, The Odonata of Sulawesi and adjacent islands. Part 5. The genus *Protosticta* Selys (Platystictidae), *Tijdschrift voor Entomologie* **143**: 221–266.

van Tol, J., 2005, Revision of the Platystictidae of the Philippines (Odonata), excluding the *Drepanosticta halterata* group, with descriptions of twenty-one new species. *Zoologische Mededelingen* **79** (2): 195–282.

van Tol, J. and Müller, R.A., 2003, Forest damselflies of the Philippines, their evolution and present status, with the description of *Drepanosticta moorei* spec. nov. from Luzon (Zygoptera: Platystictidae), *Odonatologica* **21**: 39–45.

von Rintelen, Th. and Glaubrecht, M., 2003, New discoveries in old lakes: three new species of *Tylomelania* Sarasin & Sarasin, 1897 (Gastropoda: Cerithioidea: Pachychilidae) from the Malili lake system on Sulawesi, Indonesia, *Journal of Molluscan Studies* **69**: 3–17.

Trewick, S.A., 2000, Molecular evidence for dispersal rather than vicariance as the origin of flightless insect species on the Chatham Islands, New Zealand, *Journal of Biogeography* **27**: 1189–1200.

Turner, H., Hovenkamp, P., and van Welzen, P.C., 2001, Biogeography of Southeast Asia and the West Pacific, *Journal of Biogeography* **28**: 217–230.

Ulmer, G., 1939, Eintagsfliegen (Ephemeropteren) von den Sunda-Inseln. Teil I. Einleitung und Systematik, *Archiv für Hydrobiologie* (Suppl. 16): 443–580.

Ulmer, G., 1940, Eintagsfliegen (Ephemeropteren) von den Sunda-Inseln. Teil II. Beschreibung von den Nymphen. Teil III. Zusammenfassung, *Archiv für Hydrobiologie* (Suppl. 16): 581–692.

Vane-Wright, R.I., 1990, The Philippines – key to the biogeography of Wallacea? in: Knight, W.J. and Holloway, J.D. (eds), *Insects and the Rain forests of Southeast Asia.* Royal Entomological Society of London, London, pp. 19–34.

van Oosterzee, P., 1997, *Where Worlds Collide. The Wallace line*, Cornell University Press, Ithaca, NY.

Wallace, A.R., 1860, On the zoological geography of the Malay archipelago, *Journal of the Proceedings of the Linnean Society (Zoology)* **1860**: 172–184.

Wallace, A.R., 1863, On the physical geography of the Malay Archipelago, *Journal of the Royal Geographical Society* **33**: 217–234.

Watson, J.A.L., 1982, Odonata (dragonflies and damselflies), in: Keast, A. (ed.), *Ecological Biogeography of Australia*, W. Junk, The Hague, pp. 1138–1167.

Watson, J.A.L., Theischinger, G., and Abbey, H.M., 1991, *The Australian Dragonflies. A guide to the Identification, Distributions and Habitats of Australian Odonata*, CSIRO, Canberra and Melbourne.

Wheeler, W.C., Whiting, M., Wheeler, Q.D., and Carpenter, J.M., 2001, The phylogeny of the extant hexapod orders, *Cladistics* **17**: 113–169.

Whitmore, T.C. (ed.), 1981, *Wallace's Line and Plate Tectonics*, Clarendon Press, Oxford.

Whitten, A.J., Mustafa, M., and Henderson, G.S., 1987, *The Ecology of Sulawesi*. Gadjah Mada University Press, Yogyakarta.

Wilson, K.D.P., 1997, The Platystictidae of Hong Kong and Guangdong, with descriptions of a new genus and two new species (Zygoptera), *Odonatologica* **26**: 53–63.

Wilson, M.E.J. and Moss, S.J., 1999, Cenozoic palaeogeographic evolution of Sulawesi and Borneo, *Palaeogeography, Palaeoclimatology, Palaeoecology* **145**: 303–337.

Wowor, D., 2004, A systematic revision of the freshwater prawns of the genus *Macrobrachium* (Crustacea: Decapoda: Caridea: Palaemonidae) of Sundaland, Unpublished Ph.D. thesis, National University of Singapore.

Yan, C.Y. and Kroenke, L.W., 1993, A plate tectonic reconstruction of the Southwest Pacific, 0–100 Ma. *Proceedings of the Ocean Drilling Program. Scientific Results* **130**: 697–709.

Zwick, P., 2000, Phylogenetic system and zoogeography of the Plecoptera, *Annual Review of Entomology* 45: 709–746.

Chapter 3

# Distribution and Speciation of Megapodes (Megapodiidae) and Subsequent Development of their Breeding Behaviour

RENÉ W.R.J. DEKKER
Nationaal Natuurhistorisch Museum, Naturalis, P.O. Box 9517, 2300 RA Leiden, The Netherlands, dekker@naturalis.nl



**Abstract** Megapodes (Megapodiidae), the most peculiar of all Galliformes, are of Gondwana origin. On the Australian plate they shifted from normal avian incubation to their aberrant strategy of incubating eggs in mounds of sand and leaf litter where heat is generated by microbial decomposition. From here, they spread through the Indonesian archipelago and eastwards into Polynesia, resulting in rapid speciation and the use of alternative heat sources for the incubation of their eggs.

## 1. Introduction

The distribution of megapodes (Aves, Galliformes, Megapodiidae), as well as their unique breeding behaviour, have led to conflicting theories as to their geographical origin and reproductive history. Olson (1980), and authors therein, assumed that megapodes reached Australia from southeast Asia, probably by the Miocene. He explained their current absence on the mainland of southeast Asia by competitive exclusion with pheasants (Phasianidae). However, since Olson's publication, a fossil of what was considered to be a megapode from Eocene-Oligocene deposits of France has been re-identified as representing a separate extinct family Quercymegapodiidae



*W. Renema (ed.), Biogeography, Time, and Place: Distributions, Barriers, and Islands,* 93–102

(Mourer-Chauviré, 1992), leaving no evidence for the former presence of megapodes outside Australasia. Since this publication by Mourer-Chauviré was overlooked by Newton (2003: 137–138; 160), the theory that megapodes reached Australia from southeast Asia still stands in recent literature. Alternatively, Cracraft (1973, 1980) proposed a Gondwana origin for the Galliformes and suggested that modern megapodes were present in Australia from the time of the breakup of Gondwana. Dekker (1989, 1990) considered predation by carnivores, especially Felidae and Viverridae, as a key factor explaining the absence of these birds from the mainland of southeast Asia and the Greater Sunda islands. Recent findings of a fossil (*Ngawupodius*) from the Late Oligocene at Lake Pinpa, Central Australia, now represents the earliest fossil record of the Megapodiidae (Boles and Ivison, 1999), supporting the theory of a Gondwana origin for the megapodes.

Various authors (e.g., Frith, 1962; Immelmann and Sossinka, 1986) considered the so-called burrow-nesting strategy, in which megapodes bury their eggs in holes in warm sand, as the original breeding strategy. However, Clark (1960, 1964a, b) and Dekker and Brom (1992) suggested that burrow-nesting has been derived from mound-building, in which megapodes bury their eggs in self-made mounds of decomposing, heat-generating leaflitter. The latter view was supported by recent workers. Starck (1993: 285) considered megapodes "a monophyletic taxon of Galliform birds with the superprecocial development as a derived character complex characterizing the taxon" and stated that "the breeding biology is derived from "normal" avian incubation and not from sauropsid ancestors". Later, Starck and Sutter (2000: 542) wrote "that megapode development does not differ substantially from galliform precocity". Booth and Jones (2001: 192) (unaware of Boles and Ivison, 1999) were of the opinion that "megapodes almost certainly evolved from a galliform ancestor, as early as the Pliocene ... and this probably had a ground nest with large clutches of synchronously hatching, highly precocial chicks commonly seen in extant galliforms".

Although some of these discussions are recent, all were published when no phylogenies based on nuclear and mitochondrial DNA sequences were available. These were first published by Birks and Edwards (2002). The relationships within the genus *Megapodius* might prove to be crucial to the discussion. The phylogenies of Birks and Edwards will form the basis for this discussion about the geographical origin and reproductive history of the family.

## 2. Megapode Phylogenies and Other Relevant Publications

After the publications by Mourer-Chauviré (1992), in which she no longer considered a fossil from Eocene-Oligocene deposits of France to be that of a megapode, and that of Boles and Ivison (1999), who confirmed the presence of the Megapodiidae in Australia as early as the Late Oligocene, there is general agreement that megapodes were isolated in Australo-Papua for an extended period. A Gondwana origin would in fact link the Megapodiidae with the South American cracids (Family Cracidae), which are by some considered the sister-group of the megapodes. Sibley and Ahlquist (1990) placed the megapodes and cracids together in the order Craciformes. However, research based on 102 morphological characters, including 89 based on osteology, made Dyke et al. (2003) consider a monophyletic Megapodiidae to be the sister-group to all other Galliformes and to be the most basal clade within the order. The recovery, by Sibley and Ahlquist (1990), of a single clade comprising both Megapodiidae and Cracidae is not supported by a single morphological character (Dyke et al., 2003).

Brom and Dekker (1992: 15, Fig. 5) published a phylogeny of the Megapodiidae based on traditional, mainly morphological characters (see also Jones et al., 1995). Seven years later, Mey (1999: 30, Fig. 5) published a phylogeny of the megapodes based on the phylogeny of the chewing lice found on them. Birks and Edwards (2002), who sequenced nuclear and mitochondrial DNA, included in their studies 15 of the 22 extant species representing all seven genera. Their phylogeny, which included representatives of other Galliform families as outgroups, shows an early split in megapodes, leading to two major clades: (1) *Alectura*, *Aepypodius*, *Leipoa*, *Talegalla* and *Macrocephalon* and (2) *Eulipoa* and *Megapodius*. It is largely congruent with the phylogeny by Mey (1999), but differs from the one published by Brom and Dekker (1992), in which *Leipoa*, *Talegalla*, *Macrocephalon*, and *Megapodius* (including *Eulipoa*) form a clade.

Of the 13 species within the most widely dispersed genus *Megapodius*, which includes species exhibiting all the different breeding strategies, no less than nine species (totalling 14 populations or subspecies) were sampled by Birks and Edwards (2002). Additional morphological studies by Roselaar (1994: 21) add to this and together they offer great opportunities for a renewed discussion on the historical distribution, dispersal and evolution of the unique reproductive systems within megapodes. The relationships within the genus *Megapodius* do show some remarkable results (for details, see Birks and Edwards, 2002): *M. tenimberensis* from Tanimbar and *M. cumingii* from Sulawesi and the Philippines are an early split (*M. bernsteinii* (Sula Islands), which was not included in the analyses, might fit with these (Roselaar, 1994), as might *M. nicobariensis*). *Megapodius eremita* from the Solomon Islands and *M. layardi* from Vanuatu are not close relatives as was previously suggested based on distribution: they represent descendents of different waves of dispersal. Furthermore, *M. layardi* and *M. pritchardii* from Niuafo'ou, Tonga, are sister-taxa. Some of the conclusions by Birks and Edwards (2002) are congruent with Roselaar (1994).

## 3. Possible Scenario of Current Distribution and Breeding Strategies

### 3.1. Distribution

The position of the Australian continent during the Late Oligocene (Fig. 1) is the starting point for the discussion as it was there and then that we know of the earliest occurrence of megapodes: the recently discovered, extinct, megapode *Ngawupodius minya* (Boles and Ivison, 1999).

During the Oligocene, the Australian continent was well isolated, slowly drifting north and colliding with New Guinea. In the Early Miocene, New Guinea

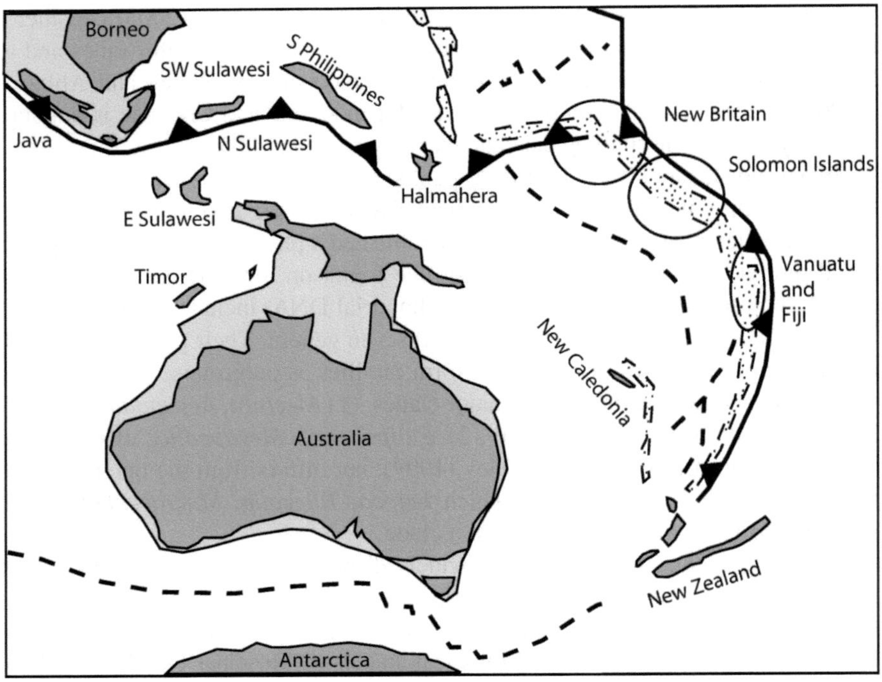


Fig. 1 Position of the Australian continent during the Oligocene (34–24 Ma) (after Hall, 1998)

was a series of islands on the northern edge of the Australian plate. During the Middle Miocene, from about 15 Ma, larger parts of New Guinea were uplifted to above sea level. Northern Australia had a warm, wet climate, which became gradually cooler and drier by the Late Miocene. During the Pliocene, 5.3–1.8 Ma there was an overall cooling of sea and land temperatures, and Australia collided with the Timor region. Active dispersal of animals, including that of megapodes, which began to gain their modern appearances, took place using islands as stepping stones as the Australian continent continued to approach the Asian continent. During the past 10,000 years the position and climate of Australia has changed little.

The earliest known fossil occurrence of the Megapodiidae was, until the discovery of the tiny *Ngawupodius minya* by Boles and Ivison (1999), from the Pliocene of Australia and relates to (what was until recently considered) a single extinct genus *Progura* (but see below). The Pleistocene fossil record of megapodes from Australia is dominated by this megapode, which was bigger than any of the extant taxa and most likely a poor flyer. Initially believed to include two species, *P. gallinacea* and *P. naracoortensis*, it was subsequently considered to represent a single species *P. gallinacea*. Differences in size were explained by sexual dimorphism (Boles and Ivison, 1999: 200; Steadman, 1999: 9) but are now seen as representing individual variation or stage of maturity (W. Boles, 2004, unpublished). Boles

(2004, unpublished) thus synonymized *Progura* with *Leipoa* stating "*Progura* is not separable from *Leipoa* and the differences between *P. gallinacea* and *L. ocellata* are mainly quantative (size), with qualitative differences so small that it is doubtful that these represent distinct species. *P. gallinacea* appears to be the megafaunal version of *L. ocellata. P. gallinacea* becomes either *Leipoa gallinacea* (if the two are kept as separate species) or *Leipoa ocellata gallinacea* (if considered temporal subspecies of a single species)". The record of the modern *L. ocellata* from the Victoria Fossil Cave, which was based on a damaged skeleton of a chick, is not valid, leaving no evidence of modern *L. ocellata* elsewhere in the Pleistocene or earlier (W. Boles, 2004, unpublished).

Outside Australia, only few fossil remains involving megapodes other than *Megapodius* species, are known. These are late Quaternary cave deposits in Irian Jaya containing the extant *Aepypodius arfakianus* (Aplin et al., 1999) and late Pleistocene-Holocene deposits in Fiji containing *Megavitiornis altirostris*, a large, quite unusual extinct megapode (Worthy, 2000). *Sylviornis neocaledoniae*, a large and flightless extinct bird from New Caledonia, once believed to be either a ratite (Poplin, 1980) or a megapode (Poplin et al., 1983), is no longer considered a megapode but better placed in a family of its own (for a summary, see Jones et al., 1995: 23). The fossil record as summarized here, with ancestral megapodes restricted to the Australian landmass (including New Guinea) and (recently) Fiji, makes it most likely that the unique incubation strategy has developed here.

No fossils are known from New Zealand, which separated from Gondwana *c.* 85 Ma. All fossil records in Polynesia are from the Holocene, representing species of the genus *Megapodius* which were exterminated from islands after the arrival of the Polynesians. The Polynesian megapode *Megapodius pritchardii*, which was long believed to have been introduced to Niuafo'ou (Tonga), is in fact a relic of a once much wider distribution. Fossil remains of this species have been found on Foa and Eua (Tonga) and it is suggested that it occurred on perhaps as many as 100 individual islands in the region (Steadman, 1999: 13). Steadman (1999) discussed the biogeography and extinction of megapodes in Oceania, from which at least five different extinct species (all *Megapodius*) are now known. He estimated that at least as many species have gone extinct as still exist today: "if not for people, the figure of 22 living species of megapodes would increase to 45–55 species" (Steadman, 1999: 16).

Although many of the present day megapodes, including all *Alectura*, *Aepypodius*, *Leipoa*, and *Talegalla* species which are restricted to Australia or New Guinea and *Macrocephalon maleo* from Sulawesi, are poor flyers, the smaller *Eulipoa wallacei* from the Moluccas and all *Megapodius* species are strong flyers. They can easily cover vast areas of open water, even at the chick stage. It is therefore no coincidence that species of *Megapodius* are widely distributed and occur on islands in the Pacific, eastern Indonesia, the Philippines, and the Mariana and Nicobar islands. *Macrocephalon*, *Eulipoa*, and *Megapodius* will have colonized these islands from the Australian plate in different waves. Also within the genus *Megapodius*, different waves of dispersal have taken place. For a map of the current distribution of the family, see Jones et al. (1995: 21).

Roselaar (1994: 21) was of the opinion that "the present-day distribution of *Megapodius* with a single species on each group of islands is apparently not due to fragmentation of the range of a once widespread species, but is more likely the result of a long history of colonizations and extinctions by a number of species". He believed that "a history of invasions and extinctions by various species of *Megapodius* may explain some of the peculiarities in their present-day distribution." Furthermore, he considered *M. reinwardt* to be a recent successfully spreading species.

## 3.2. Breeding Strategies

Megapodes exploit three different sources of heat for the incubation of their eggs. Those that build mounds use rotting vegetation for incubation, while burrow-nesters dig holes in volcanically heated soil or in sun-exposed warm sand in which they bury their eggs. All seven species of the genera *Alectura* (1), *Aepypodius* (2), *Leipoa* (1), and *Talegalla* (3), which are restricted to either Australia or New Guinea, are strict mound-builders. *Macrocephalon maleo* from Sulawesi and *Eulipoa wallacei* from the Moluccas are strict burrow-nesters and do not build mounds. Within the widely distributed genus *Megapodius* some species are strict burrow-nesters, others build mounds, while a few species can do both. Figure 2 shows the incubation strategies superimposed on the cladogram by Birks and Edwards (2002). Figure 3 shows the same for the genus *Megapodius*, superimposed on a map of their current distribution. The four *Megapodius* species not sampled by Birks and Edwards (*M. decollatus* from New Guinea, *M. geelvinkianus* from the Geelvink islands, *M. bernsteinii* from the Sula Islands and *M. nicobariensis* from the Nicobars) are all strict mound-builders.

Burrow-nesting megapodes, those using heat sources other than rotting vegetation, all live on islands and are not found in Australia and New Guinea. They occur in places which have been colonized by active dispersal (flight), as many of these islands have always been isolated. It suggests that burrow-nesting developed secondarily when new habitats, some of volcanic origin or with sun-exposed beaches, were invaded. These new, often tropical habitats led to a radiation in breeding behaviour and speciation within the genus *Megapodius*. Heat necessary for the incubation of their eggs did not have to be produced by the arduous task of raking together large mounds of leaves in which birds dig a hole and bury their eggs – heat was "simply" there. They shifted from building and maintaining mounds to the simple task of digging a hole in warm sand. The increase in relative size and yolk content of the eggs in burrow-nesting species compared to that of their mound-building relatives (Dekker and Brom, 1992), which results in even larger and more precocial (superprecocial *sensu* Starck, 1993) chicks in burrow-nesters, is an indication that burrow-nesting is derived from mound-building.

When the Australian plate moved closer to Asia and sea levels changed, islands could have acted as "stepping stones" and inter-island distances might have alternately decreased and increased. Ancestral megapodes were thus able to reach

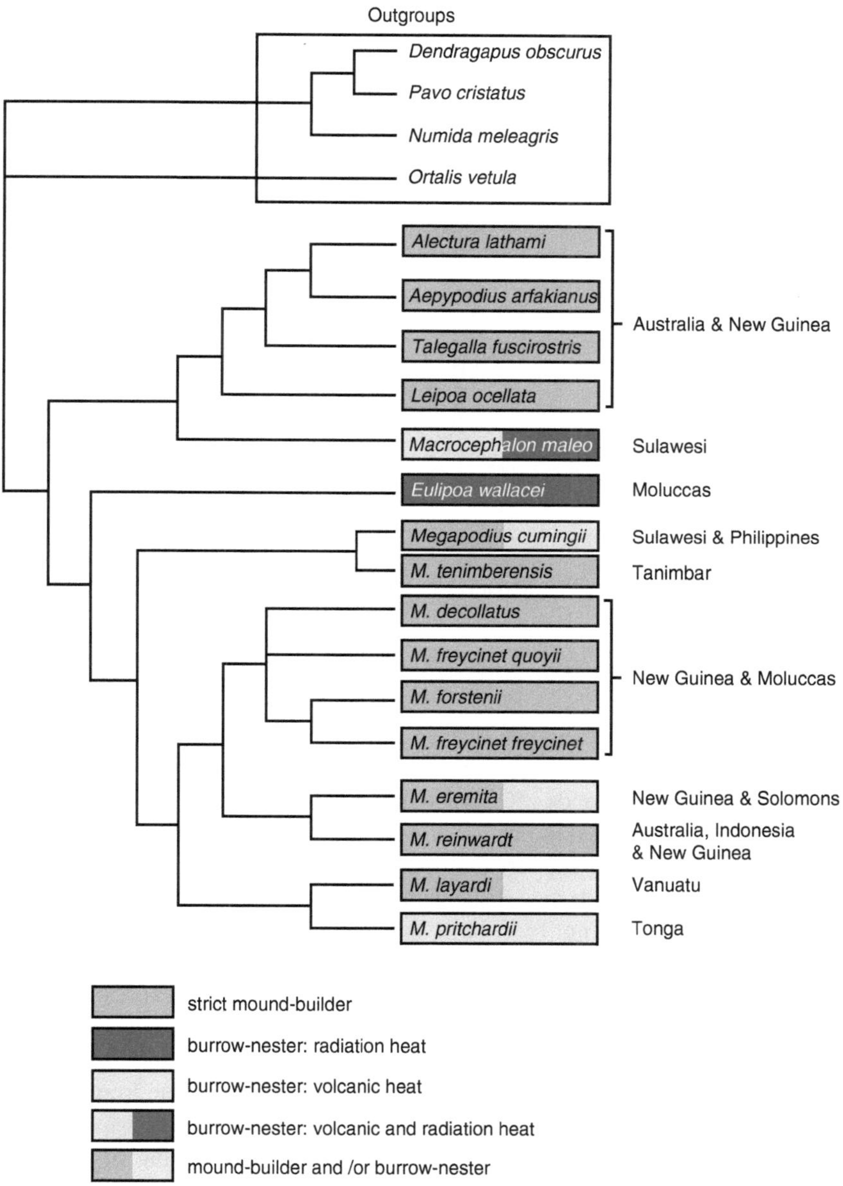


Fig. 2 Phylogenetic tree of the family Megapodiidae (redrawn from Birks and Edwards, 2002) showing the incubation strategies exhibited by the different taxa.

new habitats, passively and actively, where they became isolated. This led to a diversification in taxa as well as reproductive strategies. Their move westwards to the mainland of southeast Asia was subsequently halted by the presence of carnivore predators (Dekker, 1989) or – as suggested by Olson (1980) – by competition with

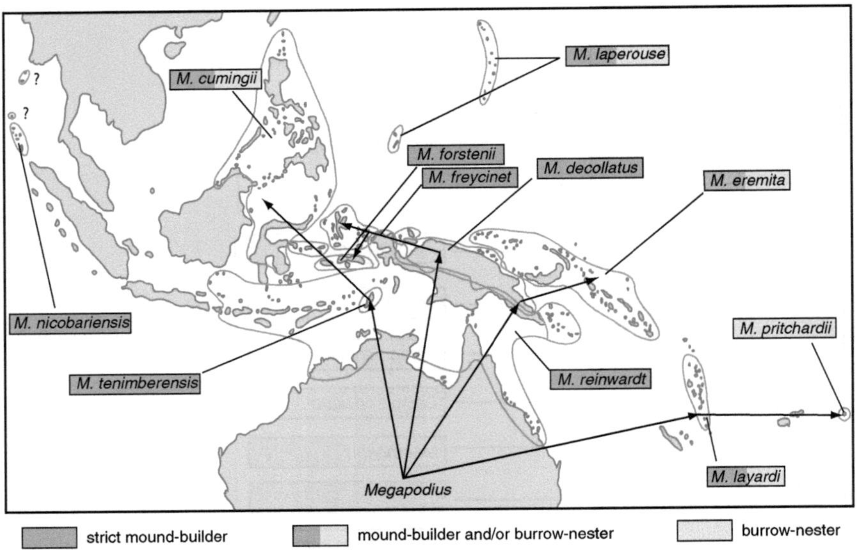


Fig. 3 Cladogram of the genus *Megapodius* superimposed upon a map of the region.

phasianids, or both. The absence of megapodes from the Greater Sunda Islands, either primarily or secondarily, can be explained in a similar way. The presence of *M. cumingii* on small islands off Borneo and of *M. nicobariensis* on the Nicobars (this species is not included in Fig. 2, but most likely closely related to *M. cumingii*) seems a relic and would suggest that megapodes once did occur on the Greater Sunda Islands from which they disappeared either through predation or competition. This has, however, not (yet) been substantiated by fossil evidence. Looking eastwards into the Pacific, we know through the work by Steadman (1999) that most islands in western Polynesia were once inhabited by megapodes and that Man is largely responsible for their absence now. The easternmost islands ever reached were Niue (Tonga) and American Samoa. The distance between the Pacific islands as well as the fact that these islands become more and more remote from the source of the "megapode origin" diminishes the chance for individual megapode taxa to reach a certain island. This might have stopped further spreading of megapodes across the Pacific.

## Acknowledgements

I would like to thank Walter Boles, Australian Museum, Sydney, and Jeremy Holloway, The Natural History Museum, London, for reviewing this paper and adding new and valuable information, which is included here.

## References


Aplin, K. P., Pasveer, J. M., and Boles, W. E., 1999, Quaternary vertebrates from the Bird's Head Peninsula, Irian Jaya, Indonesia, including descriptions of two previously unknown marsupial species, in: Baynes, A. and Long, J. A., (eds), *Papers in Vertebrate Palaeontology, Records of the Western Australian Museum,* **57** (Suppl.): 351–387.

Birks, S. M. and Edwards, S. V., 2002, A phylogeny of the megapodes (Aves: Megapodiidae) based on nuclear and mitochondrial DNA sequences, *Molecular Phylogenetics and Evolution* **23**: 408–421.

Boles, W. E., and Ivison, T. J., 1999, A new genus of dwarf megapode (Galliformes: Megapodiidae) from the Late Oligocene of central Australia, in: Olson, S. L. (ed.), *Avian Paleontology at the Close of the 20th Century: Proceedings of the 4th International Meeting of the Society of Avian Paleontology and Evolution*, Washington, DC, pp. 4–7 June 1996, *Smithsonian Contributions to Paleobiology* **89**: 199–206.

Booth, D.T., and Jones, D.N., 2001, Underground nesting in the megapodes, in: Deeming, D.C. (ed.), *Avian Incubation: Behaviour, Environment, and Evolution*, Oxford University Press, Oxford, UK, pp. 192–206.

Brom, T. G., and Dekker, R. W. R. J., 1992, Current studies on megapode phylogeny, in: Dekker, R. W. R. J. and Jones, D. N. (eds), *Proceedings of the First International Megapode Symposium, Christchurch, New Zealand*, December, 1990, *Zoologische Verhandelingen* **278**: 7–17.

Clark, G. A., 1960, Notes on the embryology and evolution of the megapodes (Aves: Galliformes), *Postilla* **45**: 1–7.

Clark, G. A., 1964a, Ontogeny and evolution in the megapodes (Aves: Galliformes), *Postilla* **78**: 1–37.

Clark, G. A., 1964b, Life histories and the evolution of megapodes, *Living Bird* **3**: 149–167.

Cracraft, J., 1973, Continental drift, paleoclimatology, and the evolution and biogeography of birds, *Journal of Zoology* **169**: 455–545.

Cracraft, J., 1980, Avian phylogeny and intercontinental biogeographic patterns, *Proceedings of the International Ornithological Congress* **17**: 1302–1308.

Dekker, R. W. R. J., 1989, Predation and the western limits of megapode distribution (Megapodiidae; Aves), *Journal of Biogeography* **16**: 317–321.

Dekker, R. W. R. J., 1990, Evolution of megapode incubation strategies, in: Dekker, R. W. R. J. (ed.), 1990, *Conservation and Biology of Megapodes (Megapodiidae, Galliformes, Aves)*, thesis, University of Amsterdam, The Netherlands, pp. 103–129.

Dekker, R. W. R. J., and Brom, T. G., 1992, Megapode phylogeny and the interpretation of incubation strategies, in: Dekker, R. W. R. J. and Jones, D. N., (eds), *Proceedings of the First International Megapode Symposium*, Christchurch, New Zealand, December, 1990, *Zoologische Verhandelingen* **278**: 19–31.

Dyke, G. J., Gulas, B. E., and Crowe, T. M., 2003, Suprageneric relationships of galliform birds (Aves, Galliformes): a cladistic analysis of morphological characters, *Zoological Journal of the Linnean Society* **137**: 227–244.

Frith, H. J., 1962, *The Malleefowl*, Angus & Robertson, Sydney.

Hall, R., 1998, The plate tectonics of Cenozoic SE Asia and the distribution of land and sea, in: Hall, R. and Holloway, J.D. (eds), *Biogeography and Geological Evolution of SE Asia*, Backhuys Publishers, Leiden, pp. 99–132.

Immelmann, K., and Sossinka, R., 1986, Parental behaviour in birds, in: Sluckin, W. and Herbert, M. (eds), *Parental Behaviour*, Blackwell, Oxford, UK, pp. 8–43.

Jones, D. N., Dekker, R. W. R. J., and Roselaar, C. S., 1995, *The Megapodes*, Oxford University Press, Oxford, UK.

Mey, E., 1999, Phylogenetic relationships of the Megapodiidae as indicated by their ischnoceran, in particular goniodid, chewing lice (Insecta: Phthiraptera), in: Dekker, R. W. R. J., Jones, D. N., and Benshemesh, J., (eds), *Proceedings of the Third International Megapode Symposium*, Nhill, Australia, December 1997, *Zoologische Verhandelingen* **327**: 23–35.

Mourer-Chauviré, C., 1992, The Galliformes (Aves) from the Phosphorites du Quercy (France): systematics and biostratigraphy, in: Cambell, K. E., (ed.), *Papers in Avain Paleontology. Honoring Pierce Brodkorb*, Science Series, LA, **36**: 67–95.

Newton, I., 2003, *The Speciation and Biogeography of Birds*, Academic Press, London.

Olson, S. L., 1980, The significance of the distribution of the Megapodiidae, *Emu* **80**: 21–24.

Poplin, F., 1980, *Sylviornis neocaledoniae* n. g., n. sp. (Aves), Ratite éteint de la Nouvelle-Calédonie, *Comptes Rendus hebdomadaires des Séances de l'Academie des Sciences*, Paris **290**: 691–694.

Poplin, F., Mourer-Chauviré, C., and Evin, J., 1983, Position systématique et datation de *Sylviornis neocaledoniae*, Mégapode géant (Aves, Galliformes, Megapodiidae) éteint de la Nouvelle-Calédonie, *Comptes Rendus hebdomadaires des Séances de l'Academie des Sciences*, Paris **297**: 301–304.

Roselaar, C. S., 1994, Systematic notes on Megapodiidae (Aves, Galliformes), including the description of five new subspecies, *Bulletin Zoologisch Museum, Universiteit van Amsterdam* **14** (2): 9–36.

Sibley, C. G. and Ahlquist, J. E., 1990, *Phylogeny and Classification of Birds. A Study in Molecular Evolution*, Yale University Press, New Haven and London.

Starck, J. M., 1993, Evolution of avian ontogenies, *Current Ornithology* **10**: 275–366.

Starck, J. M. and Sutter, E., 2000, Patterns of growth and heterochrony in moundbuilders (Megapodiidae) and fowl (Phasianidae), *Journal of Avian Biology* **31**: 527–547.

Steadman, D. W., 1999, The biogeography and extinction of megapodes in Oceania, in: Dekker, R. W. R. J., Jones, D. N., and Benshemesh, J., (eds), *Proceedings of the Third International Megapode Symposium*, Nhill, Australia, December, 1997, *Zoologische Verhandelingen* **327**: 7–21.

Worthy, T. H., 2000, The fossil megapodes (Aves: Megapodiidae) of Fiji with descriptions of a new genus and two new species, *Journal of the Royal Society of New Zealand* **30**: 337–364.

Chapter 4

# The Influence of Land Barriers on the Evolution of Pontoniine Shrimps (Crustacea, Decapoda) Living in Association with Molluscs and Solitary Ascidians

CHARLES H.J.M. FRANSEN
Nationaal Natuurhistorisch Museum Naturalis, P.O. Box 9517, 2300 RA Leiden, The Netherlands, Fransen@naturalis.nl



**Abstract** The influence of land barriers on the evolution of 50 species of pontoniine shrimps, living in association with molluscs and solitary ascidians, is studied. The interplay between the biological characteristics of the shrimps in the Atlantic/East Pacific area resulted in a biogeographical pattern best explained by a series of vicariance events. Distributional pattern in the Indo-West Pacific shows a centre of maximum diversity in eastern Indonesia, the Philippines, and New Guinea. This can be explained by the overlap of Asian, Pacific, and Australian distribuion ranges and/or the accumulation of inwardly directed, convergent dispersal, of peripheral originated species.



*W. Renema (ed.), Biogeography, Time, and Place: Distributions, Barriers, and Islands,* 103–115

## 1. Introduction

Historical biogeography tries to explain distribution patterns of taxa. These patterns are the result of the interplay between the biological characteristics of the taxa concerned (migratory capacities, habitat limitations, longevity, speciation rate, etc.) and external processes (platetectonics, climatic change, environmental change, etc.). In this paper, the influence of geological events on the evolution of pontoniine shrimps living in association with molluscs and solitary ascidians is examined (Fig. 1).

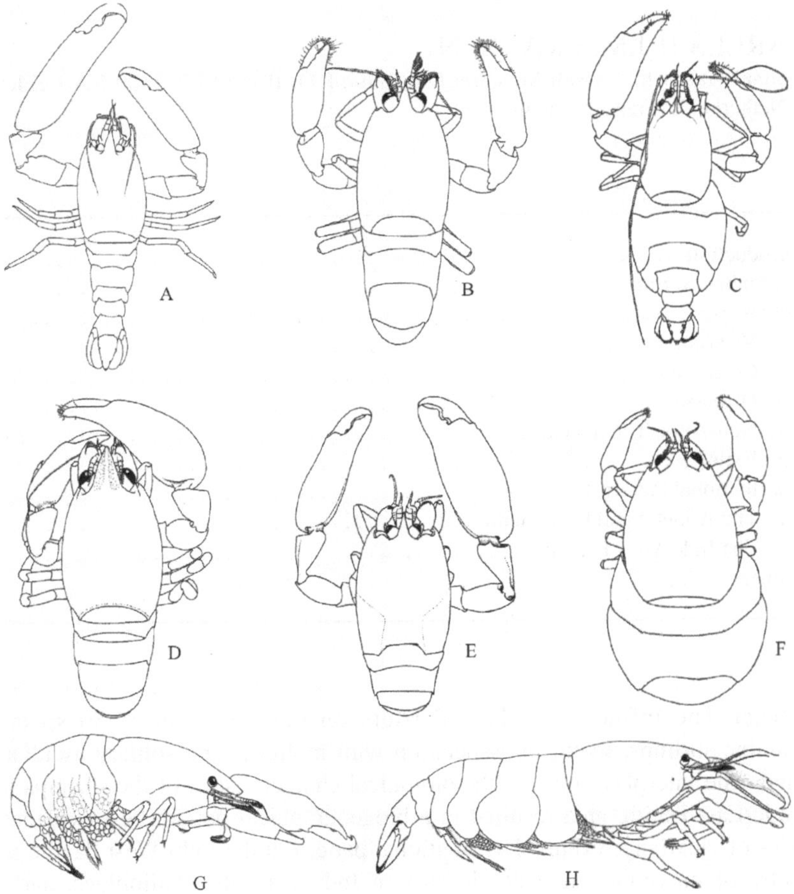


Fig. 1 Examples of species of eight pontoniine shrimp genera associated with molluscs and ascidians. (A) *Pontonia pinnae* Lockington, 1878; (B) *Ascidonia quasipusilla* (Chace, 1972); (C) *Dactylonia ascidicola* (Borradaile, 1898); (D) *Odontonia katoi* (Kubo, 1940); (E) *Chernocaris placunae* Johnson, 1967; (F) *Conchodytes pteriae* Fransen, 1994; (G) *Anchistus custos* (Forskål, 1775); (H) *Paranchistus armatus* (H. Milne Edwards, 1837).

## 2. Pontoniine Shrimps

The Pontoniinae form the most diverse subfamily of marine shrimps with over 500 recognized species, the majority of which have been recorded in the Indo-West Pacific. They occur in tropical and subtropical waters around the world, usually living in association with other invertebrates. Eighteen of the ca. 100 genera within the Pontoniinae are associated with ascidians and/or molluscs (Table 1).

The phylogeny of these genera has been analysed using morphological and, in part, molecular data (Fransen, 2002; Fransen research in progress). The genera are monophyletic except for *Anchistus* and *Paranchistus*, which ultimately will have to be redefined on the basis of apomorphic characters. The 18 genera do not form one monophyletic group. However, several monophyletic groups can be recognized in which both mollusc and ascidian hosts are recorded.

Important biological characteristics of Pontoniinae which determine their susceptibility to external processes mentioned above are: (1) preference for warm temperate and tropical coastal waters, limiting latitudinal dispersal and vicariance events; (2) the association with host-species with their own biological characteristics limiting possible distributions; and (3) a life cycle with pelagic larval stages and adult bentic stage determining dispersal capabilities.

*Table 1. Genera of pontoniine shrimps associated with ascidians and/or molluscs with number of species per host-group and geographic distribution. Asc = number of species with ascidian host; Mol = number of species with molluscan host; ? = number of species of which host is unknown; IWP = Indo-West Pacific; A = Atlantic; EP = East Pacific.*

| *Genus* | *Asc* | *Mol* | *?* | *Distribution* |
|---|---|---|---|---|
| *Amphipontonia* Bruce, 1991 | 1 | 0 | 0 | IWP |
| *Anchiopontonia* Bruce, 1992 | 0 | 1 | 0 | IWP |
| *Anchistus* Borradaile, 1898 | 0 | 7 | 0 | IWP |
| *Ascidonia* Fransen, 2002 | 5 | 0 | 0 | A, EP |
| *Bruceonia* Fransen, 2002 | 0 | 1 | 0 | IWP |
| *Cainonia* Bruce, 2005 | 0 | 1 | 0 | IWP |
| *Chernocaris* Johnson, 1967 | 0 | 1 | 0 | IWP |
| *Colemonia* Bruce, 2005 | 1 | 0 | 0 | IWP |
| *Conchodytes* Peters, 1852 | 0 | 8 | 0 | IWP |
| *Dactylonia* Fransen, 2002 | 7 | 0 | 1 | IWP |
| *Dasella* Lebour, 1945 | 3 | 0 | 0 | IWP |
| *Neoanchistus* Bruce, 1975 | 0 | 2 | 0 | IWP |
| *Odontonia* Fransen, 2002 | 7 | 0 | 0 | IWP |
| *Paranchistus* Holthuis, 1952 | 0 | 6 | 0 | IWP |
| *Platypontonia* Bruce, 1968 | 0 | 2 | 0 | IWP |
| *Pontonia* Latreille, 1829 | 0 | 9 | 1 | A, EP |
| *Pseudopontonia* Bruce, 1992 | 0 | 1 | 0 | IWP |
| *Rostronia* Fransen, 2002 | 1 | 0 | 0 | IWP |

## 3. Processes

The interplay between a taxon and its environment results in processes like vicariance, coalescence, dispersal, non-allopatric speciation, and extinction, determining their present-day biogeographical patterns. Several modes of speciation have been recognized which are non-allopatric, e.g., instantaneous through polyploidy or through ecological allopatry (but geographical sympatry) when switching to another host in taxa living in association with other organisms.

Extinction and primitive absence are commonly used as *ad hoc* explanations for otherwise unexplainable gaps in distribution patterns. Incomplete data on the distribution of certain species as well as absence of a fossil record may also cause problems in biogeographical analysis.

### 3.1. Vicariance

In relation to the biological characteristics of the group under study, several events resulting in vicariance can be recognized:

(1) In "hard" vicariance events there is complete genetic isolation between the populations concerned. An example is the separation of originally coherent water masses by a land mass. Two events are recognized to affect the studied taxa: (A) the closure of the Tethys Sea in the early Miocene, *c*. 20 Ma (Adams, 1967, 1983; Hallam, 1967; Rögl and Steiniger, 1984; Rögl, 1998); and (B) the forming of the Panama Isthmus in the early Pliocene, *c*. 3 Ma (Keigwin, 1978; Duque-Caro, 1990; Lessios, 1998). No similar effective events are known to have taken place in the Indo-West Pacific.

(2) In "soft" vicariance events no complete isolation is involved, but the barrier acts as a filter. The opening of the Atlantic in the Tertiary is a "soft" vicariance event. The rise of land masses or the lowering of sea level in the Indonesian Archipelago, with the resulting limited throughflow of water from the Pacific Ocean to the Indian Ocean, has had effects in recent geological time and can be regarded a "soft" vicariance event as well (Hoeksema, Chapter 5 of this volume).

When species are confined to hard substrate habitats, like shrimps that are associated with ascidians, the decrease of, for instance, reefs being replaced by large areas with sandy coastlines could cause small-scale vicariance effects. Events of this kind could have influenced the distribution of the group under study, partly disconnecting the western Indian Ocean from the Indo-Malayan region as seen, for example, in larger benthic foraminifera (Langer and Hottinger, 2000).

## 3.2. Coalescence

When biota, like coastal communities, merge into larger areas, the initially smaller ("ancestral") ranges disappear. Biota may react to these events partly by extinction due to species selection, and/or by increasing biodiversity due to the overlapping of originally allopatric faunas. In relation to the biological characteristics of the group under study, such an event has been recognized in the Indo-West Pacific, where Asian, Australian, and Pacific plates are merging.

## 3.3. Dispersal

Dispersal capabilities depend on the mobility of the organisms. Adults of the group under study are benthic and only capable of small-scale dispersal. Since the larvae are pelagic, it is assumed that large-scale dispersal occurs in this phase of the life cycle. Calafiore et al. (1991) were able to rear *Pontonia pinnophylax* (Otto, 1821) through eight larval stages to the first post-larva in 28–30 days. They also showed that metamorphosis from larva to first post-larva could only be induced by the presence of the host mussel, whereas larval development is retarded in the absence of the host, enhancing the dispersal capabilities of the species. This is supported by the presence of *P. pinnophylax* and in part *Ascidonia flavomaculata* (Heller, 1864) on geologically young islands like Ascension, St. Helena, Madeira, Cape Verde Islands, and the Azores. Colonization of these islands can only have happened through dispersal from continental coastal areas or along a route of stepping stones.

## 3.4. Non-Allopatric Speciation

In recent years, there is more attention for hypotheses involving sympatric speciation (Howard and Berlocher, 1998). Many of these studies are concerned with taxa that live in close association to other organisms like insects with plants (Bush, 1969, 1994; Feder, 1998; Menken and Roessingh, 1998). The mechanisms revealed in these studies could well have their parallel in pontoniine shrimps living in association with various reef organisms. Berlocher (1998) investigated how hypothesis of sympatric speciation via host or habitat shift might be tested with biogeographic and phylogenetic evidence. Sympatric distributions of sister species can be explained as allopatric speciation followed by dispersal (Mayr, 1947, 1963). However, the *ad hoc* use of dispersal can explain almost any distribution pattern and almost any speciation process and is therefore as such not very informative. One of the predictions Berlocher (1998) makes is that the frequency of sympatric sister species in the top of a phylogenetic tree will be high in the case of a predominant sympatric mode of speciation while low in a predominantly allopatric mode of speciation. This prediction can be tested in the case of the genera *Pontonia* Latreille, 1829 and *Ascidonia* Fransen, 2002. Both genera entail sister species of which one occurs on the East Pacific side and the other on the West Atlantic side of the Panama Isthmus (Fig. 2), indicating the lack of sympatric speciation during the last 3 Ma. Another indication of sympatric speciation would be the high number of host-switches in the top of the tree. Drastic host shifts from Ascidiacea to Mollusca or vice versa cannot be found in the top of the tree. The only rather drastic host-shift between bivalves and gastropods in sister species concerns the allopatric species *P. chimaera* Holthuis, 1951 and *P. domestica* Gibbes, 1850 (Fig. 2). Although sympatric speciation can still be involved, the explanation would be *ad hoc*.

## 3.5. Extinction

Extinction is an important phenomenon with regard to distributional patterns. However, there is no fossil record of the Pontoniinae that can support hypotheses on extinction events.

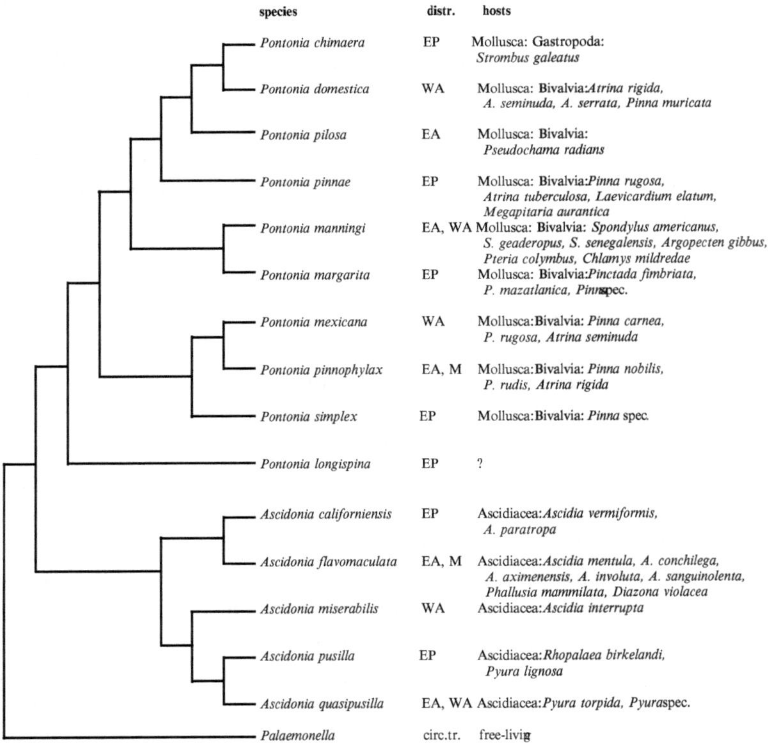


Fig. 2 Hypothesized phylogeny of *Pontonia* and *Ascidonia* with geographic distribution and host range of the species (adapted from Fransen, 2002). Circ. tr. = circumtropic; EA = East Atlantic; EP = East Pacific; M = Mediterranean Sea; WA = West Atlantic; EA = East Atlantic.

# 4. Distributional Patterns

## 4.1. The Atlantic and East Pacific

*Pontonia* and *Ascidonia* both occur in this area. Cladistic analyses based on morphological characters resulted in a robust phylogenetic hypothesis for these genera (Fransen, 2002) (Fig. 2).

Although the "soft" vicariance event of the opening of the Atlantic started much earlier than the "hard" vicariance event of the development of the Panama Isthmus, the resulting actual isolation and possible speciation might have been delayed by the dispersal capabilities of the species. If dispersal capabilities are low, it is hypothesized that speciation across the Atlantic preceded speciation across the Panama Isthmus. If dispersal capabilities are high it is hypothesized that this would be the other way around or would have hampered speciation across the Atlantic. The latter hypothesis is consistent with three clades. In the clade with *P. margarita* and *P. manningi*, the latter species occurs on both sides of the Atlantic as is also true for the clade with *A. pusilla* and *A. quasipusilla*. In the clade with *P. mexicana*, *P. pinnophylax*, and *P. simplex*, the sister species occur on both sides of the Atlantic. The colour pattern and morphology of the East Atlantic *P. pinnophylax* is very similar to the west Atlantic *P. mexicana*. It might well be that some gene-flow between these east and west Atlantic populations is still present. More material and DNA techniques could reveal the taxonomic status of these populations.

## 4.2. The Indo-West Pacific

For the group under study, the distance between the Pacific islands and the East Pacific coast has been too large to disperse along this track. In other Pontoniine genera only six species (De Grave, 2001), living in association with echinoderms and corals, have crossed the eastern Pacific filter bridge (*sensu* Glynn and Ault, 2000). Rosen (1984) showed that the coral fauna of the East Pacific became largely extinct after the rise of the Panama Isthmus and was subsequently colonized by dispersal from the Indo-West Pacific.

The distribution of land and sea, depth changes of seas, changes in climate, and in ocean circulation, all result from geological processes and have played an important role in the Cenozoic development of biogeographical patterns in the Indo-West Pacific (Hall, 1998; Hoeksema, Chapter 5 of this volume; Renema, 2002, Chapter 6 of this volume). In current vicariant biogeographical methodology, dichotomous diagrams of area relationships are generated, indicative of progressive fragmentations of previously continuous areas. As long as these breaking-up events are clear there is no methodological conflict. However, regarding the complex history of divergence and coalescence that characterizes the Indo-West Pacific, combined with the high dispersal capabilities of many marine organisms, it seems highly improbable that vicariant methods alone can clarify the observed biogeographical patterns in this area (Holloway and Hall, 1998).

From several studies on the biogeography of marine organisms of which a fossil record is available – larger benthic foraminifera (Renema, 2002, Chapter 6 of this volume), stony corals (Rosen, 1984; Wilson and Rosen, 1998), coral barnacles (Ross and Newman, 2002), and molluscs (Hallam, 1967) – it becomes apparent that high diversity shifted from the Atlanto-Mediterranean/East Pacific area, existing before the fragmentation of the Tethyan seaway, to the Indo-West Pacific centre of maximum marine diversity, including Malaysia, Indonesia, the Philippines, and Papua New Guinea (Briggs, 1974, 1992; Paulay, 1997; Hoeksema, Chapter 5 of this volume). Wilson and Rosen (1998) stressed the influence of geotectonics on biodiversity, particularly in controlling availability of suitable habitats as they observed a dramatic increase in shallow-water carbonates, which they used as a proxy for availability of coral habitats during the late Oligocene/Early Miocene (*c.* 25 Ma). This is coincident with the intensification of the collision of Australia and southeast Asia, and with the diversification of zooxanthellate corals and subsequently the fauna associated with the corals.

To explain the high diversity in "East Indies Triangle of maximum diversity", many scenarios have been postulated which can be grouped as follows (cf. Hoeksema, Chapter 5 of this volume; Wilson and Rosen, 1998):

1. *Centre of origin.* Speciation occurring inside the centre with successive outward dispersal to surrounding areas (e.g., Briggs, 1974, 1992; Stehli and Wells, 1971; Veron, 1995).
2. *Centre of overlap.* The centre of diversity consists of overlapping distribution ranges that extend to either the Asian, Pacific or Australian plate resulting from either larval dispersal or much more ancient plate tectonics (Santini and Winterbottom, 2002; Wallace, 1997).
3. *Centre of accumulation.* Speciation has occurred at the periphery of the centre after which inwardly directed convergent dispersal has caused accumulation of the ranges inside the centre (e.g., Ladd, 1960 [based on a suggestion of Fenner Chace]; McCoy and Heck, 1976; Rosen, 1984, 1988; Jokiel and Martinelli, 1992).
4. *Centre of survival.* Speciation may have occurred anywhere, but the maximum diversity is an area of survival with species extinctions in periferal area's (McCoy and Heck, 1976; Paulay, 1997).

Wilson and Rosen (1998: Table 1) summarized the evolutionary characteristics within and beyond the centre together with predicted patterns for the fossil record. Contrasting distributional patterns, of groups with similar biological characteristics, are the result of these scenarios:

(A) A low level of congruence between the distributions of the members of the group, resulting from dispersal along different tracks; a low level of endemicity in the centre as compared to peripheral areas. This pattern is consistent with the centre of overlap and centre of accumulation models.

(B) A high level of congruence between the distributions of the members of the group as the result of dispersal along the same tracks, originating from the centre; a high level of endemicity in the centre as compared to peripheral areas.

Overlapping distribution patterns, consistent with the centre of accumulation and overlap scenarios, have been observed for several scleractinian coral groups, such as fungid corals (Hoeksema, 1989, 1993), siderastereids and mussids (Pandolfi, 1992), some species of *Acropora* (Wallace et al., 1991), as well as sea urchins (Palumbi, 1996) and several other groups (Paulay, 1997). Benzie (1998) showed evidence for the centre of accumulation scenario by studying the population genetics of groups of molluscs and echinoderms. In several groups that lack a planktonic period in their development, such as certain gastropod taxa and sepioid cephalopods, closely related species have very restricted, largely allopatric distributions (e.g., Weaver and duPont, 1970; Fleminger, 1986). For these groups the dynamic Indo-West Pacific area is a centre of allopatric speciation due to vicariant isolation resulting from sea level fluctuations and tectonic movements.

Published distribution patterns of pontoniine shrimps in the Indo-West Pacific are biased. Those localities where A.J. Bruce, the foremost pontoniine specialist, resided for some time are relatively well known, e.g., Kenya, Zanzibar, Hongkong, and especially northern Australia. Recently, sampling by the present author in the Seychelles, Moluccas, Bali, Sulawesi, East Kalimantan, and the Philippines has added many records. However, many localities especially in the central Indian Ocean, western Indonesia, and west Pacific islands need to be searched more intensively to fill distributional gaps.

Another factor influencing the distributional patterns are the associations in which the shrimps live. The availability of suitable hosts in a certain area will have its effect on the distributional patterns observed (De Grave, 2001).

Nearly 8 out of 50 Indo-West Pacific taxa are widespread (Fig. 3A), occurring from eastern Africa to the western Pacific islands. Widespread taxa, however, are not informative with regards to the proposed scenarios. Nineteen species have limited known distributions indicating undersampling or actual endemicity to their distribution area (Fig. 3B). Remarkable is the absence of endemics on the major part of the Pacific plate. This could be related with the general direction of water currents going from the Pacific to the Indian Ocean (see Hoeksema, Chapter 5 of this volume). The relatively high number of endemics (6) in the western Indian Ocean is an indication that dispersal in opposite direction is hampered by the same currents or by areas in absence of suitable substrate. Several endemics are found in the centre of maximum marine diversity, which is more in favour of the centre of origin scenario. The remainder of species is distributed in two or three of the four main areas (Fig. 3C); the western Indian Ocean, Southeast Asia, Australia, and the West Pacific. Many species

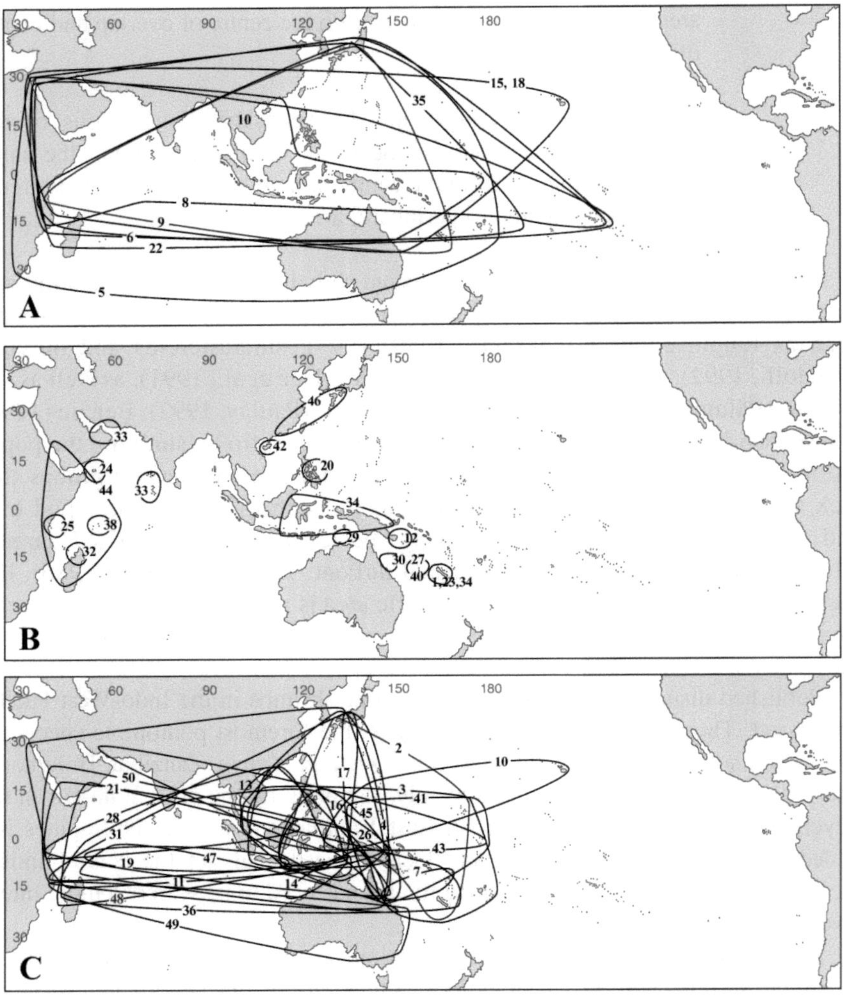


Fig. 3 Lines drawn around the extreme distributional records per species. (**A**) Widespread species; (**B**) species with small distributions; (**C**) species occurring in part of the IWP. (1) *Amphipontonia kanak*; (2) *Anchiopontonia hurri*; (3) *Anchistus australis*; (4) *A. custoides*; (5) *A. custos*; (6) *A. demani*; (7) *A. gravieri*; (8) *A. miersi*; (9) *A. pectinis*; (10) *Cainonia medipacifica*; (11), *Chernocaris placunae*; (12) *Colemonia litodactylus*; (13) *Conchodytes biunguiculatus*; (14) *C. maculatus*; (15) *C. meleagrinae*; (16) *C. monodactylus*; (17) *C. nipponensis*; (18) *C. philippinensis*; (19) *C. pteriae*; (20) *C. tridacnae*; (21) *Dactylonia anachoreta*; (22) *D. ascidicola*; (23) *D. borradailei*; (24) *D. carinicula*; (25) *D. franseni*; (26) *D. holthuisi*; (27) *D. monnioti*; (28) *D. okai*; (29) *Dasella ansoni*; (30) *D. brucei*; (31) *D. herdmaniae*; (32) *Neoanchistus cardiodytes*; (33) *N. nasalis*; (34) *Odontonia compacta*; (35) *O. katoi*; (36) *O. maldivensis*; (37) *O. rufopunctata*; (38) *O. seychellensis*; (39) *O. sibogae*; (40) *O. simplicipes*; (41) *Paranchistus armatus*; (42) *P. liui*; (43) *P. nobilii*; (44) *P. ornatus*; (45) *P. pycnodontae*; (46) *P. spondylus*; (47) *Platypontonia brevirostris*; (48) *Platypontonia hyotis*; (49) *Pseudopontonia minuta*; (50) *Rostronia stylirostris*.

have wide distributions ranging from the western Indian Ocean to eastern Indonesia and Australia, but not penetrating onto the Pacific plate. *Cainonia medipacifica* is the only species that ranges from the west Pacific to Hawaii.

The highest diversity is found in the "East Indies Triangle" and the Coral Sea (Fig. 3). Most of the distributional patterns observed can be explained according to the centre of accumulation and/or overlap scenarios (Fig. 3C). It is however evident that the other scenarios are also involved in the development of the present biogeographical patterns of this group of shrimps.

## References


Adams, C.G., 1967, Tertiary foraminifera in the Tethyan, American, and Indo-Pacific provinces, in: Adams, C.G. and Ager, D.V. (eds), *Aspects of Tethyan Biogeography*, Systematic Association Publication No. 7, pp. 195–217.

Adams, C.G., 1983, Speciation, phylogenesis, tectonism, climate and eustasy: factors in the evolution of Cenozoic larger foraminiferal bioprovinces, *In*: Sims, R.W., Price, J.H., and Whalley, P.E.S. (eds), *Evolution, Time and Space: The Emergence of the Biosphere*, Academic Press, London, pp. 255–298.

Benzie, J.A.H., 1998, Genetic structure of marine organisms and SE Asian biogeography, *In*: Hall, R. and Holloway, J.D. (eds), *Biogeography and Geological Evolution of SE Asia*, Backhuys Publishers, Leiden, pp. 197–209.

Berlocher, S.H., 1998, Can sympatric speciation via host or habitat shift be proven from phylogenetic and biogeographic evidence? in: Howard, D.J. and Berlocher, S.H. (eds), *Endless Forms: Species and Speciation*, Oxford University Press, Oxford, pp. 99–113.

Briggs, J.C., 1974, *Marine Zoogeography*, McGraw-Hill, New York.

Briggs, J.C., 1992, The marine East Indies: centre of origin? *Global Ecology and Biogeography Letters* **2**: 149–156.

Bush, G.L., 1969, Sympatric host race formation and speciation in frugivorous flies of the genus Rhagoletis (Diptera, Tephritidae), *Evolution* **23**: 237–251.

Bush, G.L., 1994, Sympatric speciation in animals – new wine in old bottles, *Trends in Ecology and Evolution* **9**: 285–288.

Calafiore, N., Costanzo, G., and Giacobbe, S., 1991, Developmental stages of *Pontonia pinnophylax* (Otto, 1821) (Decapoda, Natantia, Pontoniinae) reared in the laboratory. Mediterranean species of the genus *Pontonia* Latreille, 1829. I, *Crustaceana* **60** (1): 52–75.

De Grave, S., 2001, Biogeography of Indo-Pacific Pontoniinae (Crustacea, Decapoda): a PAE analysis, *Journal of Biogeography* **28**: 1239–1253.

Duque-Caro, H., 1990, Neogene stratigraphy, paleoceanography and paleobiology in northwest South America and the evolution of the Panama Seaway, *Palaeogeography, Palaeoclimatology, Palaeoecology* **77**: 203–234.

Feder, J.L., 1998, The apple maggot fly, *Rhagoletis pomonella*: flies in the face of conventional wisdom about speciation? *In*: Howard, D.J. and Berlocher, S.H. (eds), *Endless Forms: Species and Speciation*, Oxford University Press, Oxford, pp. 130–144.

Fleminger, A., 1986, The Pleistocene equatorial barrier between the Indian and Pacific Oceans and a likely cause for Wallace's line, *In*: Pierrot-Bults, A.C., van der Spoel, B.J.S., Zahuranec, B.J., and Johnson, R.K. (eds), *Pelagic Biogeography, UNESCO Technical Papers in Marine Science, Paris* **49**: 84–97.

Fransen, C.H.J.M., 2002, Taxonomy, phylogeny, historical biogeography, and historical ecology of the genus *Pontonia* Latreille (Crustacea: Decapoda: Caridea: Palaemonidae), *Zoologische Verhandelingen* **336**: 1–433.

Gibbes, L.R., 1850, On the carcinological collections of the cabinets of natural history in the United States: with an enumeration of the species contained therein and descriptions of new species, *Proceedings of Third Meeting of the American Association for Advancement of Science* **3**: 165–201.

Glynn, P.W. and Ault, J.S., 2000, A biogeographic analysis and review of the far eastern Pacific coral reef region, *Coral Reefs* **19**: 1–23.

Hall, R., 1998, The plate tectonics of Cenozoic SE Asia and the distribution of land and sea, in: Hall, R. and Holloway, J.D. (eds), *Biogeography and Geological Evolution of SE Asia*, Backhuys Publishers, Leiden, pp. 99–132.

Hallam, A., 1967, The bearing of certain palaeogeographic data on continental drift. *Palaeogeography, Palaeoclimatology, Palaeoecology* **3**: 201–241.

Heller, C., 1864, Horæ dalmatinae. Bericht über eine Reise nach der Ostküste des adriatischen Meeres, *Verhandlungen der Kaiserlich-königlichen Zoologisch-botanischen Gesellschaft in Wien* **14**: 17–64.

Hoeksema, B.W., 1989, Taxonomy, phylogeny and biogeography of mushroom corals (Scleractinia; Fungiidae), *Zoologische Verhandelingen* **254**: 1–295.

Hoeksema, B.W., 1993, Historical biogeography of *Fungia* (*Pleuractis*) spp. (Scleractinia: Fungiidae), including a new species from the Seychelles, *Zoologische Mededelingen* **67**: 639–654.

Holloway, J.D. and Hall, R., 1998, SE Asian geology and biogeography: an introduction, in: Hall, R. and Holloway, J.D. (eds), *Biogeography and Geological Evolution of SE Asia*, Backhuys Publishers, Leiden, pp. 1–24.

Holthuis, L.B., 1951, A general revision of the Palaemonidae (Crustacea, Decapoda Natantia) of the Americas. II. The sub-families Euryrhynchinae and Pontoniinae, *Occasional Papers Allan Hancock Foundation Publication* **11**: 1–332.

Howard, D.J. and Berlocher, S.H. (eds), 1998, Endless forms: species and speciation, Oxford University Press, New York, pp. i–xii, 1–470.

Keigwin, L.D., 1978, Pliocene closing of the isthmus of Panama, based on biostratigraphic evidence from nearby Pacific Ocean and Caribbean sea cores, *Geology* **6**: 630–634.

Ladd, H.S., 1960, Origin of the Pacific island mollusc fauna, *American Journal of Science* **258A**: 137–150.

Langer, M.R. and Hottinger, L., 2000, Biogeography of selected "larger" foraminifera, *Micropaleontology* **46** (Suppl. 1): 105–127.

Latreille, P.A., 1829, Crustacés, Arachnides et partie des Insectes, in: Cuvier, G. (ed.), *Le règne animal distribué d'après son organisation, pour servir de base à l'histoire naturelle des animaux et d'introduction à l'anatomie comparée*, édition 2, Vol. 4, Déterville Édition nouvelle, Paris, pp. i–xxvi, 1–584.

Lessios, H.A., 1998, The first stage of speciation as seen in organisms separated by the Isthmus of Panama, in: Howard, D.J. and Berlocher, S.H. (eds), *Endless Forms: Species and Speciation*, Oxford University Press, Oxford, pp. 186–201.

Mayr, E., 1947, Ecological factors in speciation, *Evolution* **1**: 263–288.

Mayr, E., 1963, *Animal Species and Evolution*, The Belknap Press of Harvard University Press, Cambridge, MA.

McCoy, E.D. and Heck K.L., Jr., 1976, Biogeography of corals, sea grasses and mangroves: an alternative to the center of origin concept, *Systematic Zoology* **25**: 201–210.

Menken, S.B.J. and Roessingh, P., 1998, Evolution of Insect–Plant Associations: sensory persection and receptor modifications direct food specialization and host shifts in phytophagous insects, in: Howard, D.J. and Berlocher, S.H. (eds), *Endless Forms: Species and Speciation*, Oxford University Press, Oxford, pp. 145–156.

Otto, A.W., 1821, *Conspectus Animalium quorundam maritimorum nondum editorum*, Typis Universitatis, Bratislava.

Palumbi, S.R., 1996, What can molecular genetics contribute to marine biogeography? An urchin's tale, *Journal of Experimental Marine Biology and Ecology* **203**: 75–92.

Pandolfi, J.M., 1992, A review of the tectonic history of New Guinea and its significance for marine biogeography, *Proceedings of the Seventh International Coral Reef Symposium Guam* **2**: 718–728.

Paulay, G., 1997, Diversity and distribution of reef organisms, *In*: Birkeland, C.E. (ed.), *Life and death of coral reefs*, Chapman & Hall, London, pp. 298–352.

Renema, W., 2002, Larger foraminifera as marine environmental indicators, *Scripta Geologica* **124**: 263 pp.

Rögl, F., 1998, Paleogeographic considerations for Mediterranean and Parathethys seaways (Oligocene to Miocene), *Annales des Naturhistorisches Museums Wien* **99A**: 279–310.

Rögl, F. and Steiniger, F.F., 1984, Neogene Paratethys, Mediterranean and Indo-Pacific seaways: implications for the paleobiogeography of marine and terrestrial biotas, *In*: Brenchley, P.J. (ed.), *Fossils and Climate*, Wiley, Chichester, pp. 171–200.

Rosen, B.R., 1984, Reef coral biogeography and climate through the Cainozoic: just islands in the sun or a critical pattern of islands? Brenchley, P.J. (ed.), *Fossils and Climate*, Wiley, Chichester, pp. 201–264.

Rosen, B.R., 1988, Progress, problems and patterns in the biogeography of reef corals and other tropical marine organisms, *Helgoländer Meeresuntersuchungen* **42**: 269–301.

Ross, A. and Newman, W.A., 2002, Coral barnacles: Cenozoic decline and extinction in the Atlantic/East Pacific versus diversification in the Indo-West Pacific, *Proceedings of the Ninth International Coral Reef Symposium*, Bali, Indonesia, pp. 179–184.

Santini, F. and Winterbottom, R., 2002, Historical biogeography of Indo-western Pacific coral reef biota: is the Indonesian region a centre of origin? *Journal of Biogeography* **29**: 189–205.

Stehli, F.G. and Wells, J.W., 1971, Diversity and age patterns in hermatypic corals, *Systematic Zoology* **20**: 115–126.

Veron, J.E.N., 1995, *Corals in Space and Time. The Biogeography and Evolution of the Scleractinia*, University of New South Wales Press, Sydney.

Wallace, C.C., 1997, The Indo-Pacific centre of coral diversity re-examined at species level, *Proceedings of the Eighth International Coral Reef Symposium, Panama* **1**: 365–370.

Wallace, C.C., Pandolfi, J.M., Young, A., and Wolstenholme, J., 1991, Indo-Pacific coral biogeography: a case study from the *Acropora selago* group, *Australian Systematic Botany* **4**: 199–210.

Weaver, C.S. and duPont, J.E., 1970, *The Living Volutes*, Delaware Museum of Natural History, Greenville, DE.

Wilson, M.E.J. and Rosen, B.R., 1998, Implications of paucity of corals in the Paleogene of SE Asia: plate tectonics or Centre of Origin? *In*: Hall, R. and Holloway, J.D. (eds), *Biogeography and Geological Evolution of SE Asia*, Backhuys Publishers, Leiden, pp. 166–195.

Chapter 5

# Delineation of the Indo-Malayan Centre of Maximum Marine Biodiversity: The Coral Triangle

BERT W. HOEKSEMA
Nationaal Natuurhistorisch Museum Naturalis, Leiden, The Netherlands,
Hoeksema@naturalis.nl



> The centre of the Malayan fauna is formed by the Malay Archipelago, joined by some neighbouring areas around it. Where its exact boundaries are located, we do not know. Also, there is no generally accepted designation; Indo-Australian, Indo-Malayan, and Insular Indian are other indications.
>
> *(Translation from the original German text; S. Ekman, 1935: 32).*

> The mechanisms giving rise to marine biodiversity in the southeast Asian region are not well-understood. *(J.A.H. Benzie, 1998: 197).*



*W. Renema (ed.), Biogeography, Time, and Place: Distributions, Barriers, and Islands,* 117–178

**Abstract** The ranges of many tropical marine species overlap in a centre of maximum marine biodiversity, which is located in the Indo-Malayan region. Because this centre includes Malaysia, the Philippines, Indonesia, and Papua New Guinea, it has been named the East Indies Triangle. Due to its dependence on the presence of coral reefs, it has recently been referred to as the Coral Triangle. Because these reefs are severely threatened by human activities, large-scale nature conservation efforts involve the establishment of a network of Marine Protected Areas (MPAs), for which it is important to know the position of this diversity hotspot. Although it is recognized where this centre is located approximately, it is unclear where its exact boundaries are. Only in a limited number of biogeographical studies, ranges and diversity centres of Indo-West Pacific (IWP) taxa have been presented. In this regard, tropical corals, marine fishes, and molluscs have received most attention. However, just for reef corals alone several different diversity centres have been proposed. The boundaries of the centre are important for reconstructing the processes that were responsible for its present shape. They may relate to the area's climatic and geological past or to the dispersal of larvae by currents in combination with ecological constraints that may prevent their settlement. Especially, in brooding organisms, without larvae or other propagules performing long-distance dispersal, isolation mechanisms may have been important for speciation and species diversity. Information on sea-level fluctuation and the past position of coastlines and data on molecular variation between and within species may help to support models that explain the present position of the centre of marine biodiversity. A detailed biogeographical study of the Fungiidae, a family of corals that disperse through larvae, is used to present a model for a diversity centre and the processes that may have caused its present position. For each species, presence–absence data were obtained from many areas in order to plot their distribution patterns. Since several species do not occur on Sunda shelf reefs, the western part of this diversity centre may have been moulded along the Sunda shelf margin since the end of the LGM (17.000–18.000 BP). Species diversity appears to be distributed unevenly among areas within this centre, which depends on habitat heterogeneity, such as cross-shelf gradients in salinity and turbidity. Eventually, the distributions of several model taxa need to be compared in a sufficiently high number of areas in order to find a more common delineation of the Coral Triangle. Many corals are widespread and have a long fossil record. Moreover, coral reefs have not always been located in their present positions. This makes it complex to find which processes have caused a present diversity maximum. Since most species are concentrated in the eastern part of the Indo-Malayan archipelago and part of the West Pacific, this may be the area where most of the youngest species have originated, but sea-level fluctuations probably have been responsible for excluding large continental shelf seas from the Coral Triangle.

## 1. Introduction

The presence of a marine latitudinal diversity gradient, i.e., an increase in species richness with decreasing latitude from the tropics to both polar regions, is commonly recognized (Ekman, 1934; Briggs, 1974). The circum-tropical belt of high species diversity within this gradient is related to high seawater temperatures and maximum solar irradiation in the proximity of the equator. These two abiotic factors are of benefit to the relatively fast growth of shallow-water corals through their photosymbiotic relation with algae (zooxanthellae), which eventually may lead to the formation of reefs. These reefs form a habitat for many other organisms, many of which live in symbiotic relationships (Paulay, 1997). Within the tropical zone, species diversity, and composition show much longitudinal variation (Ekman, 1934, 1953; Rosen, 1988c; Paulay, 1997); four major biogeographic regions can be distinguished: the Indo-West Pacific (IWP), the East Pacific (EP), the West Atlantic (WA), and the East Atlantic (EA). Among these regions, most taxa show their highest species numbers in the IWP, the largest region of the four, which extends from the Red Sea and East Africa to the Central Pacific (Ekman, 1934; Briggs, 1974; Paulay, 1997). Within the IWP region, a centre of maximum marine biodiversity can be found where most IWP species show overlaps of their distribution ranges.

For various reasons it is important to know where the highest concentration of species can be found:

(1) Coastal areas offer a great potential of natural resources for local human populations, especially in countries rich in marine species (McManus, 1997; Ablan et al., 2002; Alcala and Russ, 2002; Hughes et al., 2003; Hoeksema, 2004). For instance, an estimated 25% of the world's marine fishes inhabit coral reefs, whereas reefs only cover 0.2% of the world's oceans (McAllister, 1991; McAllister et al., 1994; Ormond and Roberts, 1997; Roberts et al., 1998). Although not directly observable, it appears that even for large oceanic predators of commercial interest, such as tuna, the proximity of coral reef areas is important (Worm et al., 2003, 2005). In addition, the strong interspecific competition in high-diversity habitats leads to a high variety of natural products that are of pharmacological value (Adey, 2000; Adey et al., 2000).

(2) For the designation of a network of Marine Protected Areas (MPAs) and for other conservation efforts (Bleakley and Wells, 1995; Green and Mous, 2004), it is important to know which areas are particularly rich in species (Gaston and Williams, 1993; Ferrier, 2002; Briggs, 2004b, 2005b), especially in endemic species that show relatively small ranges (Briggs, 1996; Myers et al., 2000; Allen, 2002; Roberts et al., 2002; Beger et al., 2003; Myers and Ottensmeyer, 2005). However, centres of high species richness are not necessarily concordant with centres of high endemicity (Hughes et al., 2002). Coral reefs in particular are in need for protection, because of their high species diversity and biological production, and because they are most severely threatened by human exploitation (Sebens, 1994; Hoeksema, 1997c; Paulay, 1997; Reaka-Kudla, 1997a, b).

(3) The marine tourism industry benefits from the high variety of marine life (Gray, 1997; Thorne-Miller, 1999; Nontji, 2002).

Several dive operators advertise the positioning of their business in the centre of marine diversity via the internet. Hence, recreational importance forms part of the economic value of high species diversity (Pet-Soede et al., 1999; Cesar, 2000; White et al., 2000; Burke et al., 2002; Cesar et al., 2003). Many small species have been discovered in coral reefs, usually well hidden, but obviously contributing even more to the species diversity in coral reef areas (Bouchet, 1997; Reaka-Kudla, 1997a, b; Carlton et al., 1999; Roberts and Hawkins, 1999; Porter and Tougas, 2001). Due to the cryptic mode of life and the camouflage of small animals, many underwater photographers find a challenge in shooting pictures of macro-life on coral reefs. They are able to spend much time on observing reef animals and therefore it is not surprising that several species illustrated in underwater field guides and diving magazines appear to be unknown to science.

(4) The position and boundaries of the centre of marine diversity are relevant for evolutionary and ecological problems (Thomas, 1997). If the boundaries of the diversity centre are known, it would be easier to understand how it was shaped, such as a centre of speciation, a centre of survival, a centre of overlap, or a centre of accumulation (Briggs, 1973, 1981, 1987b, 1999b, c, 2000, 2003, 2004a, b, 2005a, b, 2006; McCoy, 1983; Jokiel and Martinelli, 1992; Pandolfi, 1992; Santini and Winterbottom, 2002). Without knowing the patterns, one cannot understand the processes that created these patterns (Rosen, 1988a; GBDMS, 1995).

Theoretically there can only be one single area where the highest concentration of species exists, even just by stochastic processes (Osman and Whitlatch, 1978). Due to different evolutionary pasts, diversifying life history strategies, and various ecological requirements, it is unlikely that all separate higher taxa show the same area of maximum range overlap. Furthermore, the size of such focal areas also matters, considering the role of habitat heterogeneity in species–area relations (Brown, 1988; Williamson, 1988; Ricklefs and Schluter, 1993; Rosenzweig, 1995; Hubbel, 1997; Beger et al., 2003; Neigel, 2003; Mumby et al., 2004). Thus, the area size of a supposed centre of maximum diversity remains arbitrary, since larger areas are supposed to contain more species than smaller ones. Within one area, sampling effort may not be even and some of the rare species involved may appear to show disjunctive or fragmented distribution ranges (Paulay, 1997).

Hence, it is unrealistic to work with absolute maximum species numbers since this number may vary continuously depending on progressing research efforts. It is more practical to search for an area with ranges of highest species numbers instead. A fast way to achieve this goal is by the using model, key, exemplar, or focal taxa as indicator groups. These taxa should consist of a sufficiently large number of species that have reliable information on their distribution ranges, such as families and genera that are taxonomically well revised and have enough distribution records (Kohn, 1997; Meyer, 2003; Olsgard et al., 2003; Paulay and Meyer, 2006; Wallace and Rosen, 2006).

## 2. The Indo-West Pacific Region

As a result of numerous expeditions, particularly in the 19th century and the first decades of the 20th, it became evident that the highest numbers of marine animal species can be found in the (sub)tropical portions of the Indian Ocean and the Western and Central Pacific, together called the IWP (Ekman, 1934, 1935, 1953). Many families, genera and species show widespread ranges that are entirely IWP, i.e., they are endemic to the region extending from the east coast of Africa in the West Indian Ocean toward the Tuamotu Archipelago in the central Pacific (e.g., Hoeksema, 1989; Wallace, 1999).

Although Ekman (1934) was not the first to recognize the IWP as a biogeographic entity, he clearly set the IWP apart from the EP and the Atlantic and was the first to name it most appropriately (Briggs, 1974). To the southwest of the IWP, cold water around southern Africa blocks the migration of tropical species between the Indian and Atlantic Ocean and in the northwest the Red Sea is only separated from the Mediterranean by the Suez Canal. Eastward, a wide stretch of water, the East Pacific Barrier (EPB), prevents the colonization by many species of the Galapagos Islands and the western American coast.

Regardless of the long history of explorations, it is recognized that present information on the species diversity of the IWP is still poor (Hutomo and Moosa, 2005). A possible link between the high species richness of IWP biota and the occurrence of coral reefs was already determined by Forbes (1856). Because IWP coral reefs harbour a high diversity and abundance of marine habitats, they can accommodate many species of invertebrates, many of which are not yet known to science (De Fontaubert et al., 1996; Paulay, 1997; Porter and Tougas, 2001). In spite of this incomplete information, it is clear that IWP coral reefs are the world's most speciose marine ecosystems (Bellwood and Hughes, 2001; Roberts et al., 2002; Hughes et al., 2003). For certain taxa of reef-dwelling organisms, IWP coral reefs outnumber the Atlantic reefs 3–10 times in species numbers but they also measure more than ten times in reef surface area (Schuhmacher, 1976, 1988; Spalding et al., 2001).

The distribution of coral reefs is limited to tropical waters. To the north and the south, the marine tropical shelf areas are bounded by the 20°C isotherm (Briggs, 1974). In colder waters corals do no grow fast and may meet more competition by macro-algae (Johannes et al., 1983). Within the IWP tropical belt, there is a decline in species numbers to the north and south of the equator: northward most clearly from the Philippines to Japan and southward along the west and east coast of Australia and along eastern Africa (Veron and Minchin, 1992; Nishihira and Veron, 1995; Veron, 1995).

## 3. A Triangular Indo-West Pacific Biodiversity Hotspot?

Over 50% of the IWP reef area occurs in the central Indo-Pacific, the border area of the eastern Indian Ocean and the western Pacific, including the seas of East and Southeast Asia, New Guinea, and Australia (Potts, 1983; Spalding et al., 2001). Ekman (1953) considered the Malay Archipelago as the faunistic centre of the IWP from where species dispersed to peripheral areas. Although this area between Asia and Australia is recognized as one of the world's major centres of diversity, there is little agreement about its name (Heads, 2005). Subsequent authors since Ekman (1953) have presented their own terminology and definitions for this centre, especially in recent years (Table 1). Most of these authors have given the same explanations as Ekman (1953) for the positioning of this centre, i.e., a centre of speciation (Briggs, 2004b, 2005a, b, 2006), possibly in combination with a relatively low extinction here (Vermeij, 1989). These and other theories on the origin of the centre of maximum marine biodiversity are discussed in section 5.5.

*Table 1. Terminology for hypothetical centres of marine diversity in the Indo-West Pacific*

| *Term* | *Focus area* | *Reference* |
|---|---|---|
| The Malay Archipelago | Malaysia, Indonesia, Philippines, Papua New Guinea | Ekman (1953) |
| A faunistic centre within a triangle with apices in South Japan, Sumatra, and New Guinea | Indonesia | Kohn (1967) |
| Indo-Malayan centre shaped like a triangle | Philippines, the Malay Peninsula, and New Guinea as corners | Briggs (1974) |
| Indo-Malayan region | Around the Philippines, Indonesia, New Guinea, Solomon Islands | Vermeij (1978, 1989) |
| East Indies central triangle area | cf. Briggs (1974) | Briggs (1987a) |
| Coral Triangle, Malayan triangle | Philippines, North Borneo | Paine (1988) |
| Highest species diversity in a triangle | Between the Philippines, New Guinea, and Sumatra | Williams (1993) |
| Indo-Malaysia | India–Southeast Asia | Ricklefs and Latham (1993) |
| Australasia | New Guinea, Australia, West Pacific islands | Ricklefs and Latham (1993) |
| Coral Triangle | cf. Briggs (1974) | McAllister et al. (1994) |
| East Indies Triangle | cf. Briggs (1974) | Briggs (1995) |
| Indonesian–Philippines region | Richest area within a coral triangle, comprising northern Australia, the Malay-Indonesian Archipelago, Philippines, and western Melanesia | Werner and Allen (1998, 2000) |
| Indo-Australian region | 120–170° E | Bellwood and Hughes (2001) |
| Southeast Asian centre of diversity | A triangle encompassing the Philippines and central and eastern Indonesia | Spalding et al. (2001) |

(continued)

*Table 1.* (continued)

| *Term* | *Focus area* | *Reference* |
|---|---|---|
| Coral Triangle | Indonesia, Philippines, Malaysia, Papua New Guinea, Japan, and Australia | McKenna et al. (2002) |
| Coral Triangle | Between the Ryukyu Islands, the east coast of Malaysia, and western New Guinea | Allen, 2002, 2003 |
| Indo-Malayan Triangle, Epicenter of Global Marine Biodiversity | Philippines–South Indonesia,Java–New Guinea | Burke et al. (2002) |
| Indo-Malay triangle | Reference to Brigss (1992, 2000) | Wörheide et al. (2002) |
| Western Pacific diversity triangle | Philippines, Indonesia, Papua New Guinea | Gosliner (2002) |
| Central Indo-Pacific (CI-P) biodiversity hot spot | between 100–140° E, and 10°S–10°N | Hughes et al. (2002)<br>Connolly et al. (2005) |
| Indo-Malay–Philippines Archipelago (IMPA) | Indonesia, Malaysia, Philippines | Bellwood and Wainwright (2002) |
| A centre of high diversity in the Indonesian–Philippine region (IPR) | Philippines, Indonesia | Mora et al. (2003) |
| Indo-Australian Archipelago (IAA) | A triangle bounded by Sumatra, Philippines, Papua New Guinea, Solomon Islands | Connolly et al. (2003)<br>Bellwood et al. (2005)<br>Barber and Bellwood (2005) |
| IWP diversity triangle | Reference to Briggs' (1999) "Indo-Malay triangle" | Kirkendale and Meyer (2004) |
| Philippines–South China Sea–Indonesia triangle | Philippines, central Indonesia | Kulbicki et al. (2004) |
| West Pacific diversity triangle, Malesia | Reference to Indo-Australian Archipelago (IAA), the triangle formed by Sumatra, the Philippines, and New Guinea | Heads (2005) |
| Global center of marine biodiversity in Indo-Malaya | No description, reference to a map | Paulay and Meyer (2005) |

Kohn (1967) appears to be the first author who referred to a triangular shape in connection to a centre of marine diversity (Table 1). The concept of a triangular centre was adopted and illustrated by Briggs (1974; see Fig. 1A). His triangle is slightly smaller than the one indicated by Kohn (Table 1). The positions of the triangle sides appear not to have significance as biogeographical boundaries but should be considered indicative, since Briggs (1987a) mentioned that the area of highest species diversity for most tropical marine animal groups is located 'within' the East Indies central triangle area. The word 'central' refers here to its position inside the IWP.

Later on Briggs (1995, 2003, 2004b, 2005a, b) referred to this 'East Indies Triangle' itself as a center of evolutionary radiation. Again, reference to the East Indies Triangle or Indo-Malayan Triangle as a centre of biodiversity indicates its approximate position

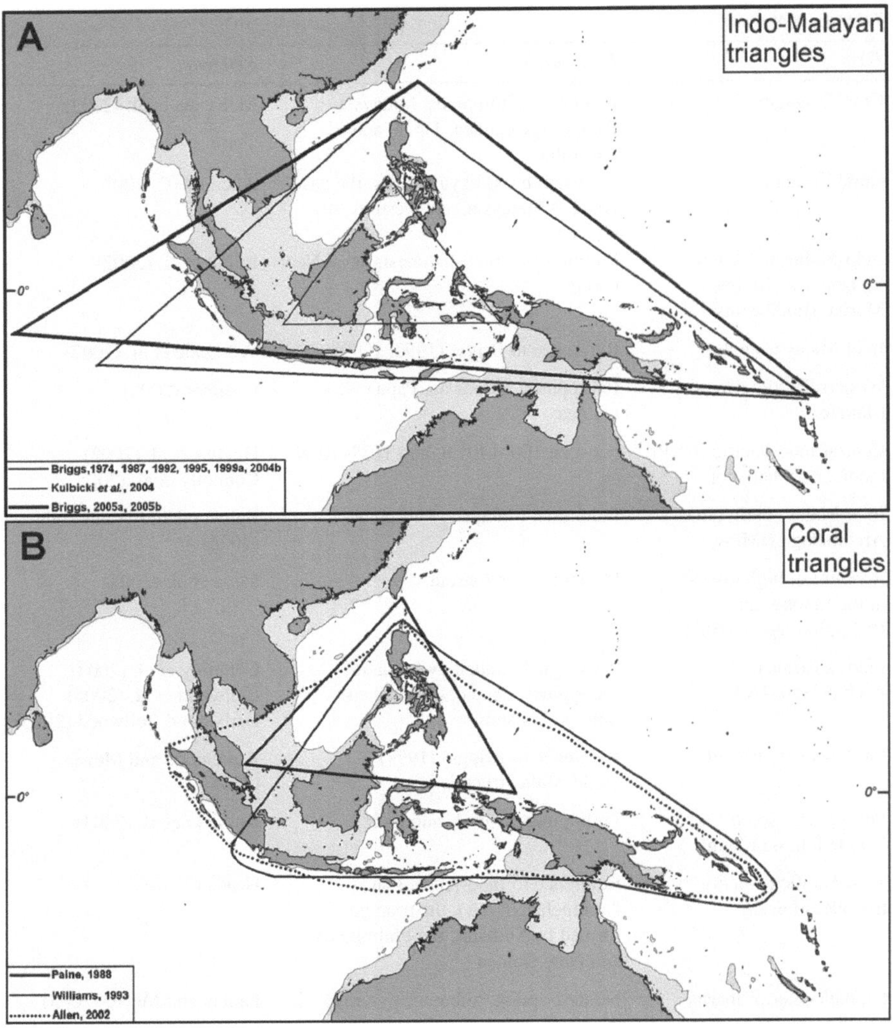


Fig. 1(**A**) Briggs' (1974) Indo-Malayan centre of marine biodiversity depicted as the "East Indies Triangle" (Briggs, 1987). His later version (Briggs, 2005a) is slightly larger, including all of Sumatra, and therefore more similar to the Coral Triangle indicated by Allen (2002; Fig. 1B). Kulbicki et al. (2004) refer to a centre of fish diversity, which they call "the Philippines–South China Sea–Indonesia triangle"; (**B**) The centre of maximum diversity presented as coral triangles (Paine, 1988; Allen, 2002). The centre of reef-associated pennatulacean octocorals is also presented as a triangle (Williams, 1993).

(e.g., Donaldson, 1986; Santini and Winterbottom, 2002; Williams and Reid, 2004), but it is clear that no specific meaning is given to its delineation (Mironov, 2006).

The first use of the name 'Coral Triangle' has probably occurred in a television documentary (Paine, 1988), which explained that the seas within an imaginary Malayan triangle are so rich in marine life forms that they can be considered the evolutionary origin of numerous marine animals (Table 1; Fig. 1B).

A map of a larger Coral Triangle (Fig. 1B) is presented by Allen (2002, 2003), who collaborated with Wells (2002) and Roberts et al. (2002). It resembles a triangle of octocoral diversity previously illustrated by (Williams 1993; Fig. 1). However, the first use of 'coral triangle' in scientific literature is by McAllister et al. (1994: 165) with reference to the triangle of Briggs (1974). Werner and Allen (1998, 2000), Allen and McKenna (2001), and McKenna et al. (2002) have proposed a coral triangle, in which the Indonesian–Philippines region is indicated as probably the most diverse (Table 1). The smallest triangular diversity centre of all refers to fishes in 'the Philippines–South China Sea–Indonesia triangle' (Kulbicki et al., 2004; see Fig. 1A).

Spalding et al. (2001) also refer to a centre of diversity that does not include the Malay Peninsula and Papua New Guinea, but Burke et al. (2002) and Gosliner (2002) have also included Papua New Guinea (Table 1). Bellwood and Hughes (2001) do not point out a particular shape but refer to the Indo-Australian region, whereas Bellwood (1997) refers to a centre of diversity in the western Pacific (Table 1). It has also been called the Central Indo-Pacific (CI-P) biodiversity hotspot (Hughes et al., 2002) and been referred to as the Indo-Malay–Philippines Archipelago (IMPA) (Bellwood and Wainwright, 2002; Carpenter and Springer, 2005), a centre of high diversity in the Indonesian and Philippine region (IPR) (Mora et al., 2003; Gaston, 2003), and the Indo-Australian Archipelago (IAA) (Connolly et al., 2003), all with various positions and sizes (Table 1).

## 4. Marine Biodiversity Patterns Among Various Taxa

The hypothetical Indo-Pacific biodiversity hotspots treated in the previous section are generally not congruent. Studies that present accurate data at species level are scarce. Their boundaries are not based on presence/absence records of species. Many localities need to be studied for quantitative comparisons of faunas and therefore the subject of sampling is usually restricted to large, relatively easy to find animals. The following sections aim to give an overview of marine biodiversity studies for various taxa or functional groups of organisms. Since these groups represent different life histories, also various diversity patterns may exist. A major distinction is made between broadcasting and brooding organisms: broadcasting organisms are lifelong pelagic or have a long-lived pelagic stage, which results in a high-dispersion capacity with wide-ranged interconnected populations that are less likely to undergo fast speciation than brooding species with a low-dispersion capacity, limited and isolated geographic ranges, and a high degree of endemism (Paulay, 1997).

### 4.1. Pelagic Biodiversity

In contrast to other marine environments, the open ocean contains relatively small numbers of species in diverse groups of organisms. This relatively low diversity of pelagic species (plankton and nekton) is due to the scarcity of obvious barriers

(Boxshall, 1981; Angel, 1993, 1997; Pierrot-Bults, 1997, 2003; Pierrot-Bults and Van der Spoel, 2003). However, it is usually overlooked that many benthic species also posses a pelagic stage in their life cycle (Scheltema and Hall, 1975; Scheltema, 1988; Sinclair, 1988; Scheltema et al., 1996).

Pelagic species have widespread distributions through dispersal and allopatric speciation seems to have little importance. Their populations are also quite well capable of surviving at the periphery of their ranges. Nevertheless, the fossil record suggests that species turnover is very rapid compared to coastal ecosystems and this is most likely the result of sympatric speciation (Norris, 2000). Hence, it is not surprising that no clear centre of pelagic diversity has been distinguished so far, although planktonic Foraminifera show some indication of a large circum-tropical belt of maximum diversity (Stehli, 1968; Tokeshi, 1999). Due to their fossil record, the potential importance of planktonic foraminifera in (paleo-)biogeographical studies may need more attention in studies determining diversity patterns (see, e.g., Kennett et al., 1985). Diversity of large oceanic predators is generally highest in subtropical zones where tropical and temperate species ranges overlap with individual hotspots close to coral reef areas, seamounts, and oceanic islands, coinciding with concentrations of zooplankton and coral reef species near shores (Worm et al., 2003, 2005).

The Indo-Malayan area is considered as only one of 40 planktonic faunal centres of the world oceans, without sharp boundaries and further distinction with regard to diversity (Van der Spoel and Heyman, 1983). Longhurst (1998) classifies this area into three categories: the northern part of the Australia–Indonesia Coastal Province of the Indian Ocean Coastal Biome, the Sunda–Arafura (i.e., Sahul) Shelves Province as part of the Pacific Coastal Biome and the Archipelagic Deep Basins Province as part of the Pacific Trade Wind Biome. In general, there is a latitudinal increase in pelagic species richness from the poles to the equator and most biogeographic studies appear to concentrate on oceanic waters and not on coastal waters (Van der Spoel and Pierrot-Bults, 1979; McGowan and Walker, 1993; Angel, 1997; Pierrot-Bults, 1997). This focus on oceanic patterns appears not to be a bias, since various ranges of separate Indo-Pacific planktonic taxa show a disjunction in the shelf-based waters of Southeast Asia (Van der Spoel and Heyman, 1983). Hence, many low-latitude species have tropical ocean-wide or even circum-equatorial ranges (Johnson, 1986; Norris, 2000). These pelagic biogeographic zones reflect main oceanographic features and large-scale oceanic circulation patterns of the present and the geological past (Backus, 1986; McGowan, 1986; Olson, 1986; Angel, 1997).

Unresolved taxonomic problems and indistinct distribution boundaries may hinder a fast progress in further analysis of species richness patterns. Hence, supposedly single, very widely distributed species may actually represent species complexes obscuring cryptic species that may have smaller ranges (Sinclair, 1998; Norris, 2000; Pierrot-Bults and Van der Spoel, 2003). A large potential for unknown species is distinctly the case for the deep basins of eastern Indonesia (Othman et al., 1990). The deep-sea basins between the land areas of the Sunda and Sahul shelves of the Indo-Malayan region probably played a role as dispersal barriers during

Pleistocene glacial low eustatic sea-level stands in combination with extensive cold upwelling, which may have provided suitable conditions for speciation and causing high local diversity in pelagic taxa (Fleminger, 1986). An example for fast Holocene evolution in IWP marine zooplankton in isolated water basins is presented by Dawson and Hamner (2005), who studied jellyfish in marine lakes.

## 4.2. Marine Plants: Mangroves, Seagrasses, and Algae

Biogeographically, mangroves and corals show similar distributions that are latitudinally determined by the 20°C winter isotherm (McCoy and Heck, 1976; Woodroffe and Grindrod, 1991; Hogarth, 2001). Ecologically they are different, since mangroves occur in relatively warm tidal surface water, they are exposed to variable air temperatures, they need soft bottom substrata for their roots and can tolerate water with low salinity from river discharge (Van Steenis, 1962; Ricklefs and Latham, 1993), whereas corals may be subject to lower subsurface temperatures, cannot tolerate long-lasting exposure to air, and cannot live on muddy substrates near river outlets with freshwater discharge. Although mangroves can dwell on sandy reef flats, they cannot live on steep rocky cliffs, like corals (Van Steenis, 1962; Woodroffe and Grindrod, 1991). Due to these different ecological requirements, mangrove and coral reef distributions in Southeast Asia are not similar on a local scale but usually adjacent to each other (WCMC, 1997; Groombridge and Jenkins, 2002). Because of their buoyant propagules, mangrove ranges are determined by drift dispersal through surface currents. Mangrove species diversity is higher in the IWP than in the Atlantic and East Pacific. Within the IWP, Ricklefs and Latham (1993) recognize two relatively rich subregions, i.e., Indo-Malesia (from the Indian subcontinent to Southeast Asia) and Australasia (from New Guinea and Australia to the West Pacific oceanic islands). The mangrove floras of these two subregions are relatively similar (Hogarth, 1999, 2001). Within these two subregions together, comprising the East Indian Ocean and the West Pacific, a centre of diversity has been distinguished at the Indo-Pacific convergence, ranging from the seas around Sumatra and the southern half of peninsular Malaysia, to the easternmost point of New Guinea (Groombridge and Jenkins, 2000, 2002; see also Fig. 2).

The associated fauna in mangroves also shows a concentration of species in the Asian–Indonesian part of the IWP, although there may be a collector bias in the comparison of different regions involved (Ricklefs and Latham, 1993; Hogarth, 1999, 2001).

Seagrasses occur in a wide latitudinal range from tropical to temperate-boreal shallow seas (Den Hartog, 1970; McCoy and Heck, 1976; Phillips and Meñez, 1988; Hemminga and Duarte, 2000; Spalding et al., 2003). Basically, there are nine seagrass floras, including an Indo-Pacific and a West Pacific flora, whose ranges overlap in the Indo-Malayan area (Hemminga and Duarte, 2000; Duarte, 2001). In this area of overlap, also called Malesia, the highest concentration of seagrass species is found (Mukai, 1993; Duarte, 2001). Spalding et al. (2003) recognize five areas of high seagrass species diversity (11–15 species), all in the tropical IWP:

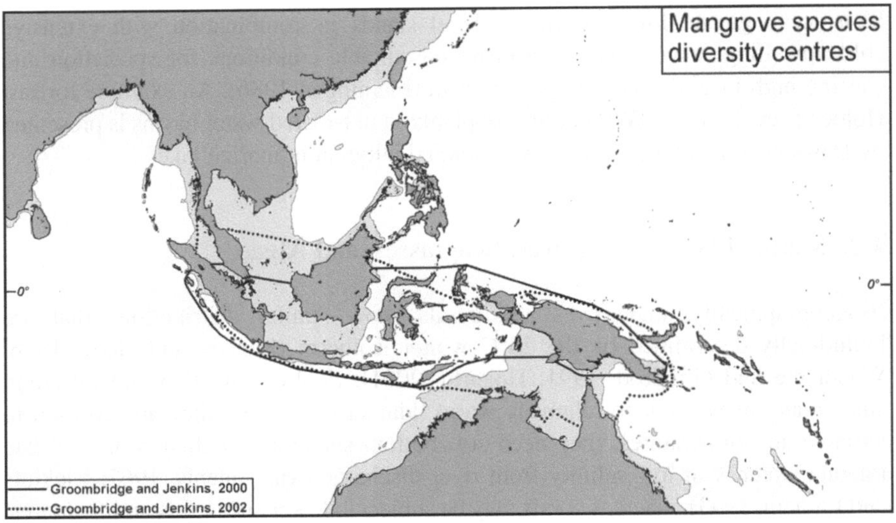


Fig. 2 Centres of diversity for mangrove species (Groombridge and Jenkins, 2000, 2002).

insular Southeast Asia, Japan with Korea, southwest Australia, East Africa, and Southeast India. They also consider the Philippines and New Guinea, together with Indonesia to be the centre of global seagrass biodiversity (Fig. 3).

Seagrass meadows may consist of multi-species stands. In addition there are many species of epiphytic macro-algae and invertebrates that live on seagrasses (Verheij and Erftemeijer, 1993). Various species of animals live among the seagrass leaves and in the sediment held by the seagrass roots (Duarte, 2001). In the tropics, seagrass meadows occur for a large part on coral reef flats, using coral sand as

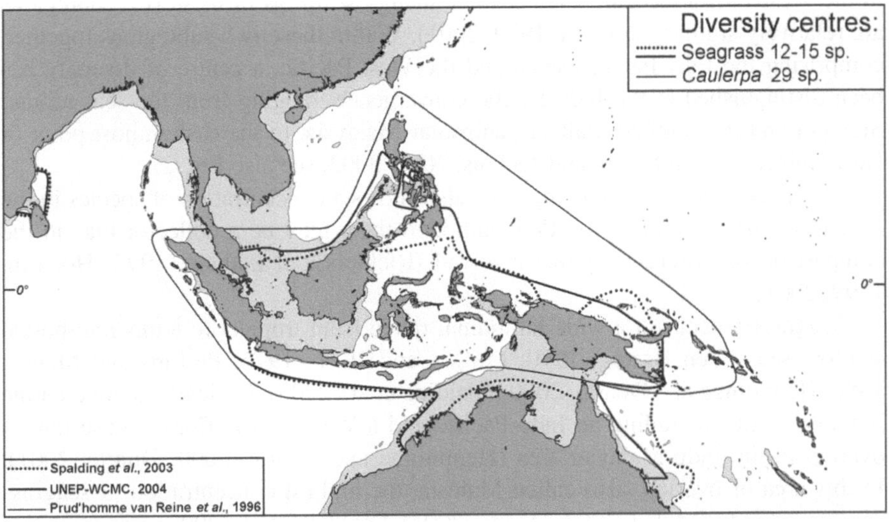


Fig. 3 Centres of diversity for seagrass species (Spalding et al., 2003; UNEP-WCMC, 2004) and *Caulerpa* spp. (Prud'homme van Reine et al., 1996).

substrate, and therefore species that are part of such tropical seagrass communities may also be considered to be part of coral reef-associated biota.

Apart from these epiphytic algae, seaweeds also play an important role on the hard substrata of coral reefs and many species of calcareous algae even take part in the calcification process (Dawes, 1981). Seaweeds occur in all seas. The ranges of individual species very much depend on how temperature regulates their life histories, in particular with regard to lethal and reproductive boundaries (Van den Hoek, 1982). In the Indo-Pacific, a tropical IWP seaweed region has been distinguished with regard to red algae, but its easternmost boundaries are not clearly defined (Joosten and Van den Hoek, 1986; Lawson, 1988).

Apparently, the most diverse seaweed floras do not occur in the Indo-Malayan region but in subtropical areas, notably southern Australia (Silva, 1992; Bolton, 1994; Phillips, 2001). This southern Australian algal flora is composed of four elements, comprising endemics, widely distributed temperate species, tropical species (carried by the southward Leeuwin Current along the Western Australian coast) and polar species (Phillips, 2001).

Exceptional algal diversity patterns are formed by taxa of reef-dwelling green algae, such as the genus *Caulerpa* (Fig. 3) and the order Bryopsidales, which both show maximum species concentrations in Malesia (Indo-Malayan archipelago) due to their preference for warm water (Prud'homme van Reine et al., 1996; Kerswell, 2006). Southern Australia still has the highest number of endemics in *Caulerpa* (Prud'homme van Reine et al., 1996; Phillips, 2001), but this should not be a main criterion for defining marine biogegraphic areas (Adey and Steneck, 2001). Since Prud'homme van Reine et al. (1996) and De Senerpont Domis (2004) discuss the importance of *Caulerpa* as a model taxon for biodiversity assessments of tropical algae, it may be relevant to study the biogeographic species richness patterns of other algal taxa that are well represented in warm water, such as the closely related genus *Halimeda* (order Caulerpales) and coralline red algae (order Corallinales). Recent information concerning *Halimeda* indicates that Malesia is its centre of diversity with nearly all IWP species represented in Indonesia and Papua New Guinea (Dargent and Coppejans, 1998).

## 4.3. Molluscs

Among marine animals, molluscs are third in species numbers after the arthropods and nematodes (Briggs, 1994). They are very diverse in life form, life history, and ecology (Kohn, 1983; Kohn and Perron, 1994). Many marine species have long-lived veliger larvae that may travel with the currents whereas others produce eggs that release far-advanced individuals with very little dispersal (Shuto, 1983; Scheltema and Williams, 1983; Perron and Kohn, 1985; Vermeij, 1987; Scheltema et al., 1996; Paulay, 1997). Modes of mobility (free-swimming, crawling, buried, or attached) and life history strategy (brooders or broadcasters) strongly affect the range of individual mollusc species.

The results of most revisionary studies confirm high concentrations of species in the Indo-Malayan region or a larger part of the Central Indo-West Pacific (CI-P)

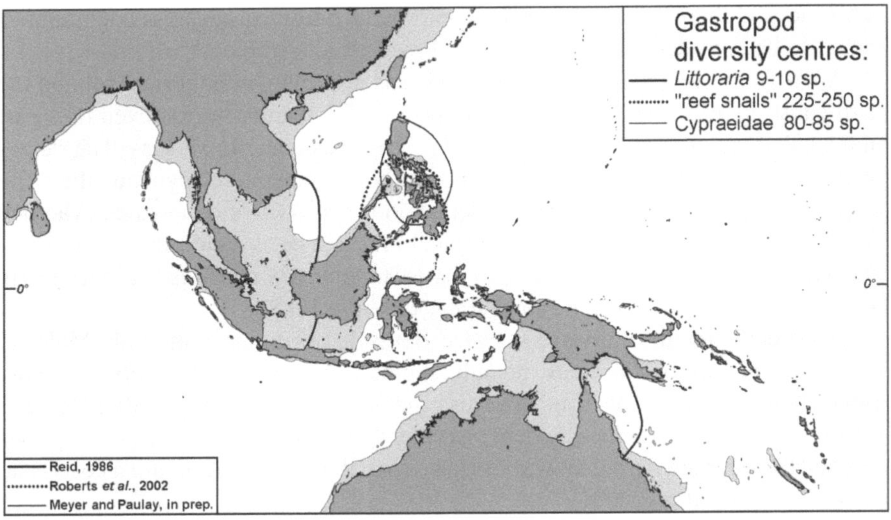


Fig. 4 Centres of diversity for gastropod snails: *Littoraria* spp. (Reid, 1986), reef gastropods (Roberts et al., 2002), and Cypraeidae (Paulay and Meyer, 2006).

(Table 2). A classic example of an Indo-Malayan diversity centre is presented in a map of *Strombus* (Abott, 1960; Briggs, 1995, 1996, 1999a). In an update on Strombidae given by Roy et al. (2001a) the highest diversity is indicated to occur in the Philippines and in eastern Indonesia. Additional studies usually confirm such centres (Table 2). In most gastropod families or genera only a few species show widespread Indo-Pacific distributions, while the majorities have more restricted ranges. As another example, cowries (Cypraeidae) constitute a gastropod family that has been subject to detailed geographical studies during several decades, which has resulted in the distinction of various centres, the Philippines in particular (Schilder and Schilder, 1938; Schilder 1965, 1969; Foin, 1976; Meyer, 2003, 2004; Meyer and Paulay, 2005; Paulay and Meyer, 2006; Table 2; Fig. 4).

In a comparison of species numbers of selected gastropod families in northern Australia, New Guinea, and Indonesia, Wells (1990) found the highest numbers for New Guinea and the second highest for northern Australia. There may be a sampling bias in such patterns, depending on where authors have performed most of their field research. In a later study, Wells (2002) has included records from Indonesia, which consequently is considered part of the "coral triangle" centre of maximum diversity. Reef-associated species of three well-studied gastropod families have been used in a recent study on marine biodiversity hotspots for establishing geographic conservation priorities (Roberts et al., 2002). In that study, which referred to the coral triangle, the highest concentration of species was found in the Philippines (Fig. 4).

Regarding marine bivalve molluscs, three large regions of over 500 species have been identified, i.e., East Indian Ocean–West Pacific, the Southern Caribbean and the tropical EP (Stehli et al., 1967; Clarke and Crame, 1997). Within the first, there are two smaller adjacent high-diversity hotspots of over 1,000 species, among

*Table 2.* *Terminology for examples of centres of marine diversity used for various taxa in the Indo-West Pacific*

| *Taxon* | *Centre* | *Reference* |
|---|---|---|
| **Mollusca: Gastropoda** | | |
| Cerithiidae: *Cerithium* | Philippines–Indonesia–East Australia | Houbrick (1992) |
| Cerithiidae: *Clypeomorus* | Philippines–Indonesia | Houbrick (1985) |
| Conidae | Indo-Malayan region | Kohn (1967, 1985, 1997) |
| Conidae | Indo-West Pacific island arc | Vallejo (2005) |
| Cypraeidae | Philippines–South Japan | Foin (1976) |
| Cypraeidae | Philippines | Meyer and Paulay (in preperation) |
| Haliotidae | Central Indo-Pacific | Geiger and Groves (1999) |
| Haliotidae | Indo-Malayan area | Geiger (2000) |
| Littorinidae: *Littoraria* | West Sunda shelf | Reid (1986, 2001) |
| Muricidae: *Murex-Haustellum* | South Japan–Southeast Asia–Phillipines, Indonesia–Papua New Guinea | Shuto (1983), Ponder and Vokes (1988) |
| Olividae: *Oliva* | Philippines–East Indonesia–Papua New Guinea–Solomons | Tursch and Greifeneder (2001) |
| Strombidae: *Strombus* | Philippines–Central Indonesia–New Guinea | Abbott (1960) |
| Strombidae | Philippines–East Indonesia | Roy et al. (2001a) |
| **Mollusca: Bivalvia** | Indonesia–Philippines | Crame (2000) |
| **Sipuncula** | Indo-Malay Archipelago | Murina (1975) |
| **Crustacea: Cirripedia** | | |
| Coral-associated barnacles | Malaysian Triangle | Ross and Newman (2002) |
| **Crustacea: Decapoda** | | |
| Pontoniinae | Indo-Philippine region | De Grave (2001) |
| Pontoniinae | Indo-Malayan region | Fransen chapter 4, this publication |
| **Chordata: Pisces** | | |
| Marine fishes | East Indian region | Randall (1998) |
| Reef fishes | East Indonesia–South Philippines | Allen (2002) |
| Reef fishes | Philippines–South China Sea–Indonesia triangle | Kulbicki et al. (2004) |
| Marine shore fishes | Central Philippines, Malacca Strait | Carpenter and Springer (2005) |
| Chaetodontidae | South Philippines, Northwest Borneo, East Indonesia, and Micronesia | Roberts et al. (1998) |
| Siganidae | Andaman Sea to the Solomon Islands, and to southern Japan | Woodland (1983) |

which "Indonesia-the Philippines" is the richest and most clearly associated with coral reefs (Crame, 2000).

The effect of sampling effort on species numbers has been demonstrated by Bouchet et al. (2002), who found extremely high numbers of mollusc species at New Caledonia after intensive collecting. Nearly 30% of these were only represented by empty shells, which usually represent animals that live hidden as endo- or ectoparasite in association with a host species. Systematic checks of potential host species on the presence of parasitic molluscs usually result in the discovery of many cryptic species, which demonstrates that our perception of species richness strongly depends on sampling strategy (e.g., Gittenberger et al., 2000; Massin, 2000; Goud and Hoeksema, 2001; Hoeksema and Kleemann, 2002; Kleemann and Hoeksema, 2002; Gittenberger, 2003; Massin and Dupont, 2003; Gittenberger and Gittenberger, 2005). This also implies that many invertebrate species that do not leave behind empty shells or other calcareous body parts will remain unnoticed and their numbers underestimated (Bouchet et al., 2002). The effect of increasing sampling effort in time has been demonstrated on opisthobranch molluscs in Papua New Guinea, where each survey has added new records to the growing list of species (Gosliner, 1993; Gosliner and Draheim, 1996). Consequently, more species have been documented from Papua New Guinea than from the Philippines, but unpublished data suggests that species numbers are highest in Indonesia (Gosliner, 2002).

Vermeij (1973) noticed a peak in species richness observed among mangrove gastropod assemblages in the Indo-Malayan region. Reid (1986, 2001) indicated a more specific centre of diversity within the CI-P among mangrove-associated littorinid molluscs of the genus *Littoraria* (Fig. 4; Table 2). He considers this to be correlated with high mangrove habitat diversity on the shores of continental land masses. Large land masses possess longer coastlines than oceanic islands, which may contain more habitat variation, such as environmental gradients: inshore–offshore perpendicular to the coastline and alongshore in between river mouths. This habitat dependence becomes more apparent when the diversity pattern in *Littoraria* is compared with the distribution of the littorinid genus *Echinolittorina*, whose species are confined to rocky substrates and whose diversity is highest to East and South of the Sunda Shelf (Williams and Reid, 2004).

In conclusion, although many molluscan taxa show a diversity maximum in the Indo-Malayan region (Table 2), differences in IWP distribution maxima show that there is a clear dependency on habitat diversity. The diversity patterns of reef-dwelling molluscs may differ from those of mangrove-associated molluscs, whereas others taxa may occur in both kinds of environments. Many molluscan taxa are found on coral reefs, where much more habitat heterogeneity occurs (due to depth gradients and variable distance offshore) than in the shallow mangroves fringing the coastlines. Especially in parasitic snails, the observed numbers also depend on a systematic sampling strategy, in which knowledge on the host taxa is required, and on sampling intensity as demonstrated in all-species inventories (Bouchet, 1997; Bouchet et al., 2002).

### 4.4. Arthropods: Aquatic Insects and Crustaceans

Semi-aquatic bugs (Heteroptera, Gerromorpha) are a generally much overlooked group of animals in studies related to marine biodiversity. On the other hand, very few entomologists have recognized the Indo-Pacific as a biogeographical region (Andersen, 1991). Water striders and sea skaters in particular are believed to have originated in estuaries and mangrove swamps and to have diversified further toward intertidal reef flats and the open ocean surface (Andersen, 1991, 1999; Andersen et al., 2000). They are wingless and therefore depend on surface winds and currents for their dispersal. All of the more than 170 species of these marine insects are known to occur in subtropical and tropical water. By linking the geographical areas for these bugs with those of mangroves, Andersen (1999) reports the highest number of species from Asia with the Indo-Malayan Archipelago (77), and the second highest from the West Pacific (68). In the total IWP, the number of species occurring in mangroves and estuaries (79) is double the amount of species living on coral reef flats (38), whereas outside the IWP only one reef-dwelling species has been recorded in contrast to 36 species in mangroves and estuaries. In other words, the high concentration of IWP species is related to a high number of species in estuaries and mangroves, but even more so by coral reefs acting as an additional habitat.

The subfamily Pontoniinae (Decapoda) is among the most diverse shrimp taxa associated with Indo-Pacific coral reefs, only outnumbered by the Alpheidae (De Grave, 2001). The highest diversity is found in the Indo-Philippine or the Indo-Malayan region (Table 2), where most range overlap occurs (De Grave, 2001; Fransen, chapter 4, this publication). Its species show highly specific ecological requirements by being associated with host species that belong to various kinds of reef-dwelling sessile invertebrates (De Grave, 2001; Fransen, chapter 4, this publication). Hence, the ranges and diversity patterns shown by the shrimps depend completely on those of their respective hosts (Fransen, 2002). Similarly, the alpheid (snapping) shrimp genus *Alpheus* consisting of more than 100 species that contribute to reef communities, also shows a wide range of reef-dwelling invertebrate hosts, especially sponges and crinoids (VandenSpiegel et al., 1998; Didderen et al., 2006).

Amphipod crustaceans have been proposed as a biodiversity indicator group for species-rich tropical ecosystems such as coral reefs (Thomas, 1993; Myers, 1997). As an example, the Madang Lagoon is mentioned at the north coast of Papua New Guinea, which appears to have exceptionally high numbers of amphipods and other organisms (Thomas, 1997). Comparison of species richness of reef-dwelling isopods has started but too few sites have been studied so far in order to find a distinct biogeographical diversity pattern (Kensley, 1998).

In a study on coral-associated barnacles, it has been argued that although the centre of species richness was once thought to be the Malaysian Triangle, this centre has become less distinct since research on these animals shifted to areas outside this traditional centre (Ross and Newman, 2002). However, since coral-associated barnacles may be extremely host-specific (Achituv and Hoeksema, 2003) their diversity centre is expected to be linked to that of corals.

## 4.5. Fishes

Reef fishes are spectacular animals with regard to their variation in shape, colour, and behaviour. Since they have commercial value in the aquarium business their diversity is important to fish collectors. Areas famous for reef fish diversity also attract diving tourists and underwater photographers. The richest marine fish fauna of the world is found in the "East Indian region" (eastern Indonesia, New Guinea, and the Philippines), which is related to a high habitat diversity (Randall, 1998). This Indo-Malayan area has also been highlighted by Allen (2002), who considers eastern Indonesia–southern Philippines as the most likely diversity centre (Table 2; Fig. 5). This centre has also been indicated by Roberts et al. (2002) and a slightly larger one by Mora et al. (2003), who use species counts of 13 fish families (Fig. 5). In contrast, Carpenter and Springer (2005) indicate two peaks for marine shore fishes that are more to the north and west (Table 2).

An east Indonesian diversity centre, based on over 40% of the species of coral reef fishes, is indicated by Roberts et al. (1998), in a study on the westernmost tip of New Guinea. They presented a map of butterfly fishes (Chaetodontidae) indicating a centre extending eastward from the Indo-Malayan region (Table 2). Another family of tropical IWP fishes, the Siganidae, is represented by a larger diversity centre (Woodland, 1983; Table 2). The special biogeographic position of Indonesia within this centre is also emphasized by a high concentration of endemic and rare fish species (Erdmann et al., 1998; Forey, 1998; Randall, 1998; Allen, 2002; Roberts et al., 2002; Mora et al., 2003).

The northernmost boundary of the diversity centre of shore fish indicated by Allen (2002; see Fig. 5) is concordant with the boundary between the Indo-Malayan tropical subregion and the Sino-Japanese subtropical subregion, based on the 20°C winter surface isotherm (Qiyong and Yazhi, 1986). The range limits of species restricted to tectonic crustal plates, so-called plate endemics, have received special attention in fish biogeography (Springer, 1982; Myers, 1989; Springer and Williams, 1990). Species ranges may be confined to tectonic plates. When these plates move to or from each other, so do their faunas, which may cause them to

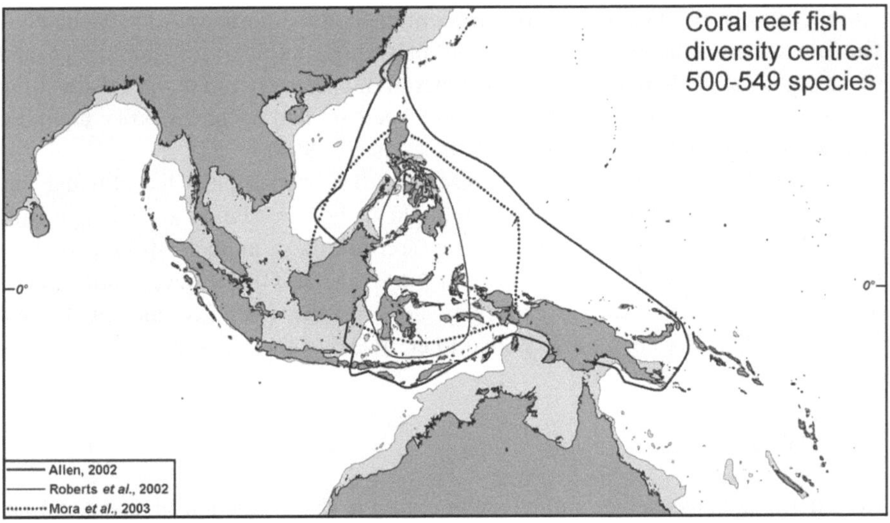


Fig. 5 Centres of diversity for coral reef fishes (from Allen, 2002; Roberts et al., 2002; Mora et al., 2003).

join with each other or to split up from each other. The collision of Australia with Southeast Asia may have caused an influx of West Pacific species in the latter area, increasing the species number of the Indo-Malayan fish fauna (Woodland, 1986).

## 4.6. Larger Benthic Foraminifera

Larger Foraminifera do not constitute a taxonomic unit but they form a functional group of tropical, symbiont-bearing, shallow-living, benthic marine unicellular organisms (Renema, 2002, 2006). Most species of larger Foraminifera that harbour algal symbionts live in the proximity of coral reefs or other carbonate depositional environments. The highest concentration of species is found in the Indo-Malayan area (Hallock, 1988), where a core area has been distinguished from southern Japan to the Sahul shelf, including the Philippines and most of the Indonesian Archipelago (Langer and Hottinger, 2000). At the generic level, a modelled diversity centre extends from Borneo to the northern coast of New Guinea (Belasky, 1996). In conclusion, the centre of diversity for larger benthic diversity may not only include the Philippines, eastern Indonesia, and Japan (Renema, 2002) but probably also the northern coast of New Guinea.

## 4.7. Stony Corals: Scleractinians, Hydrocorals, and Octocorals

The patterns of reef coral diversity have been a subject for study in tropical marine biogeography by biologists and geologists for some decades already. The earliest analyses concentrated on generic diversity patterns due to the initial lack of species records from a sufficient number of localities. This lack of sufficient data was partly related to taxonomic errors and uncertainties (cf. Jokiel and Martinelli, 1992; Sheppard, 1998). Maps indicating areas of similar generic diversity, which are outlined by "isopleths" or indicated by symbols, usually indicate a focus of maximum diversity in the CI-P (Wells, 1954, 1966; Rosen, 1984, 1988c; Best et al., 1989; Veron, 1985, 1986, 1993–1995; Hoeksema, 1993b, 1997a; Fraser and Currie, 1996; Wilson and Rosen, 1998; Veron, 2000), occasionally with a second centre in the western Indian ocean (Rosen, 1971; Stehli and Wells, 1971; Coudray and Montaggioni, 1982), or a combined West Indian Ocean–Indo-Malayan centre (Belasky, 1992, 1996), and a third centre in the Red Sea (McCoy and Heck, 1976; Schuhmacher, 1976, 1988). The former secondary centres of generic diversity in the western Indian Ocean resulted from expeditions and surveys in that region around 1980, but dissolved after subsequent research on coral diversity in the West Pacific (e.g., Veron, 1986; Best et al., 1989). Hence, the latest opinions on reef dwelling Scleractinia concerns a single major centre of generic diversity of 80–90 genera in the Indo-Pacific convergence, including parts of Southeast Asia and the West Pacific (Paulay, 1997; Karlson, 1999; Veron, 2000). For comparison, the highest concentration of Caribbean coral reef genera is less than 25 (McCoy and Heck, 1976; Porter and Tougas, 2001).

In 1995, worldwide scleractinian reef coral species diversity patterns started to become published next to patterns at genus level (Fig. 6A, B). Veron's (1995;

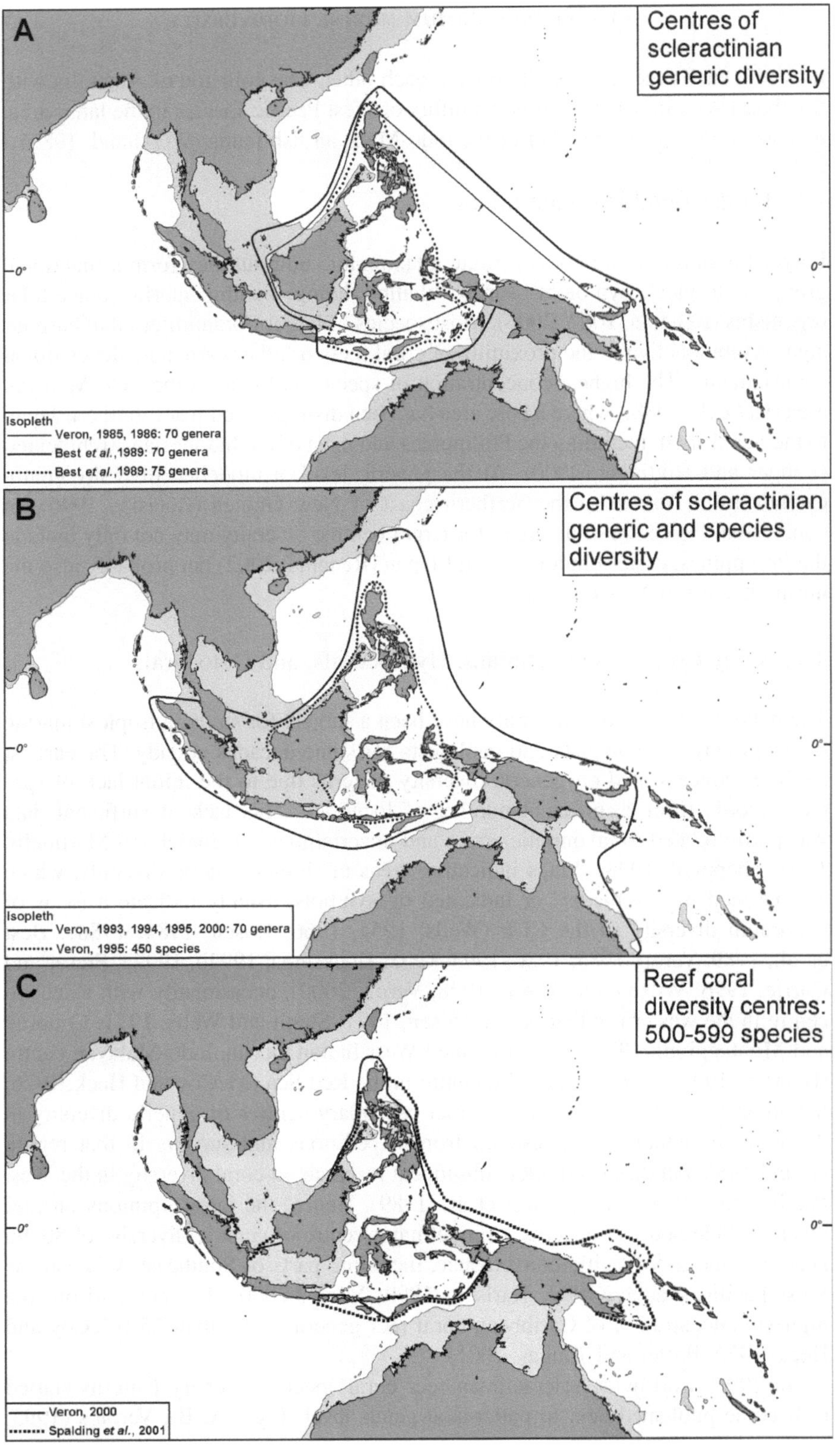

A
Centres of scleractinian generic diversity
0°
0°
Isopleth
Veron, 1985, 1986: 70 genera
Best et al.,1989: 70 genera
Best et al.,1989: 75 genera
B
Centres of scleractinian generic and species diversity
0°
0°
Isopleth
Veron, 1993, 1994, 1995, 2000: 70 genera
Veron, 1995: 450 species
C
Reef coral diversity centres: 500-599 species
0°
0°
Veron, 2000
Spalding et al., 2001

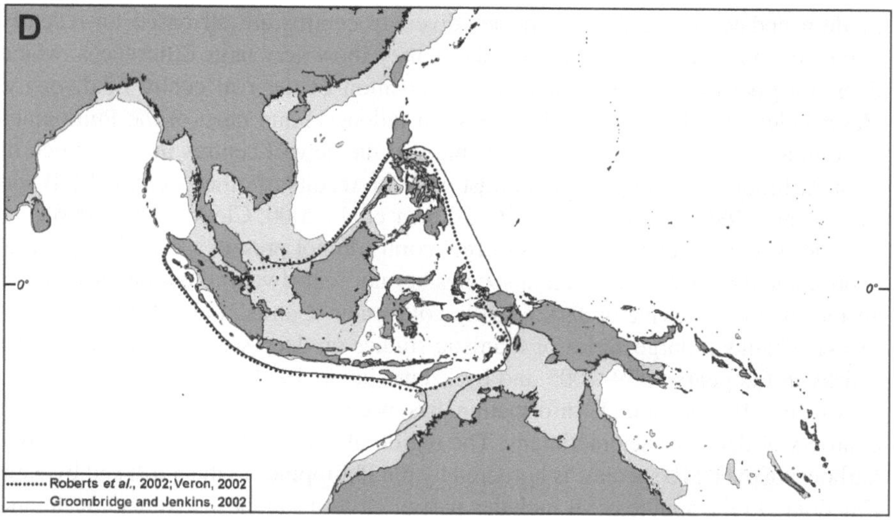


Fig. 6 Various centres of diversity for reef coral genera and species after (**A**) Veron (1985, 1986), Best et al. (1989); (**B**) Veron (1993, 1994, 1995, 2000). Various centres of diversity for reef coral species from; (**C**) Veron (2000), Spalding et al. (2001); (**D**) Roberts et al. (2002), Veron (2002), Groombridge and Jenkins (2002).

Fig. 5) overview of coral species diversity contours does not represent real species counts but is a reconstruction based on multiplying generic distribution diversity data by the species numbers per genus (e.g., Veron and Minchin, 1992; Veron, 1993). Therefore, his areas of similar species diversity reflect very much areas of similar generic diversity. He added a centre of maximum coral diversity (contour $n = 450$ species), which contains Borneo, the Philippines, and much of Wallacea (i.e., the area between the Sunda and the Sahul shelves. This centre of scleractinian reef coral diversity is very similar to the centre of reef coral genera presented by Best et al. (1989) (Fig. 6A, B).

Successive reef coral surveys and analyses resulted in various additional centres of reef coral species diversity (Fig. 6), which show a remarkable variation to the west and east ($n = 500$–599 species). Veron's (2000) latest centre differs from his earlier one (Veron, 1995) by the inclusion of the Java Sea and the westernmost point of New Guinea (Fig. 6C). Veron's (2000) maps on species diversity of some selected scleractinian families and genera show very similarly shaped centres with an inclusion of the Java Sea, such as the Acroporidae ($n > 150$), *Montipora* ($n > 50$), *Acropora* ($n > 100$), Faviidae ($n > 75$), the Poritidae ($n > 50$), *Porites* ($n > 20$), and also for numbers of regional endemic species ($n = 31$). The scleractinian species diversity centre presented by Spalding et al. (2001) shows similar contours to Veron's (2000) but with a major extension eastward, including the northern part of New Guinea and the Bismarck Sea (Fig. 6C). The centres presented by Roberts et al. (2002) and Groombridge and Jenkins (2002) both differ from Veron's (1995) by the exclusion of northern Luzon (Philippines) and the inclusion of Sumatra and the latter also by the inclusion of Western Australia (Fig. 6D). Although these modern updates on reef coral species diversity centres are all based on recently obtained data (maybe even the same data), they show very large differences, which does not give much more insight in the position of the real centre of diversity. Nevertheless, all these centres have eastern Indonesia and most of the Philippines in common (Fig. 6). It is also remarkable that the newest centres include much of western Indonesia, which is in contrast with the results of other studies (Moll and Suharsono, 1986; Van Woesik, 1996; Edinger et al., 2000; Cleary et al., 2006).

The species diversity contours of reef corals do not present a complete picture. More than 40% of the scleractian species are not reef-dwelling but deep-sea coral species that can survive in dark, usually cold water (Cairns et al., 1999; Cairns, in press). Thanks to large series of faunistic studies on deep-sea corals published by Cairns in the years 1979–2005 and by Zibrowius in 1974–1997 (e.g., Cairns and Zibrowius, 1997), enough information has been gathered for plotting diversity contours of deep-sea scleractinians. The region of highest diversity, with localities harbouring over 110 species, is bordered by the Philippines to the north and by New Caledonia to the southeast; it includes part of eastern Indonesia, New Guinea, and northeastern Australia (Cairns, in press).

Like scleractinian deep-water corals, stylasterid corals (Hydrozoa: Stylasteridae) are not restricted to shallow environments in the photic zone (Cairns et al., 1999; Lindner et al., 2004). Most of the 250 species live in deep water and some of them do not occur in warm water and can even be found in Antarctica (Cairns, 1992; Cairns et al., 1999). In comparison, shallow-water stylasterids appear less well known at species level than their deep-water counterparts and new species are still being discovered (Cairns and Hoeksema, 1998; Lindner et al., 2004). Although Boschma (1957) concluded that the most diverse faunas of Pacific Stylasteridae occur in the East Indies, off Japan, and around the Aleuthian islands, he did not discuss which factors would affect the distribution of these corals. According to Cairns (1992), stylasterids do not occur near continental landmasses and show a predominant insular distribution pattern. He attributes this to the absence of suspended sediment and fluctuating salinities otherwise found near river outlets and to the presence of vertical hard substrates where they can prosper as slow-growing, fragile animals. Despite the presence of large landmasses, the Indo-Malayan region also harbours offshore insular habitats, like atolls, seamounts, and submarine ridges. Here, these marine environments are usually sufficiently remote from the influence of terrestrial runoff and therefore the occurrence of stylasterids here does not contradict the explanation given by Cairns (1992).

Fire corals (Milleporidae) are reef-dwelling hydrocorals that need sunlight for growth. Since these stony corals are generally not included in reef coral surveys and faunistic reef coral reports, they are usually also excluded from biodiversity studies. The number of species may only be about 10–12, with about twice as many species in the Indo-Pacific than in the Atlantic (Boschma, 1948; Cairns et al., 1999; Razak and Hoeksema, 2003). Most Indo-Pacific species appear to have been reported from eastern Indonesia, only six in total, including a species that previously was supposed to be endemic to the EP (Razak and Hoeksema, 2003). Therefore, addition of fire corals to the faunistic inventories of stony corals will not significantly affect any differences in total numbers of stony corals between areas.

Only two octocoral species are known to produce massive carbonate skeletons on IWP coral reefs, whereas the others are either soft or rigid, depending on the concentration and the sizes of their sclerites, i.e., carbonate particles imbedded in the coral's soft tissue (Fabricius and Alderslade, 2001; Hoeksema and Van Ofwegen, 2004). Without massive skeleton, specimens need to be stored in ethanol, whereas stony corals only need to be kept dry for examination. Since taxonomic research of octocoral sclerites is specialistic and time-consuming work, this impedes the comparison of species numbers for obtaining good overviews of species diversity patterns in octocorals. Like in stony corals, the identification of specimens of closely related species is difficult when species show much intraspecific variability. Octocoral biodiversity patterns that have been published are far from conclusive due to the poor taxonomy (Williams, 1993; Van Ofwegen, 2002). By use of selected model taxa in the study of biogeographic diversity patterns, fewer species are necessary and therefore it will become less difficult to obtain complete data sets that will enable further analysis.

## 4.8. A Model Taxon: The Mushroom Coral Family (Scleractinia, Fungiidae)

A major risk of applying data sets including all scleractinian species in diversity analyses is the incompleteness of data for many areas. An alternative approach in investigating species diversity patterns is by only using selected species groups, for which taxonomic revisions based on museum collections and thorough field surveys have been completed, such as the scleractinian mushroom coral family Fungiidae ($n = 45$; Hoeksema, 1989, 1993b, 1997b; Hoeksema and Moka, 1989; Hoeksema and Dai, 1991; Hoeksema and Putra, 2002) or the staghorn coral genus *Acropora* ($n > 120$; Wallace et al., 1991, 2002; Wallace, 1997, 1999a, b, 2002; Wallace and Wolstenholme, 1998; Wallace and Rosen, 2006). By setting up sampling programmes using model taxa as target groups (Kohn, 1985, 1997) reliable data sets can be obtained on species presence or absence. Per area surveyed, data gathering should cover the complete habitat range of each model taxon to prevent sampling bias. Cumulative species curves based on numbers of species observed per unit area or unit sampling effort should indicate whether sampling has been sufficient for each area included. By concentrating monitoring surveys in and around the supposed diversity centre, and by using various model taxa, more precise diversity centres can be reconstructed than using many species, because also reliable species absence can be obtained (Hoeksema, 1997b; Hoeksema and Putra, 2002).

A taxonomic revision of mushroom corals predominantly based on museum specimens initially showed a concentration of species in the area comprising Indonesia, the Philippines, and northern New Guinea (Fig. 7). The selection of sampling areas has only been based on topography, with units just consisting of island clusters and portions of continents with no distinct biogeographical significance. By the addition of detailed presence/absence data and the use of bathymetry, a more reliable centre for mushroom corals has been hypothesized (Fig. 8). In addition to an overall similarity in mushroom coral fauna of the areas investigated

(Hoeksema and Putra, 2002), the shape of this centre is also determined by the position of continental shelf margins. During low sea-level stands these shelf margins represented continental coastlines along which coral reefs occurred (Hoeksema and Putra, 2002; Hoeksema, 2004).

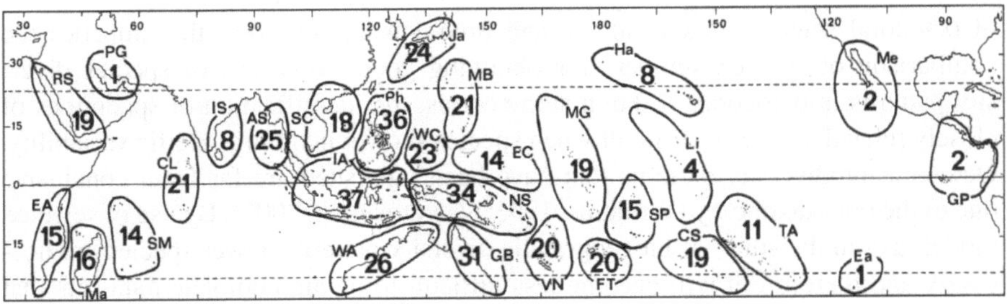


Fig. 7 Pattern of mushroom coral species diversity (Fungiidae) in the entire tropical Indo-Pacific, mainly based on records obtained from museum collections. The species numbers are not up to date and the sample areas are arbitrary (from Hoeksema, 1989).

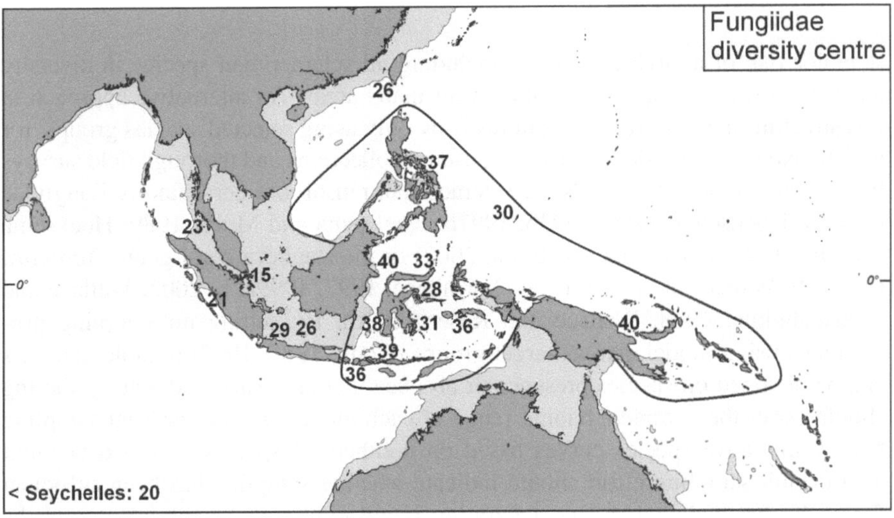


Fig. 8 Centre of mushroom coral diversity. Species numbers are based on field observations. Localities visited outside the centre: Seychelles, West Indian Ocean (20); Phuket, Andaman Sea (23); West Sumatra (21); Singapore (15); Kepulauan Seribu (e.g., Thousand Islands) and Jakarta Bay, Java Sea (29), Karimunjawa, Java Sea (26), Southern Taiwan (26). Localities visited inside the centre: Cebu Strait, Visayas, Philippines (37); Palau, West Pacific (30); Madang, Papua New Guinea, Bismarck Sea (40); Berau Islands, East Kalimantan, Sulawesi Sea (40); Bunaken and Lembeh Strait, North Sulawesi (33); Togian Islands, Tomini Bay, Central Sulawesi (28); Wakatobi Islands (e.g., Tukang Besi Islands), Southwest Sulawesi, Flores Sea (31); Ambon, Moluccas (36); Spermonde Archipelago, Makassar Strait, South Sulawesi (38); Komodo (39); Bali, Lombok Strait (36). Centre outline based on continental shelf contours and island clusters, compare with Fig. 9 (from Hoeksema, 2004).

## 5. Processes Affecting Marine Biodiversity

The processes that determine the diversity patterns of marine taxa are either historical or ecological. Plate tectonics and eustatic sea-level fluctuations are examples of processes that affected species distributions millions and thousand years ago, respectively (Potts, 1984, 1985; Rosen, 1998b; Rosen and Smith, 1988; Veron, 1995; Wilson and Rosen, 1998). The effects of Quaternary sea-level fluctuations on present-day distribution ranges can relatively easily be reconstructed.

For a good understanding of processes regulating species diversity we need to understand how species disperse and which factors restrict their settlement and survival. We may need to know more about oceanic currents but also about the genetic similarity of populations in order to learn how distribution ranges are generated and how they are maintained. In addition, we need to know about the ecological factors that regulate species diversity. Although various hypothetical models have been proposed for explaining how the centre of maximum diversity originated, a combination of various models offers the most satisfying solution as clarification for the position and shape of the most likely centre of maximum marine diversity.

### 5.1. Holocene Recolonization of Southeast Asian Coral Reefs

During the Pleistocene, the alternation of glacials and interglacials caused large-scale eustatic sea-level movements, which had a remarkable effect on the shore lines and coral reefs of the Indo-Malayan region, especially on and along the continental shelves (Chappell and Thom, 1997; Chappell, 1983; Potts, 1983–1985; McManus, 1985; Veron, 1995; Voris, 2000). Due to the effects of changing ice volume and distributions, the sea-level fluctuated world-wide through a range of 100–120 m, with vertical changes of up to 10–20 m per 1,000 years (Chappell, 1983; Masse and Montaggioni, 2001; Siddall et al., 2003), and horizontal shoreline migrations across the broad shelves exceeding far over 100 m per 10 years (Chappell and Thom, 1977). At the last glacial maximum (LGM) (17–18 ka), the sea reached an estimated level 120 m lower than at present (Fig. 9), which exterminated the emerged coral reefs on the continental shelves for several thousand years (Potts, 1983; Meyers, 1989; Voris, 2000). Coral communities had to move downward with the descending sea level along the relatively steep slopes of continental margins and volcanic islands in the open seas (Potts, 1984).

The water exchange between the Pacific and the Indian Ocean was more obstructed than at present due to narrower and shallower sea passages. The oceanic basins in the Indo-Malayan region may also have become more isolated from each other during the low sea-level stands (McManus, 1985; Longhurst, 1998). In addition, the river discharge into these narrow straits and nearly enclosed basins was also much more concentrated because water and silt was carried from much larger continental catchment areas than at present (Moolengraaff, 1922, 1929; Umbgrove, 1929, 1947; Voris, 2000; see Fig. 10). This increase of turbid, hyposaline conditions

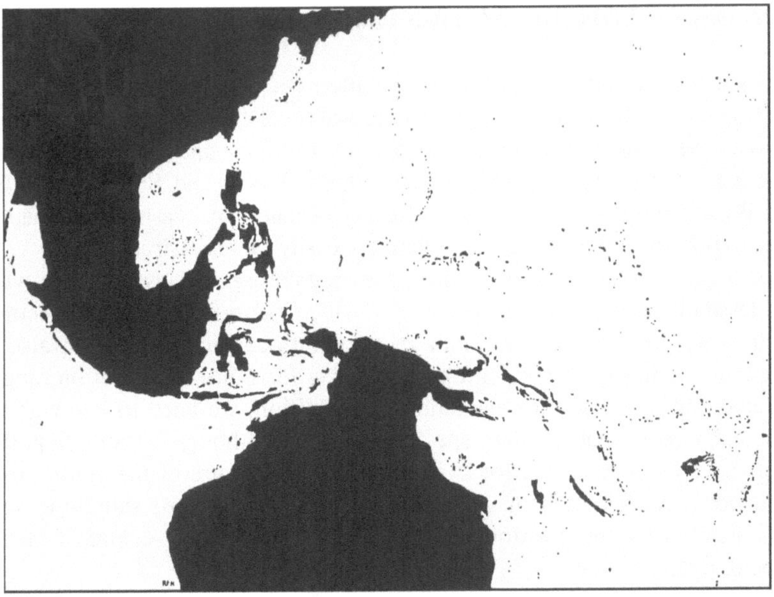

Fig. 9 Australasia and the West Pacific during the lowest sea-level stand of the last glacial maximum (18–17 ka) 120 m lower than now. Modern coral reefs on continental shelves (e.g., Sunda shelf and Sahul shelf) were emerged and dry, with nearest coral reef life remaining along the continental margins. Interoceanic water circulation between the two continents was nearly blocked, which limited gene flow, and possibly caused isolation in oceanic basins between the East Indian Ocean and the West Pacific, and also between the two oceans themselves. Terrestrial run-off from the continents may have lowered salinity in the isolated basins, thus increasing a barrier effect to predominantly offshore species. Present-day shoals had emerged and become oceanic islands; some islands had become larger, and those arranged in island chains on oceanic ridges became connected to each other (reproduced with permission from Myers, 1989).

may have caused additional habitat loss for stenohaline species (Garvine, 1986; Smetacek, 1986; Springer and Williams, 1990). With nearly no shallow seas left, the remaining coral reefs consisted mainly of fringing reefs along the steep slopes of continental margins and volcanic islands (Fig. 9). During the LGM, the surface seawater in, for instance, the Sulu Sea was less saline and 2.3°C±0.5°C cooler than at present (Rosenthal et al., 2003). Furthermore, large areas of dry land may have served as a source for dust that became transported to the sea by wind, after which it became suspended silt, which hindered reef development (Montaggioni, 2005).

Lower seawater temperatures during the last glacial and upwelling of cool water along and between the land masses may also have acted as a barrier blocking the dispersal of species through the Indonesian seaways. This may have split geographic ranges of species that became divided into disconnected, isolated populations, which may have caused subsequent speciation in both pelagic and benthic taxa (Umbgrove, 1947; Fleminger, 1986; Blum, 1989; Springer and Williams, 1990).

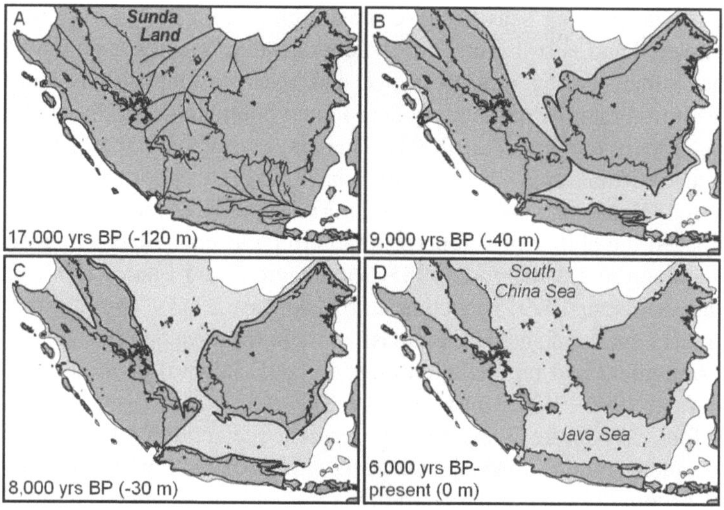


Fig. 10 The Sunda continental shelf during various sea-level stands (from Moolengraaff, 1922; Van Weel, 1923; Umbgrove, 1929, 1947; Tomascik et al., 1997; Voris, 2000). (A) During the last glacial maximum (18–17 ka) with the course of major riverbeds shown. (B) Borneo and Java still connected to Sundaland (southeast Asian mainland) during the post-glacial sea level rise. (C) Borneo and Java disconnected from Sundaland; water circulation between the Java Sea and South China Sea was possible, but the sea water level was still lower than the present-day coral reefs.

At present, the shallow seas of the Sunda and Sahul continental shelves show sparse and scattered development of coral reefs (Moolengraaff, 1922, 1929; Umbgrove 1929, 1947). In comparison to the deep straits and basins of eastern Indonesia, there are hardly any fringing reefs in the shelf seas, i.e., the Java Sea, the southern part of the South China Sea, and the Arafura Sea (Moolengraaff, 1922; Map IV; Spalding et al., 2001). During the latest transgression, the shelf seas expanded very rapidly while they remained rather shallow. This rapid transgression over shallow muddy sea bottoms did not offer ideal conditions for coral reef development. Even now, the river discharge and sediment yield per drainage area is extremely high along the coastlines of these seas (Umbgrove and Verwey, 1929; Milliman and Meade, 1983; Longhurst and Pauly, 1987; Edinger and Browne, 2000). Hence, water turbulence may cause frequent occasions of sediment resuspension in the water, which is unfavourable for reef coral growth. During deglaciation, periods of high-frequency storms may have led to recurrent sediment resuspension and may also have hindered the formation of reef structures due to coral fragmentation (Montaggioni, 2005).

What matters for the biodiversity of these thinly distributed shelf reefs is whether they offer suitable environmental conditions for these reef organisms with planktonic larval dispersal into these seas, and whether they have lasted long enough for species to invade the newly inundated reefs. According to Moolengraaff (1922, 1929), the reefs of the Sunda shelf may be relatively poor in corals compared to those of surrounding seas because their immigration and recruitment has not been completed and is still going on. In general, some species must have migrated faster than others during transgressions, and species compositions and numbers probably shifted rapidly, particularly near faunal boundaries (Hellberg et al., 2001; Roy et al., 2001b; Hughes et al., 2003). This may imply that not all species present in the centre of maximum diversity may have had the opportunity yet to invade the Java Sea (Hoeksema and Putra, 2002). The rapid Holocene sea-level rise may have outpaced the ability of submerged reefs to build upward as so-called give-up reefs (Neumann and Macintyre, 1985; Montaggioni and Faure, 1997; Grigg and Epp, 1989; Karlson, 1999; Blanchon and Blakeway, 2003). This did not concern the previously emerged shelf-based fossil reefs in the Java Sea that served as basis for new settlement and repopulation after a glacial interval of thousands of years (Moolengraaff 1992, 1929; Umbgrove, 1947).

Therefore, it is more likely that the present reef communities on the Sunda shelf are much younger than 10,000 years (Fig. 10). Although most of the area started to become flooded 9–10 ka, the Java Sea and the southern part of the South China Sea were only deep inland bays (Fig. 10B) containing shallow, hyposaline, and turbid water unfit for coral growth. Areas were present-day coral reefs occur became in contact with seawater 8 ka, when the Java Sea and the South China Sea connected with each other and with the Indian Ocean (Fig. 10C). However, the bottom of the Sunda shelf seas was still shallow and only 5–6 ka, the water level reached its present position (Fig. 10D) enabling the reefs to develop with a habitat heterogeneity related to the same depth gradients as at present.

After a rapid sea-level rise from 18 to 6 ka, the sea level started to stabilize, but local differences in the IWP may have occurred due to variation in vertical crustal adjustments related to the ice caps melting at higher latitudes, with stabilizations starting 7.5–2.5 ka (Adey, 1978; Cabioch et al., 1995, 1999a, b; Bard et al., 1996; Montaggioni and Faure, 1997; Ota and Chappell, 1999; Montaggioni, 2000; Masse and Montaggioni, 2001; Yamano, 2002). Of all Holocene sea-level curves presented, one reconstructed for Singapore is geographically the most relevant to the Sunda shelf deglaciation history (Hesp et al., 1998). In this curve, the post-glacial marine transgression reached the present level at 7–6.5 ka, which compares well with similar curves for the Strait of Malacca (Geyh et al., 1979; Tjia and Fuji, 1992; Tomascik et al., 1997).

During the Pleistocene ice ages, reef corals remained in eastern Indonesia, even just along the Sunda shelf (Moolengraaff, 1929; Umbgrove, 1947). Pre-Holocene reef coral settlement along the continental margins and around oceanic islands had to keep pace with the rising sea level. It is not likely that all species present in the centre of maximum diversity were able to migrate upward on the steep slopes of the continental margins during the fast transgression. Since the stabilization of the last 6,500–7,000 years, they may also not have been able to colonize the remote shallow-based shelf reefs that show little habitat heterogeneity. In other words, the boundaries of the centre of maximum diversity have partly been determined by extinction on the continental shelf reefs during the last glaciation and selective recolonization of these continental seas during the last deglaciation.

## 5.2. Oceanic Currents and Dispersal Barriers

Especially for tropical marine bottom-dwelling invertebrates, long-distance transport of pelagic larvae by currents is important for their dispersal (Scheltema and Williams, 1983; Scheltema, 1988; Scheltema and Rice, 1990; Jokiel and Martinelli, 1992; Scheltema et al., 1996; Vermeij, 2004). This explains why some species of coral (e.g., *Fungia* (*Cycloseris*) *curvata*, Hoeksema, 1989; and *F.* (*C.*) *distorta*, Michelin, 1842) and coral-associated organisms show pan-Indo-Pacific ranges from the eastern coast of Africa to the west coast of Central America (Ekman, 1953; Hoeksema, 1989; Lessios et al., 2003; see also Fig. 7). Although the EP ocean is well known for its high degree of endemism (e.g., Briggs, 1974; Heck and McCoy, 1978; Veron, 1995; Glynn and Ault, 2000; Roberts et al., 2002; Mora and Robertson, 2005), the existence of pan-Indo-Pacific species (Cairns et al., 1999; Bowen et al., 2001), implies that some of these species have been able to cross the East Pacific Barrier (EPB, Scheltema, 1988) or in the case of pantropical species evolved before the closure of the Panama Isthmus. Therefore, for shallow-water, bottom-dwelling species, deep ocean basins are considered porous or semi-permeable barriers. The position and effectiveness of dispersal barriers have changed in time as result of plate tectonics and changing seawater temperatures (Vermeij, 2004). Ocean basins can be large but when the eustatic sea level was very low, volcanic islands could have emerged and acted as stepping stones over bridging long stretches of seas that separated suitable habitats (Rosen, 1984, 1988b; Rosen and Smith, 1988; see also Fig. 9).

Species that show such wide pan-Pacific distributions are exceptional: they are either old species that have had more time to cross the EPB, taking advantage of emerging stepping stones during low sea-level stands, or species that have long larval stages. The duration of the larval stage may vary strongly between and among various taxa, which has great implications for speciation (Paulay and Meyer, 2002, 2006). The average duration for coral reef fish species is 26 days (Brothers and Thresher, 1985) and for reef corals only 7–10 days (Fadlallah, 1983). Among 250 Indo-Pacific species of the gastropod genus *Conus*, species with a greater potential for larval dispersal show the broadest geographic ranges (Perron and Kohn, 1985; Kohn and Perron, 1994).

The distribution ranges within the Indo-Malayan centre are also subject to currents and their outermost range limits may have been determined by how far currents have been able to carry the larvae away from their place of origin (Fig. 11). Because the average sea level is higher at the western equatorial Pacific than in the eastern Indian Ocean, Pacific water flows through the Indo-Malayan region into the Indian Ocean (Wyrtki, 1961; Gordon and Fine, 1996; Ilahude and Gordon, 1996; Gagan et al., 2004). The North Equatorial Current (NEC) and the South Equatorial Current (SEC) carry water from the West Pacific to the Indo-Malayan region (Myers, 1989; Gordon and Fine, 1996). The SEC turns around and becomes part of the Equatorial Counter Current (ECC). The NEC splits into the northward-flowing Kuroshio and the southward Mindanao Current (Nitani, 1972; Qiu et al., 1986). The latter continues into the Indonesian Throughflow (ITF), through the Makassar

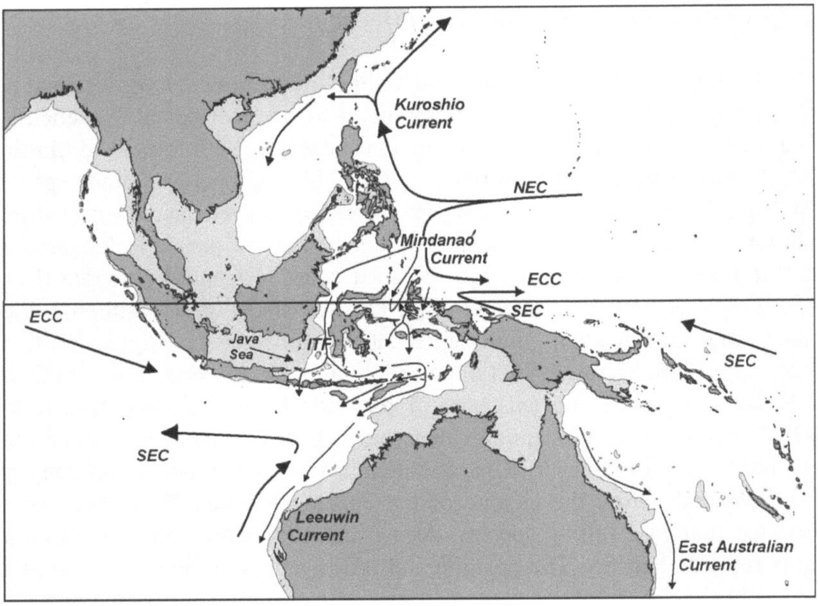


Fig. 11 Main patterns of oceanic and interoceanic currents (after Myers, 1989; Gordon and Fine, 1996; Gordon, 2001).

Strait, to which water from the Java Sea is added (Gordon and Fine, 1986; Gordon et al., 2003).

The current passing through Indonesia is subject to seasonal variation and tidal mixing (Hatayama et al., 1996; Miyama et al., 1996a, b), which may have some scattering effect on larval dispersal. To the south, water from eastern Indonesia flows southward via the Leeuwin Current along the west coast of Australia (Cresswell and Golding, 1980; Cresswell, 1986) where the number of benthic species declines in southward direction with decreasing seawater temperature (Marsh and Marshall, 1983; Veron and Marsh, 1988; Veron, 1995). Meanwhile, part of the water masses carried by the SEC turns southward along the Great Barrier Reef and moves with the East Australian Current where species numbers also decline in southward direction (Mukai, 1993; Veron, 1995). To the north, the warm water of the Kurioshio Current passes east of the Philippines to southern Japan and also into the South China Sea via a side branch south of Taiwan (Fan and Yu, 1981; Qiu et al., 1986; Chen, 1999). Here, species numbers decline in northward direction, also with lower temperatures (Veron, 1995; Chen, 1999). Thus, considering the present inter-oceanic currents, one can conclude that the Indo-Malayan area has a potential of receiving species from the West Pacific, and that from here species may disperse to the north, south, and back to the east again, but not likely to the west. In fact, the ITF acts as a semi-permeable barrier for species dispersal from the Pacific to the Indian Ocean with the exception of Indian Ocean coastlines along eastern Indonesia and Western Australia.

Many coral species cannot live close to river outlets and show a restricted offshore distribution across continental shelves (Done, 1982; Moll, 1983; Hoeksema, 1990). River discharge, which decreases salinity and increases suspended sediment in marine coastal environments, may form a physical barrier by hindering the dispersal of certain species (Potts, 1983). Continental Southeast Asia and the Indo-Malayan archipelago have the highest rates of accumulated sediment discharge into the ocean as compared to other continents (Milliman and Meade, 1983; Longhurst and Pauly, 1987). Salinity is particularly low in the western part of the Indo-Malayan region, i.e., around Borneo, the Malay Peninsula, and northern Sumatra (Fig. 12). Coral reefs occur predominantly outside the low salinity areas, e.g., at West Thailand, West Sumatra, North Java, Sabah (northern Borneo), and Northeast Kalimantan, at the entrance of the Makassar Strait. This implies that the Strait of Malacca, between peninsular Malaysia and Sumatra, may act as a dispersal barrier for many reef-dwelling species. Therefore, the Java Sea and the South China Sea are likely the westernmost part of the border area between the Pacific and the Indian Oceans, with very little input from the Indian Ocean. Another physical barrier may be formed by cold-water upwelling along the Indian Ocean coasts in Indonesia, which may hinder the dispersal of planktonic species intolerant to low temperatures (Fleminger, 1986) and benthic animals with short-lived larvae (Samyn and Tallon, 2005).

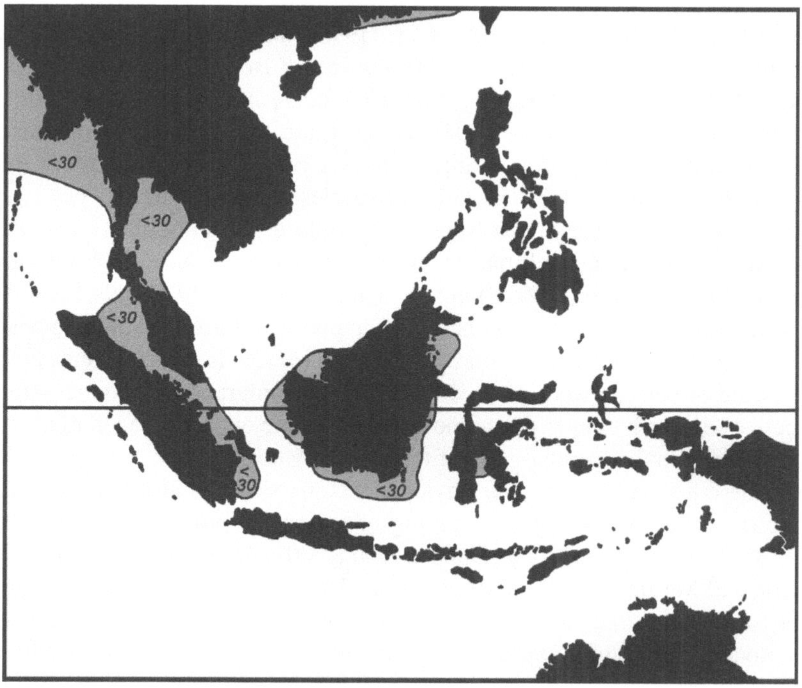


Fig. 12 Coastal areas of low salinity (<30%) due to river discharge (after Van Weel, 1923; Wyrtki, 1961).

## 5.3. Speciation and Connectivity

Since species boundaries depend on the dispersal capacity of species, and the cumulative species range overlaps constitute the Indo-Malayan centre of maximum marine biodiversity, it is important to get insight in species dispersal patterns for finding justifications for the position of such a centre. The way in which populations are linked to each other by the exchange of migrating larvae, recruits, juveniles, and adults, is termed connectivity (Ablan et al., 2002; Burke et al., 2002; Palumbi, 2003). Larval dispersal is difficult to study, because larvae cannot easily be identified at species level and both the direction and distance of larval transport need to be monitored for knowing the connectivity between populations.

Thanks to the application of molecular techniques in biogeographic research, species boundaries, and phylogenetic affinities between populations within species can be studied with more certainty (Palumbi, 1994, 1996; Knowlton, 2000; Vermeij 2004). Molecular techniques have also started to become used in research on marine species with high-dispersal capacities and wide ranges that can be found in the IWP or even the entire Indo-Pacific (Lessios et al., 1998). Hence, studies on genetic flow help to support explanations for species boundaries and the concentration of species in the Indo-Malayan region. The boundaries of separated population ranges give much insight in past and present dispersal barriers. They confirm a past disjunction along the eastern side of the Sunda Shelf, which may be related to the Holocene sea-level fluctuations (McMillan and Palumbi 1995; Palumbi, 1996, 1997; Benzie, 1998, 1999a, b, 2001; Chenowith et al., 1998a, b; Barber et al., 2000, 2002; McCartney et al., 2000; Williams, 2000; Perrin and Borsa, 2001; Chenoweth and Hughes, 2003; Uthicke and Benzie, 2003; Kirkendale and Meyer, 2004; Lourie and Vincent, 2004; Ovendale et al., 2004). These results agree with the importance of the latest sea-level lowstand as an important factor for determining present species ranges and consequently for boundaries of species diversity patterns.

In a marine context, comparisons with Wallace's Line (Collette and Russo, 1984; Woodland, 1986; Helfman et al., 1997; Barber et al., 2000; Lourie and Vincent, 2004; Collette, 2005) may be a bit confusing. Wallace's Line in the Indo-Malayan region is about limits in the separation of terrestrial biota in between the Sunda and the Sahul shelves during low sea-level stands. Species range boundaries of land organisms were extended as a result of migration over emerging land bridges, whereas the ranges of marine species became cut off at the margins of the continents due to extinction on top of the emerging shelves.

On the other hand, molecular phylogenetic analyses can also help to discover cryptic or sibling species and the timing of speciation processes (Lessios et al., 1999; McCartney et al., 2000; Bowen et al., 2001; Dawson and Jacobs, 2001; Dawson and Martin 2001; McCafferty et al., 2002; Landry et al., 2003). If sibling species are discovered, their biogeographic ranges need to be re-evaluated and their biodiversity patterns need to be corrected. Knowing the time of particular speciation events will help to get insight in the development of species ranges, for instance whether the ranges might have been affected by sea-level fluctuations. This has direct impact on species richness as sibling species are able to coexist in the same area, especially if they are co-occurring as parasites on the same host (Gittenberger and Gittenberger, 2005).

Molecular studies of genetic patterns in relation to isolation by distance also have a potential to get insight in the geographic scales on which populations interact, which can be applied in the planning of marine reserves (Barber et al., 2000, 2002; Hellberg et al., 2002; Gerber et al., 2003; Kirkendale and Meyer, 2004). Theoretically, populations are connected by either (1) an "Island model", in which propagules migrate equally between populations within a species range, resulting in a genetically even dispersion pattern, or (2) a "Stepping-stone model", in which mainly nearest neighbour populations are connected and genetic distance is directly related to geographic distance (Palumbi, 2003). In the case of disjunct species ranges, the stepping-stone model may apply and absence of species in a certain area may be related to the unavailability of suitable habitats here. Habitat heterogeneity is important here. If the populations appear well connected despite a disjunction, this may be indicative of the island model and species ranges and diversity have been affected by a recent local extinction. In relation to mushroom coral diversity patterns (Fig. 8), connectivity studies may give insight in why some areas within the centre of diversity show relatively low species numbers, while they are surrounded by areas with higher species numbers.

Direct geographic distance appears to be most important at a large scale. Although wide-ranged Indo-Pacific species show much genetic uniformity, a major distinction is not only seen between IWP and Central-East Pacific clades (Williams et al., 1999; Landry et al., 2003) but also between Indian and Pacific ocean populations within the IWP (Palumbi, 1997; Duke et al., 1998; Williams and Benzie, 1996–1998; Duda and Palumbi, 1999; Benzie 1999a, b, 2000; Andersen et al., 2000; Lessios et al., 2001; McCafferty et al., 2002; Williams et al., 2002; Reid et al., 2006). However, in these biogeographic inter-oceanic comparisons, the closure of the Panama Isthmus as a major structuring factor of the EP tropical fauna should not be overlooked (Bowen et al., 2001; Muss et al., 2001).

At a smaller scale, the position of populations in relation to the flow direction of currents has to be considered as an explanation for distinct differences in genetic variation between populations (Macaranas et al., 1992; Ablan et al., 2002; Ovenden et al., 2004). It can also be used to explain why populations of the Philippines (West Pacific) and Western Australia (East Indian Ocean) are related more closely to each other and those of Fiji (West Pacific) and the Great Barrier Reef (Williams and Benzie, 1997; Benzie, 1999).

The impact of Quaternary ice ages on the genetic structure and distribution ranges of terrestrial species has been well studied (Hewitt, 2000), but less attention has been paid to tropical marine species. Sharp genetic breaks at much closer distance than may be predicted on the base of current flow direction can be related to sea-level fluctuations. Populations may have become separated by the emergence of the Sunda and Sahul continental shelves during the latest low glacial sea-level stand, after which post-glacial recolonization of these shelf areas decreased their geographic distance (Barber et al., 2000, 2002; Perrin and Borsa, 2001; Benzie et al., 2003; Lourie and Vincent, 2004; Collette, 2005;

Lourie et al., 2005; Reid et al., 2006). The course of the present currents (Fig. 11) probably prevents a remixing of these separated populations.

Geographic patterns in genetic diversity suggest that extinction during the last glacial may have caused low genetic variation in peripheral isolated populations with a high extinction rate and that higher diversity remained in the centre of species diversity, where populations had lower risks to become totally extinct (Palumbi, 1996). Genetic differentiation patterns also indicate that recolonization of areas that were emerged during the last glacial period and became submerged during the Holocene marine sea-level rise may explain a relatively high variation in phylogeographical patterns found in a teleost fish species on the Sahul Shelf (Chenoweth and Hughes, 2003). Some authors suggest that genetic differences within species cannot be attributed to Plio-Pleistocene sea-level fluctuations alone and may be of much older origin (Uthicke and Benzie, 2003; Kirkendale and Meyer, 2004). In general, the major population genetic structures detected are likely to be originated from before the last glaciation, with indications for high dispersal of recent origin, which can be attributed to the last sea-level rise and the inundation of the Sunda and Sahul continental shelves, and probably also by easier water transport via various seaways, including the Indonesian Throughflow. Such pulsed dispersal events may also have caused range expansions in previous interglacial periods, but not for all species in the same way, which may explain why species may show differences in gene flow patterns that cannot be directly related to present-day ocean currents (Benzie and Williams, 1995, 1997; Benzie, 1998, 1999a, 2001).

## 5.4. Habitat Heterogeneity and Environmental Constraints

Species diversity on coral reefs strongly depends on habitat heterogeneity (Hoeksema and Moka, 1989; Cornell and Karlson, 1996; Karlson, 1999; Karlson and Cornell, 1999). Similar observations on the importance of habitat diversity concern mangrove-associated fauna (Vermeij, 1973; Reid, 1986, 2001; Ricklefs and Latham, 1993; Hogarth, 1999, 2001). Areas with relatively low species numbers in the middle of species-rich areas usually represent relatively low-habitat heterogeneity, as indicated by mushroom coral distributions (Fig. 8: Togian Islands and Wakatobi Islands). Outside the periphery of a high-diversity centre, areas with low environmental variability may hinder the outward expansion of this centre.

Among areas showing contrasting diversity patterns, poor areas usually lack clear environmental gradients, such as those related to increasing distance from river outlets (onshore–offshore or alongshore). Oceanic island groups without large rivers may for instance lack low-salinity habitats, gently declining reef slopes, or variation in exposure to wave action along their coastlines.

Cross-shelf reef systems present a large variety of environmental gradients, from inshore to oceanic conditions. Together with fringing reefs, they have variation in salinity, nutrients, and turbidity in common; in addition, they show much more variation in depth ranges and wave exposure (depending on increasing shelf depth), and temperature (cold-water upwelling near barrier reefs). Examples of the effect of onshore–offshore gradients on species composition and diversity are known from a small number of shelf areas (Table 3), most of which are shelf-based barrier-reef systems and one representing an island chain (Thousand Islands off Jakarta, northwest Java). Species diversity of many taxa is generally highest in the mid-section of each cross-shelf reef system, where terrigenous impact (river discharge) is less severe than near the shore and oceanic effects (wave action, upwelling) are also weak (Hoeksema, 1990; references therein). Due to a deeper shelf bottom here as compared to close to the shore, the reef slopes have greater depth ranges with more reef surface area and habitat heterogeneity (light penetration, sediment). Distributions of species that prefer nearshore conditions may overlap with those that prefer offshore conditions (references in Table 3). Consequently, entire cross-shelf reef systems contain a lot of habitat diversity along gradients perpendicular to the coastline. The total reef surface area along such gradients with increasing distance offshore accumulates a rising habitat heterogeneity, which reflects a species–area relationship (Williamson, 1988; Ricklefs and Schluter, 1993, Rosenzweig, 1995; Neigel, 2003). In contrast, fringing reefs along coastlines and reefs that are arranged in rows parallel to the coastline, such as in relatively small barrier reefs systems (Spermonde Archipelago, Madang Lagoon), show less variation in habitats (references in Table 3).

*Table 3. Indo-West Pacific areas and taxa studied for onshore–offshore gradients*

| *Area/Taxon* | *References* |
|---|---|
| **Thousand Islands, Northwest Java, Indonesia** | |
| Scleractinia | Moll and Suharsono (1986); Cleary et al. (2006) |
| **Spermonde Archipelago, Southwest Sulawesi, Indonesia** | |
| Scleractinia | Moll (1983) |
| Scleractinia: Fungiidae | Hoeksema (1990), Cleary et al. (2005), Becking et al. (2006) |
| Porifera | De Voogd et al. (1999, 2006), Cleary et al. (2005), Becking et al. (2006) |
| Echinodermata: Echinoidea | De Beer (1990a, b), Cleary et al. (2005), Becking et al. (2006) |
| Seagrasses, Macroalga | Verheij and Erftemeier (1993) |
| Larger benthic Foramifera | Renema et al. (2001), Renema and Troelstra (2001) Cleary et al. (2005), Becking et al. (2006) |
| **Berau Archipelago, East Kalimantan, Indonesia** | |
| Larger benthic Foramifera | Renema (2006) |
| **Madang Lagoon, Bismarck Sea, Papua New Guinea** | |
| Scleractinia: Fungiidae | Hoeksema (1993a) |
| Larger benthic Foramifera | Langer and Lipps (2003) |
| **Great Barrier Reef, Eastern Australia** | |
| Scleractinia | Done (1982, 1983) |
| Alcyonacaea | Dinesen (1983) |
| Porifera | Wilkinson and Trott (1985), Wilkinson and Cheshire (1989, 1990) |
| Reef fishes | Williams (1982), Williams and Hatcher (1983) |
| Copepods and fish larvae | Willams et al. (1988) |

Parasite–host, predator–prey, and other interspecific relationships between organisms form additional biotic factors increasing species diversity. These biotic habitat constraints are clearly demonstrated by families of gastropod and bivalve molluscs and crustaceans (see sections 4.3 and 4.4). Many marine parasites and commensals are very host-specific. However, a single host species may harbour several different species of associated organisms, e.g., species of mushroom corals that offer housing to gastropods (Gittenberger et al., 2000; Massin, 2000; Massin and Dupont, 2003, Gittenberger and Gittenberger, 2005) and mytilid bivalves (Hoeksema and Achituv, 1993; Hoeksema and Kleemann, 2002; Kleemann and Hoeksema, 2002). Many species of parasitic snails remain undiscovered until a systematic survey of their host species is performed (Bouchet, 1997; Bouchet et al., 2002; Schiaparelli et al., 2005). Consequently, these specific associations result in a centre of coral diversity that involves many other reef-dwelling organisms depending on corals. Therefore, the name "Coral Triangle" is correct with regard to its relation to corals.

## 5.5. Models for the Development of Marine Biodiversity Centres

The tropical Indo-Pacific biota are characterized by a high species diversity, which includes many species with wide distribution ranges (Kohn, 1983, 1990). Although species with widespread IWP distributions occur in the Indo-Malayan centre of diversity and form part of the cumulative range overlaps that constitute this centre, they do not determine the centre's boundaries. The centre outlines are determined by species with range boundaries restricted inside or along the centre's limits. By knowing the causes that have restricted these boundaries, the processes that have shaped the centre's position and outline are also understood much better. There are at least five theoretical mechanisms that are used as explanatory models for the centre of diversity, although most authors only emphasize some of them (Rosen, 1981, 1988c; Potts, 1985; Jokiel and Martinelli, 1992; Paulay, 1997; Wilson and Rosen, 1998; Briggs, 2005b, 2006; Mironov, 2006):

(1) Centre of speciation. Many authors consider the centre of diversity to be a centre of origin, with speciation occurring inside the centre with successive outward dispersal to surrounding areas (Van Steenis, 1962; Briggs, 1981, 1992, 1999, 2003, 2004b, 2005b, 2006). Successive periods of glaciations and low sea-level stands caused the emergence of barriers and the isolation of populations in deep-sea basins in between the island arcs of Indonesia and the Philippines (McManus, 1985; Vallejo, 2001, 2005). However, this does not explain the high species diversity demonstrated by many taxa along the northern coastline of New Guinea. Many Recent coral reef species have fossil records dating from periods before the glaciations, hampering the detection of speciation patterns

(Hoeksema, 1989). Molecular clock reconstructions in other marine taxa also indicate that the role of Pleistocene sea level fluctuations may be less important as a speciation mechanism than expected (Uthicke and Benzie, 2003; Kirkendale and Meyer, 2004; Bellwood et al., 2004; Barber and Bellwood, 2005).

(2) Centre of overlap. The centre of diversity consists of overlapping distribution ranges that extend into either the Pacific or the Indian Ocean, resulting from either larval dispersal or ancient plate tectonics (Santini and Winterbottom, 1992; Wallace, 1997). Overlapping ranges among sister taxa as demonstrated in phylogenetic studies would be most suitable for testing this model (Barber and Bellwood, 2005; Lourie et al., 2005).

(3) Centre of accumulation. Speciation has occurred outside the periphery of the centre (fringe isolation) after which inwardly directed convergent dispersal has caused accumulation of the ranges inside the centre (Ladd, 1960; Kay, 1984, 1990; Rosen, 1984, 1988a, b, c; Jokiel and Martinelli, 1992; Pandolfi, 1992; Landry et al., 2003). Due to the large size of the Indo-Pacific more species evolved here (Rosen, 1981) and during glaciations a smaller proportion became extinct than in the Atlantic (Mironov, 2006).

(4) Centre of survival. Speciation may have occurred anywhere. The centre of diversity is just an area of survival with species extinctions outside the centre's boundaries. Areas that have maintained high levels of habitat heterogeneity during Cenozoic low sea-level stands, may have acted as refugia, especially in the West Pacific (Paulay, 1990, 1996, 1997). An important condition in this model is the maintenance of habitat diversity for many species and the availability of sufficient abundance for each species (Barber and Bellwood, 2005).

(5) Centre of mid-domain overlaps. This model is a variety of the centre of overlap (2). A recent hypothesis that has been proposed to explain the position of the centre of Indo-Pacific coral reef biodiversity is the theory of the mid-domain effect (MDE). A MDE stands for a maximum in species richness in the middle of a geographical area that has been formed by overlaying randomized distributions of the locations of individual geographical ranges within species groups (Connolly et al., 2003; Bellwood et al., 2005). This peak is predicted where the probability of maximum species range overlaps is highest. Deviations may have resulted from ocean currents and historical processes. It is affected by the disproportional abundance of suitable coral reef habitats in the Indo-Malayan area (Bellwood and Hughes, 2001; Bellwood et al., 2005). In order to test whether local extinctions during the LGM have had long-lasting effect on the present position of the centre of maximum marine biodiversity, it might be interesting to reconstruct a MDE exclusively for reef areas outside the large Sunda and Sahul shelves.

In the case of the mushroom coral diversity patterns (Figs. 7–8), an analysis of individual species ranges will be presented separately (Hoeksema, in preperation). The ranges of most predominantly West Pacific mushroom coral species show a westernmost boundary at the Sunda Shelf. The presence of this boundary suggests either (1) local extinction of such species to the west of this boundary during ice ages; or (2) a dispersal limit to species that cannot tolerate low salinity and high turbidity in shelf seas surrounded by continental landmasses with much river discharge, preferring oceanic conditions instead. In the first case, the centre of diversity would be a centre of survival (model 4). Since many mushroom coral species are widespread, showing an IWP distribution, and an age of 2–5 My (Hoeksema, 1989), detection of a centre of speciation (model 1) is not feasible. A mid-domain overlapping (model 5) is also not obvious due to the large proportion of species with an IWP distribution range. The presence of much coral reef area in the Central IWP, with exclusion of reefs on top of continental shelves, might be important for maintaining high species diversity in the Coral Triangle of maximum marine biodiversity. With regard to mushroom corals, the Coral Triangle consists for about 50% of IWP range species and for 30% of species with predominantly West Pacific ranges. Although direct indications for a centre of overlap (model 2) or a centre of accumulation (model 3) cannot be discerned, either one or both may be applicable, but due to the Holocene sea-level fluctuations and local extinctions, it will be hard to find direct evidence for any model. Perhaps molecular studies may help to discover past speciation events and dispersal patterns.

## 6. Conclusions

One model is not enough to explain the centre of maximum diversity centre (Wilson and Rosen, 1998; Fransen, 2006). Habitat heterogeneity as an ecological variable in coral reef areas may be an additional factor that has to be taken into account as a condition related to biodiversity patterns (Hallock, 1988; Cornell and Karlson, 1996; Paulay, 1997; Kohn, 1997; Taylor, 1997; Karlson and Cornell, 1999; Renema and Troelstra, 2001; Renema, 2002).

Based on most diversity patterns, continental shelf bathymetry, oceanic currents and phylogeographic patterns, an Indo-Malayan centre of benthic biodiversity extending from East Indonesia to the Philippines and the Solomon Islands, can be explained by:

1. Most speciation took place before the Plio-Pleistocene sea-level fluctuations.
2. Complete extinction of species on the Sunda and Sahul continental shelves occurred during low sea-level stands.
3. Selective migration of species took place from West Pacific refugia to relatively few reefs based on the continental shelves, which limited the diversity on the Sunda and Sahul continental shelves.

4. Sufficient habitat heterogeneity occurred on coral reefs along the present coastlines with gradients in nutrient input, salinity, siltation, wave exposure, and turbidity, predominatly varying from nearshore to offshore.

## Acknowledgements

I want to thank the reviewers and editors for improving comments on the manuscript. This paper is partly a result of the project "Climate change and Indonesian coral reef biotas" within the research theme "Biodiversity in relation to Global Change" of the Council for Earth and Life Sciences of the Netherlands Organisation for Scientific Research (ALW-NWO grant 852.000.50). Some of the ideas presented in this paper were inspired by results obtained during field surveys for which various sponsors helped in getting research permits, such as the Indonesian Institute of Sciences (LIPI), in particular the Research Centre for Oceanography at Jakarta (director Dr. Suharsono) and Hasanuddin University at Makassar (especially through the help given by Professor Dr. Alfian Noor and Drs. Willem Moka). Financial support for the field research was given by various funds, such as the Netherlands Foundation for the Advancement of Tropical research (WOTRO grants W01-60, W77-96, WK84-354, WT87-299), the Schure Beijerinck Popping fund (KNAW), the Alida Buitendijkfonds (Naturalis), and the Jan Joost ter Pelkwijkfonds (Naturalis), The Nature Conservancy (Indonesia) and the WWF (Netherlands). Various institutes and organizations acted as host, such as field stations at Pulau Pari and Ambon (PPO-LIPI), UNHAS (Makassar), Bung Hatta University (Padang), The Nature Conservancy (Komodo and Wakatobi), Christensen Research Institute (Madang), the Coral Reef Research Foundation (Palau), and several dive resorts, such as Bali Hai Diving Adventures, Bali Blue Dive, Derawan Dive Resort, and Sangalaki Dive Lodge. Dr. Charles Fransen designed the basic maps used for showing the various diversity centers. I want to thank Dr. David G. Reid (The Natural History Museum, London), Dr. Christopher P. Meyer (Florida Museum of Natural History, Gainesville), Dr. Stephen D. Cairns (National Museum of Natural History, Smithsonian Institution, Washington, D.C.), and my colleagues of the Naturalis Marine Research Team (Dr. Charles H.J.M. Fransen, Dr. Leen P van Ofwegen, Dr. Nicole J. de Voogd) for their constructive remarks on the manuscript.

## References

Abbott, R.T., 1960, The genus *Strombus* in the Indo-Pacific, *Indo Pacific Mollusca* **1**: 33–146.

Ablan, M.C.A., McManus, J.W., Chen, C.A., Shao, K.T., Bell, J., Cabanban, A.S., Tuan, V.S., and Arthana, I.W., 2002, Meso-scale transboundary units for the management of coral reefs in the South China Sea area, *Naga Worldfish Center Quarterly* **25**: 4–9.

Achituv Y. and Hoeksema, B.W., 2003, *Cantellius cardenae* spec. nov. (Cirripedia: Pyrgomatinae) from *Acropora (Isopora) brueggemanni* (Brook, 1893) (Anthozoa: Acroporidae), a case of host specificity in a generalist genus, *Zoologische Mededelingen Leiden* **77**: 1–8.

Adey, W.H., 1978, Coral reef morphogenesis: a multidimensional model, *Science* **202**: 831–837.

Adey, W.H., 2000, Coral reef ecosystems and human health: biodiversity counts! *Ecosystem Health* **6**: 227–236.

Adey, W.H., McConnaughey, T.A., Small, A.M., and Spoon, D.M., 2000, Coral reefs: endangered biodiverse, genetic resources, in: Sheppard, C.R.C. (ed.), *Seas at the Millennium: An Environmental Evaluation, Volume III, Global Issues and Processes*, Elsevier, Amsterdam, pp. 33–42.

Adey, W.H. and Steneck, R.S., 2001, Thermogeography over time creates biogeographic regions: a temperature/space/time-integrated model and an abundance-weighted test for benthic marine algae, *Journal of Phycology* **37**: 677–698.

Alcala, C. and Russ, G.R, 2002, Role of socioeconomic factors in coral reef protection and management, *Proceedings Ninth International Coral Reef Symposium*, Bali, 2000, **1**: 29–32.

Allen, G.R., 2002, Indo-Pacific coral-reef fishes as indicators of conservation hotspots, *Proceedings Ninth International Coral Reef Symposium*, Bali, 2000, **2**: 921–926.

Allen, G.R., 2003, Reef fishes of Milne Bay Province, Papua New Guinea, in: Allen, G.R., Kinch, J.P., McKenna, S.A., and Seeto, P. (eds), *A Rapid Marine Biodiversity Assessment of Milne Bay Province, Papua New Guinea – Survey II (2000). RAP Bulletin of Biological Assessment* 29, Conservation International, Washington, DC pp. 46–55.

Allen, G.R. and McKenna, S.A., 2001, *A Marine Rapid Assessment of the Togean and Banggai Islands, Sulawesi, Indonesia. RAP Bulletin of Biological Assessment 20*, Conservation International, Washington, DC.

Andersen, N.M., 1991, Cladistic biogeography of marine water striders (Insecta, Hemiptera) in the Indo-Pacific, *Australian Systematic Botany* **4**: 151–163.

Andersen, N.M., 1999, The evolution of marine insects: phylogenetic, ecological and geographical aspects of species diversity in marine water striders, *Ecography* **22**: 98–111.

Andersen, N.M., Cheng, L., Damgaard, J., and Sperling, F.A.H., 2000, Mitochondrial DNA variation and phylogeography of oceanic insects (Hemiptera: Gerridae: *Halobates* spp.), *Marine Biology* **136**: 421–430.

Angel, M.V., 1993, Biodiversity of the pelagic ocean, *Conservation Biology* **7**: 760–772.

Angel, M.V., 1997, Pelagic biodiversity, in: Ormond, R.F.G., Gage, J.D., and Angel, M.V. (eds), *Marine Biodiversity: Patterns and Processes*, Cambridge University Press, Cambridge, UK, pp. 35–68.

Backus, R.H., 1986, Biogeographic boundaries in the open ocean, *UNESCO Technical Papers in Marine Science* **49**: 9–13.

Barber, P.H. and Bellwood, D.R., 2005, Biodiversity hotspots: evolutionary origins of biodiversity in wrasses (*Halichoeres*: Labridae) in the Indo-Pacific and new world tropics, *Molecular Phylogenetics and Evolution* **35**: 235–253.

Barber, P.H., Palumbi, S.R., Erdmann, M.V., and Moosa, M.K., 2000, A marine Wallace's line? *Nature* **406**: 392–693.

Barber, P.H., Palumbi, S.R., Erdmann, M.V., and Moosa, M.K., 2002, Sharp genetic breaks among populations of *Haptosquilla pulchella* (Stomatopoda) indicate limits to larval transport: patterns, causes, and consequences, *Molecular Ecology* **11**: 659–674.

Bard, E., Hamelin, B., Arnold, M., Montaggioni, L.F., Cabioch, G., Faure, G., and Rougerie, F., 1996, Deglacial sea level record from Tahiti corals and the timing of global meltwater discharge, *Nature* **382**: 241–244.

Becking, L.E., Cleary D.F.R., Voogd, N.J. de, Renema, W., Beer, M. de, Soest, R.W.M. van, and Hoeksema, B.W., 2006, Beta diversity of tropical marine benthic assemblages in the Spermonde Archipelago, Indonesia, *Marine Ecology PSZNI* **27**: 76–88.

Beer, M. de, 1990a, A marine biological study on the cross-shelf distribution patterns of regular sea urchins (Echinodermata: Echinodea) in the Spermonde Archipelago, SW Sulawesi, (Indonesia), *Netherlands Journal of Zoology* **40**: 544.

Beer, M. de, 1990b, Distribution patterns of regular sea urchins (Echinodermata: Echinodea) across the Spermonde Shelf, SW Sulawesi, (Indonesia), in: Ridder, C. de, Dubois, P., Lahaye, M.C., and Jangoux, M. (eds), *Echinoderm Research*, Balkema Publishers, Rotterdam, pp. 165–169.

Beger, M., Jones, G.P., and Munday, P.L., 2003, Conservation of coral reef biodiversity: a comparison of reserve selection procedures for corals and fishes, *Biological Conservation* **111**: 53–62.

Belasky, P., 1992, Assessment of sampling bias in biogeography by means of a probabilistic estimate of taxonomic diversity: application to modern Indo-Pacific reef corals, *Palaeogeography, Palaeoclimatology, Palaeoecology* **99**: 243–270.

Belasky, P., 1996, Biogeography of Indo-Pacific larger foraminifera and scleractinian corals: a probabilistic approach to estimating taxonomic diversity, faunal similarity, and sampling bias, *Palaeogeography, Palaeoclimatology, Palaeoecology* **122**: 119–141.

Bellwood, D.R., 1997, Reef fish biogeography: habitat associations, fossils and phylogenies, *Proceedings Eighth International Coral Reef Symposium*, Panama **1**: 379–384.

Bellwood, D.R. and Hughes, T.P., 2001, Regional-scale assembly rules and biodiversity of coral reefs, *Science* **292**: 1532–1534.

Bellwood, D.R., Hughes, T.P., Connolly, S.R., and Tanner, J., 2005, Environmental and geometric constraints on Indo-Pacific coral reef biodiversity, *Ecology Letters* **8**: 643–651.

Bellwood, D.R., Herwerden, L. van, and Konow, N., 2004, Evolution and biogeography of angelfishes, *Molecular Phylogenetics and Evolution* **33**: 140–155.

Bellwood, D.R. and Wainwright, P.C., 2002, The history and biogeography of fishes on coral reefs, in: Sale, P.F. (ed.), *Coral Reef Fishes. Dynamics and Diversity in a Complex Ecosystem*, Academic Press, NY, pp. 5–32.

Benzie, J.A.H., 1998, Genetic structure of marine organisms and Southeast Asian biogeography, in: Hall, R. and Holloway, D. (eds), *Biogeography and Geological Evolution of Southeast Asia*, Backhuys Publishers, Leiden pp. 197–209.

Benzie, J.A.H., 1999a, Genetic structure of coral reef organisms: ghosts of dispersal past, *American Zoologist* **39**: 131–145.

Benzie, J.A.H., 1999b, Major genetic differences between Crown-of-Thorns starfish (*Acanthaster planci*) populations in the Indian and Pacific oceans, *Evolution* **53**: 1782–1795.

Benzie, J.A.H., 2000, The detection of spatial variation in widespread marine species: methods and bias in the analysis of population structure in the crown of thorns starfish (Echinodermata: Asteroidea), *Hydrobiologia* **420**: 1–14.

Benzie, J.A.H., 2001, Ocean structure, permeable barriers, rare events and patterns of genetic diversity in the Indo-Pacific, *American Zoologist* **41**: 1389–1390.

Benzie, J.A.H., Smith, C., and Sugama, K., 2003, Mitochondrial DNA reveals genetic differentiation between Australian and Indonesian pearl oyster *Pinctada maxima* (Jameson 1901) populations, *Journal of Shellfish Research* **22**: 781–787.

Benzie, J.A.H. and Williams, S.T., 1995, Gene flow among giant clam (*Tridacna gigas*) populations in Pacific does not parallel ocean circulation, *Marine Biology* **123**: 781–787.

Benzie, J.A.H. and Williams, S.T., 1997, Genetic structure of Giant Clam (*Tridacna maxima*) populations in the West Pacific is not consistent with dispersal by present-day ocean currents, *Evolution* **51**: 768–783.

Best, M.B., Hoeksema, B.W., Moka, W., Moll, H., Suharsono and Sutarna, I.N., 1989, Recent scleractinian coral species collected during the Snellius-II Expedition in eastern Indonesia, *Netherlands Journal of Sea Research* **23**: 107–115.

Blanchon, P. and Blakeway, D., 2003, Are catch-up reefs an artefact of coring? *Sedimentology* **50**: 1271–1282.

Bleakley, C. and Wells, S. (eds), 1995, Marine region 13: East Asian Seas, in: Kelleher, G.G., Bleakley, C., and Wells, S (eds), *A Global Representative System of Marine Protected Areas* 3 Great Barrier Reef Park Authority, Canberra, The World Bank, Washington, DC, The World Conservation Union (IUCN), Gland and Cambridge, pp. 107–140.

Blum, S.D., 1989, Biogeography of the Chaetodontidae: an analysis of allopatry among closely related species, *Environmental Biology of Fishes* **25**: 9–31.

Bolton, J.J., 1994, Global seaweed diversity: patterns and anomalies, *Botanica Marina* **37**: 241–245.

Boschma, H., 1948, The species problem in *Millepora*, *Zoologische Verhandelingen Leiden* **1**: 1–115.

Boschma, H., 1957, List of the described species of the Order Stylasterina, *Zoologische Verhandelingen Leiden* **33**: 1–72.

Bouchet, P., 1997, Inventorying the molluscan diversity of the world: what is our rate of progress? *Veliger* **40**: 1–11.

Bouchet, P., Lozouet, P., Maestrati, P., and Heros, V., 2002, Assessing the magnitude of species richness in tropical marine environments: exceptionally high numbers of molluscs at a New Caledonia site, *Biological Journal of the Linnean Society* **75**: 421–436.

Bowen, B.W., Bass, A.L., Garcia-Rodriguez, A.I., Rocha, L.A., and Robertson, D.R., 2001, Phylogeography of the trumpetfish (*Aulostomus* spp.): a ring species complex on a global scale, *Evolution* **55**: 1029–1039.

Boxshall, G.A., 1981, Community structure ands resource partitioning – the plankton, in: Forey, P.L. (ed.), *The Evolving Biosphere. Chance Change and Challenge*, British Museum (Natural History), London, and Cambridge University Press, Cambridge, UK, pp. 143–156.

Briggs, J.C., 1973, Operation of zoogeographic barriers, *Systematic Zoology* **23**: 248–256.

Briggs, J.C., 1974, *Marine Biogeography*, McGraw-Hill, New York.

Briggs, J.C., 1981, Do centers of origin have a center, *Paleobiology* **7**: 305–307.

Briggs, J.C., 1987a, *Biogeography and Plate Tectonics*, Elsevier, Amsterdam.

Briggs, J.C., 1987b, Antitropical distribution and evolution in the Indo-West Pacific Ocean, *Systematic Zoology* **36**: 237–247.

Briggs, J.C., 1992, The marine East Indies: centre of origin? *Global Ecology and Biogeography Letters* **2**: 149–156.

Briggs, J.C., 1994, Species diversity: land and sea compared, *Systematic Biology* **43**: 30–135.

Briggs, J.C., 1995, *Global Biogeography*, Elsevier, Amsterdam.

Briggs, J.C., 1996. Tropical diversity and conservation, *Conservation Biology* **10**: 713–718.

Briggs, J.C., 1999a, Coincident biogeographic patterns: Indo-West Pacific Ocean, *Evolution* **53**: 326–335.

Briggs, J.C., 1999b, Extinction and replacement in the Indo-West Pacific Ocean, *Journal of Biogeography* **26**: 777–783.

Briggs, J.C., 1999c, Modes of speciation: marine Indo-West Pacific, *Bulletin of Marine Science* **65**: 645–656.

Briggs, J.C., 2000, Centrifugal speciation and centres of origin, *Journal of Biogeography* **27**: 1183–1188.

Briggs, J.C., 2003, Marine centres of origin as evolutionary engines, *Journal of Biogeography* **30**: 1–18.

Briggs, J.C., 2004a, Older species: a rejuvenation on coral reefs? *Journal of Biogeography* **31**: 525–530.

Briggs, J.C., 2004b, A marine center of origin: reality and conservation? in: Lomolino M.V. and Heaney, L.R. (eds), *Frontiers of Biogeography. New Directions in the Geography of Nature*, Sinauer Associates, Sunderland, UK, pp. 255–269.

Briggs, J.C., 2005a, The marine East Indies: diversity and speciation, *Journal of Biogeography* **32**: 1517–1522.

Briggs, J.C., 2005b, Coral reefs: conserving the evolutionary sources, *Biological Conservation* **126**: 297–305.

Briggs, J.C., 2006, Proximate sources of marine diversity, *Journal of Biogeography* **33**: 1–10.

Brothers, E.B. and Thresher, R.E., 1985, Pelagic duration, dispersal and the distribution of Indo-Pacifc coral reef fishes, in: Reaka, M.L. (ed.), *The Ecology of Coral Reef Symposia Series for Undersea Research*, NOAA's Undersea Research Program **3** (1): 53–69.

Brown, J.H., 1988, Species diversity, in: Meyers, A.A. and Giller, P.S. (eds), *Analytical Biogeography. An Integrated Approach to the Study of Animal and Plant Distributions*, Chapman & Hall, London, pp. 57–89.

Burke, L., Selig, E., and Spalding, M., 2002, *Reefs at Risk in Southeast Asia*, World Resources Institute, Washington, DC.

Cabioch, G., Camoin, C.F., and Montaggioni, L.F., 1999a, Postglacial growth history of a French Polynesian barrier reef tract, Tahiti, Central Pacific, *Sedimentology* **46**: 985–1000.

Cabioch, G., Montaggioni, L.F., and Faure, G., 1995, Holocene initiation and development of New Caledonian fringing reefs, South-West Pacific, *Coral Reefs* **14**: 131–140.

Cabioch, G., Montaggioni, L.F., Faure, G., and Ribaud-Laurenti, A., 1999b, Reef coralgal assemblages as recorders of paleobathymetry and sea level changes in the Indo-Pacific province, *Quaternary Science Reviews* **18**: 1681–1695.

Cairns, S.D., 1992, Worldwide distribution of of the Stylasteridae (Cnidaria: Hydrozoa), *Scientia Marina* **56**: 125–130.

Cairns, S.D., in press, Deep-water corals: an overview with special reference to diversity and distribution of deep-water scleractinian corals, *Bulletin of Marine Science* **80**.

Cairns, S.D. and Hoeksema, B.W., 1997, *Distichopora vervoorti*, a new shallow-water stylasterid coral (Cnidaria : Hydrozoa : Stylasteridae) from Bali, Indonesia, *Zoologische Verhandelingen Leiden* **323**: 311–318.

Cairns, S.D., Hoeksema, B.W., and Land, J. van der, 1999, List of extant stony corals, *Atoll Research Bulletin* **459**: 13–46.

Cairns, S.D. and Zibrowius, H., 1997, Cnidaria Anthozoa: azooxanthellate Scleractinia from the Philippine and Indonesian regions, *Mémoirs du Muséum National d'Histoire Naturelle Paris* **172**: 27–243.

Carlton, J.T., Geller, J.B., Reaka-Kudla, M.L., and Norse, E.A., 1999, Historical extinctions in the sea, *Annual Review of Ecology and Systematics* **30**: 515–538.

Carpenter, K.E. and Springer, V.G., 2005, The center of the center of marine shorefish biodiversity: the Philippine Islands, *Environmental Biology of Fishes* **72**: 467–480.

Cesar, H.S.J. (ed.), 2000, *Collected Essays on the Economics of Coral Reefs*, CORDIO, Kalmar, Sweden.

Cesar, H., Burke, L., and Pet-Soede, L., 2003, *The Economics of Worldwide Coral Reef Degradation*, Cesar Environmental Economics Consulting (CEEC), Arnhem, The Netherlands.

Chappell, J., 1983, Sea level changes and coral reef growth, in: Barnes, D.J. (ed.), *Perspectives on Coral Reefs*, Brian Clouston, Manuha, Australia, pp. 46–55.

Chappell, J. and Thom, B.G., 1977, Sea levels and coasts, in: Allen, J., Golson, J., and Jones, R. (eds), *Sunda and Sahul: Prehistoric Studies in Southeast Asia, Melanesia and Australia*, Academic Press, London, pp. 275–291.

Chen, C.A., 1999, Analysis of scleractinian distribution in Taiwan indicating a pattern congruent with sea surface temperatures and currents: examples from *Acropora* and Faviidae corals, *Zoological Studies* **38**: 119–129.

Chenoweth, S.F. and Hughes, J.M., 2003, Oceanic interchange and nonequilibrium population structure in the estuarine dependent Indo-Pacific tasselfish, *Polynemus sheridani, Molecular Ecology* **12**: 2387–2397.

Chenoweth, S.F., Hughes, J.M., Keenan, C.P., and Lavery, S., 1998a, Concordance between dispersal and mitochondrial gene flow: isolation by distance in a tropical teleost, *Lates calcarifer* (Australian baramundi), *Heredity* **80**: 187–197.

Chenoweth, S.F., Hughes, J.M., Keenan, C.P., and Lavery, S., 1998b, When oceans meet: a teleost shows secondary intergradation at an Indian-Pacific interface, *Proceedings of the Royal Society B* **265**: 415–420.

Clarke, A. and Crame, J.A, 1997, Diversity, latitude and time: patterns in the shallow sea, in: Ormond, R.F.G., Gage, J.D., and Angel, M.V. (eds), *Marine Biodiversity: Patterns and Processes*, Cambridge University Press, Cambridge, UK, pp. 122–147.

Cleary D.F.R., Becking, L.E., Voogd, N.J. de, Renema, W., Beer, M. de, Soest, R.W.M. van, and Hoeksema, B.W., 2005, Variation in the diversity and composition of benthic taxa as a function of distance offshore, depth and exposure in the Spermonde Archipelago, Indonesia, *Estuarine and Coastal Shelf Science* **65**: 557–570.

Cleary, D.F.R, Suharsono, and Hoeksema, B.W., 2006, Coral diversity across a disturbance gradient in the Pulau Seribu reef complex off Jakarta, Indonesia, *Biodiversity and Conservation*. **15**: 3653–3676.

Collette, B.B., 2005, Is the east-west division of haplotypes of the three-spot seahorse along Wallace's Line novel among marine organisms? *Journal of Biogeography* **32**: 1286.

Collete, B.B. and Russo, J.L., 1984, Morphology, systematics, and biology of the Spanish mackerels (Scomberomorus, Scombridae), *Fishery Bulletin* **82**: 545–692.

Connolly, S.R., Bellwood, D.R., and Hughes, T.P., 2003, Indo-Pacific biodiversity of coral reefs: deviation from a mid-domain model, *Ecology* **84**: 2178–2190.

Connolly, S.R., Hughes, T.P., Bellwood, D.R., and Karlson, R.H., 2005, Community structure of corals and reef fishes at multiple scales, *Science* **309**: 1363–1365.

Cornell, H.V. and Karlson, R.H., 1996, Species richness of reef-building corals determined by local and regional processes, *Journal of Animal Ecology* **65**: 233–241.

Coudray, J., and Montaggioni, L., 1982, Coraux et récifs coralliens de la province indo-pacifique: répartition géographique et altitudinale en relation avec la tectonique globale, *Bulletin de la Society Géoloqique de France* **7**: 981–993.

Crame, J.A., 2000, Evolution of taxonomic diversity gradients in the marine realm: evidence from the composition of recent bivalve faunas, *Paleobiology* **26**: 188–214.

Cresswell, G., 1986, The role of the Leeuwin current in the life cycles of several marine creatures, *UNESCO Technical Papers in Marine Science* **49**: 60–64.

Cresswell, G.R. and Golding, T.J., 1980, Observations of a south south-flowing current in the southeastern Indian Ocean, *Deep-sea Research* **27A**: 449–466.

Dargent, O. and Coppejans, E., 1998, The species diversity of the genus *Halimeda* (Caulerpales, Bryopsidophyceae) in Indonesia and Papua New Guinea, *Biologisch Jaarboek Dodonaea Gent* **65**: 119–120.

Dawes, C.J., 1981, *Marine Botany*, Wiley, New York.

Dawson, M.N. and Hamner, W.M., 2005, Rapid evolutionary radiation of marine zooplankton in peripheral environments, *Proceedings of the National Academy of Sciences USA* **102**: 9235–9240.

Dawson, M.N. and Jacobs, D.K., 2001, Molecular evidence for cryptic species of *Aurelia aurita* (Cnidaria, Scyphozoa), *Biological Bulletin* **200**: 92–96.

Dawson, M.N. and Martin, L.E., 2001, Geographic variation and ecological adaptation in *Aurelia* (Scyphozoa, Semaeostomeae): some implications from molecular phylogenetics, *Hydrobiologia* **451**: 259–273.

Didderen, K., Fransen, C.H.J.M., and Voogd, N.J. de, 2006, Observations of sponge-dwelling colonies of *Synalpheus* (Decapoda, Alpheidae) of Sulawesi, Indonesia, *Crustaceana* **79**: 961–975.

Dinesen, Z.D., 1983, Patterns in the distribution of soft corals across the central Great Barrier Reef, *Coral Reefs* **1**: 229–236.

Donaldson, T.J., 1986, Distribution and species richness of Indo-West Pacific Cirrhitidae: support for Woodland's hypothesis, in: Uyeno, T., Arai, R., Taniuchi, T., and Matsuura, K. (eds), *Indo-Pacific Fish Biology. Proceedings of the Second International Conference on Indo-Pacific Fishes*, Ichtyological Society of Japan, Tokyo, pp. 623–628.

Done, T.J., 1982, Patterns in the distribution of coral communities across the Central Great Barrier Reef, *Coral Reefs* **1**: 95–107.

Done, T.J., 1983, The distribution of coral communities on the Great Barrier Reef, *Proceedings of the Inaugural Great Barrier Reef Conference*, Townsville, 1983, pp. 197–202.

Duarte, C.M., 2001, Seagrasses, in: Levin, S.A. (ed.), *Encyclopedia of Biodiversity*, Vol. 5, Academic Press, San Diego, CA, pp. 255–268.

Duda, T.F. and Palumbi, S.R., 1999, Population structure of the black tiger prawn, Penaeus monodon, among western Indian Ocean and western Pacific populations, *Marine Biology* **34**: 705–710.

Duke, N.C., Benzie, J.A.H., Goodall, J.A., and Ballment, E.R., 1998, Genetic structure and evolution of species in the mangrove genus *Avicennia* (Avicenniaceae) in the Indo-West Pacific, *Evolution* **52**: 1612–1626.

Edinger, E. and Browne, D.R., 2000, Continental seas of western Indonesia, in: Sheppard, C.R.C. (ed.), *Sea at the Millenium: An Environmental Evaluation, Volume II. Regional Chapters: The Indian Ocean to the Pacific*, Elsevier Science, Amsterdam, pp. 381–404.

Edinger, E.N., Kolasa, J., and Risk, J.M., 2000, Biogeographic variation in coral species diversity on coral reefs in three regions of Indonesia, *Diversity and Distributions* **6**: 113–127.

Ekman, S., 1934, Indo-Westpazifik und Atlanto-Ostpazifik, eine tiergeographische Studie, *Zoogeographica* **2**: 320–374.

Ekman, S., 1935, *Tiergeographie des Meeres*. Akademische Verlagsgesellschaft M.B.H. Leipzig.

Ekman, S., 1953, *Zoogeography of the Sea*, Sidgwick & Jackson, London.

Erdmann, M.V., Caldwell, R.L., and Kasim Moosa, M., 1998, Indonesian 'king of the sea' discovered, *Nature* **395**: 335–336.

Fabricius, K. and Alderslade, P., 2001, *Soft Corals and Sea Fans. A Comprehensive Guide to the Tropical Shallow-Water Genera of the Central-West Pacific, the Indian Ocean and the Red Sea*, Australian Institute of Marine Science, Townsville.

Fadlallah, Y.H., 1983, Sexual reproduction, development and larval biology in scleractinian corals, *Coral Reefs* **2**: 129–150.

Fan, K.L. and Yu, C.Y., 1981, A study of water masses in the seas of southernmost Taiwan, *Acta Oceanographica Taiwanica* **12**: 94–111.

Ferrier, S., 2002, Mapping spatial pattern in biodiversity for regional conservation planning: where to from here? *Systematic Biology* **51**: 331–363.

Fleminger, A., 1986, The Pleistocene equatorial barrier between the Indian and Pacific oceans and a likely cause for Wallace's Line, *UNESCO Technical Papers in Marine Science* **49**: 84–97.

Foin, T.C., 1976, Plate tectonics and the biogeography of the Cypraeidae (Mollusca: Gastropoda), *Journal of Biogeography* **3**: 19–34.

Fontaubert, A.C. de, Downes, D.R., and Agardy, T.S., 1996, *Biodiversity in the Seas: Implementing the Convention of Biological Diversity in Marine and Coastal Habitats*, IUCN, Gland and Cambridge.

Forbes, E., 1856, Map of the distribution of marine life, in: Johnston, A.K. (ed.), *The Physical Atlas of Natural Phenomena* (new edition), W. Johnston and A.K. Johnston, London, p. 31.

Forey, P., 1998, A home away from home for coelacanths, *Nature* **395**: 319.

Fransen, C.H.J.M., 2002, Taxonomy, phylogeny, historical biogeography, and historical ecology of the genus *Pontonia* Latreille (Crustacea: Decapoda: Caridea: Palaemonidae), *Zoologische Verhandelingen Leiden* **336**: 1–433.

Fransen, C.H.J.M., 2007, The influence of land barriers on the evolution of pontoniine shrimps (Crustacea, Decapoda) living in association with molluscs and solitary ascidians, in Renema, W. (ed.), *Biogeography, Time and Place: Distributions, Barriers and Islands*, Springer, Dordrecht, pp. 103–116. (Chapter 4 this book).

Fraser, R. H. and Currie, D.J., 1996, The species richness-energy hypothesis in a system where historical factors and thought to prevail: coral reefs, *American Naturalist* **148**: 138–159.

Garvine, R.W., 1986, The role of brackish plumes in open shelf waters, in: Skreslet, S. (ed.), *The Role of Freshwater Outflow in Coastal Marine Ecosystems*, Springer, Berlin, pp. 47–65.

Gagan, M.K., Hendy E.J., Haberle, S.G., and Hantoro, W.S., 2004, Post-glacial evolution of the Indo-Pacific Warm Pool and El Niño-Southern oscillation, *Quaternary International* **118–119**: 127–143.

Gaston, K.J., 2003, The how and why of biodiversity, *Nature* **421**: 900–901.

Gaston, K.J. and Wiliams, P.H., 1993, Mapping the world's species – the higher taxon approach, *Biodiversity Letters* **1**: 2–8.

GBDMS (Committee on Biological Diversity in Marine Systems), 1995, *Understanding Marine Biodiversity*, National Academy Press, Washington, DC.

Geiger, D.L., 2000, Distribution and biogeography of the recent Haliotidae (Gastropoda: Vetigastropoda) world-wide, *Bolletino Malacologico* **35**: 57–120.

Geiger, D.L. and Groves, L.T., 1999, Review of fossil abalone (Gastropoda: Vetigastropoda: Haliotidae) with comparison to recent species, *Journal of Paleontology* **73**: 872–885.

Gerber, L.R., Botsford, L.W., Hastings, A., Possingham, H.P., Gaines, S.D., Palumbi, S.R., and Andelman, S., 2003, Population models for marine reserve design: a retrospective and prospective synthesis, *Ecological Applications* **13** (Suppl.): S47–S64.

Geyh, M.A., Kudrass, H.R., and Streiff, H., 1979, Sea-level changes in the late Pleistocene and Holocene in the Strait of Malacca, *Nature* **278**: 441–443.

Gittenberger, A., 2003, The wentletrap *Epitonium hartogi* spec. nov. (Gastropoda: Epitoniidae), associated with bubble coral species, *Plerogyra* spec. (Scleractinia: Euphyllidae), off Indonesia and Thailand, *Zoologische Verhandelingen Leiden* **345**: 139–150.

Gittenberger, A., Goud, J., and Gittenberger, E., 2000, *Epitonium* (Gastropoda: Epitoniidae) associated with mushroom corals (Scleractinia: Fungiidae) from Sulawesi, Indonesia, with the description of four new species, *Nautilus* **114**: 1–13.

Gittenberger, A. and Gittenberger, E., 2005, A hitherto unnoticed adaptive radiation: epitoniid species (Gastropoda: Epitoniidae) associated with corals (Scleractinia), *Contributions to Zoology* **74**: 125–203.

Glynn, P.W. and Ault, J.S., 2000, A biogeographic analysis and review of the far eastern Pacific coral reef region, *Coral Reefs* **19**: 1–23.

Gordon, A.L., 2001, Interoceanic exchange, in: Siedler, G., Church, J., and Gould, J. (eds), *Ocean Circulation and Climate*, Academic Press, San Diego, CA, pp. 303–314.

Gordon, A.L. and Fine. R.A., 1996, Pathways of water between the Pacific and Indian Oceans in the Indonesian seas, *Nature* **379**: 146–149.

Gordon, A.L., Susanto, R.D., and Vranes, K., 2003, Cool Indonesian throughflow as a consequence of restricted surface layer flow, *Nature* **425**: 824–828.

Gosliner, T.M., 1993, Biodiversity of tropical opisthobranch gastropod faunas, *Proceedings Seventh International Coral Reef Symposium, Guam*, 1992, **2**: 702–709.

Gosliner, T.M., 2002, Biodiversity, endemism, and evolution of opisthobranch gastropods on Indo-Pacific coral reefs, *Proceedings Ninth International Coral Reef Symposium*, Bali, 2000, **2**: 937–940.

Gosliner, T.M. and Draheim, R., 1996, Indo-Pacific opisthobranch gastropod biogeography: how do we know what we don't know, *American Malacological Bulletin* **12**: 37–43.

Goud, J. and Hoeksema, B.W., 2001, *Pedicularia vanderlandi* spec. nov., a symbiotic snail (Caenogastropoda: Ovulidae) on the hydrocoral *Distichopora vervoorti* Cairns & Hoeksema, 1998 (Hydrozoa: Stylasteridae), from Bali, Indonesia, *Zoologische Verhandelingen Leiden* **334**: 77–97.

Gray, J.S., 1997, Marine biodiversity: patterns, threats and conservation needs, *Reports and Studies GESAMP* **62**: 1–24.

Grave, S. de, 2001, Biogeography of Indo-Pacific Pontoniinae shrimps (Crustacea, Decapoda): a PAE analysis, *Journal of Biogeography* **28**: 1239–1253.

Green, A. and Mous, P.J., 2004, Delineating the Coral Triangle, its ecoregions and functional seascapes. *Report on an expert workshop, held at the Southeast Asia Center for Marine Protected Areas*, Bali, Indonesia (April 30–May 2, 2003). Version 1.1. The Nature Conservancy, Southeast Asia Center for Marine Protected Areas, Bali, Indonesia.

Grigg, R.W. and Epp, D., 1989, Critical depth for the survival of coral islands: effects on the Hawaiian Archipelago, *Science* **243**: 638–641.

Groombridge, B. and Jenkins, M.D, 2000, *Global Biodiversity. Earth's Living Resources in the 21st Century*, World Conservation Press, Cambridge.

Groombridge, B. and Jenkins, M.D, 2002, *World Atlas of Biodiversity. Earth's Living Resources in the 21st Century*, University of California Press, Berkeley, CA.

Hallock, P., 1988, Diversification in algal symbiont-bearing Foraminifera: a response to oligotrophy? *Revue de Paléobiologie Vol. Spéc.* **2**: 789–797.

Hartog, C. den, 1970, The sea-grasses of the world, *Verhandelingen van de Koninklijke Nederlandse Akademie van Wetenschappen, Afdeling Natuurkunde, Reeks 2*, **59**: 1–275, pls. 1–31.

Hatayama, T., Awaji, T., Akitomo, K., and Imasato, N., 1996, Tidal current in the Indonesian seas and its effect on the transport and mixing processes, *Proceedings of the IOC-WESTPAC Third International Science Symposium*, Bali, 1994, pp. 636–649.

Heads, M., 2005, Towards a panbiogeography of the seas, *Biological Journal of the Linnean Society* **84**: 675–723.

Heck, K.L. and McCoy, E.D., 1978, Long-distance dispersal and the reef-building corals of the eastern Pacific, *Marine Biology* **48**: 349–356.

Helfman, G.S., Collette, B.B., and Facey, D.E., 1997, *The Diversity of Fishes*, Blackwell Science, Malden, MA.

Hellberg, M., Balch., D.P., and Roy, K., 2001, Climate-driven range expansion and morphological evolution in a marine gastropod, *Science* **292**: 1707–1710.

Hellberg, M.E., Burton, R.S., Neigel, J.E., and Palumbi, S.R., 2002, Genetic assessment of connectivity among marine populations, *Bulletin of Marine Science* **70** (Suppl.): 273–290.

Hemminga, M.A. and Duarte, C.M., 2000, *Seagrass Ecology*, Cambridge University Press, Cambridge, UK.

Hesp, P.A., Chang, C.H., Hilton, M., Chou, M.L., and Turner, I.M., 1998, A first tentative Holocene sea-level curve for Singapore, *Journal of Coastal Research* **14**: 308–314.

Hewitt, G., 2000, The genetic legacy of the Quaternary ice ages, *Nature* **405**: 907–913.

Hoeksema, B.W., 1989, Taxonomy, phylogeny and biogeography of mushroom corals (Scleractinia: Fungiidae), *Zoologische Verhandelingen Leiden* **254**: 1–295.

Hoeksema, B.W., 1990, *Systematics and ecology of mushroom corals (Scleractinia: Fungiidae)*. Ph.D. thesis, Leiden University, Leiden, The Netherlands.

Hoeksema, B.W., 1993a, Mushroom corals (Scleractinia: Fungiidae) of Madang Lagoon, northern Papua New Guinea: an anonotated check-list with the description of *Cantherellus jebbi* spec. nov., *Zoologische Mededelingen Leiden* **67**: 1–19.

Hoeksema, B.W., 1993b, The position of northern New Guinea in the center of marine benthic diversity: a reef coral perspective, *Proceedings Seventh International Coral Reef Symposium*, Guam, 1992, **2**: 710–717.

Hoeksema, B.W., 1997a, Generic diversity of Scleractinia in Indonesia, in: Tomascik, T., Mah, A.J., Nontji, A., and Moosa M.K. (eds), *The Ecology of the Indonesian Seas,* Part I, Periplus, Singapore, pp. 308–311.

Hoeksema, B.W., 1997b, Diversity of mushroom corals (Scleractinia: Fungiidae in: Tomascik, T., Mah, A.J., Nontji, A., and Moosa M.K. (eds), *The Ecology of the Indonesian Seas*, Part I, Periplus, Singapore, pp. 311–313.

Hoeksema, B.W., 1997c, Conservation problems in coelenterates with emphasis on coral reef communities, in: Hartog, J.C. den, Ofwegen, L.P. van, Spoel, S. van der (eds), *Proceedings of the Sixth International Conference on Coelenterate Biology*, Noordwijkerhout, 1995, Nationaal Natuurhistorisch Museum, Leiden, The Netherlands, pp. 253–264.

Hoeksema, B.W., 2004, Biodiversity and the natural resource management of coral reefs in Southeast Asia, in: Visser, L.E. (ed.), *Challenging Coasts. Transdisciplinary Excursions into Integrated Coastal Zone Development*, Amsterdam University Press, Amsterdam, pp. 49–71.

Hoeksema, B.W. and Achituv, Y., 1993, First Indonesian record of *Fungiacava eilatensis* Goreau et al., 1968 (Bivalvia: Mytilidae), endosymbiont of *Fungia* spp. (Scleractinia: Fungiidae), *Basteria* **57**: 131–138.

Hoeksema, B.W. and Dai, C.F., 1991, Scleractinia of Taiwan II: family Fungiidae (including a new species), *Bulletin of the Institute of Zoology Academia Sinica* **30**: 203–228.

Hoeksema, B.W. and Kleemann, K., 2002, New records of *Fungiacava eilatensis* Goreau et al., 1968 (Bivalvia, Mytilidae) boring into Indonesian mushroom corals (Scleractinia, Fungiidae), *Basteria* **66**: 25–30.

Hoeksema, B.W. and Moka, W., 1989, Species assemblages and phenotypes of mushroom corals (Fungiidae) related to coral reef habitats in the Flores Sea, *Netherlands Journal of Sea Research* **23**: 149–160.

Hoeksema, B.W. and Putra, K.S., 2002, The reef coral fauna of Bali in the centre of marine diversity, *Proceedings Ninth International Coral Reef Symposium*, Bali, 2000, **1**: 173–178.

Hoeksema, B.W. and Ofwegen, L.P. van (eds), 2004, *Indo-Malayan Reef Corals: A Generic Overview, World Biodiversity Database CD-Rom Series*, ETI, Amsterdam.

Hogarth, P.J., 1999, *The Biology of Mangroves*, Oxford University Press, Oxford, NY.

Hogarth, P.J., 2001, Mangrove ecosystems, in: Levin, S.A. (ed.), *Encyclopedia of Biodiversity*, Vol. 3, Academic Press, San Diego, CA, pp. 853–870.

Houbrick, R.S., 1985, Genus *Clypeomorus* Jousseaume (Cerithiidae: Prosobranchia), *Smithsonian Contributions to Zoology* **403**: 1–131.

Houbrick, R.S., 1992, Monograph of the genus *Cerithium* Bruguière in the Indo-Pacific (Cerithiidae: Prosobranchia), *Smithsonian Contributions to Zoology* **510**: 1–211.

Hubbel, S.P., 1997, A unified theory of biogeography and relative species abundance and its application to tropical rain forests and coral reefs, *Coral Reefs* **16** (Suppl.): S9–S21.

Hughes, T.P., Bellwood, D.R., and Connolly, S.R., 2002, Biodiversity hotspots, centres of endemicity, and the conservation of coral reefs, *Ecology Letters* **5**: 775–784.

Hutomo, M. and Moosa, M.K., 2005, Indonesian marine and coastal biodiversity: present status, *Indian Journal of Marine Science* **34**: 88–97.

Ilahude, A.G. and Gordon, A.L., 1996, Water masses of the Indonesian seas throughflow, *Proceedings of the IOC-WESTPAC Third International Science Symposium*, Bali, 1994, pp. 572–587.

Johannes, R.E., Wiebe, W.J., Crossland, C.J., Rimmer, D.W., and Smith, S.V., 1983, Latitudinal limits to coral reef growth, *Marine Ecology Progress Series* **11**: 105–111.

Johnson, R.K., 1986, Polytypy, boundary zones and the place of broadly-distributed species in mesopelagic zoogeography, *UNESCO Technical Papers in Marine Science* **49**: 156–165.

Jokiel, P. and Martinelli, F.J., 1992, The vortex model; of coral reef biogeography, *Journal of Biogeography* **19**: 449–458.

Karlson, R.H., 1999, *Dynamics of Coral Communities*, Kluwer, Dordrecht, The Netherlands.

Karlson, R.H. and Cornell, H.V., 1999, Integration of local and regional perspectives on the species richness of coral assemblages, *American Zoologist* **39**: 104–112.

Kay, E.A., 1984, Patterns of speciation in the Indo-West Pacific, *B.P. Bishop Museum Special Publications* **72**: 15–31.

Kay, E.A., 1990, Cypraeidae of the Indo-Pacific: cenozoic fossil history and biogeography, *Bulletin of Marine Science* **47**: 23–34.

Kensley, B., 1998, Estimates of species diversity of free-living marine isopod crustaceans on coral reefs, *Coral Reefs* **17**: 83–88.

Kennett, J.P., Keller, G., and Srinivasan, M.S., 1985, Miocene planktonic foraminiferal biogeography and paleoceanographic development of the Indo-Pacific region, *Geological Society of America Memoirs* **163**: 197–236.

Kerswell, A.P., 2006, Global biodiversity patterns of benthic marine algae, *Ecology* **87**: 2479–2488.

Kirkendale, L. and Meyer, C.P., 2004, Phylogeography of the *Patelloida profunda* group (Gastropoda: Lottidae): diversification in a dispersal-driven marine system, *Molecular Ecology* **13**: 2749–2762.

Kleemann, K. and Hoeksema, B.W., 2002, *Lithophaga* (Bivalvia, Mytilidae), including a new species, boring into mushroom corals (Scleractinia, Fungiidae) off South Sulawesi, Indonesia, *Basteria* **66**: 11–24.

Knowlton, N., 2000, Molecular genetic analyses of species boundaries in the sea, *Hydrobiologia* **420**: 73–90.

Kohn, A.J., 1967, Environmental complexity and species diversity in the gastropod genus *Conus* on Indo-West Pacific reef platforms, *American Naturalist* **101**: 251–259.

Kohn, A.J., 1983, Marine biogeography and evolution in the tropical pacific: zoological perspectives, *Bulletin of Marine Science* **33**: 528–535.

Kohn, A.J., 1985, Evolutionary ecology of *Conus* on Indo-Pacific coral reefs, *Proceedings Fifth International Coral Reef Congress*, Tahiti **4**: 139–144.

Kohn, A.J., 1990, Biogeography and evolution of Indo-Pacific marine mollusca: patterns, progress, problems and prospect, *Bulletin of Marine Science* **47**: 2–9.

Kohn, A.J., 1997, Why are coral reef communities so diverse? in: Ormond, R.F.G., Gage, J.D., and Angel, M.V. (eds), *Marine Biodiversity: Patterns and Processes*, Cambridge University Press, Cambridge, UK, pp. 201–215.

Kohn, A.J. and Perron, F.E., 1994, *Life History and Biogeography Patterns in Conus*, Clarendon Press, Oxford, UK.

Kulbicki, M., Labrosse, P., and Ferraris, J., 2004, Basic principles underlying research projects on the links between the ecology and the uses of coral reef fishes in the Pacific, in: Visser, L.E. (ed.), *Challenging Coasts. Transdisciplinary Excursions into Integrated Coastal Zone Development*, Amsterdam University Press, Amsterdam, pp. 119–158.

Ladd, H.S., 1960, Origin of the Pacific Island molluscan fauna, *American Journal of Science* **258A**: 137–150.

Landry, C., Geyer, L.B., Arakaki, Y., Uehara, T., and Palumbi, S.R., 2003, Recent speciation in the Indo-West Pacific: rapid evolution of gamete recognition and sperm morphology in cryptic species of sea urchin, *Proceedings of the Royal Society B* **270**: 1839–1847.

Langer, M.R. and Hottinger, L., 2000, Biogeography of selected "larger" foraminifera, *Micropaleontology* **46** (Suppl. 1): 105–126.

Langer, M.R. and Lipps, J.H., 2003, Foraminiferal distribution and diversity, Madang Reef and Lagoon, Papua New Guinea, *Coral Reefs* **22**: 143–154.

Lessios, H.A., Kane, J., and Robertson, D.R., 2003, Phylogeography of the pantropical sea urchin *Tripneustes*: contrasting patterns of population structure between oceans, *Evolution* **57**: 2026–2036.

Lessios, H.A., Kessing, B.D., and Pearse, J.S., 2001, Population structure and speciation in tropical seas: global phylogeography of the sea urchin *Diadema*, *Evolution* **55**: 955–975.

Lessios, H.A., Kessing, B.D., and Robertson, D.R., 1998, Massive gene flow across the world's most potent marine biogeographic barrier, *Proceedings of the Royal Society B* **265**: 583–588.

Lessios, H.A., Kessing, B.D., Robertson, D.R., and Paulay, G., 1999, Phylogeny of the pantropical sea urchin *Eucidaris* in relation to land barriers and ocean currents, *Evolution* **53**: 806–817.

Lindner, A., Cairns, S.D., and Guzman, H.M., 2004, *Distichopora robusta* sp. nov., the first shallow-water stylasterid (Cnidaria: Hydrozoa: Stylasteridae) from the tropical eastern Pacific, *Journal of the Biological Association UK* **84**: 943–947.

Longhurst, A., 1998, *Ecological Geography of the Sea*, Academic Press, San Diego, CA.

Longhurst, A. and Pauly, D., 1987, *Ecology of Tropical Oceans*, Academic Press, San Diego, CA.

Lourie, S.A., Green, D.M., and Vincent, A.C.J., 2005, Dispersal, habitat differences, and comparative phylogeography of Southeast Asian seahorses (Syngnathidae: *Hippocampus*), *Molecular Ecology* **14**: 1073–1094.

Lourie, S.A. and Vincent, A.C.J., 2004, A marine fish follows Wallace's Line: the phylogeography of the three-spot seahorse (*Hippocampus trimaculatus*, Syngnathidae, Teleostei) in Southeast Asia, *Journal of Biogeography* **31**: 1975–1985.

Macaranas, J.M. Ablan, C.A., Pante, M.J.R., Benzie, J.A.H., and Williams, S.T., 1992, Genetic structure of giant clam (*Tridacna derasa*) populations from reefs in the Indo-Pacific, *Marine Biology* **113**: 231–238.

Marsh, L.M. and Marshall, J.I., 1983, Some aspects of the zoogeography of northwestern Australian echinoderms (other than holothurians), *Bulletin of Marine Science* **33**: 671–687.

Masse, J.P. and Montaggioni, L.F., 2001, Growth history of shallow-water carbonates: control of accommodation on ecological and depositional processes, *International Journal of Earth Sciences* **90**: 452–469.

Massin, C., 2000, Ecology of the *Leptoconchus* spp. (Gastropoda, Coralliophilidae) infesting Fungiidae (Anthozoa, Madreporaria), *Bulletin de l'Institut Royal des Sciences Naturelles de Belgique Biologie* **70**: 235–252.

Massin, C. and Dupont, S., 2003, Study on *Leptoconchus* species (Gastropoda, Coralliophilidae) infesting Fungiidae (Anthozoa: Scleractinia). 1. Presence of nine Operational Taxonomic Units (OTUs) based on anatomical and ecological characters, *Belgian Journal of Zoology* **133**: 121–126.

McAllister, D.E., 1991, What is the status of the world's coral reef fishes? *Sea Wind* **5**: 14–18.

McAllister, D.E., Schueler, F.W., Roberts, C.M., and Hawkins, J.P., 1994, Mapping and GIS analysis of the global distribution of coral reef fishes on an equal-area grid, in: Miller, R.I. (ed.), *Mapping the Diversity of Nature*, Chapman & Hall, London, pp. 155–175.

McCafferty, S., Bermingham, E., Quenouille, B., Planes, S., and Hoelzer, G., 2002, Historical biogeography and molecular systematics of the Indo-Pacific genus *Dascyllus* (Teleostei: Pomacentridae), *Molecular Ecology* **11**: 1377–1392.

McCartney, M.A., Keller, G., and Lessios, A., 2000, Dispersal barriers in tropical oceans and speciation in atlantic and eastern pacific sea urchins of the genus *Echinometra*, *Molecular Ecology* **9**: 1391–1400.

McCoy, E.D., 1983, Centers of origin revisited, *Paleobiology* **9**: 17–19.

McCoy, E.D. and Heck, K.L., 1976, Biogeography of corals, seagrasses and mangroves: an alternative to the center of origin concept, *Systematic Zoology* **25**: 201–210.

McGowan, J.A., 1986, The biogeography of pelagic ecosystems, *UNESCO Technical Papers in Marine Science* **49**: 191–200.

McGowan, J.A. and Walker, P.W., 1993, Pelagic diversity patterns, in: Ricklefs, R.E. and Schluter, D. (eds), *Species Diversity in Ecological Communities. Historical and Geographical Perspectives*, Chicago University Press, Chicago, pp. 203–214.

McKenna, S.A., Allen, G.R., and Suryadi, S., 2002, *A marine rapid assessment of the Raja Ampat Islands, Papua Province, Indonesia. RAP Bulletin of Biological Assessment 22*, Conservation International, Washington, DC.

McManus, J.W., 1985, Marine speciation, tectonics and sea-level changes in southeast Asia, *Proceedings Fifth International Coral Reef Congress*, Tahiti **4**: 133–138.

McManus, J.W., 1997, Tropical marine fisheries and the future of coral reefs: a brief review with emphasis on Southeast Asia, *Proceedings Eighth International Coral Reef Symposium*, Panama, 1996, **2**: 129–134.

McMillan, W.O. and Palumbi, S.R., 1995, Concordant evolutionary patterns among Indo-West Pacific Butterfly fish, *Proceedings of the Royal Society B* **260**: 229–236.

Meyer, C., 2003, Molecular systematics of cowries (Gastropoda: Cypraeidae) and diversification patterns in the tropics, *Biological Journal of the Linnean Society* **79**: 401–459.

Meyer, C., 2004, Toward comprehensiveness: increased molecular sampling within Cypraeidae and its phylogenetic implications, *Malacologia* **46**: 127–156.

Meyer, C. and Paulay, G., 2005, DNA barcoding: error rates based on comprehensive sampling, *PLOS Biology* **3** (12): e422 (1–10).

Milliman, J.D. and Meade, R.H., 1983, World-wide delivery of river sediment to the oceans, *Journal of Geology* **91**: 1–21.

Mironov, A.N., 2006, Centers of marine fauna redistribution, *Entomological Review* **86** (Suppl. 1): 32–44.

Miyama, T., Awaji, T., Akitomo, K., and Imasato, N., 1996a, Seasonal transport variations in the Indonesian Seas, *Proceedings of the IOC-WESTPAC Third International Science Symposium*, Bali, 1994, pp. 650–664.

Miyama, T., Awaji, T., Akitomo, K., and Imasato, N., 1996b, Seasonal variation of salinity in the mixed layer of the Indonesian Seas, *Proceedings of the IOC-WESTPAC Third International Science Symposium*, Bali, 1994, pp. 772–784.

Moll, H., 1983, Zonation and diversity of Scleractinia on reefs off SW Sulawesi, Indonesia, Ph.D. thesis, Leiden University, Leiden, The Netherlands.

Moll., H. and Suharsono, 1986, Distribution, diversity and abundance of reef corals in Jakarta Bay and Kepulauan Seribu, *UNESCO Reports in Marine Science* **40**: 112–125.

Montaggioni, L.F., 2000, Postglacial reef growth, *Comptes Rendus de l'Academie des Sciences Paris, Sciences de la Terre et des Planètes* **331**: 319–330.

Montaggioni, L.F., 2005, History of Indo-Pacific coral reef systems since the last glaciation: development patterns and controlling factors, *Earth-Science Reviews* **71**: 1–75.

Montaggioni, L.F. and Faure, G., 1997, Response of coral reef communities to sea-level rise: a Holocene model from Mauritius (Western Indian Ocean), *Sedimentology* **44**: 1053–1070.

Moolengraaff, G.A.F., 1922, Geologie, in: Stok, J.P. van der (ed.), *De Zeeën van Nederlandsch Oost-Indië*, Brill, Leiden, pp. 272–357, maps I–VII.

Moolengraaff, G.A.F., 1929, The coral reefs in the East Indian Archipelago: their distribution and mode of development, *Proceedings Fourth Pacific Science Congress* Java, pp. 1–35.

Mora, C., Chittaro, P.M., Sale, P.F., Kritzer, J.P., and Ludsin, S.A., 2003, Patterns and processes in reef fish diversity, *Nature* **421**: 933–936.

Mora, C. and Robertson, D.R., 2005, Factors shaping the range-size frequency distribution of the endemic fish fauna of the Tropical Eastern Pacific, *Journal of Biogeography* **32**: 277–286.

Mukai, H., 1993, Biogeography of the tropical seagrasses in the western Pacific, *Australian Journal of Marine and Freshwater Research* **44**: 1–17.

Mumby, P.J., Edwards A.J, Arias-González, J.E., Lindeman, K.C., Blackwell, P.G., Gall, A., Gorczynska, M.I., Harborne A.R., Pescod, C.L., Renken, R., Wabnitz, C.C.C., and Llewellyn, G., 2004, Mangroves enhance the biomass of coral reef fish communities in the Caribbean, *Nature* **427**: 533–536.

Murina, V.G., 1975, The geographical distribution of marine worms of the phylum Sipuncula of the world ocean, in: Rice, M.E. and Todorovic, M. (eds), *Proceedings of the International Symposium on the Biology of the Sipuncula and Echira*, Vol. I, Institute for Biologica Research "Sinisa Stankovic", Belgrade, and National Museum of Natural History, Smithsonian Institution, Washington, D.C, pp. 3–18.

Muss, A., Robertson, D.R., Stepien, C.A., Wirtz, P., and Bowen, B.W., 2001, Phylogeography of *Ophioblennius*: the role of ocean currents and geography in reef fish evolution, *Evolution* **55**: 561–572.

Myers, A.A., 1997, Biogeographic barriers and the development of marine biodiversity, *Estuarine and Coastal Shelf Science* **44**: 241–248.

Myers, N., Mittermeier, R.A., Mittermeier, C.G., da Fonseca, G.A.B., and Kent, J., 2000, Biodiversity hotspots for conservation priorities, *Nature* **403**: 853–858.

Myers, R.A. and Ottensmeyer, C.A., 2005, Extinction risk in marine species, in: Norse, E.A. and Crowder, L.B. (eds), *Marine Conservation Biology: The Science of Maintaining the Sea's Biodiversity*, Island Press, Washington, DC, pp. 126–174.

Myers, R.F., 1989, *Micronesian Reef Fishes. A Comprehensive Guide to the Coral Reef Fishes of Micronesia*, Coral Graphics, Guam.

Neigel, J.E., 2003, Species-area relationships and marine conservation, *Ecological Applications* **13** (Suppl.): S138–S145.

Neumann, A.C. and Macintyre, I., 1985, Reef response to sea-level rise: keep-up, catch-up or give-up, *Proceedings Fifth International Coral Reef Congress*, Tahiti, 1985, **3**: 105–110.

Nishihira, M. and Veron, J.E.N., 1995, *Corals of Japan*, Kaiyusha Publishers, Tokyo.

Nitani, H., 1972, Beginning of the Kurioshio, in: Stommel, H. and Yoshida, K. (eds), *Kuroshio: Its Physical Aspects*, University of Tokyo Press, Tokyo, pp. 129–163.

Nontji, A., 2002, Coral reefs of Indonesia: past, present and future, *Proceedings Ninth International Coral Reef Symposium*, Bali, 2000, **1**: 17–27.

Norris, R.D., 2000, Pelagic species diversity, biogeography, and evolution, in: Erwin, D.H. and Wing, S.L. (eds), *Deep Time: Paleobiology's Perspective*, The Paleontological Society, Lawrence, *Paleobiology* **26** (4) (Suppl.), pp. 236–258.

Ofwegen, L.P. van, 2002, Status of the knowledge of the Indo-Pacific soft coral genus *Sinularia* May, 1989 (Anthozoa: Octocorallia), *Proceedings Ninth International Coral Reef Symposium*, Bali, 2000, **1**: 167–171.

Olsgard, F., Brattegard, T., and Holthe, T., 2003, Polychaetes as surrogates for marine biodiversity: lower taxonomic resolution and indicator groups, *Biodiversity and. Conservation* **12**: 1033–1049.

Olson, D.B., 1986, Transition zones and faunal boundaries in relationship to physical properties of the ocean, *UNESCO Technical Papers in Marine Science* **49**: 219–225.

Ormond, R.F.G. and Roberts, C.M., 1997, The biodiversity of coral reef fishes, in: Ormond, R.F.G., Gage, J.D., and Angel, M.V. (eds), *Marine Biodiversity: Patterns and Processes*, Cambridge University Press, Cambridge, UK, pp. 216–257.

Osman, R.W. and Whitlatch, R.B., 1978, Patterns of species diversity: fact or artefact? *Paleobiology* **4**: 41–54.

Ota, Y. and Chappell, J., 1999, Holocene sea-level rise and coral reef growth on a tectonically rising coast, Huon Peninsula, Papua New Guinea, *Quaternernary International* **55**: 51–59.

Othman, B.H.R., Greenwood, J.G., and Rothlisberg, P.C., 1990, The copepod fauna of the Gulf of Carpentaria, and its Indo-West Pacific affinities, *Netherlands Journal of Sea Research* **25**: 561–572.

Ovenden, J.R., Salini, J., O'Connor, S., and Street, R., 2004, Pronounced genetic population structure in a potentially vagile fish species (*Pristipomoides multidens*, Teleostei; Perciformes; Lutjanidae) from the East Indies triangle, *Molecular Ecology* **13**: 1991–1999.

Paine, B., 1988, Coral Triangle, in: Neal, A. (ed.), *The Natural World*, BBC, Bristol (45 minutes documentary for television on behalf of UNEP).

Palumbi, S.R., 1994, Genetic divergence, reproductive isolation and speciation in the sea, *Annual Review of Ecology and Systematics* **25**: 547–572.

Palumbi, S. R., 1996, What can molecular genetics contribute to marine biogeography? An urchin's tale, *Journal of Experimental Marine Biology and Ecology* **203**:75–92.

Palumbi, S.R., 1997, Molecular biogeography of the Pacific, *Coral Reefs* **16** (Suppl.): S47–S52.

Palumbi, S.R., 2003, Population genetics, demographic connectivity and the design of marine protected areas, *Ecological Applications* **13** (Suppl.): S146–S158.

Pandolfi, J.M., 1992, Successive isolation rather than evolutionary centres for the origination of Indo-Pacific reef corals, *Journal of Biogeography* **19**: 593–609.

Paulay, G., 1990, Effects of late Cenozoic sea-level fluctuations on the bivalve faunas of tropical oceanic islands, *Paleobiology* **16**: 415–434.

Paulay, G., 1996, Dynamic clams: changes in the bivalve fauna of Pacific islands as a result of sea-level fluctuations, *American Malacological Bulletin* **12**: 45–57.

Paulay, G., 1997, Diversity and distribution of reef organisms, in: Birkeland, C.E. (ed.), *Life and Death of Coral Reefs*, Chapman & Hall, NY, pp. 298–353.

Paulay, G. and Meyer, C., 2002, Diversification in the Tropical Pacific: Comparisons between marine and terrestrial systems and the importance of founder speciation, *Integrative and Comparative Biology* **42**: 922–934.

Paulay, G. and Meyer, C., 2006, Dispersal and divergence across the greatest ocean region: Do larvae matter? *Integrative and Comparative Biology* **46**: 269–281.

Perrin, C. and Borsa, P., 2001, Mitochondrial DNA analysis of the geographic structure of Indian scad mackerel in the Indo-Malay archipelago, *Journal of Fisheries Biology* **59**: 1421–1426.

Perron, F.E. and Kohn, A.J., 1985, Larval dispersal and geographic distribution in coral reef gastropods of the genus *Conus*, *Proceedings Fifth International Coral Reef Congress*, Tahiti, 1985, **5**: 95–100.

Pet-Soede, L., Cesar, H.S.J., and Pet, J.S., 1999, The economics of blast fishing on Indonesian coral reefs, *Environmental Conservation* **26**: 83–93.

Phillips, J.A., 2001, Marine macroalgal biodiversity hotspots: why is there high species richness and endemism in southern Australian marine benthic flora? *Biodiversity and Conservation* **10**: 1555–1577.

Phillips, R.C. and Meñez, E.G., 1988, Seagrasses, *Smithsonian Contributions to Marine Science* **34**: 1–104.

Pierrot-Bults, A.C., 1997, Biological diversity in macrozooplankton: more than counting species, in: Ormond, R.F.G., Gage, J.D., and Angel, M.V. (eds), *Marine Biodiversity: Patterns and Processes*, Cambridge University Press, Cambridge, UK, pp. 69–93.

Pierrot-Bults, A.C., 2003, Pelagic biodiversity and biogeography around the Southamerican continent, *Gayana* **67**: 161–167.

Pierrot-Bults, A.C. and Spoel, S. van der, 2003, Macrozooplankton diversity: how much do we really know? *Zoologische Verhandelingen Leiden* **345**: 297–312.

Ponder, W.F. and Vokes, E.H., 1988, A revision of the Indo-West Pacific fossil and recent species of *Murex* s.s. and *Haustellum* (Mollusca: Gastropoda: Muricidae), *Records of the Australian Museum* (Suppl.) **8**: 1–160.

Porter, J.W. and Tougas, J.I., 2001, Reef ecosystems: threats to their biodiversity, in: Levin, S.A. (ed.), *Encyclopedia of Biodiversity*, Vol. 5, Academic Press, San Diego, CA, pp. 73–93.

Potts, D.C., 1983, Evolutionary disequilibrium among Indo-Pacific corals, *Bulletin of Marine Science* **33**: 619–632.

Potts, D.C., 1984, Generation times and the Quaternary evolution of reef-building corals, *Paleobiology* **10**: 48–58.

Potts, D.C., 1985, Sea-level fluctuations and speciation in the Scleractinia, *Proceedings Fifth International Coral Reef Congress*, Tahiti, 1985, **4**: 127–132.

Prud'homme van Reine, W.F., Verheij, E., and Coppejans, E., 1996, Species and ecads of *Caulerpa* (Ulvophyceae, Chlorophyta) in Malesia (South-East Asia): taxonomy, biogeography and biodiversity, *Netherlands Journal of Aquatic Ecology* **30**: 83–98.

Qiu, B., Potemra, J.T., and Lukas, R., 1996, On the seasonal and interannual bifurcation of the North Equatorial Current along the Pacific western boundary, *Proceedings of the IOC-WESTPAC Third International Science Symposium*, Bali, 1994, pp. 772–784.

Randall, J.E., 1998, Zoogeography of shore fishes of the Indo-Pacific region, *Zoological Studies* **37**: 227–268.

Razak, T.B. and Hoeksema, B.W., 2003, The hydrocoral genus *Millepora* (Hydrozoa: Capitata: Milleporidae) in Indonesia, *Zoologische Verhandelingen Leiden* **345**: 313–336.

Reaka-Kudla, M.L., 1997a, The global biodiversity of coral reefs: a comparison with rain forests, in: Reaka-Kudla, M.L., Wilson, D.E., and Wilson, E.O. (eds), *Biodiversity II. Understanding and Protecting our Biological Resources*, Joseph Henry Press, Washington, DC, pp. 83–108.

Reaka-Kudla, M.L., 1997b, An estimate of known and unknown biodiversity and potential for extinction on coral reefs, *Reef Encounter* **17**: 8–12.

Reid, D.G., 1986, *The littorinid molluscs of mangrove forest in the Indo-Pacific region. The genus Littoraria (Mollusca, Gastropoda, Littorinidae)*, British Museum (Natural History), London.

Reid, D.G., 2001, New data on the taxonomy and distribution of the genus *Littoraria* Griffith and Pidgeon, 1834 (Gastropoda: Littorinidae) in Indo-West Pacific mangrove forests, *Nautilus* **115**: 115–139.

Reid, D.G., Lal, K., Mackenzie-Dodds, J., Kaligis, F., Littlewood, D.T.J., and Williams, S.T., 2006, Comparative phylogeography and species boundaries in *Echinolittorina* snails in the central Indo-West Pacific, *Journal of Biogeography* **33**: 990–1006.

Renema, W., 2002, Larger foraminifera as marine environmental indicators, *Scripta Geologica* **124**: 1–260.

Renema, W., 2006, Large benthic foraminifera from the deep photic zone of a mixed siliciclastic-carbonate shelf off East Kalimantan, Indonesia, *Marine Micropaleontology* **58**: 73–82.

Renema, W., Hoeksema, B.W., and Hinte, J.E. van, 2001, Larger benthic foraminifera and their distribution patterns on the Spermonde shelf, South Sulawesi, *Zoologische Verhandelingen Leiden* **334**: 115–149.

Renema, W. and Troelstra, S.R., 2001, Larger foraminifera distribution on a mesotrophic carbonate shelf in SW Sulawesi, *Paleogeography, Paleoclimatology, Paleoecology* **175**: 125–146.

Ricklefs, R.E. and Latham, R.E., 1993, Global patterns of diversity in mangrove floras, in: Ricklefs, R.E. and Schluter, D. (eds), *Species Diversity in Ecological Communities. Historical and Geographical Perspectives*, Chicago University Press, Chicago, pp. 215–229.

Ricklefs, R.E. and Schluter, D., 1993, Species diversity: regional and historical influences, in: Ricklefs, R.E. and Schluter, D. (eds), *Species Diversity in Ecological Communities. Historical and Geographical Perspectives*, Chicago University Press, Chicago, pp. 350–363.

Roberts, C.M. and Hawkins, J.P., 1999, Extinction risk in the sea, *Trends in Ecology and Evolution* **14**: 241–246.

Roberts, C.M., Hawkins, J., Strong, A.E., Schueler, F.W., and McAllister, D.E., 1998, The distribution of coral reef fish biodiversity: the climate-biodiversity connection, *Sea Wind* **12**: 14–28.

Roberts, C.M., McClean, C.J., Veron, J.E.N., Hawkins, J.P., Allen, G.R., McAllister, D.E., Mittermeier, C.G., Schueler, F.W., Spalding, M., Wells, F., Vynne, C., and Werner, T.B., 2002, Marine conservation hotspots and conservation priorities for tropical reefs, *Science* **295**: 1280–1284.

Rosen, B.R., 1971, The distribution of reef coral genera in the Indian Ocean, in: Stoddart, D.R. and Yonge, C.M. (eds), *Regional Variation in Indian Ocean Coral Reefs*, Academic Press, London, *Symposium Zoological Society London* **28**: 263–299.

Rosen, B.R., 1981, The tropical high diversity enigma – the corals'-eye view, in: Forey, P.L. (ed.), *The Evolving Biosphere. Chance Change and Challenge*, British Museum (Natural History), London, and Cambridge University Press, Cambridge, UK, pp. 103–129.

Rosen, B.R., 1984, Reef coral biogeography and climate through the late Cainozoic: just islands in the sun or a critical pattern of islands? in: Brenchley, P.J. (ed.), *Fossils and Climate*, Wiley, Chichester, UK, pp. 201–262.

Rosen, B.R., 1988a, Biogeographic patterns: a perceptual overview, in: Meyers, A.A. and Giller, P.S. (eds), *Analytical Biogeography. An Integrated Approach to the Study of Animal and Plant Distributions*, Chapman & Hall, London, pp. 23–55.

Rosen, B.R., 1988b, From fossils to earth history: applied historical biogeography, in: Meyers, A.A. and Giller, P.S. (eds), *Analytical Biogeography. An Integrated Approach to the Study of Animal and Plant Distributions*, Chapman & Hall, London, pp. 437–481.

Rosen, B.R., 1988c, Progress, problems and patterns in the biogeography of reef corals and other tropical marine organisms, *Helgoländer Meeresuntersuchungen* **42**: 269–301.

Rosen, B.R. and Smith, A.B., 1988, Tectonics from fossils? Analysis of reef coral and sea urchin distributions from Late-Cretaceous to Recent, using a new method, in: Audley-Charles, M.G. and Hallam, A. (eds), *Gondwana and Tethys*, Oxford University Press, Oxford, NY, pp. 275–306.

Rosenthal, Y., Oppo, D.W., and Linsley B.K., 2003, The amplitude and phasing of climate change during the last deglaciation in the Sulu Sea, western equatorial Pacific, *Geophysical Research Letters* **30** (8), 1428, doi:10.1029/2002GL016612.

Rosenzweig, M.L., 1995, *Species Diversity in Space and Time*, Cambridge University Press, Cambridge, UK.

Ross, A. and Newman, W.A., 2002, Coral barnacles: Cenozoic decline and extinction in the Atlantic/East Pacific versus diversification in the Indo-West Pacific, *Proceedings Ninth International Coral Reef Symposium*, Bali, 2000, **2**: 179–184.

Roy, K., Balch, D.P., and Hellberg, M.E., 2001a, Spatial patterns of morphological diversity across the Indo-Pacific: analyses using strombid gastropods, *Proceedings of the Royal Society B* **268**: 2503–2508.

Roy, K., Jablonsky, D., and Valentine, J.W., 2001b, Climate change, species range limits and body size in marine bivalves, *Ecology Letters* **4**: 366–370.

Samyn, Y. and Tallon, I., 2005, Zoogeography of the shallow-water holothuroids of the western Indian Ocean, *Journal of Biogeography* **32**: 1523–1538.

Santini, F. and Winterbottom, R., 2002, Historical biogeography of Indo-western Pacific coral reef biota: is the Indonesian region a centre of origin? *Journal of Biogeography* **29**: 189–205.

Scheltema, R.S., 1988, Initial evidence for the transport of teleplanic larvae of benthic invertebrates across the East Pacific Barrier, *Biological Bulletin* **174**: 145–152.

Scheltema, R.S. and Hall, J.R., 1975, The dispersal of pelagosphaera larvae by ocean currents and the geographical distribution of sipunculans, in: Rice, M.E. and Todorovic, M. (eds), *Proceedings of the International Symposium on the Biology of the Sipuncula and Echira*, Vol. I, Institute for Biologica Research "Sinisa Stankovic", Belgrade, and National Museum of Natural History, Smithsonian Institution, Washington, DC, pp. 103–115, pl. 1.

Scheltema, R.S. and Rice, M.E., 1990, Occurrence of teleplanic pelagosphera larvae of sipunculans in tropical regions of the Pacific and Indian Oceans, *Bulletin of Marine Science* **47**: 159–181.

Scheltema, R.S. and Williams, I.P., 1983, Long-distance dispersal of planktonic larvae and the biogeography and evolution of some Polynesian and western Pacific mollusks, *Bulletin of Marine Science* **33**: 545–565.

Scheltema, R.S., Williams, I.P., and Lobel, P.S., 1996, Retention around and long-distance dispersal between oceanic islands by planktonic larvae of benthic gastropod Mollusca, *American Malacological Bulletin* **12**: 67–75.

Schiaparelli, S., Barucca, M., Olmo, E., Boyer, M., and Canapa, A., 2005, Phylogenetic relationships within Ovulidae (Gastropoda: Cypraeoidea) based on molecular data from the 16S rRNA gene, *Marine Biology* **147**: 411–420.

Schilder, F.A. and Schilder, M., 1938, Prodrome of a monograph on living Cypraeidae, *Proceedings of the Malacological Society of London* **23**: 119–231.

Schilder, F.A., 1965, The geographical distribution of cowries, *The Veliger* **7**: 171–183.

Schilder, F.A., 1969, Zoogeographical studies on living cowries, *The Veliger* **11**: 367–377.

Schuhmacher, H., 1976, *Korallenriffe: Verbreitung, Tierwelt, Ökologie*. 1st Edition, BLV Verlagsgesellschaft, München.

Schuhmacher, H., 1988, *Korallenriffe: Verbreitung, Tierwelt, Ökologie*. 3rd Edition, BLV Verlagsgesellschaft, München.

Sebens, K.P., 1994, Biodiversity of coral reefs: what are we losing and why? *American Zoologist* **34**: 115–133.

Senerpont Domis, L.N. de, 2004, Assessing macrophyte diversity on coral reefs in the Philippines. Evaluation of the use of indicator species, functional groups and other surrogates, Ph.D. thesis, Leiden University, Leiden, The Netherlands.

Sheppard, C.R.C., 1998, Biodiversity patterns in Indian Ocean corals, and effects of taxonomic error in data, *Biodiversity and Conservation* **7**: 847–868.

Shuto, T., 1983, Larval development and geographical distribution of the Indo-West Pacific *Murex*, *Bulletin of Marine Science* **33**: 536–544.

Siddall, M., Rohling, E.J., Almogi-Labin, A., Hemleben, C., Meischner, D., Schmelzer, I., and Smeed, D.A., 2003, Sea-level fluctuations during the last glacial cycle, *Nature* **423**: 853–858.

Silva, P.C., 1992, Geographic patterns in diversity in benthic marine algae, *Pacific Science* **46**: 429–437.

Sinclair, M., 1988, *Marine Populations: An Essay on Population Regulation and Speciation*, University of Washington Press, Seattle, WA.

Smetacek, V.S., 1986, Impact of freshwater discharge on production and transfer of materials in the marine environment, in: Skreslet, S. (ed.), *The Role of Freshwater Outflow in Coastal Marine Ecosystems*, Springer, Berlin, pp. 85–106.

Spalding, M.D., Ravilious, C., and Green, E.P., 2001, *World Atlas of Coral Reefs*, University of California Press, Berkeley, CA.

Spalding, M.D., Taylor, M., Ravilious, C., Short, F., and Green, E., 2003, Global overview. The distribution and status of seagrasses, in: Green, E.P. and Short, F.T. (eds), *World Atlas of Seagrasses* University of California Press, Berkeley, CA, pp. 5–26.

Spoel, S. van der, and Heyman, R.P., 1983, *A Comparative Atlas of Zooplankton. Biological Patterns in the Oceans*, Wetenschappelijke Uitgeverij Bunge, Utrecht.

Spoel, S. van der, and Pierrot-Bults, A.C. (eds), 1979, *Zoogeography and Diversity in Plankton*, Bunge Scientific Publishers, Utrecht.

Springer, V.G., 1982, Pacific plate biogeography, with special reference to shore fishes, *Smithsonian Contributions to Zoology* **367**: 1–182.

Springer, V.G. and Williams, J.T., 1990, Widely distributed Pacific plate endemics and lowered sea-level, *Bulletin of Marine Science* **47**: 631–640.

Steenis, C.G.G.J. van, 1962, The distribution of mangrove plant genera and its significance for palaeography, *Proceedings van de Koninklijke Nederlandse Academie van Wetenschappen C* **65**: 164–169.

Stehli, F.G., 1968, Taxonomic diversity gradients in pole location: the recent model, in: Drake, E.T. (ed.), *Evolution and Environment*, Yale University Press, New Haven, CT, pp. 163–227.

Stehli, F.G. and Wells, J.W., 1971, Diversity and age patterns in hermatypic corals, *Systematic Zoology* **20**: 115–126.

Stehli, F.G., McAlester, A.L., and Helsley, C.E., 1967, Taxonomic diversity of recent bivalves and some implications for geology, *Bulletin of the Geological Society of America* **78**: 455–466.

Taylor, J.D., 1997, Diversity and structure of tropical Indo-Pacific benthic communities: relations to regimes of nutrient input, in: Ormond, R.F.G., Gage, J.D., and Angel, M.V. (eds), *Marine Biodiversity: Patterns and Processes*, Cambridge University Press, Cambridge, UK, pp. 178–200.

Thomas, J.D., 1993, Biological monitoring and tropical biodiversity in marine environments: a critique with recommendations, and comments on the use of amphipods as bioindicators, *Journal of Natural History* **27**: 795–806.

Thomas, J.D., 1997, Using marine invertebrates to establish research and conservation priorities, in: Reaka-Kudla, M.L., Wilson, D.E., and Wilson, E.O. (eds), *Biodiversity II. Understanding and Protecting our Biological Resources*, Joseph Henry Press, Washington, DC, pp. 357–369.

Thorne-Miller, B., 1999, *The Living Ocean. Understanding and Protecting Marine Biodiversity*, Island Press, Washington, DC.

Tjia, H.D. and Fuji, S., 1992, Late Quarternary shorelines in Peninsula Malaysia, in: Tjia, H.D. and Abdullah, S.M.S. (eds), *The Coastal Zones of Peninsular Malaysia*, Penerbit Universiti Kebangsaan Malaysia, Bangi, pp. 28–41.

Tokeshi, M., 1999, *Species Coexistence: Ecological and Evolutionary Perspectives*, Blackwell Science, Oxford, UK.

Tomascik, T, Mah, A,J., Nontji, A., and Moosa, M.K., 1997, *The Ecology of the Indonesian Seas*, 2 volumes, Periplus, Singapore.

Tursh, B. and Greifeneder, D., 2001, *Oliva Shells. The Genus Oliva and the Species Problem*, L'Informatore Piceno, Ascona (Italy), and Bosque BMT, Costa Rica.

Umbgrove, J.H.F, 1929, De koraalriffen der Duizend Eilanden (Java-Zee), *Wetenschappelijke Mededelingen van de Dienst Mijnbouw in Nederlandsch-Indië* **12**: 1–47, pls. I–VI.

Umbgrove, J.H.F, 1947, Coral reefs of the East Indies, *Bulletin of the Geological Society of America* **58**: 729–778.

Umbgrove, J.H.F. and Verwey, J., 1929, The coral reefs in the Bay of Batavia, *Fourth Pacific Science Congress Java, Excursion* **A2**: 1–30.

UNEP-WCMC, 2004, *World Atlas of Seagrasses*, http://www.unep-wcmc.org/marine/seagrassatlas/

Uthicke, S. and Benzie, J.A.H., 2003, Gene flow and population history in high dispersal marine invertebrates: mitochondrial DNA analysis of *Holothuria nobilis* (Echinodermata: Holothuroidea) populations from the Indo-Pacific, *Molecular Ecology* **12**: 2635–2648.

Vallejo, B.M., 2001, The biogeography of Philippine marine molluscs, *Loyola Schools Reviews* **1**: 21–42.

Vallejo, B.M., 2005, Inferring the mode of speciation in Indo-Pacific *Conus* (Gastropoda: Conidae), *Journal of Biogeography* **32**: 1429–1439.

VandenSpiegel, D., Eeckhaut, I., and Jangoux, M., 1998, Host selection by *Synalpheus stimpsoni* (De Man), an ectosymbiotic shrimp of comatulid crinoids, inferred by a field survey and laboratory experiments, *Journal of Experimental Marine Biology and Ecology* **225**: 185–196.

Verheij, E. and Erftemijer, P.L.A., 1993, Distribution of seagrasses and associated macroalgae in South Sulawesi, Indonesia, *Blumea* **38**: 45–64.

Vermeij, G.J., 1973, Molluscs in mangrove swamps: physiognomy, diversity, and regional differences, *Systematic Zoology* **22**: 609–624.

Vermeij, G.J., 1978, *Biogeography and Adaptation. Patterns of Marine Life*, Harvard University Press, Cambridge, MA.

Vermeij, G.J., 1987, The dispersal barrier in the tropical Pacific: implications for molluscan speciation and extinction, *Evolution* **41**: 1046–1058.

Vermeij, G.J., 1989, Interoceanic differences in adaptation: effects of history and productivity, *Marine Ecology Progress Series* **57**: 293–305.

Vermeij, G.J., 2004, Island life: a view from the sea, in: Lomolino, M.V. and Heaney, L.R. (eds), *Frontiers of Biogeography. New Directions in the Geography of Nature*, Sinauer Associates, Sunderland, UK, pp. 239–254.

Veron, J.E.N., 1985, Aspects of the biogeography of hermatypic corals, *Proceedings Fifth International Coral Reef Congress*, Tahiti, 1985, **4**: 83–88.

Veron, J.E.N., 1986, *Corals of Australia and the Indo-Pacific*, Angus & Robertson, Sydney.

Veron, J.E.N., 1993, A biogeographic database of hermatypic corals: species of the central Indo-Pacific, genera of the world, *Australian Institute of Marine Science Monograph Series* **10**: 1–433.

Veron, J.E.N., 1994, Biodiversity of reef corals: is there a problem in the Indo-Pacific centre of diversity? in: Ginsburg, R.N. (ed.), *Proceedings of the Colloquium on Global Aspects of Coral Reefs: Health, Hazards, and History*, University of Miami, Florida, 1993, Rosenstiel School of Marine and Atmospheric Science, University of Miami, Miami, pp. 365–370.

Veron, J.E.N., 1995, *Corals in Space and Time: the Biogeography and Evolution of the Scleractinia*, Cornell University Press, Ithaca.

Veron, J.E.N., 2000, *Corals of the World*, 3 volumes, Australian Institute of Marine Science, Townsville, MC.

Veron, J.E.N., 2002, Reef corals of the Raja Ampat Islands, Papua Province, Indonesia, in: McKenna, S.A., Allen, G.R., and Suryadi, S. (eds), *A Marine Rapid Assessment of the Raja Ampat Islands, Papua Province, Indonesia. RAP Bulletin of Biological Assessment 22*, Conservation International, Washington, DC, pp. 26–28.

Veron, J.E.N. and Marsh, L.M., 1988, Hermatypic corals of Western Australia: records and annotated species list, *Records of the West Australian Museum* **29** (Suppl.): 1–136.

Veron, J.E.N. and Minchin, P.R., 1992, Correlations between sea surface temperature, circulation patterns and the distribution of hermatypic corals of Japan, *Continental Shelf Research* **12**: 835–857.

Voogd, N.J. de, Cleary, D.F.R., Hoeksema, B.W., Noor, A., and Soest, R.W.M. van, 2006, Sponge beta diversity in the Spermonde Archipelago, SW Sulawesi, Indonesia, *Marine Ecology Progress Series* **309**: 131–142.

Voogd, N.J. de, Soest, R.W.M. van, and Hoeksema, B.W., 1999, Cross-shelf distribution of southwest Sulawesi reef sponges, *Memoirs of the Queensland Museum* **44**: 147–154.

Voris, H.K., 2000, Maps of Pleistocene sea levels in Southeast Asia: shorelines, river systems and time durations, *Journal of Biogeography* **27**: 1153–1167.

Wallace, C.C., 1997, The Indo-Pacific centre of coral diversity re-examined at species level, *Proceedings Eighth International Coral Reef Symposium*, Panama, 1996, **1**: 365–370.

Wallace, C.C., 1999a, The Togian Islands: coral reefs with a unique coral fauna and a hypothesised Tethys Sea signature, *Coral Reefs* **18**: 162.

Wallace, C.C., 1999b, *Staghorn Corals of the World: A Revision of the Coral Genus Acropora (Scleractinia; Astrocoeniina; Acroporidae) Worldwide, with Emphasis on Morphology, Phylogeny and Biogeography*, CSIRO Publishing, Melbourne.

Wallace, C.C., 2002, Journey to the heart of the centre - Origins of high marine faunal diversity in the central Indo-Pacific from the perspective of an acropologist, *Proceedings Ninth International Coral Reef Symposium*, Bali, 2000, **1**: 33–39.

Wallace, C.C., Pandolfi, J.M., Young, A. and Wolstenholme, J., 1991, Indo-Pacific coral biogeography: a casestudy from the *Acropora selago* group, *Australian Systematic Botany* **4**: 199–210.

Wallace, C.C., Paulay, G., Hoeksema, B.W., Bellwood, D.R., Hutchings, P.A., Barber, P.H., Erdmann, M., and Wolstenholme, J., 2002, Nature and origins of unique high diversity reef faunas in the Bay of Tomini, Central Sulawesi: The ultimate "centre of diversity? *Proceedings Ninth International Coral Reef Symposium*, Bali, 2000, **1**: 185–192.

Wallace, C.C. and Rosen, B.R., 2006, Diverse staghorn corals (*Acropora*) in high-latitude Eocene assemblages: implications for the evolution of modern diversity patterns of reef corals, *Proceedings of the Royal Society B* **273**: 975–982.

Wallace, C.C. and Wolstenholme, J., 1998, Revision of the coral genus *Acropora* (Scleractinia: Astrocoeniina: Acroporidae) in Indonesia, *Zoological Journal of the Linnean Society* **123**: 199–384.

WCMC, 1997, *Coral Reefs and Mangroves of the World*, http://www.unep-wcmc.org/marine/data/coral_mangrove/marine_maps_main.html.

Weel, K.M. van, 1923, Meteorological and hydrographical observations made in the western part of the Netherlands East Indian Archipelago, *Treubia* **4**: 1–559.

Wells, F.E., 1990, Comparative zoogeography of marine molluscs from northern Australia, New Guinea, and Indonesia, *The Veliger* **33**: 140–144.

Wells, F.E., 2002, Centres of species richness and endemism of shallow-water marine molluscs in the tropical Indo-West Pacific, *Proceedings Ninth International Coral Reef Symposium*, Bali, 2000, **2**: 941–945.

Wells, J.W., 1954, Recent corals of the Marshall Islands, *US Geological Survey Professional Papers* **260**: 609–615.

Wells, J.W., 1966, Evolutionary development in the scleractinian family Fungiidae, in: Rees, W.J. (ed.), *The Cnidaria and their Evolution*, Academic Press, London, *Symposium Zoological Society London* **16**: 223–246.

Werner, T.B. and Allen, G.R., 1998, *A rapid biodiversity assessment of the coral reefs of Milne Bay Province, Papua New Guinea. RAP Bulletin of Biological Assessment* 11, Conservation International, Washington, DC.

Werner, T.B. and Allen, G.R., 2000, *A Rapid Marine Biodiversity Assessment of the Calamianes Islands, Palawan Province, Philippines. RAP Bulletin of Biological Assessment 17*, Conservation International, Washington, DC.

White, A.T., Vogt, H.P., and Arin, T., 2000, Philippine coral reefs under threat: the economic losses caused by reef destruction, *Marine Pollution Bulletin* **40**: 598–605.

Wilkinson, C.R. and Cheshire, A.C., 1989, Patterns in the distribution of sponge populations across the central Great Barrier Reef, *Coral Reefs* **8**: 127–134.

Wilkinson, C.R. and Cheshire, A.C., 1990, Comparisons of sponge populations across the barrier reefs of Australia and Belize: evidence for higher productivity in the Caribbean, *Marine Ecology Progress Series* **67**: 285–294.

Wilkinson, C.R. and Trott, L.A., 1985, Light as a factor determining the distribution of sponges across the central Great Barrier Reef. *Proceedings Fifth International Coral Reef Congress*, Tahiti, 1985, **5**: 125–130.

Williams, D.M., 1982, Patterns in the distribution of fish communities across the central Great Barrier Reef, *Coral Reefs* **1**: 35–43.

Williams, D.M. and Hatcher, A.I., 1983, Structure of fish communities on outer slopes of inshore, mid-shelf and outer shelf reefs of the Great Barrier Reef, *Marine Ecology Progress Series* **10**: 239–250.

Williams, D.M., Dixon, P., and English, S., 1988, Cross-shelf distribution of copepods and fish larvae across the central Great Barrier Reef, *Marine Biology* **99**: 577–589.

Williams, G.C., 1993, Biotic diversity, biogeography, and phylogeny of Pennatulacean Octocorals associated with coral reefs in the Indo-Pacific, *Proceedings Seventh International Coral Reef Symposium*, Guam, 1992, **2**: 729–735.

Williams, S.T., 2000, Species boundaries in the starfish genus *Linckia*, *Marine Biology* **136**: 137–148.

Williams, S.T. and Benzie, J.A.H., 1996, Genetic uniformity of widely separated populations of the coral reef starfish *Linckia laevigata* from the East Indian and West Pacific Oceans, revealed by allozyme electrophoresis, *Marine Biology* **126**: 99–107.

Williams, S.T. and Benzie, J.A.H., 1997, Indo-West Pacific patterns of genetic differentiation in the high-dispersal starfish *Linckia laevigata*, *Molecular Ecology* **6**: 559–573.

Williams, S.T. and Benzie, J.A.H., 1998, Evidence of a biogeographic break between populations of a high dispersal starfish: congruent regions within the Indo-West Pacific defined by color morphs, mtDNA, and allozyme data, *Evolution* **52**: 87–99.

Williams, S.T., Jara, J., Gomez, E., and Knowlton, N, 2002, The marine Indo-West Pacific break: contrasting the resolving power of mitochondrial and nuclear genes, *Integrative and Comparative Biology* **42**: 941–952.

Williams, S.T., Knowlton, N., and Weigt, L.A., 1999, Indo-Pacific molecular biogeography of the coral-dwelling snapping shrimp *Alpheus lottini* (Decapoda: Caridea: Alpheidae), in: Sheppard, C.R.C. and Seaward, M.R.D. (eds), *Ecology of the Chagos Archipelago, Occassional Publications of the Linnean Society*, 2, pp. 195–206.

Williams, S.T. and Reid, D.G., 2004, Speciation and diversity on tropical rocky shores: a global phylogeny of snails of the genus *Echinolittorina*, *Evolution* **58**: 2227–2251.

Williamson, M., 1988, Relationship of species number to area, distance and other variables, in: Meyers, A.A. and Giller, P.S. (eds), *Analytical Biogeography. An Integrated Approach to the Study of Animal and Plant Distributions*, Chapman & Hall, London, pp. 91–115.

Wilson, E.J. and Rosen, B.R., 1998, Implications of paucity of corals in the Paleogene of SE Asia: plate tectonics or Centre of Origin, in: Hall, R. and Holloway, D. (eds),

*Biogeography and Geological Evolution of Southeast Asia*, Backhuys Publishers, Leiden, pp. 165–195.

Woesik, R. van, 1996, Geographic location influences coral composition: a comparison between Bali and Sulawesi, Indonesia, *Proceedings of the IOC-WESTPAC Third International Science Symposium*, Bali, 1994, pp. 280–290.

Woodland, D.J., 1983, Zoogeography of the Siganidae (Pisces): an interpretation of distribution and richness patterns, *Bulletin of Marine Science* **33**: 713–717.

Woodland, D.J., 1986, Wallace's Line and the distribution of marine inshore fishes, in: Uyeno, T., Arai, R., Taniuchi, T., and Matsuura, K. (eds), *Indo-Pacific Fish Biology. Proceedings of the Second International Conference on Indo-Pacific Fishes*, Ichtyological Society of Japan, Tokyo, pp. 453–460.

Woodroffe, C.D. and Grindrod, J., 1991, Mangrove biogeography: the role of Quaternary environmental and sea-level change, *Journal of Biogeography* **18**: 479–492.

Wörheide, G., Hooper, J.N.A., and Degnan, B.M., 2002, Phylogeography of western Pacific *Leucetta 'chagosensis'* (Porifera: Calcarea) from ribosomal DNA sequences: implications for population history and conservation of the Great Barrier Reef World Heritage Area (Australia), *Molecular Ecology* **11**: 1753–1768.

Worm, B., Lotze, H.K., and Myers, R.A., 2003, Predator diversity hotspots in the blue ocean, *Proceedings of the National Academy of Sciences, USA* **100**: 9884–9888.

Worm, B., Sandow, M., Oschlies, A., Lotze, H.K., and Myers, R.A., 2005, Global patterns of predator diversity in the open oceans, *Science* **309**: 1365–1369.

Wyrtki, K., 1961, Scientific results of marine investigations of the South China Sea and the Gulf of Thailand. Physical oceanography of the Southeast Asian Waters, *Naga Reports* **2**: 1–195.

Yamano, H., 2002, Sensitivity of reef flats and reef islands to sea-level change, *Proceedings Ninth International Coral Reef Symposium*, Bali, 2000, **2**: 1193–1197.

Zhang, Q. and Zhang, Y., 1986, The fish fauna of Taiwan Strait, in: Uyeno, T., Arai, R., Taniuchi, T., and Matsuura, K. (eds), *Indo-Pacific Fish Biology. Proceedings of the Second International Conference on Indo-Pacific Fishes*, Ichtyological Society of Japan, Tokyo, pp. 465–470.

Chapter 6

# Fauna Development of Larger Benthic Foraminifera in the Cenozoic of Southeast Asia

WILLEM RENEMA
Nationaal Natuurhistorisch Museum Naturalis, P.O. Box 9517, 2300 RA Leiden, The Netherlands, Renema@naturalis.nl



**Abstract** The central part of the Indo-West Pacific (IWP) harbours the highest marine biodiversity, but only few data of its origin are available. In this paper the Cenozoic records of large benthic foraminifera are reviewed. Since many groups need species level revision, the occurrences of genera were reviewed based on records from literature. It is acknowledged that using genus level data is susceptible



*W. Renema (ed.), Biogeography, Time, and Place: Distributions, Barriers, and Islands,* 179–215

to differences in interpretation between authors, but the observed patterns are tested against morphological disparity and the number of taxa observed in a single locality. The analyses show a longitudinal shift in diversity from the western Tethys in the Eocene to the central IWP in the Miocene and younger. Oligocene faunas were relatively homogenous between these two regions. During the Eocene radiation occurred in the western Tethys, and abundant and often stratigraphically long-ranging species migrated east and reached Indonesia. This pattern reversed from the Late Eocene onwards, continuing in the Oligocene, and especially the Miocene. However, although the same genera were observed in Oligocene deposits from the western and eastern Tethys lineages showing similar morphological trends evolved independently and at different rates in the two regions. The observed patterns compare very well with the availability of suitable habitat in shallow tropical seas, and are fine-tuned by environmental parameters such as temperature and nutrient availability.

## 1. Introduction

The history of diversity patterns in the Cenozoic fauna is poorly known for the Indo-West Pacific (IWP). Wilson and Rosen (1998) showed that high diversity assemblages of zooxanthellate corals did not appear in the Indo-Malayan region until the start of the Neogene. During most of the Paleogene, zooxanthellate coral diversity was much higher in western Tethys, especially in Europe. However, Paleogene, and especially pre-Chattian, corals are rare in the IWP and unevenly spread through time. They are found in several (Middle) Eocene assemblages, but the Late Eocene to Early Oligocene is poorly represented in their data set. Consequently, Wilson and Rosen (1998) argued that the IWP was not a centre of diversity for zooxanthellate corals during the Paleogene. Instead, this centre was positioned within the western Tethyan realm in Europe. With the Australian plate moving northward into warmer areas during the Oligocene to Miocene, the isolation of the Indo-Pacific area decreased and diversity of zooxanthellate corals increased (Wilson and Rosen, 1998). Meanwhile in Europe, zooxanthellate coral diversity increased until the Burdigalian and then gradually decreased. At the end of the Messinian zooxanthellate corals became extinct in the Mediterranean as the result of loss of habitat and climatic cooling (Wilson and Rosen, 1998; Rosen, 1999).

Scleractinian corals, several families of benthic molluscs and other taxa living in shallow marine, tropical conditions have their highest current diversity in the IWP region (Paulay, 1997; Hoeksema, Chapter 5 of this volume and references therein). Diversity within the IWP region peaks in Indo-Malaysia (IM), is relatively even across the Indian Ocean, and falls gradually eastward across the western and central Pacific (Paulay, 1997). Other, less diverse, faunal provinces are the East Pacific, West Atlantic (including the Caribbean), and the East Atlantic. Of these, the West Atlantic is the most diverse (Paulay, 1997). Larger symbiont-bearing foraminifera also show this biogeographical pattern (Hottinger, 1983; Langer and Hottinger, 2000). The main faunal provinces are the Caribbean and the IWP. Some genera/species seem to have a circum-tropical distribution (e.g., *Heterostegina depressa*, *Sorites orbiculus*), while others are restricted to the Caribbean (e.g., *Cyclorbiculina compressa*; Langer and Hottinger, 2000) or IWP (e.g., *Calcarina* spp. and *Alveolinella quoyii*). The Caribbean Province houses only about 60% of the number of genera compared to the IWP. Within the IWP, faunal composition differs between the West Pacific (including southern Japan, the Philippines, and Indonesia) and East Africa (including the Red Sea and the East African coast). For example, the genus *Calcarina* occurs in the IWP, but is absent from East Africa and the Red Sea, whilst for *Heterocyclina tuberculata* the opposite is the case.

Three parameters determine the occurrence of larger foraminifera; temperature, oligotrophy, and life history. Symbiosis is only profitable (for both host and symbiont) in high temperatures (Hallock, 1987; Hollaus and Hottinger, 1997). The distribution of larger symbiont-bearing foraminifera is restricted to a worldwide climatic belt bordered by the 15–20°C minimum temperature range (Adams et al., 1990; Langer and Hottinger, 2000), similar temperature limits have been observed for zooxanthellate corals (Adams et al., 1990; Veron, 1995). All species have a different sea surface temperature (SST) limit in the coldest winter month, the area showing highest diversity is the area with the highest SST; (Langer and Hottinger, 2000). *Amphistegina* has the largest tolerance (>14°C in the coldest month), while *Operculina heterosteginoides* and *A. quoyii* have the most limited temperature tolerance (>26°C in the coldest month). The biogeographic distribution of living larger foraminifera is controlled by the extent of oligotrophic water masses (Langer and Hottinger, 2000). The geometric delimitation of faunal provinces does not reflect an important role of the surface currents as the main agents for the distribution and propagation of shallow benthic species. Moreover, the main faunal provinces are bounded by the extent of land masses and upwelling zones (Langer and Hottinger, 2000). On a smaller scale, the input of terrestrial run-off affects the distribution patterns of species on a regional or local scale (Langer and Hottinger, 2000; Renema and Troelstra, 2001; Renema, 2006a, b).

Biogeographical patterns are the result of three categories of biological processes, related to, respectively, maintenance, distributional change, and origination (Rosen, 1988). Maintenance refers to processes by which an organism maintains its presence in a particular area: extinction can be seen as the failure of an organism to maintain itself. Distributional change refers to those processes by which an organism survives in response to changing environmental conditions by shifting its geographical range, for example, in following the most suitable habitat. Origination is related to the evolution of new taxa (Rosen, 1988). The larger the geographical and stratigraphical scale, the more influence tectonic, climatologic, and oceanographic changes have on the observed patterns. Changing positions of land masses can create or remove barriers, whilst climatologic changes directly affect the range of latitudes at which larger foraminifera can be expected to occur. However, within this zone differences in seasonality and rainfall (among other parameters) influence the distribution of taxa. Therefore, palaeogeographic and palaeoceanographic changes in the geologic history should be considered as well in order to understand present-day biogeographical patterns.

## 1.1. Sea Surface Temperature (SST)

Plate configurations, and the presence or absence of either Antarctic or Arctic polar ice-caps during the Cenozoic resulted in changing ocean circulation patterns and global climatic conditions, i.e., in latitudinal contractions and expansions of the high SST belt (e.g., Adams et al., 1990). Throughout the Cenozoic, ocean circulation patterns and basin configurations led to high temperatures at the western sides of oceans. IM was always positioned on or near the equator, within the warmest regions on earth. Due to the collision of Australia with Eurasia and the narrowing of the Indo-Pacific gateway, the Pacific Warm Pool moved gradually eastward during the Miocene (Wang, 1994; Hall, 1998). Europe, positioned at the northern margin of the Western Tethys, lay at considerably higher latitudes, around 30°N, and was thus more affected by changes in the temperature gradient. All other parameters remaining constant, a prediction based on SST would show a high diversity in Indonesia throughout the Cenozoic, and at all times a lower and decreasing diversity from the Early Eocene onwards in Europe.

## 1.2. Palaeoceanography and Ocean Circulation Rates

Several studies have linked the diversity of larger foraminifera to that of planktonic foraminifera by the Trophic Resource Continuum (TRC) model. The TRC is the spectrum of conditions from the richest run-off and upwelling to the most nutrient deficient subtropical seas (Hallock, 1987). Though at first developed to predict larger foraminiferal diversity and distribution in recent oceans, the model can also be used to explain diversity trends through time (Hallock, 1987, 1988; Hallock et al., 1991; Boersma et al., 1998; McGowran and Li, 2000).

Changes in ocean circulation rates influence the global nutrient gradient in surface waters. Reduced rates of circulation result in an expanded nutrient gradient and a higher potential diversity, while higher mixing rates contract the nutrient gradient and have a lower potential diversity (Hallock, 1987; Hallock et al., 1991). The deep photic zone assemblage, with species adapted to living in very low light levels, disappear resulting in decreased habitat fractionation during periods of strong oceanic mixing. Thus, if tropical diversity is driven by ocean circulation rates, and associated changes in nutrient availability, a stepwise decrease diversity during the Cenozoic would be expected.

## 1.3. Eustasy

Apart from the effect of direct causes of eustatic sea level change (predominantly produced by the waxing and waning of ice caps and changes in rate of sea floor spreading), an obvious result of high sea levels is the flooding of continental margins leaving more potential habitats for shallow marine fauna like larger foraminifera. During periods with low sea levels the base level of erosion is lowered and, accordingly, erosion rates will be higher, thus supplying more nutrients and sediment to the coastal zone, further deteriorating the environment for larger benthic foraminifera.

### 1.4. Outline

In this paper the diversity development of larger benthic foraminifera in the Cenozoic of IM is discussed and compared with coeval patterns in Europe. Outcrops and stratigraphy, occurrence and taxonomy of individual genera will be discussed. Section 2 discusses some tectonic and oceanographic patterns affecting the distribution patterns, which will form the basis of a discussion that focuses on whether the patterns we see are real, or just an artefact of our understanding of the fossil record.

European faunas include localities from Spain, France, Italy, former Yugoslavia, and Greece, but exclude Turkey (Fig. 1). IM consists of records from Indonesia (without Irian Jaya), Sabah, Sarawak, and the Philippines (Fig. 1), but does not include New Guinea or other areas on the Australian plate. The denomination IM is chosen in preference to Southeast Asia, since the region studied is smaller than that used by Wilson and Rosen (1998). However, it is emphasized that it is used as a geographical term and that no biogeographical implications are intended.

## 2. Genera Included

Larger foraminifera are a polyphyletic group, recognized mainly by their size and complex internal structure. Only those families showing characters indicative of algal symbiosis were included in the present analysis. Such characteristics include stolon systems to facilitate the transportation of symbionts responding to irradiation levels; flosculization of the lateral wall; the development of cups and chamberlets

Fig. 1 Palaeogeography of the Tethys Ocean during the Lutetian. The map is based on reconstructions by Dercourt et al. (1993), with additional information from Hall (1996) concerning the IWP. The areas of upwelling (curved arrows) are indicated after Parrish and Curtis (1982). IM = Indo-Malaysia, E = Europe, C = Caribean, J = Java, B = Borneo, WS = West Sulawesi, ES = East Sulawesi, M = Mindoro. Black = land, dark grey = shallow marine carbonates, light grey = shallow marine clastics.

within the chambers; and the development of lateral chamberlets and pillars to regulate the amount of light getting to the symbionts. In general, all genera belonging to the superfamilies Nummulitoidea, Soritoidea, and Alveolinoidea are included (Table 1). In the superfamilies Rotaloidea, Asterigerinoidea, and Orbitoidoidea (Table 1)

*Table 1. Superfamilies and families included in the database on larger foraminifera occurrences. Note that the nominate taxa of the families Lepidocyclinidae, Meandropsinidae, and Lepidorbitoididae are not included since they have not been recorded in the studied area or time interval. Genera marked by a dot have not been observed in situ in Indo-Malaysia*

| *Superfamily* | *Families included* | *Genera included* |
|---|---|---|
| Milioloidea Ehrenberg, 1839 | Austrotrillinidae Loeblich and Tappan, 1986 | *Austrotrillina* Parr, 1942 |
| Alveolinoidea Ehrenberg, 1839 | Alveolinidae Ehrenberg, 1839 | *Alveolinella* Douvillé, 1907 |
| | | • *Rhabdorites* Fleury, 1996 |
| | | • *Glomalveolina* Hottinger, 1962 |
| | | *Flosculinella* Schubert, 1910 |
| | | • *Bullalveolina* Reichel, 1936 |
| | | *Borelis* de Montfort, 1808 |
| | | *Alveolina* d'Orbigny, 1826 |
| | | • *Praebullalveolina* Sirel and Acar, 1982 |
| | Fabulariidae Ehrenberg, 1839 | • *Pseudolacazina* Caus, 1979 |
| | | • *Periloculina* Munier-Chalmas and Schlumberger, 1885 |
| | | • *Lacazina* Munier-Chalmas, 1882 |
| | | • *Fabularia* Defrance, 1820 |
| | | • *Lacazinella* Crespin, 1962 |
| Soritoidea Ehrenberg, 1839 | Meandropsinidae Henson, 1948 | • *Hottingerina* Drobne, 1975 |
| | Peneroplidae Schultze, 1854 | • *Reulina* Lamarck, 1804 |
| | | • *Archiacina* Munier-Chalmas, 1878 |
| | | *Spirolina* Lamarck, 1804 |
| | | *Dendritina* d'Orbigny, 1826 |
| | | *Peneroplis* de Montfort, 1808 |
| | Soritidae Ehrenberg, 1839 | • *Opertorbitolites* Nutall, 1925 |
| | | *Orbitolites* Lamarck, 1801 |
| | | *Praerhapydionina* van Vessem, 1943 |
| | | *Pseudotaberina* Eames, 1971 |
| | | *Archaias* de Montfort, 1808 |
| | | *Sorites* Ehrenberg, 1839 |
| | | • *Globoflarina* Fleury, 1982 |
| | | *Marginopora* Quoy and Gaimard, 1830 |
| | | • *Cyclorbiculina* Silvestri, 1937 |
| | | *Parasorites* Seiglie and Rivera, 1977 |
| | | *Amphisorus* Ehrenberg, 1839 |
| Asterigerinoidea D'Orbigny, 1839 | Amphisteginidae Cushman, 1927 | *Amphistegina* d'Orbigny, 1826 |
| | Boreloididae Reiss, 1963 | • *Boreloides* Cole and Bermudez, 1947 |
| | Lepidocyclinidae Scheffen, 1932 | *Nephrolepidina* Douvillé, 1911 |
| | | *Eulepidina* Douvillé, 1911 |

(*continued*)

*Table 1.* (continued)

| *Superfamily* | *Families included* | *Genera included* |
|---|---|---|
| Rotalicea Ehrenberg, 1839 | Calcarinidae Schwager, 1876 | *Silvestriella* Hanzawa, 1952 |
| | | *Calcarina* d'Orbigny, 1826 |
| | | *Baculogypsina* Sacco, 1893 |
| | | *Sclumbergerella* Yabe and Hanzawa, 1930 |
| | | *Baculogypsinoides* Yabe and Hanzawa, 1930 |
| | Miogypsinidae Vaughan, 1928 | *Miogypsina* Sacco, 1893 |
| | | *Miogypsinoides* Yabe and Hanzawa, 1928 |
| | | *Miolepidocyclina* Silvestri, 1907 |
| | | *Miogypsinoidella* Boudagher-Fadel et al., 2000 |
| | | *Lepidosemicyclina* Rutten, 1911 |
| | Rotaliidae Ehrenberg, 1839 (pars)* | *Neorotalia* Bermudez, 1952 |
| | | *Pararotalia* Le Calvez, 1949 |
| Orbitoidoidea Schwager, 1876 | Linderinidae Loeblich and Tappan, 1974 | *Linderina* Schlumberger, 1893 |
| | Eoannularidae Ferràndez-Cañadell and Serra-Kiel, 1999 | • *Eoannularia* Cole and Bermudez, 1944 |
| | Lepidorbitoididae Vaughan, 1933 | • *Actinosiphon* Vaughan, 1929 |
| | | • *Daviesina* Smout, 1954 |
| Nummulitoidea de Blainville, 1827 | Asterocyclinidae Brönnimann, 1951 | *Orbitoclypeus* Silvestre, 1907 |
| | | *Asterocyclina* Gümbel, 1870 |
| | Discocyclinidae Galloway, 1928 | *Discocyclina* Gümbel, 1870 |
| | | • *Nemkovella* Less, 1987 |
| | Miscellaneidae Sigal, 1952 | • *Miscellanea* Pflender, 1935 |
| | Nummulitidae de Blainville, 1827 | *Assilina* d'Orbigny, 1839 |
| | | *Cycloclypeus* Carpenter, 1856 |
| | | *Grzybowskia* Bieda, 1950 |
| | | *Heterostegina* d'Orbigny, 1826 |
| | | *Katacycloclypeus* Tan, 1932 |
| | | *Nummulites* Lamarck, 1801 |
| | | • *Nummulitoides* Abrard, 1956 |
| | | *Operculina* d'Orbigny, 1826 |
| | | *Operculinella* Yabe, 1918 |
| | | *Palaeonummulites* Schubert, 1908 |
| | | *Planocamerinoides* Cole, 1957 |
| | | *Radiocycloclypeus* Tan, 1932 |
| | | • *Ranikothalia* Caudri, 1944 |
| | | *Spiroclypeus* Douvillé, 1905 |
| | | *Tansinhokella* Banner and Hodgkinson, 1991 |
| | Pellatispiridae Hanzawa, 1937 | *Pellatispira* Boussac, 1906 |
| | | *Biplanispira* Umbgrove, 1937 |
| | | *Vaculispira* Tan, 1936a |

**Only the subfamily Pararotalinae, with the genera* Neorotalia *and* Pararotalia*. The extant genus* Neorotalia *houses symbionts. Loeblich and Tappan (1987) regarded these genera as synonymous, but Hottinger et al. (1991) have shown sufficient morphological difference to warrant generic distinction.*

not all families house symbionts. Austrotrillinidae (Milioloidea) are included because their alveolate walls indicating a zooxanthellate life habit (Adams, 1968).

Although the use of higher-level taxonomy increases the problems of interpretation (Hottinger, 1990), the generic level is used herein since the systematics at the level of species are not yet resolved for some important taxa (e.g., *Nephrolepidina*, *Eulepidina*, Miogypsinidae, *Spiroclypeus*). The most basic taxonomical problem is that even within the same family, different concepts of generic classification have been used. This problem cannot be resolved completely at present. For example, historically there have been researchers using mainly numerical data for taxonomy (e.g., in Lepidocyclinidae, Discocyclinidae, and Miogypsinidae), while others used typological defined species (e.g., in Nummulitidae and Alveolinidae). Also, there is no consensus about the value of coiling patterns for generic classification.

It is evident that future revisions will affect several of the genera mentioned. Some of the smaller genera might turn out to be synonymous, whilst others might have to be split. The latter is most likely to happen in the larger genera like *Nummulites* and *Nephrolepidina*. Both genera can be divided into several lineages (with side branches) of species. For *Nummulites*, seven species groups were described by Schaub (1981), of which only four have been found in southeast Asia (Renema et al., 2003). At present, only the *Nummulites fichteli* group is sufficiently distinct on morphological characters to warrant a separate count. If the other species groups were given generic status too, the Ta3 interval would have an additional two genera in southeast Asia as opposed to six in Europe. Although turnover rate at the Eocene–Oligocene boundary interval would increase, the overall pattern remains identical. Thus, it is believed that taxonomic revisions will only affect the numbers, but not the observed patterns.

## 3. The East Indian Letter Classification

The Letter stages used in the biostratigraphy of the Cenozoic of southeast Asia are assemblage zones defined by the occurrences of larger benthic foraminifera. The ranges have been used as the basis for the "letterklassificatie" in the then Netherlands East Indies since the 1920s (Van der Vlerk and Umbgrove, 1927; Leupold and Van der Vlerk, 1931). Originally the letter stages replaced the stage names (Ta for Middle Eocene, Tb Late Eocene, etc.). In the earliest publications the number of stages varied considerately and some range adjustments had to be made. For example, even before the letter classification was formalized, the stratigraphic range of *Spiroclypeus* (now *Tansinhokella* and *Spiroclypeus*) was used to separate two stages within the post-Eocene (Van der Vlerk, 1925), equivalent to Te in Van der Vlerk and Umbgrove (1927) and Te1–5 in Leupold and Van der Vlerk (1931). Tan (1930a, 1937) found that *Spiroclypeus* also occurred in the Eocene, which was one of his arguments against the further subdivision of lower Te (Tan, 1936a, 1939). However, the main stages defined by the overlap of ranges of some key genera, were fairly consistent (Ta–Tg or Th). Most of the boundaries are defined by the first (FO) or last (LO) occurrences of important genera (Fig. 2; Table 2).

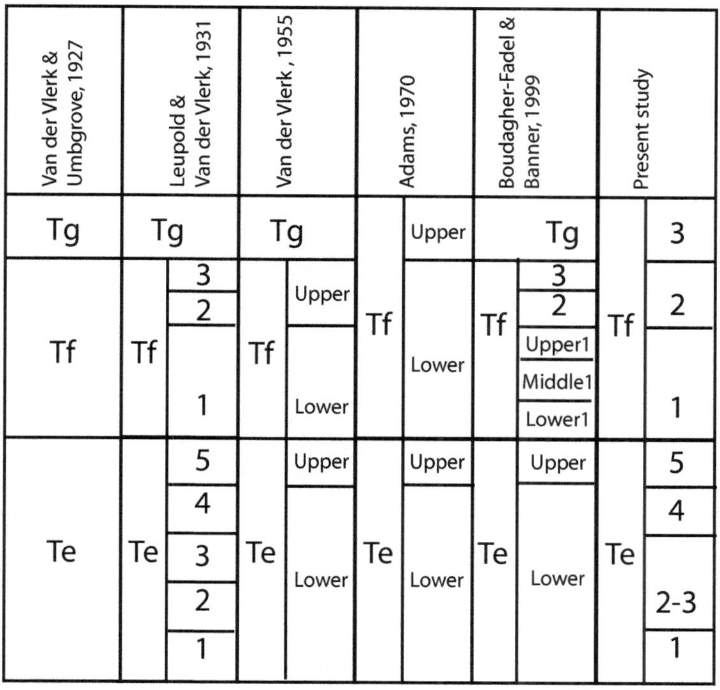


Fig. 2 Development of the Te–Tf part of the letter classification since 1927.

Deposits older than Tb are rare in IM, and sections in which the fauna changes over boundary intervals are absent or not studied. The revision of this part of the letter stages by Adams (1970) was mainly based on sections in India. Based on the ranges of nummulitids without secondary chamberlets, Ta3, the only stage unambiguously present, can be separated into two parts, Ta3–1 (i.e., upper part of the Lutetian), and Ta3–2 (i.e., lower part of the Bartonian). Ta3–2 is the range zone of *N. javanus* and *N. boninensis*, Ta3–1 the range of *N. djokdjokartae* (Renema et al., 2003).

The occurrence of *Planocamerinoides/Assilina* and *Alveolina* are used to indicate Ta, followed by a short period of overlap with *Pellatispira* and *Biplanispira*, and only the latter two genera in Tb (Lunt, 2003). *Alveolina* is rare in Ta3–2, and, especially in East Indonesia, more abundant in Ta3–1; its absence is thus not unambiguous proof of Tb age. The last occurrence (LO) of *Planocamerinoides/Assilina* falls together with the LO of large, non-reticulate *Nummulites* (Renema et al., 2003). This combination of events can be traced in other parts of the Tethys as well (LFA6-7 in India, Govindan, 2003; SBZ17-18 in the western Tethys, Serra-Kiel et al., 1998). The Ta3–2/Tb boundary is better defined by the extinction of large *Nummulites* (*N. javanus* and *N. boninensis*, see Renema et al., 2003).

Tc and Td together span the range of *N. fichteli*. In the upper part of its range this species is rare, and the end of Td can better be defined by the FO of *Heterostegina (Vlerkina) borneensis* than the LO of *N. fichteli* (Tan, 1936a). The Te biozone spans

*Table 2.* *Boundary definitions of the stages and substages of the East Indian Letter classification as proposed in this article*

| | *Lower boundary* | *Upper boundary* | *Subdivision* | |
|---|---|---|---|---|
| Ta3 | | LO of *Alveolina*, large *Nummulites* | | |
| Tb | | LO of *Pellatispira* | | |
| Tc | FO of *N. fichteli* | | | |
| Td | FO of *Eulepidina* | | | |
| Te1–4 | FO of *H.(V.) borneensis* | LO of *H.(V.) borneensis* | Te1 | Lower boundary: FO of *H.(V.) borneensis* |
| | | | Te2–3 | Lower boundary: FO of *Tansinhokella* |
| | | | Te4 | Lower boundary: FO of *Miogypsinoides* |
| Te5 | FO of *Miogypsina* | LO of *Eulepidina*. *Spiroclypeus* | | |
| Tf | | LO of Nephrolepidina | Tf1 | Upper boundary LO of *Austrotrillina* |
| | | | Tf2 | Upper boundary LO of *Flosculinella borneensis* |
| | | | Tf3 | Upper boundary: LO of *Nephrolepidina* |
| Tg | | Tg/Th boundary: not defined on LBF occurrences | | |
| Th | | Th/Pleistocene boundary: not defined on LBF occurrences | | |

a long and important stratigraphical interval, including the Oligo–Miocene transition. Leupold and Van der Vlerk (1931) subdivided Te into 5 subzones, mainly based on occurrences of species of *Eulepidina*, *Nephrolepidina*, and *Spiroclypeus*. This subdivision was rejected by Tan (1939) and Mohler (1949). Mohler only accepted a subdivision in lower Te or Te1–4 and upper Te or Te5. This was used in later revisions as well (Fig. 3; Van der Vlerk, 1955; Adams, 1970; Boudagher-Fadel and Banner, 1999). Evaluation of occurrences in drill hole samples collected for the "Bataafse Petroleum Maatschappij" (BPM) and the "Nederlandsche Pacifische Petroleum Maatschappij" (NPPM) shows that a subdivision generally can be made, but that all diagnostic species are rarely present. Te1 (*H. borneensis* without *Tansinhokella* or *Spiroclypeus*) can be recognized, as *Spiroclypeus* and *Tansinhokella* are generally abundant within their range (Te2–Te4). In Tawoen and Kahajan drill holes, *Miogypsinoides* is observed together with *Heterostegina (Vlerkina)*, typical of Te1–4. In the original definition of Te4, Leupold and Van der Vlerk (1931) also used the FO of *Mygypsinoides*, their *Miogypsina* with lateral chamberlets. This is reinstated in the present study. Because *Miogypsinoides* is rare in the early part of its range, its absence is not always indicative for Te2–3 (when *Spiroclypeus* is present). In Te4, however *Tansinhokella* is rare and *Spiroclypeus* is abundant, whilst in Te2–3 *Tansinhokella* is the most abundant genus.

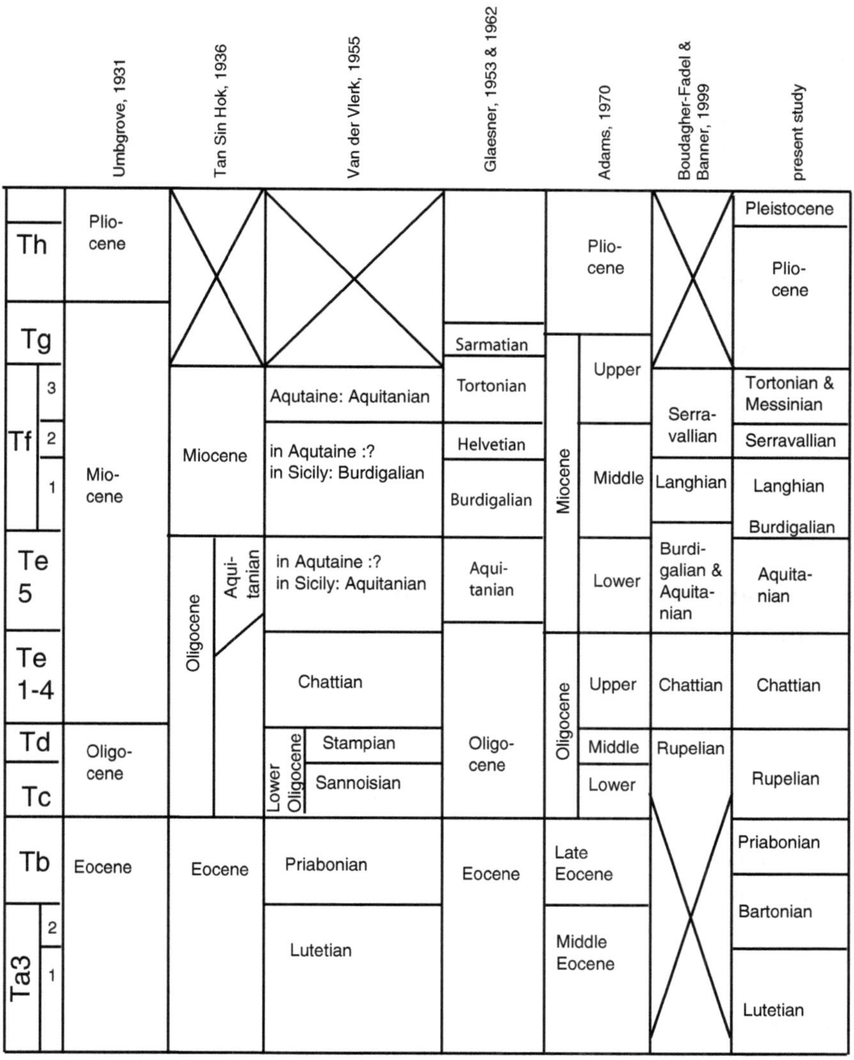


Fig. 3 Changes in the correlation of the East Indian Letter Classification since 1931.

In the definition of Tf1–3 (Leupold and Van der Vlerk, 1931), the upper boundary of Tf was defined as the LO of *Nephrolepidina* and *Miogypsina*. During subsequent work, it was found that *Nephrolepidina* ranges beyond *Miogypsina* and that the division between Tf2 and Tf3 could not be traced. This resulted in various ways of grouping the substages in Tf. Successive divisions of Tf varied from two (Adams, 1970) to five substages (Fig. 3; Boudagher-Fadel and Banner, 1999). The division of Tf1 into three parts is based on the extinctions of *Miogypsinoides* and *Miolepidocyclina* in the lower part of Tf1 (Boudagher-Fadel and Banner, 1999). Both genera are rare and facies–dependent, and the age of many samples cannot be assessed up to this level. In the present study a division into three substages of Tf was used (Table 2). The lower boundary of Tf3 is characterized by the LO of *F. bontangensis*; miogypsinids are absent.

Tg and Th, the Tertiary without orbitoids according to Van der Vlerk and Umbgrove (1927), was separated by Leupold and Van der Vlerk (1931) on the basis of the proportion extant mollusc genera, and were equivalent to the Late Miocene and Pliocene respectively. Adams (1970) showed that *Nephrolepidina* extended beyond Tf2 and ranges through most, if not all, of the Late Miocene (also Adams, 1984; Van Vessem, 1978). In the present study, Adams (1970) is followed in extending Tf3 until the Miocene/Pliocene boundary, and thus in omitting Tg. The Late Miocene is subdivided into Tortonian and Messinian in Europe. Most deposits are from deeper water carbonate platforms or shelves, equivalent to or slightly deeper than the recent Spermonde shelf, with assemblages including both planktonic and larger foraminifera. Most assemblages can thus be characterized at least to the above-mentioned resolution.

## 4. Correlation to Plankton Foraminifera Zonal Schemes and European Stage Names

The letter classification was developed because correlation with Europe turned out to be very difficult and larger benthic foraminifera are abundant components in Cenozoic carbonates. Van der Vlerk and Umbgrove (1927) used the open nomenclature for their biozones, because they were not confident direct correlation of bioevents between Europe (where most of the stage names were defined) and Indonesia.

Calibration of the letter classification with independent stratigraphical methods is the only way to confidently correlate with a chronostratigraphical scheme. Problems arise because, in many type-sections of European stages, no larger foraminifera occur in common with Far Eastern sections. Also, several of the type-sections, and many of the larger benthic foraminifera-rich deposits in Indonesia, contain only a few planktonic foraminifera. The increasing use and understanding of planktonic foraminifera biostratigraphy has led to an increasing understanding of the correlation between Europe and the Far East (Fig. 3).

The FO of *Orbulina universa*, a very characteristic planktonic foraminifer, was used by Le Roy (1948, 1952) to correlate the top of the Telissa Formation with strata in Europe, Australia, and New Zealand. The FO of *Orbulina universa* was used by Boudagher-Fadel and Banner (1999) to discriminate between middle and upper TF1. Glaessner (1943, 1953, 1959) critically reviewed the letter stages, and all subsequent modifications. Glaessner (1959) acknowledged the great practical value of a simple system of recognizing the major subdivisions of the Tertiary, while criticizing certain details. Glaessner (1959) continued the discussion on correlating the letter stages with the European stage-names and thought that the letter classification might have outlived its usefulness. In his reconsideration of the letter stages Adams paid little attention to correlation with plankton stratigraphy, although he concludes the problem part at the end of his paper by stating that "Accurate correlation ... with the planktonic zonal scheme is a most important aim, and one that undoubtedly be realized in the not too distant future" (Adams, 1970: 127).

Adams (1984) discussed many of the important datum planes on which the letter stages were based and related these to the plankton correlation schemes, focussing on Tc and younger. Later studies included the whole Cenozoic (Haak and Postuma, 1975; van Gorsel, 1988), and those of Boudagher-Fadel and Banner (1999) for Td and younger built on these results.

In the present study, the results of Renema et al. (2003) and Boudagher-Fadel and Banner (1999) have been used for the correlation with the planktonic zonal scheme (Fig. 4). The Tc–Td transition is at *c*. 32 Ma, approximately the P18/P19 boundary (see discussion in Renema et al., 2003). As Renema et al. (2003) have shown, the youngest rocks containing *N. fichteli* have been dated as 29.4–28.4 Ma (slightly older or equivalent to the Rupelian–Chattian boundary).

The substages Te1, Te2–3, and Te4 were not included in the above-mentioned studies. At the moment only incomplete data on planktonic foraminifera are available, and the resolution of the zonation in the plankton scheme is also not as good. The Te1 to Te2–3 boundary has been taken as the P21a–P22 boundary, the Te2–3 to Te4 boundary falls within P22. Future work should confirm or detail these findings.

## 5. Remarks on Some Stratigraphic Occurrences

The stratigraphical ranges of genera have been summarized in Fig. 5. Some of the ranges will be discussed below.

### 5.1. Miogypsinidae

The generic classification of Boudagher-Fadel et al. (2000) has been used. The stratigraphical data have been adjusted with own observations where necessary.

### 5.2. Calcarinidae

Six genera of calcarinids have been found in IM. *Silvestriella* has been found on Sulawesi (Osimo, 1908) and Java (Sangiran; Renema, unpublished record) from Tb assemblages (Renema, 2002). Of the extant genera *Schlumbergerella*, *Baculogypsina*, and *Baculogypsinoides* all have their FO in the latest Pliocene. *Neorotalia* ranges from Tc to Te4 (Lunt and Allan, 2004). Whittaker and Hodgkinson (1979) found middle Pliocene *Neorotalia*.

*Calcarina* occurs form the Late Miocene to recent, but occasionally older reports have been published in both IM and Europe. A revision is needed to establish the supraspecific relations within this group.

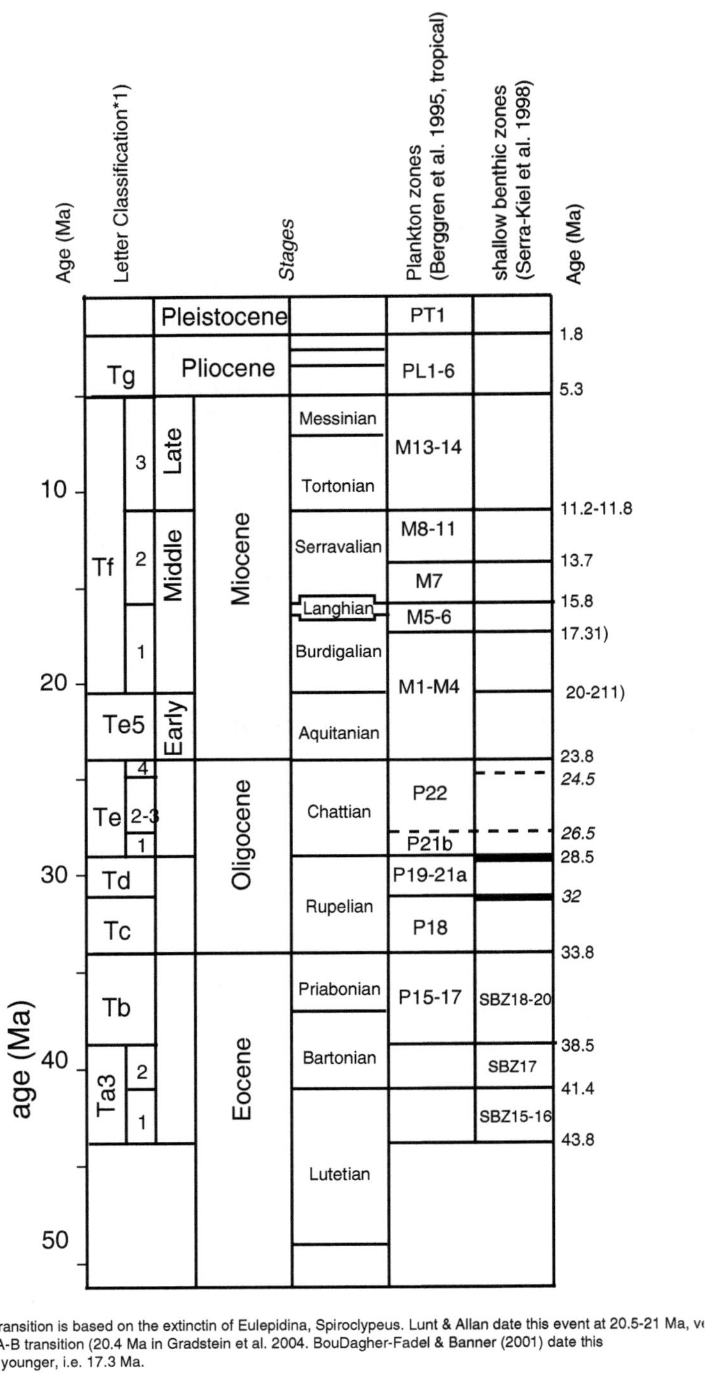


Fig. 4 Correlation of the letter classification with plankton zonal schemes and chronostratigraphy. The fat lines at the Tc/Td and Td/Te boundaries indicate uncertainties based on Sr-isotope data discussed in Renema et al. (2003).

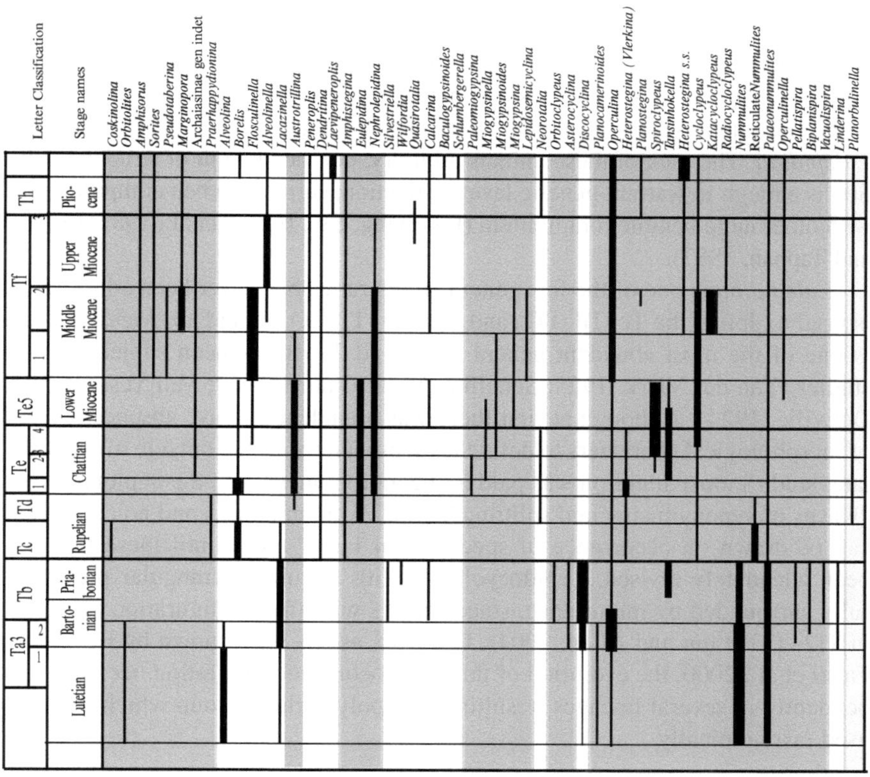


Fig. 5 Range chart of the Indo-Malayan larger foraminifera genera.

## 5.3. *Eulepidina* and *Nephrolepidina*

The southeast Asian Lepidocyclinidae have been classified into two or three (sub)genera (Tan, 1936b; Eames et al., 1982; Adams, 1987); *Eulepidina* and *Nephrolepidina*, to which some authors add *Trybliolepidina*. Megalospheric specimens regularly can be recognized up to this level, but microspheric specimens, especially in random thin section often cannot. Because of the similarity of the microspheric generation they have often been regarded subgenera within the genus *Lepidocyclina*.

The generic classification of the *Lepidocyclinidae* proposed by Sirotti (1982) is used, in which *Nephrolepidina* and *Lepidocyclina* are considered to have evolved in separate lineages from a common ancestor, and therefore are separate genera, with *Eulepidina* evolving from *Lepidocyclina*. Boudagher-Fadel and Banner (1997) proposed a different classification, which summarized all B-forms as *Lepidocyclina*. The latter is a genus occurring only in the Caribbean, and it would create confusion to use *Lepidocyclina* for B-forms of either *Eulepidina* or *Nephrolepidina* (Renema, 2006c).

The principal difference between *Eulepidina* and *Lepidocyclina* is that the former possesses adauxilary chambers while the other does not. *Lepidocyclina*, on the other hand always possesses a complete periembryonic ring of primary auxiliary and interauxillary chambers, a feature missing in known representatives of *Eulepidina*. The rank of these units is arbitrary, but the amount of structural variation is enough to warrant generic level distinction, certainly when comparing them with other large benthic foraminifera (following, e.g., Eames et al., 1982; Loeblich and Tappan, 1987).

*Eulepidina* is one of the few genera which range has never been adjusted, and is used to define the Tc–Td (FO) and the Te5–Tf (LO) boundary. *Nephrolepidina* is one of the most abundant genera in Te and Tf. It has been subject to many studies (Van der Vlerk, 1928; Scheffen, 1932; Caudri, 1939; Van Vessem, 1978; Douvillé, 1924). Although at first there was a tendency to use absence/presence of morphological characters to define taxa, especially on the outside of the test, in later studies most attention was paid to the morphometrics of the nepiont. Several "waves of synonymising and splitting" left a taxonomic mess and no conclusions can be drawn on occurrence of species over larger areas until these taxa have been adequately revised. Lepidocyclinids with a (sub)quadrangular protoconch fully surrounded by importance, since species with this configuration only occur in Tf2–Tf3 (Lunt and Allan, 2004). However, as has been shown by Boudagher-Fadel et al., 2000, the evolution of this nucleoconch configuration occurred independently in several lineages, resulting in a polyphyletic group which cannot be used taxonomically.

## 5.4. Nummulitidae

### *5.4.1. Nummulites Sensu Lato*

The ranges of the species of *Nummulites* and *Palaeonummulites* were summarized in Renema et al. (2003). Important is the LO of large *Nummulites*, i.e., *N. javanus* and *N. boninensis* at the Ta–Tb boundary.

*Nummulites* (reticulate). The reticulate *Nummulites* are a species group that on morphological grounds can be separated from the true *Nummulites*. This group was already treated as a separate group for stratigraphical reasons to define Tc and Td. The traces of the marginal cord are straight and parallel, instead of anastomizing as in *Palaeonummulites* and *Nummulites*, and the very even height of the alar prolongations (as opposed to a gradually thinning alar prolongation towards the centre), intersected by pillars, creating structures that resemble lateral chamberlets, but are not. These two characters would warrant inclusion in a separate genus. This will have to wait on a revision of the other species groups of *Nummulites* for which no defining characters have been found. In the present study the reticulate *Nummulites* have been counted as if they were a separate genus.

*Operculinella*. This involute genus is very similar to *Palaeonummulites* and involute *Operculina*. It is distinguished from *Palaeonummulites* by the absence of trabeculae. The trabeculae are only visible in loose specimens, and never in thin sections. It is possible that *Operculinella* has been overlooked. The oldest certain record is from the Prupuh beds, Te5 or lower Miocene. It is distinguished from *Operculina* by the alar prolongations that protrude over the previous whorl.

*5.4.2. Nummulitidae with Secondary Chamberlets*

Banner and Hodgkinson (1991) revised this group of genera. In their classification they considered the mode of coiling, and the presence of lateral chamberlets or cubicula to be most important. Their classification has been used here as well. However, *Heterostegina* (*Vlerkina*) (Eames et al., 1968) has been included as a separate entity because of its distinct morphological characters. Banner and Hodgkinson (1991) merely maintained subgeneric status so that unidentifiable immature specimens could be referred to as *Heterostegina* s.l. As both groups are clearly distinct in their stratigraphical distribution, I prefer to use both subgenera as separate entities in this count.

*Heterostegina* (*Heterostegina*) (d'Orbigny, 1826). So far this subgenus is only known from Late Miocene to recent (Lunt and Allan, 2004; Banner and Hodgkinson, 1991).

*Heterostegina* (*Vlerkina*) (Eames et al., 1968; emended Banner and Hodgkinson, 1991). *Heterostegina (V.) borneensis*, the type species of this subgenus, is of high stratigraphical value as an indicator of Te1–4. The number of occurrences is highest in samples with assemblages referable to Te1, and rare in samples of Te2–3 and Te4 (number of abundances counted in boreholes by BPM of which the larger benthic foraminifera containing samples were sent to Van der Vlerk for identification and age determination).

*Planostegina* (Banner and Hodgkinson, 1991). This genus occurs in two stratigraphic intervals in Indonesia. *Planostegina praecursor* (Tan, 1930) and *P. bantamensis* (Tan, 1932) have been found in typical Tc–Td assemblages of Java and Borneo. Van der Vlerk (1922, 1925a) mentions *P. reticulata* Rütimeyer from Tb to Tc beds, but so far I have been unable to trace specimens of *P. reticulata* from samples with a Tb-age from the region. Specimens, some identified as *P. reticulata*, from Tc to Td assemblages from Kalimantan in the Van der Vlerk collection so far all turned out to be *P. praecursor*.

*P. heterosteginoides* (Hofker, 1927) occurs from Pliocene to recent (Whittaker and Hodgkinson, 1979; Hofker, 1927).

*Grzybowskia* (Bieda, 1950). Lunt (2003) reported an isolated occurrence of *Grybowskia* from the Late Eocene of Christmas Island. No other occurrences of this genus are known from IM. The nearest record from this otherwise boreal genus is from Bonin Island (South Japan; Matsumaru, 1996).

*Tansinhokella* (Banner and Hodgkinson, 1991) and *Spiroclypeus* (Douvillé, 1905). *Tansinhokella* has its FO in Tb of several localities (Kutei, South Kalimantan; Sarawak; Boeton, East Sulawesi), but has not been found in Tc–Te1 assemblages in the region. In sections and drill holes in Java and Borneo on top of beds with *N. fichteli* first some beds occur with *H. borneensis* (indicative of Te), but without *Spiroclypeus* and *Tansinhokella*. Te2–3 is defined at the FO of *Tansinhokella*.

*5.4.3. Nummulitidae with Annular Chambers*

This group evolved from *P. praecursor* (Tan, 1930). In 1932, Tan Sin Hok made a detailed study of the cycloclypei, describing two new subgenera, being defined on the presence of one or more concentric thickenings, a structure he called annuli (*Katacycloclypeus*) or the presence of radial ribs (*Radiocycloclypeus*). Both subgenera have a very restricted stratigraphical range in the Middle Miocene (Tf). Subsequently, there has been some discussion as to whether the presence of annuli is a valid character with which to define a genus. Tan (1932) already realized that some species in *Cycloclypeus* s.s. could develop uncommon annuli as well. The most important reason for Cole (1975) to abandon the subgenus *Katacycloclypeus* was that specimens of the same species could be assigned to two different subgenera. Adams and Frame (1979) reinvestigate the status of *Katacycloclypeus*, concluding that *Katacycloclypeus* should be assigned at least subgenus rank, but that the mere presence or absence of annuli in the megolospheric form is not diagnostic at the genus level, although it has some value at the species level. They stated that "if reduction of the nepionic spiral in *Cycloclypeus* is accepted as an indication of the grade of evolution, then *C. (K) annulatus* Martin, 1880, is the most highly evolved member of the genus, notwithstanding the fact that it is extinct and that a more 'primitive' member lives on today" (Adams and Frame, 1979: 15). Thus, *Katacycloclypeus* can be regarded as a separate lineage, next to the lineage that eventually led to *C. carpenteri* (Brady, 1861; Adams and Frame, 1979). Within *Katacycloclypeus* no trends in reduction of nepionic spiral, nor in the number of annuli has been proven. Both subgenera have been given genus rank in this study (following Loeblich and Tappan (1987) for *Radiocycloclypeus*).

*Cycloclypeus* (Carpenter, 1861). Tc-recent. The oldest records come from Cimanggu (South Bantam, Java), in an association with *N. fichteli*, and *P. praecursor*.

*Radiocycloclypeus* (Tan, 1932). The only known species of this genus, *R. stellatus* (Tan, 1930) occurs together with *Nephrolepidina rutteni* in Kalimantan and can thus not be defined more precisely than Tf, but the absence of other *Nephrolepidina* species might hint at a Tf2 or Tf3 age.

## 5.5. Porcelaneous Foraminifera

*Praerhapydionina*. The youngest record is from the lower part of the Tonasa limestone, where *P. delicata* is found with *N. fichteli* and *Eulepidina* in Td deposits (Crotty and Engelhardt, 1993). Eames et al. (1962) found this species in deposits as young as Te5 in the Asmari limestone in Iran and Iraq, but I do not know of any Indo-Malayan record younger than Td.

*Austrotrillina*. The FO occurrence of the genus *Austrotrillina* is within the Early Oligocene in the Asmari limestone in Iran and Iraq (Henson, 1950; Adams, 1968). The oldest record from the Indo-Malayan region is from Td, for example, from the Tonasa limestone in southwest Sulawesi, where *A. striata* occurs together with *Eulepidina* and *N. fichteli*. The LO of *Austrotrillina* indicates the end of Tf1, but records in the youngest part of its range are rare (see also Adams, 1984).

*5.5.1. Borelis, Flosculinella, and Alveolinella*

*Borelis pygmea* (Hanzawa, 1930). Specimens identified as *B. pygmea* have been encountered in boreholes in Borneo from Tc to Te5; and it was most abundant in Te1 (archives of Van der Vlerk, Leiden). Bakx (1932) gave a range of uppermost Ta–Tc and Te. The Ta/Tb record is based on a single record from Loh Oeloh published as *Alveolina* sp. by Verbeek and Fennema (1896). Of these specimens, the initial chambers were missing (Bakx, 1932), and, although the later chambers match with *B. pygmea*, more proof is needed to extend the range of this genus into the Eocene. Other Eocene records are by Henrici (1934), but these specimens can be referred to as *Alveolina* (Adams, 1970, and own research on specimens deposited in Technische Hogeschool Delft, now in Nationaal Natuurhistorisch Museum Naturalis (NNM), Leiden, and Caudri (1934) from Sumba (specimens restudied in NNM). It has been recorded from Tb strata in the West-Pacific (Cole, 1957), thus it is possible that it occurs very rarely in Tb in the area. No records from Tf are known. Recently *Borelis* was discovered in Pliocene (Kalimantan) and Pleistocene deposits (southeast Sulawesi; unpublished personal observation), and it has been reported in recent sediments (McCulloch, 1977; unpublished personal observation). In drill holes in West-Pacific atolls, *B. schlumbergeri* also occurs in Pliocene and younger deposits (Cole, 1957; Adams, 1970).

*Flosculinella* (Schubert, 1910). *Flosculinell*a is differentiated by having two rows of secondary chamberlets instead of only one, as in *Borelis*, or three, as in *Alveolinella*. Representatives of *Flosculinella* form a morphogenetic series differentiated by their size and length/width ratio. Restudying the type material and additional specimens from the type locality of *F. globulosa* (Rutten, 1913) shows that these populations are perfectly spherical and of similar size as the type material of *F. reicheli* (Mohler, 1949), and the latter is thus considered a junior synonym of *F. globulosa*. The oldest species, *F. globulosa* (Rutten), is found in the upper part of Te and the lowest part of Tf1. Specimens from the type locality of *F. bontangensis* are about 2–2.5× as long and as thick, but in many samples transitional specimens have been found. The youngest species, *F. borneensis* (Tan, 1936a), is found in the Tawoen (East Java) drill holes in Tf2, together with *N. rutteni* and *Katacycloclypeus*. This species is even more elongated (3–5× as long and as wide) and towards the poles one or more additional rows of chamberlets appear. This is very similar to *A. praequoyi* (Wonders and Adams, 1991) with which it is possibly synonymous.

*Alveolinella* (Douvillé, 1907). The inclusion of *Flosulinella borneensis* remains ambiguous. The oldest unambiguous species of *Alveolinella*, i.e., *A. quoyii*, is recorded from Tf3 onwards.

*5.5.2. Soritidae, Peneroplidae, Calcarinidae*

The taxonomy of the Soritoidea is not yet resolved. Hottinger (2001) (re)described several Caribbean genera showing the importance of internal morphological characters of the test. He also showed that the generic and higher taxonomy in this group is far from resolved in the Tethys region, for a large part because of lack of well-preserved, free specimens. Furthermore, specimens of the Soritoidea usually make up only a very small fraction of the total assemblage and occur patchily.

Most members of the above families are extremely rare and known from only a few records per stage. This makes the effect of "occasional" finds of one the genera relatively large. Within the Soritidae I have recognized *Orbitolites* in the Eocene, and single-layered *Sorites*, double-layered *Amphisorus*, and multilayered *Marginopora*. Occasional loose finds from Java show that *Sorites* might have to be split into at least two, but until a good descriptions of these specimens is produced, they will not be available for worthwhile analysis.

Involute Soritoidea with an endoskeleton have a very fragmented occurrence. The records of *Pseudotaberina* are evaluated in Renema (in press) and occur predominantly in Tf1, with one record from Tf2. The BPM collections contain fragments or sections of other taxa have been found as well. These specimens show no annular passage and are thus different from *Pseudotaberina*, and have been included as Archaiasinae gen. indet. These specimens occur in samples from Te2–3 to Tf3 age.

## 6. Generic Diversity of the Indo-West Pacific as Compared to Europe

In the IM late Lutetian 9 and the early Bartonian 13, genera have been recorded, compared to 16 in the late Bartonian/Priabonian (Fig. 6). After a drop in diversity at the Eocene–Oligocene boundary, the generic diversity continued to increase until a maximum was reached in Te5–Tf2 (18–22 genera). During the Middle Miocene the number of genera decreased rapidly from 22 to 18 in the Late Miocene (Fig. 6).

The occurrences of larger benthic foraminifera genera in Europe have been summarized in Renema (2002), and those data are included herein. The oldest Cenozoic occurrences of larger symbiont-bearing foraminifera in Europe are from the Selandian (Middle Paleocene), but these are not considered here. Twenty genera have been recorded from the late Lutetian, while the diversity maximum was reached in the Priabonian with 24 genera. Most of the genera that have been found in the IM Ypresian to Lutetian interval are represented by the largest number of species in Europe in the same period (*Assilina*, *Nummulites*, *Alveolina*,*Orbitolites*, *Planocamerinoides*, *Discocyclina*, and *Asterocyclina*; Schaub, 1981; Less, 1987; Renema, 2002).

The faunal character changed markedly at the Eocene–Oligocene transition (Adams, 1973). The genera mentioned above were the most important genera during the Eocene, while from the Oligocene onwards *Nephrolepidina*, *Eulepidina*, *Cycloclypeus* and miogypsinids dominated the fauna. The extinction at the end of the Eocene was not rapid, but stepwise, starting in the Bartonian and with the last extinctions at the Priabonian–Rupelian boundary. In Europe, at both the Bartonian–Priabonian and the Priabonian–Rupelian boundaries, about half of the existing genera went extinct; the number of FO in the succeeding period is similar to those in earlier stages, resulting in a net loss in generic diversity. In IM the extinction at the Early–Late Bartonian boundary was less pronounced. At the end of the early Bartonian four genera went extinct, but six genera had their FO in Tb, resulting in a net increase in generic diversity over the Bartonian–Priabonian boundary (Figs. 6, 7). At the end of Tb 13 genera went extinct. However, nine genera are restricted

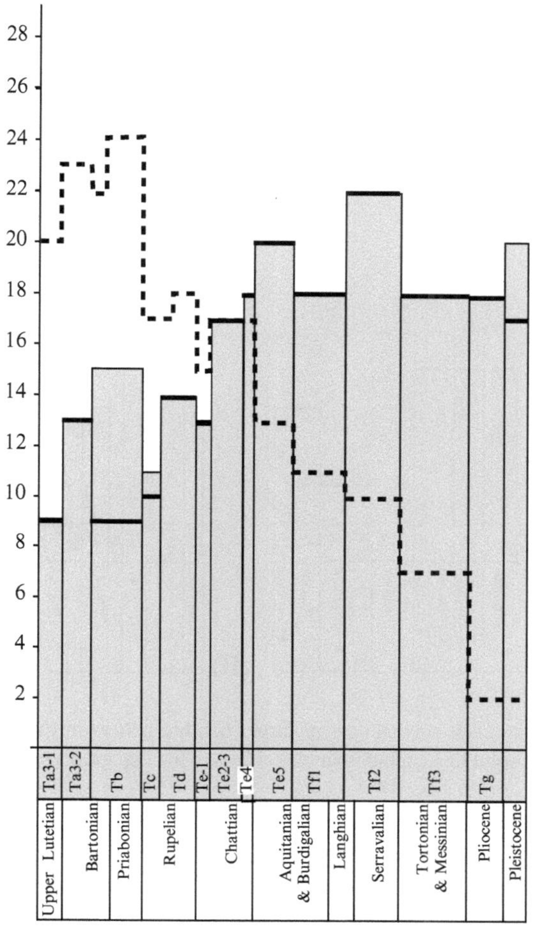


Fig. 6 Diversity of larger benthic foraminifera in Indo-Malaysia including (grey bars) and excluding (solid lines) genera who have their first and last occurrence within the same stage, as compared to Europe (dashed line). The European diversity data are taken from Renema (2002).

to Tb, whilst *Tansinhokella* had its FO in the Priabonian, but went locally extinct at the end of the Priabonian and did not return until the Te2–3. This increased the number of genera, FO and LO in Tb by nine.

## 6.1. Palaeogeographic Events Influencing the Distribution of Larger Foraminifera

The distribution of larger foraminifera is influenced by changes in the geography and distribution of habitats through time. During the Paleocene and Eocene, the Tethys Ocean still existed and formed a more or less continuous passageway at

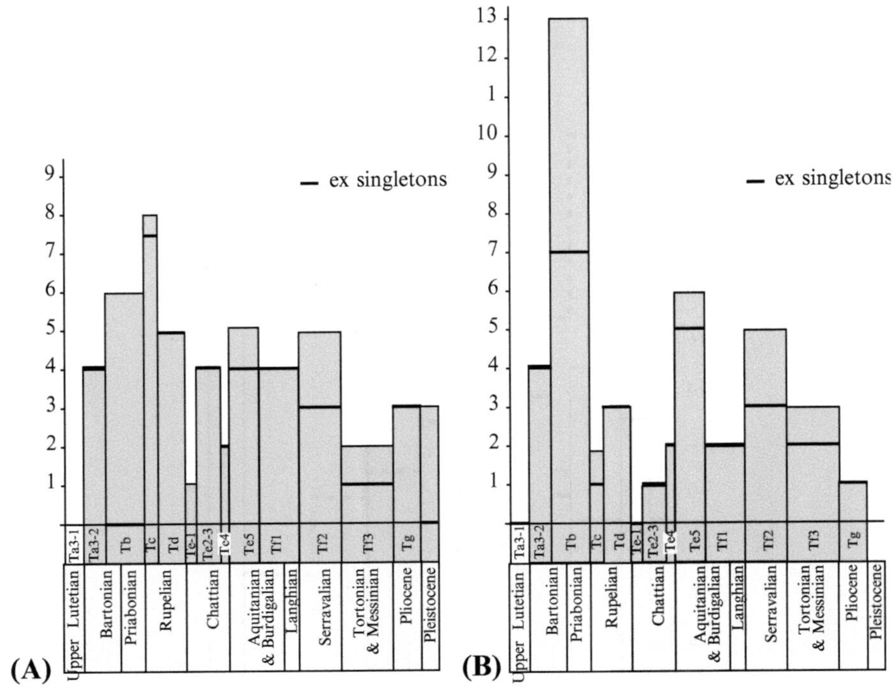


Fig. 7A, B First and last occurrences of larger benthic genera in Indo-Malaysia. Grey bar represents total, the solid line excludes singletons. (A) First occurrences, (B) Last occurrences.

low latitudes (Fig. 1). Currents flowed unhindered from the Pacific into the Indian Ocean between Australia and Eurasia. The main currents connecting the Indian and Atlantic Ocean, certainly the deep water flow was south of Africa, and then through the Caribbean back into the Pacific Ocean. Extensive shallow marine areas, including large carbonate platforms, existed in the central Tethys and flanked the northern coast of the African continent and the southern shores of Europe. The IM region was formed largely by oceanic crust with island arc at the plate boundaries (Fig. 8; Hall, 1996). Carbonate platforms did not cover large areas and shelf edges were comparatively steep, leaving only few shallow margins in IM. With the opening of the Makassar Strait large areas of shallow water habitats became available between southwest Sulawesi and Kalimantan in Indonesia and at Mindoro in the Philippines. Around the equator some shallow seas and islands existed on crustal fragments that nowadays belong to the Philippines. The position of these shallow carbonate platforms in Fig. 8 is inferred from the relative position of parts of the Philippines and palaeoenvironmental information (Hall, 1996, 1998; Wilson and Rosen, 1998; Moss and Wilson, 1998; Wilson, 2002), and own observations on southern Luzon, Mindoro and Mindanao in the Philippines.

With the continuing rotation of Africa, and the northward movement of India and Australia, the palaeogeographic configuration changed markedly. The most

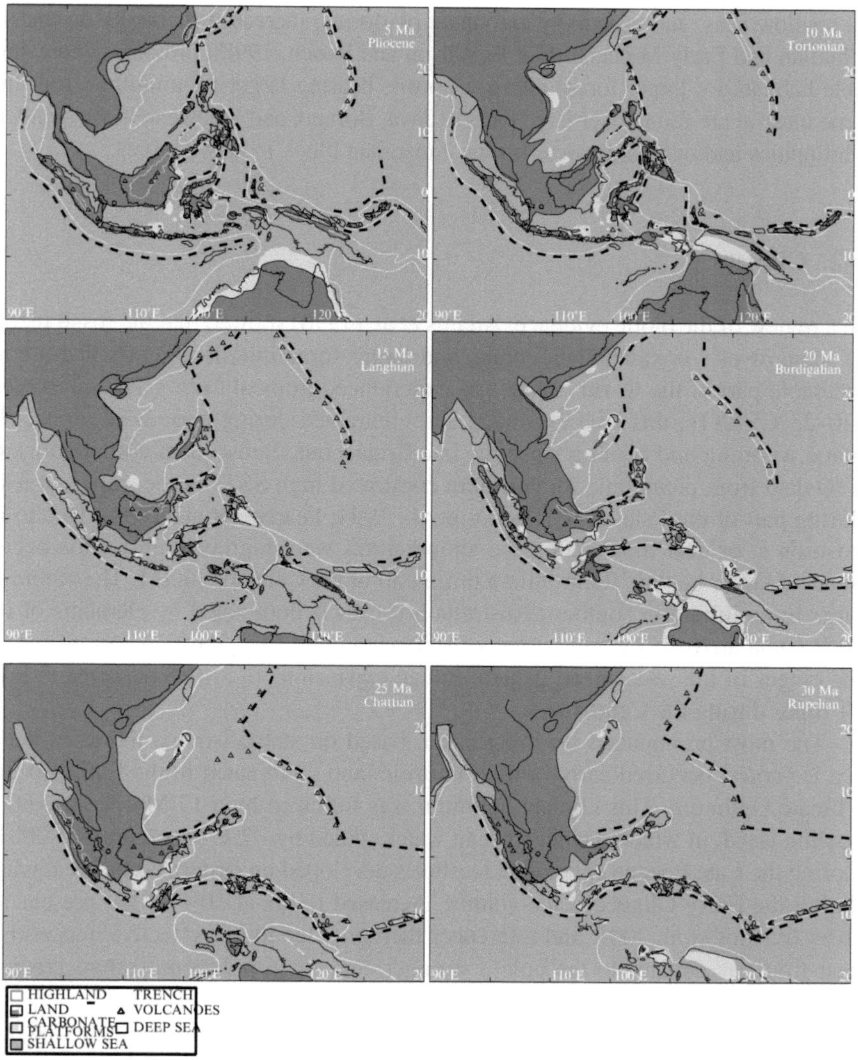


Fig. 8 Paleogeographic maps, showing mountainous areas, land, shallow seas, and deep sea during the Rupelian, Chattian, Burdigalian, Langhian, Tortonian, and Pliocene (after Hall, 1998).

important change was the collision of Africa with Eurasia, resulting in the initial closure of the connection between the Indian Ocean and the Mediterranean in the Burdigalian, after which connections were re-established several times until the definite closure during the Burdigalian (Jones, 1999; Rögl, 1998).

During the Chattian to Burdigalian, the extent of shallow seas and carbonate platforms in the western Tethys decreased. The reverse was the case for the IWP. Australia moved northwards and started to collide with IM in the Chattian. The extent of shallow seas, and especially carbonate platforms, increased markedly during the Chattian and Early Miocene (Fig. 8; Wilson and Rosen, 1998), providing more suitable habitats for larger foraminifera. Deposits bearing larger foraminifera from this time interval are found over large parts of Java, Borneo, and Sulawesi, in the southern Philippines and over large areas on the Australian block in New Guinea.

## 6.2. Climate

In a review of the biotic evidence, Adams et al. (1990) showed that the fossil record of mangroves, zooxanthellate corals, and larger foraminifera suggests that a considerable part of the world oceans has experienced tropical (>25°C) or subtropical (20–25°C) SSTs; this area expanded and contracted during periods of global climatic warming and cooling since the late Cretaceous. Subsequent compilations of $\delta^{18}O$ data from planktonic foraminifera confirmed high SSTs in the tropics at least during part of the Paleogene (Zachos et al., 1994; Pearson et al., 2001). Due to its position at or near the equator, the temperatures were high enough for the occurrence of larger benthic foraminifera during all of the Cenozoic in IM. The record of large foraminifera in southern Australia however is dominated by elements of the IWP fauna arrive via warm currents (McGowran and Lee, 2000) as expansions of the ranges of the widespread genera. Range expansions to >40°S occurred at least 11 times during the Cenozoic.

The most pronounced warming trend, based on stable isotope trends in deep-sea ice cores, occurred early in the Cenozoic, and culminated in the Early Eocene Climate Optimum. This climate optimum was followed by a 17 My period with a cooling trend, in which the deep ocean water cooled by ~7°C (Zachos et al., 2001). During the Late Eocene ephemeral icesheets developed on eastern Antarctica, while during the Early Oligocene ice-volume increased to about 50% of the present ice sheet (Zachos et al, 2001, and references therein). Due to less effective transport of heat from the equator to the pole a stronger equator to pole temperature gradient developed (Zachos et al., 1994).

In general, the onset of permanent glaciation in Antarctica, around the Priabonian–Rupelian boundary, marked a period of changing ocean circulation patterns. Circum-tropical transport became increasingly difficult with the collision of India and Eurasia, and even more so because of Africa approaching Eurasia. A complete circum-Antarctic seway did not open until the opening of the Tasman Sea and Drake Passage for deep water currents. The former can be dated at *c.* 32 Ma, the latter at 29–33 Ma (Lawver and Gahagan, 2003). Ocean circulation was driven by differences in temperature and to a lesser extent by salinity during the Palaeocene–Eocene, whilst in the new (post-Oligocene) situation salinity became more important, resulting in larger density differences between water masses and thus faster circulation, at least in the Southern Ocean (Diester-Haass, 1995; Salamy and Zachos, 1999).

Faster ocean circulation rates result in increased upwelling, thus increasing surface water productivity (Salamy and Zachos, 1999). The Paleogene thermal isolation of Antarctica and mountain building in the Himalayas also resulted in seasonally more variable conditions; the first evidence for a monsoonal climate around the equator comes from this time (Morley, 2000). In a monsoonal climate the seasonality increased, decreasing predictability of the environment and reducing the availability of suitable conditions for long living specialists. Especially the deeper living larger benthic foraminifera specialized at low light intensities disappear, whilst shallower species still occur (Renema and Troelstra, 2001). The Eocene–Oligocene boundary and the Middle Miocene extinction are worldwide characterized by the disappearance of large, flat species (*Discocyclina*, large *Nummulites*, large *Assilina* in the Eocene, lepidocyclinids and *Cycloclypeus* in the Miocene), whilst more eurytypic species dominate the fauna after the extinctions.

## 6.3. Facies and Outcrop

During most of the Cenozoic, Southeast Asia consisted of several small, tectonically controlled basins. Many of these basins experienced rapid subsidence or uplift. Often, the occurrence of larger foraminiferal species is not determined by stratigraphical events, but by facies changes (Racey, 1995). Composite sections and isolated samples cover most of the Ypresian to Priabonian in southeast Asia. For his revision of the letter classification, Adams (1970) used reference sections in India for the Palaeocene–Eocene in his revision of the letter stages, as outcrops of this age are rare in Indonesia.

Unfortunately, no reliable Palaeocene to lowermost Ypresian records of larger foraminifera are known from IM. At the Kudat Peninsula (Sabah), a mixed assemblage of reworked Late Palaeocene and *in situ* Early Miocene foraminifera has been found (van der Vlerk, 1951). Hashimoto and Matsumaru (1981a) interpreted the interval without Miocene foraminifera as *in situ* Upper Palaeocene sedimentary rocks. Although undoubtedly Thanetian (or possibly Ypresian) sedimentary rocks were deposited in this area, these have now been reworked into younger deposits and an exact age for them cannot be determined (van der Vlerk, 1951).

Ypresian deposits are also rare and can only be confirmed on Halmahera. These consist of deep-water marine sedimentary rocks and yield no larger foraminifera. The oldest Cenozoic deposits found on Sumba could not be dated, since no age-diagnostic fossils have been found in these strata. The basal Lutetian is rare, and has only been found on Timor, Sumba, East Sulawesi, Mindoro and Luzon. No age diagnostic genera are present in these deposits and they are merely characterized by the presence of *Palaeonummlites* sp.1 of Renema et al. (2003) in the absence of other taxa such as *P. beaumonti*.

Middle Lutetian to Bartonian sedimentary rocks are the most widespread Eocene deposits in IM. These have been found at ten to fourteen localities throughout the archipelago (Renema et al., 2003). All assemblages found in Ta are typical of deeper water, soft bottom sediments and characterized by the occurrence of nummulitids. In east Indonsia *Alveolina* is a more abundant element of assemblages this age, but its low diversity in IM might be explained by a reduced diversity of sedimentary facies types. However, also within the same range of facies diversity on the species level is smaller in IM than in Europe. Far fewer species of *Nummulites* (*s.l.*) have been recorded, and the size variation is also smaller than in European basins (Renema, 2002).

Deposits of top Barthonian/Priabonian (Tb) age are more common. On Java and southwest Sulawesi these are of similar marly facies, and contain numerous large (up to 30 mm diameter) *Discocyclina*, *Operculina* and small *Palaeonummulites*. Sedimentary rocks of Priabonian age are, however, rare and have only been found in the Melinau Limestone (Sarawak), a carbonate platform deposited from the Bartonian to Early Miocene (Adams, 1965). The boundary interval between the uppermost Priabonian and Rupelian is present, but does not contain larger benthic foraminifera. Other Tb-deposits have been recorded from Sumba (Caudri, 1934), Kalimantan (Tan, 1930a, 1937), Central and Southeast Sulawesi (Osimo, 1908), and Luzon (Renema, unpublished data) and a reworked boulder on Java (Lunt, personal communication). These beds all contain a very characteristic assemblage of several genera that have not been recorded anywhere else, like *Orbitoclypeus*, *Linderina*, *Silvestriella*, and *Tansinhokella vermicularis*. These deposits probably are younger than the *Discocyclina* beds in Tb, and might correspond to the Priabonian. No sections have been found and this correlation cannot be proven in Indonesia as yet. In south Japan, the same faunal transition has been recorded (Matsumaru, 1996).

The Priabonian fauna, characterized by the occurrence of large *Discocyclina* and *Pellatispira*, in the absence of typical Middle Eocene genera like *Assilina*, *Alveolina*, and large representatives of *Nummulites*, has been recognized as a separate chronofauna (Hottinger, 1997). However, the distinct nature of the Priabonian fauna could instead be a response to environmental conditions causing the presence of more algal overgrowth on the substrate (Renema, 2002). The presence and extinction of large discocyclinids (with a marked adult dimorphism) can be seen in the line of "progressive" extinctions of the largest species of genera (or complete genera) at the end of Ta3–1, *Assilina* and *Pseudocamerinoides*; and Ta3–2, *Nummulites*. The morphology of the characterizing genera of Tb can be compared to the exposed slope assemblages at the Spermonde Archipelago (*T. vermicularis* vs. *H. depressa*, *Sylvestriella* vs. *Baculogypsinoides*, and *Linderina* vs. *Amphisorus*). Rubble and reef rock inhabited by this assemblage of larger foraminifera is usually covered by coralline-algae, and an increase in genera preferring this kind of substrate in the Priabonian thus hints at an increase in algal dominance. Algae dominate reefal systems, for example, when nutrient availability is higher, or when temperature is lower.

The post-Eocene is much better represented at many places in the region. Sections covering Tc–Td or Te4 are present in southwest Java (e.g., Tan, 1932; Doornink, 1934), southwest Sulawesi (Wilson, 1995; Renema, unpublished data), and Sarawak (Adams, 1965). The Java (composite) sections are sandstones with interbedded limestones and shales. The Tb/Tc boundary has not been reported in these deposits, which mostly comprise deep water facies. The sections in the Tonasa limestone document mainly carbonates and shales. Although typical Tb faunas with large *Discocyclina* and *Pellatispira* have been found here, no continuous Eocene–Oligocene section has been reported so far. The Tc–Td beds range from deep to shallow photic zone and are mainly carbonates (Crotty and Engelhardt, 1993; Wilson, 1995; Renema, unpublished data).

Chattian to Middle Miocene (Te1–Tf3) larger benthic foraminifera are abundant throughout the region. For example, the BPM/NPPM material contains at least 50 cores or sections yielding larger benthic foraminifera covering rocks of this age range. These samples were taken (mostly) at Borneo, Java, Sulawesi, and Sumatra. Additional material is available form Sumba (Caudri, 1934). Samples from many different facies types are present. The end of Tf3 is often not represented due to a big hiatus, caused by a sharp drop in sea level at the end of the Serravallian. Due to this hiatus, it is impossible to determine whether the extinction events marking the absolute end of the Oligocene–Miocene fauna occur simultaneously or not.

Pliocene data are rare, so the diversity in this period might be too low. Little attention has been paid to long ranging taxa in publications dealing with stratigraphy. Important publications covering (parts of) the Pliocene are Boudagher-Fadel (2002) from southwest Sulawesi, Whittaker and Hodgkinson (1979) from Sabah and Van Marle (1991; East Indonesia). The later publications mainly involve relatively deep facies for larger benthic foraminifera, so most species will have been washed into them. Unpublished records come from Cabaruyan and Cagayan Valley (both Luzon, Phillipines).

## 7. Regional Distribution and Fauna Provinces

The Cenozoic tropics can be divided into three bioprovinces: Central American, Mediterranean, and the IWP (Adams, 1967, 1983). The Central American Province reaches from southern California and Florida to Peru and Brazil. The IWP Province extends from the east coast of Africa to the central Pacific islands and from Japan to Australia and New Zealand. The Mediterranean Province covers most of the present-day Mediterranean coasts. The border zones are characterized by faunal influences of adjacent provinces (Adams, 1983).

This division is present in most of the Cenozoic, but all three provinces cannot be recognized at all times. During the Eocene there are several genera restricted to Central America, like *Helicostegina*, *Lepidocyclina*, *Nephrolepidina*, and *Eulepidina*. However, there is only a limited difference between the Mediterranean and IWP Province. Genera like *Assilina*, *Planocamerinoides*, *Discocyclina*, and *Alveolina* dominate the faunas. However, diversity of all four of these genera is highest in the Mediterranean area, and only the most abundant, usually longer ranging species reach the eastern part of the Tethys (Renema, 2002). Nevertheless, Lunt (2003) has shown that during the Eocene there is a distinction between faunas associated with the Sundashelf craton, which is essentially Mediterranean–IWP, and faunas found on the Australian plate. The latter fauna is typified by the occurrence of *Lacazinella*, and the absence of *Assilina*, *Planocamerinoides*, and pellatispirids.

Of the Late Eocene fauna, the pellatispirids have their oldest record within the IWP, and also show a higher species, generic and morphologic diversity in the east part of its range. For example, the large and flat *Biplanispira mirabilis* and the almost globular *Vacuolispira* spp. (Hottinger, et al. 2001; Renema, unpublished data) are restricted to the eastern part of the Tethys.

The Early Oligocene is characterized by a limited provincialism. First *Eulepidina* and later *Nephrolepidina* migrated to the Mediterranean and IWP provinces, and *N. fichteli* migrated from Europe to the IWP. The *N. fichteli* group radiated in India and Indonesia into some short ranging, but distinct species, including a large and tightly coiled species, *N. subbrongniarti* Verbeek (Renema, 2002) and *N.* cf. *fichteli* with intercalary whorls (Sengupta, 2000, 2002).

During the Oligocene to Miocene, and eventually recent, the Mediterranean and IWP provinces showed an increasing taxonomic separation. *Miogypsina* appears to be limited to the Mediterranean in the Chattian, whilst it has its first appearance in the Aquitanian in the IWP. Other faunal differences are subtler, as was shown by O'Herne (1972), who evaluated the morphological evolution of *Cycloclypeus* in Europe and Indonesia. He found that there is a single trend in the IWP, but that the Spanish material follows another trend (Fig. 9). This shows that although *Cycloclypeus* occurred both in the Mediteranean and IWP, they evolved as separate lineages and might even have a polyphyletic origin (MacGillavry, 1962).

Especially in this time interval, lineages in several groups have been used for circum-tropical correlation, such as the miogypsinids (Drooger, 1963; 1993; Raju, 1974), and the lepidocyclinids (Van der Vlerk, 1968; Van der Vlerk and Postuma, 1967; Van Vessem, 1978). However, independent unidirectional evolution on either side of major barriers, provided they share a common ancestor, could result in morphological similar members in lineages. This would make the repeated crossing of transoceanic or other barriers unnecessary. Regional sides branches are expected and produced (Fig. 10; Adams, 1983). Lineages need not be initiated at the same time and stage, as can be seen by the longer initial spire of *Miogypsina* in the Americas than in the Mediterranean and IWP (Fig. 10). Once started, extinctions can occur at any moment, and not all lineages necessarily proceed to completion in all areas (Fig. 10; Adams, 1983).

It cannot be discounted that at least the *Miogypsinella* lineages are polyphyletic, as the oldest records occur in the Americas, put intermediate forms have been found in much younger strata in Borneo and South Japan (Boudagher-Fadel et al., 2000; Matsumaru, 1996). Apparently, the new appearance of equatorial chambers is a relatively "easy" evolutionary step in rotalines, as can also be seen in recent specimens of *Neoeponides* on Bali, which are developing equatorial chambers (Renema, unpublished data). Repeated evolution of rotalines with equatorial chambers could be a way to explain relatively young occurrences of very primitive miogypsinids as found by Adams and Belford (1974), Raju (1974), and Glaessner (in Adams, 1983). If this is the case, the barrier between the faunal provinces might be harder to cross than previously thought. In the Burdigalian to Serravallian, the faunas of the IWP start to differentiate more from those of the Mediterranean. Typical faunal elements are *Flosculinella*, *Alveolinella*, and *Katacycloclypeus*. This definite separation occurs at the same time as the closure of the Tethys by the collision of Africa with Eurasia.

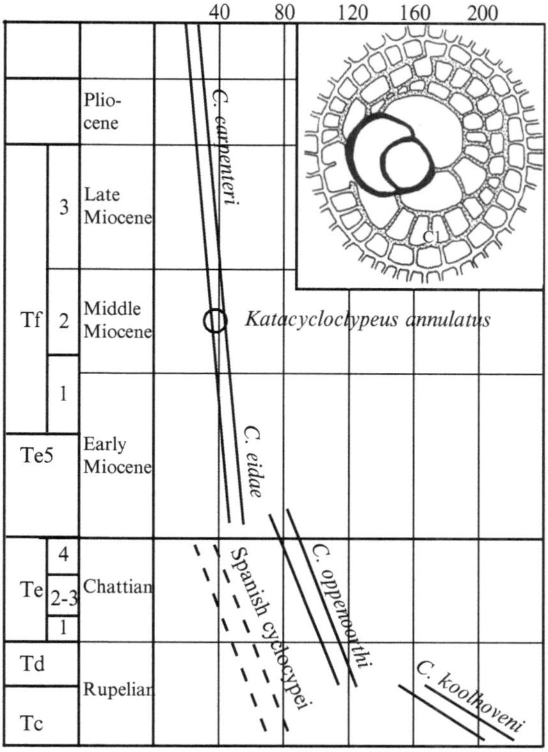


Fig. 9 Numbers of secondary chamberlets of the first neanic chamber plotted against time, for the Indo-Malaysian and European cycloclypei. The inset shows the nepiont of a megalospheric Cycloclypeus. C1 indicates the first neanic chamber (after figure 14 of O'Herne, 1972).

The northward movement of Africa continued during the Late Miocene to recent and severely reduced the extent of shallow carbonate platforms in Europe. Together with lower temperatures this led to the almost complete extinction of larger symbiont-bearing foraminifera in Europe. In the IWP many typical Oligocene–Miocene genera went extinct. In the Late Miocene–Pleistocene of the IWP several new genera have their FO, for example, *Calcarina*, *Baculogypsinoides*, and *Schlumbergerella*.

## 8. Conclusions

During the Eocene, the diversity of larger benthic foraminifera was lower in the IWP than in Europe. Although part of the pattern might be explained by a predominance of deeper water, soft bottom communities in the IWP, within these facies the number of species and the morphological diversity is lower.

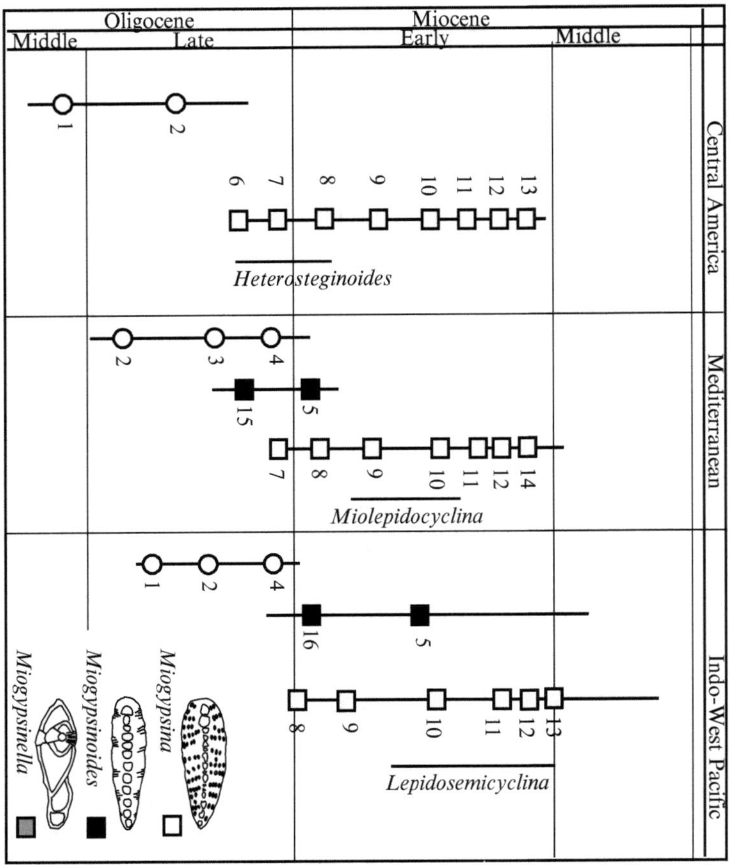


Fig. 10 Stratigraphical distribution of some miogypsinids. Figure drawn after Adams (1983), but with the inclusion of *Miogypsinella* for the primitive *Miogypsinoides* species (*M. bermudezi–M. bantamensis*). In Europe *Miogypsinoides mauretanius* and *M. dehaarti* have been added (see Cahuzac, 1984). (1) *Miogypsinella bermudezi*, (2) *M. complanatus*, (3) *M. fomosensis*, (4) *M. bantamensis*, (5) *Miogypsinoides dehaarti*, (6) *Miogypsina thalmanni*, (7) *Miogypsina basraensis*, (8) *Miogypsina gunteri*, (9) *Miogypsina tani*, (10) *Miogypsina globulina*, (11) *Miogypsina intermedia*, (12) *Miogypsina cushmanni*, (13) *Miogypsina antillea*, (14) *Miogypsina mediterranea*, (15) *Miogypsinoides mauretanicus*, (16) *Miogypsinoides ubaghsi*.

The genera which are most characteristic of the Lutetian fauna attained their highest diversity in the Mediterranean faunal province and only long ranging, widely distributed taxa reached the IWP and may or may not evolve in sister species. This pattern started to change at the end of the Bartonian. In the Rupelian some (sub)families evolved in India and/or Indonesia and only the most abundant species reached the Mediterranean, although at the same time new taxa evolved in the Mediterranean area and migrated east. With the closure of the Tethys in the Burdigalian, the

Mediterranean, and IWP faunas were definitely separate. Due to habitat loss and climate change no refreshment of the Mediterranean fauna occurred, and most larger benthic foraminifera genera gradually became extinct in Europe.

In the IWP the availability of suitable habitats increased and the taxonomic diversity increased with it. Thus, coincidentally or not, the diversity compares very well with the presence and abundance of suitable habitats. However, species and morphological diversity indicate that the diversity in the IWP in the Eocene, and in the Mediterranean during the (Middle to Late) Miocene, were also restricted by other environmental parameters such as temperature and nutrient availability.

## References

Adams, C.G., 1965, The foraminifera and stratigraphy of the Melinau Limestone, Sarawak, and its importance in Tertiary correlation, *Quarterly Journal of the Geological Society, London* **121**: 283–338.

Adams, C.G., 1967, Tertiary foraminifera in the Tethyan, American and Indo-Pacific Provinces, in: Adams, C.G. and Ager, D.V. (eds), *Aspects of Tehyan Biogeography*, Systematics Association, London, pp. 195–217.

Adams, C.G., 1968, A revision of the foraminiferal genus *Austrotrillina* Parr, *Bulletin of the British Museum (Natural History) Geology* **16**: 73–97.

Adams, C.G., 1970, A reconsideration of the East Indian letter classification of the Tertiary, *Bulletin of the British Museum (Natural History) Geology* **19**: 1–137.

Adams, C.G., 1973, Some Tertiary foraminifera, in: Hallam, A. (ed.), *Atlas of Palaeobiology*, Elsevier, Amsterdam, pp. 453–471.

Adams, C.G., 1983, Speciation, phylogenesis, tectonism, climate and eustasy: factors in the evolution of Cenozoic larger foraminiferal bioprovinces, in: Sims, R.W., Price, J.H., and Whalley, P.E.S. (eds), *Evolution, Time and Space: The Emergence of the Biosphere, Systematics Association Special Volume* **23**: 255–289.

Adams, C.G., 1984, Neogene larger foraminifera, evolutionary and geological events in the context of datum planes, in: Ikebe, N. and Tsuchi, R. (eds), *Pacific Neogene Datum Planes*, University of Tokyo Press, Tokyo, pp. 47–67.

Adams, C.G. and Belford, 1974, Foraminiferal biostratigraphy of the Oligocene–Miocene limestones of Christmas Island (Indian Ocean), *Palaeontology* **17**: 475–506.

Adams, C.G. and Frame, P., 1979, Observations on *Cycloclypeus* (*Cycloclypeus*) Carpenter and *Cycloclypeus (Katacycloclypeus)* Tan (Foraminferida), *Bulletin of the British Museum (Natural History) Geology* **32**: 3–17.

Adams, C.G., Lee, D.E. and Rosen, B.R., 1990, Conflicting isotopic and biotic evidence for tropical sea-surface temperatures during the Tertiary, *Palaegeography, Palaeoclimatology, Palaeoecology* **77**: 289–313.

Bakx, L.A.J., 1932, De genera *Fasciolites* en *Neoalveolina* in het Indo-Pacifische gebied, *Verhandelingen van het Koninklijk Nederlandsch Geologisch en Mijnbouwkundig Genootschap voor Nederland en Koloniën* **9**: 205–267.

Banner, F.T. and Hodgkinson, R.L., 1991, A revision of the foraminiferal subfamily Heterosteginінae, *Revista Española de Micropaleontología* **23**: 101–140.

Boersma, A., Premoli Silva, I., and Hallock, P., 1998, Trophic models for the well-mixed and poorly mixed warm oceans across the Paleocene/Eocene Epoch boundary, in: Aubry, M.-P.,

Lucas, S.G., and Berggren, W.A. (eds), *Late Paleocene–Early Eocene Climatic and Biotic Events in the Marine and Terrestrial Records*. Columbia University Press, New York, pp. 204–213.

Boudagher-Fadel, M.K., 2002, The stratigraphical relationship between planktonic and larger benthic foraminifera in Middle Miocene to Lower Pliocene carbonate facies of Sulawesi, Indonesia, *Micropaleontology* **48**: 153–176.

Boudagher-Fadel, M.K. and Banner, F., 1997, The revision of some genus-group names in Tethyan Lepidocyclininae, *Paleopelagos* **7**: 3–16.

Boudagher-Fadel, M.K. and Banner, F., 1999, Revision of the stratigraphic significance of the Oligocene–Miocene "Letter-Stages", *Revue de Micropaléontologie* **42**: 93–97.

Boudagher-Fadel, M.K., Lord, A.R., and Banner, F.T., 2000, Some Miogypsinidae (Foraminiferida) in the Miocene of Borneo and nearby countries, *Revue Paléobiology* **19**: 137–156.

Cahuzac, B.C., 1984, *Les faunes de Miogypsinidae d'Aquitaine Méridionale (France)*, Benthos'83, Pau France, pp. 117–129.

Caudri, C.M.B., 1934, *Tertiary Deposits of Sumba*, H.J, Paris, Amsterdam.

Caudri, C.M.B., 1939, Lepidocyclinen von Java, *Verhandelingen van het Geologisch en Mijnbouwkundig Genootschap voor Nederland en Koloniën* **12**: 135–237.

Cole, W.S., 1957, Larger foraminifera from Eniwetok drill holes, *U.S. Geological Survey Professional Paper* **260**-V: 743–784.

Cole, W.S., 1975, Concordant age determination by larger and planktonic foraminifera in the Tertiary of the Indo-Pacific region, *Journal of Foraminiferal Research* **5**: 21–39.

Crotty, K.J. and Engelhardt, D.W., 1993, Larger foraminifera and palynomorphs of the upper Malawa and lower Tonasa Formations, southwestern Sulawesi Island, Indonesia, *Proceedings of the International Symposium on Biostratigraphy of Mainland Southeast Asia: Facies and Paleontology*, Indonesian Petroleum Association, Jakarta, pp. 71–82.

Dercourt, J., Ricou, L.E. and Vrielynck, B. (eds), 1993, *Atlas Tethys Palaeoenvironmental Maps*, Gauthier-Villars, Paris.

Diester-Haass, L., 1995, Middle Eocene to early Oligocene paleooceanography of the Antarctic Ocean (Maud Rise, ODP leg 113, site 689): change from a low to a high productivity ocean, *Palaegeography, Palaeoclimatology, Palaeoecology* **113**: 311–334.

Doornink, H.W., 1932, Tertiary Nummulitidae from Java, *Verhandelingen van het Geologisch en Mijnbouwkundig Genootschap voor Nederland en Koloniën* **9**: 267–315.

Douvillé, H., 1924, Revision des Lepidocyclines, *Mémoires de la Société Géologique de France (N.S.)* **2**: 115 pp, 7 plates.

Drooger, C.W., 1963, Evolutionary trends in Miogypsinidae, in: Koenigswald, G.H.R. von, Emeis, J.D., Buning, W.L., and Wagner, C.G. (eds), *Evolutionary Trends in foraminifera*, Elsevier, Amsterdam, pp. 315–349.

Drooger, C.W., 1993, Radial foraminifera; morphometrics and evolution, *Proceedings of the Royal Academy of Sciences and Arts of the Netherlands, Science Series* **41**: 242 pp.

Eames, F.E., Banner, F.T., Blow, W.H. and Clarke, W.J., 1962, *Fundamentals of Mid-Tertiary Stratigraphical Correlation*, Cambridge University Press, Cambridge.

Eames, F.E., Clarke, W.J., Banner, F.T., Smout, A.W.H. and Blow, W.H., 1968, Some larger foraminifera from the Tertiary of Central America, *Palaeontology* **11**: 283–305.

Glaessner, M.F., 1943, Problems of stratigraphic correlation in the Indo-Pacific Region, *Proceedings of the Royal Society of Victoria* (n.s.) **55**: 41–80.

Glaessner, M.F. 1953, Time-stratigraphy and the Miocene epoch, *Bulletin of the Geological Society of America* **64**: 647–658.

Glaessner, M.F., 1959, Tertiary stratigraphic correlation in the Indo-Pacific region and Australia, *Journal of the Geological Society of India* **1**: 53–67.

van Gorsel, J.T., 1988, Biostratigraphy in Indonesia: methods, pitfalls and new directions, *Proceedings of the Indonesian Petroleum Association, 16th Annual Convention*, IPA, Jakarta, pp. 275–300.

Govindan, A., 2003, Tertiary larger foraminifera in Indian basins: correlation with standard planktic zones and "letter stages", *Gondwana Geological Magazine* **6**: 45–78.

Gradstein, F.M., Ogg, J.G. and Smith, A.G., 2005. A geologic timescale 2004, Cambridge University Press, Cambridge, p. 589.

Haak, R. and Postuma, J.A., 1975, The relation between the tropical planktonic foraminiferal zonation and the Tertiary Far East letter classification, *Geologie en Mijnbouw* **54**: 195–198.

Hall, R., 1996, Reconstructing Cenozoic SE Asia, in: Hall, R. and Blundell, D. (eds), *Tectonic Evolution of Southeast Asia*, Geological Society Special Publication **106**: 153–185.

Hall, R., 1998, The plate tectonics of Cenozoic SE Asia and the distribution of land and sea, in: Hall, R. and Holloway, J.D., (eds), *Biogeography and Geological Evolution of Southeast Asia*, Backhuys Publishers, Leiden, pp. 99–131.

Hallock, P., 1987, Fluctuations in the trophic resource continuum: a factor in global diversity cycles? *Paleoceanography* **2**: 457–471.

Hallock, P., 1988, Interoceanic differences in foraminifera with symbiotic algae: a result of nutrient supplies? *Proceedings Sixth International Coral Reef Symposium, Volume* **3**: 251–255.

Hallock, P., Premoli Silva, I., and Boersma, A., 1991, Similarities between planktonic and larger foraminiferal evolutionary trends through Paleogene paleoceanographic changes, *Palaegeography, Palaeoclimatology, Palaeoecology* **83**: 49–64.

Hashimoto, W. and Matsumaru, K., 1981, Larger foraminifera from Sabah, Malaysia, part 1, Larger foraminifera from the Kudat Peninsula, the Gomanton Area, and the Semporna Peninsula, *Geology and Palaeontology of Southeast Asia* **22**: 49–54.

Henrici, H., 1934, Foraminiferen aus dem Eocän und Altmiocän von Timor, *Palaeontographica* **4**: 56 pp.

Henson, F.R.S., 1950, *Middle Eastern Tertiary Peneroplidae (Foraminfera), with Remarks on the Phylogeny and Taxonomy of the Family*, West Yorkshire, Wakefield.

Hofker, J.S.R., 1927, The foraminifera from the Siboga expedition; uitkomsten op zoologisch, botanisch, oceanografisch en geologisch gebied, verzameld in Nederlandsch Oost-Indië 1899–1900 aan boord van H.M. Siboga, E.J., Brill, Leiden.

Hollaus, S.-S. and Hottinger, L., 1997, Temperature dependence of endosymbiotic relationships? Evidence from the depth range of Mediterranean *Amphistegina lessonii* (foraminiferida) truncated by the thermocline, *Ecologae Geologicae Helvetiae* **90**: 591–597.

Hottinger, L., 1983, Processes determining the distribution of larger Foraminifera in space and time, *Utrecht Micropaleontological Bulletins* **30**: 239–253.

Hottinger, L., 1990, Significance of diversity in shallow benthic foraminifera, *Atti del Quattro Simposio di Ecologia e Paleoecologia delle Comunita Bentoniche*, Museo Regionale de Scienze, Naturali, Torino, pp. 35–51.

Hottinger, L., 1997, Shallow benthic foraminiferal assemblages as signals for depth of their deposition and their limitations, *Bulletin de la Société Géologique de France* **168**: 491–505.

Hottinger, L., 2001, Archiasinids and related porcelaneous larger foraminifera from the late Miocene of the Dominican Republic, *Journal of Paleontology* **75**: 475–512.

Hottinger, L., Halicz, E. and Reiss, Z., 1991, The foraminiferal genera *Pararotalia*, *Neorotalia* and *Calcarina*: taxonomic revision, *Journal of Paleontology* **65**: 18–33.

Hottinger, L., Romero, J., and Caus, E., 2001, Architecture and revision of the Pellatispirines, planispiral canaliferous foraminifera from the Late Eocene Tethys, Micropaleontology **47** (Suppl. 2): 35–77.

Jones, R.W., 1999, Marine invertebrate (chiefly foraminiferal) evidence for the palaeogeography of the Oligocene–Miocene of western Eurasia, and consequences for terrestrial vertebrate migration, in: Agustí, J., Rook, L. and Andrews, P. (eds), *The Evolution of Neogene Terrestrial Ecosystems in Europe*, Cambridge University Press, Cambridge, pp. 274–309.

Langer, M.R. and Hottinger, L., 2000, Biogeography of selected "larger" foraminifera, *Micropaleontology* **46** (Suppl. 1): 105–127.

Lawver, L.A., and Gahagen, L.M., 2003, The development of palaeo seaways around Antarctica, *Palaegeography, Palaeoclimatology, Palaeoecology* **198**: 11–37.

Less, G., 1987, Paleontology and stratigraphy of the European Orthophragminae, *Geologica Hungarica Series Palaeontologica* **51**: 373 pp.

Le Roy, L.W., 1948, The foraminifer *Orbulina universa* d'Orbigny, a suggested middle Tertiary time indicator, *Journal of Paleontology* **22**: 500–508.

Le Roy, L.W., 1952, *Orbulina universa* d'Orbigny in Central Sumatra, *Journal of Paleontology* **26**: 576–584.

Leupold, W. and van der Vlerk, I.M., 1931, The Tertiary, *Leidsche Geologische Mededelingen* **5**: 54–78.

Loeblich, A.R., Jr. and Tappan, H., 1987, *Foraminiferal genera and their classification*, Van Nostrand Reinhold, New York.

Loeblich, A.R., Jr. and Tappan, H., 1994, Foraminifera from the Sahul shelf and Timor Sea, *Cushman Foundation for Foraminiferal Research Special Publication* **13**: 1–661.

Lunt, P., 2003, Biogeography of some Eocene larger foraminifera, and their application in distinguishing geological plates, *Palaeontologica Electronica* **6** (1): 1–22.

Lunt, P. and Allan, T., 2004, *Larger foraminifera in Indonesian biostratigraphy, calibrated to isotopic dating*. GRDC Museum Workshop on Micropaleontology, June 2004.

MacGillavry, H.J., 1962, Lineages in the genus *Cycloclypeus* Carpenter (Foraminifera), *Proceedings Konijnklijke Nederlandse Academie voor Wetenschappen* **B-65**: 430–458.

van Marle, L.J., 1991, Eastern Indonesian, Late Cenozoic smaller benthic foraminifera, Verhandelingen der Koninklijke Nederlandse Akademie van Wetenschappen, afd Natuurkunde, *Eerste Reeks* **34**: 1–328.

Matsumaru, K., 1996, Tertiary larger foraminifera (foraminiferida) from the Ogasawara Islands, Japan, *Paleontological Society of Japan, Special Papers* **36**: 239 pp.

McGowran, B. and Li, Q., 2000, Evolutionary palaeoecology of Cainozoic Foraminfera: Tethys, Indo-Pacific, Southern Australia, *Historical Biology* **15**: 3–27.

Morley, R.J., 2000, *Origin and Evolution of Tropical Rain Forests*, Wiley, London.

Mohler, W.A., 1949, *Flosculinella reicheli* sp. nov. aus dem Tertiär von Borneo, *Eclogae Geologiae Helvetiae* **42**: 521–527.

Moss, S.J. and Wilson, M.E.J., 1998, Biogeographic implactions of the Tertiary palaeogeographic evolution of Sulawesi and Borneo, in: Hall, R. and Holloway, J.D. (eds), *Biogeography and Geological Evolution of Southeast Asia*, Backhuys, Leiden, pp. 133–163.

O'Herne, L. 1972, Secondary chamberlets in *Cycloclypeus*, *Scripta Geologica* **7**: 1–38.

Osimo, G., 1908, Di alcuni foraminiferi dell 'Eocene superiore di Celebes, *Rivista Italiana di Paleontologia* **14**: 28–54.

Parrish, J.T., and Curtis, R.L., 1982, Atmospheric circulation, upwelling, and organic rich rocks in the Mesozoic and Cenozoic eras, *Palaeogeography, Palaeoclimatology, Palaeoecology* **40**: 31–66.

Paulay, G., 1997, Diversity and distribution of reef organisms, in: Birkeland, C. (ed.), *Live and Death of Coral Reefs*, Chapman & Hall, New York, pp. 298–352.

Pearson, P.N., Ditchfield, P.W., Singano, J., Harcourt-Brown, K.G., Nicholas, C.J., Olsson, R., Shackleton, N.J., and Hall, M., 2001, Warm tropical sea surface temperatures in the Late Creatateceous and Eocene epochs, *Nature* **413**: 481–487.

Racey, A., 1995, Lithostratigraphy and larger foraminiferal (nummulitid) biostratigraphy of the Tertiary of northern Oman, *Micropaleontology* **41**: 1–123.

Raju, D.S.N., 1974, Study of Indian Miogypsinidae, *Utrecht Micropaleontological Bulletins* **9**: 1–148.

Renema, W., 2002, Larger foraminifera as marine environmental indicators, *Scripta Geologica* **124**: 263 pp.

Renema, W., 2006a, Large benthic foraminifera from the deep photic zone of a mixed siliciclastic-carbonate shelf off East Kalimantan, Indonesia, *Marine Micropaleontology* **58**: 73–82.

Renema, W. 2006b, Habitat variables determining the occurrence of large benthic foraminifera in the Berau area (East Kalimantan, Indonesia), *Coral Reefs* **25**:351–359.

Renema, W., 2006c, Comment on "Significant Miocene larger foraminifera from South Central Java" by M.K. Boudagher and S.W. Lokier, *Revue de Paléobiologie* **25**: 205–206.

Renema, W., in press, the internal architecture of the miocene *Pseudotaberina* and its relation to caribbean archiasinids, *Palaeontology*.

Renema, W., Racey, A., and Lunt, P., 2003, Paleogene nummulitid foraminifera from the Indonesian Archipelago: a review, *Cainozoic Research* **2** (1–2): 23–78.

Renema, W. and Troelstra, S.R., 2001, Larger foraminifera distribution on a mesotrophic carbonate shelf in SW Sulawesi (Indonesia), *Palaeogeography, Palaeoclimatology, Palaeoecology* **175**: 125–146.

Rögl, F., 1998, Paleogeographic considerations for Mediterranean and Paratethys seaways (Oligocene to Miocene), *Annales des Naturhistorisches Museums Wien* **99A**: 279–310.

Rosen, B.R., 1988, From fossils to earth history; applied historical biogeography, in: Myers, A.A. and Giller, P. S. (eds), *Analytical Biogeography; An Integrated Approach to the Study of Animal and Plant Distributions*, Chapman & Hall, London, pp. 437–481.

Rosen, B.R., 1999, Paleoclimatic implications of the energy hypothesis from Neogene corals of the Mediterranean region, in: Agustí, J., Rook, L., and Andrews, P. (eds), *The Evolution of Neogene Terrestrial Ecosystems in Europe, Homonoid Evolution and Climatic Change in Europe*, Vol. 1, Cambridge University Press, Cambridge, pp. 309–327.

Salamy, K.A. and Zachos, J., 1999, Latest Eocene–Early Oligocene climate change and Southern Ocean fertility: inferences from sediment accumulation and stable isotope data, *Palaeogeography, Palaeoclimatology, Palaeoecology* **145**: 61–77.

Schaub, H., 1981, Nummulites et Assilines de la Tethys paleogène, Taxonomie, phylogenèse et biostratigraphie, *Schweizerische Palaeontologische Abhandlungen* **104**: 1–238.

Scheffen, W., 1932, Ostindische Lepidocyclinen. Part 1, Wetenschappelijke Mededeelingen, *Dienst Mijnbouw Bandoeng* **19**: 5–76.

Sengupta, S., 2000, Problems of classifying Early Oligocene Reticulate *Nummulites* (Foraminiferida) from Southwestern Kutch, Gujarat, *Journal Geological Society of India* **56**: 673–677.

Sengupta, S., 2002, A note on test shape of *Nummulities* cf. *fichteli* Michelloti form A (Foraminiferida) from the early Oligocene deposit of southwestern Kutch, India, *Journal of the Geological Society of India* **60**: 223–224.

Serra-Kiel, J., Hottinger, L., Caus, E., Drobne, K., Fernandez, C., Jauhri, A.K., Less, G., Pavlovec, R., Pignatti, J., Samso, J.M., Schaub, H., Sirel, E., Strougo, A., Tambareau, Y., Tosquella, Y., and Zakrevskaya, E., 1998, Larger foraminiferal biostratigraphy of the Tethyan Paleocene and Eocene, *Bulletin de la Société Géologique de France* **169**: 281–299.

Sirotti, A., 1982, A tentative phylogenetic interpretation of Lepidocyclinidae, in: Gallitelli, E.M. (ed.), *Proceedings of the First International Meeting on "Paleontology, Essential of Historical Geology"* STEM, Mucchi, Modena, pp. 467–480.

Tan, S.H., 1930a, Over *Spiroclypeus* met opmerkingen over zijn stratigraphische verspreiding, *De Mijningenieur* **11**: 180–184.

Tan, S.H. 1930b, Over *Cycloclypeus*: voorlopige resultaten eener biostratigraphischer studie, *De Mijningenieur* **11**: 233–242.

Tan, S.H., 1932, In the genus *Cycloclypeus* Carpenter, *Wetenschappelijke Mededeelingen Dienst van den Mijnbouw in Nederlands-Indië* **19**: 3–194.

Tan, S.H. 1936a, Over verschillende palaeontologische criteria voor de geleiding van het Tertiair, *De ingenieur in Nederlands Oost-Indië* **3** (4): 35–45.

Tan, S.H. 1936b, Zur Kenntnis der Lepidocycliniden, *Natuurkundig Tijdschrift voor Nederlandsch-Indië* **96**: 235–280.

Tan, S.H., 1937, On the genus *Spiroclypeus* H.Douvillé with a description of the Eocene *Spiroclypeus vermicularis* nov. sp. from Koetai in East Borneo, *De Mijningenieur in Nederlands Oost-Indië* **4** (4): 177–193.

Tan, S.H., 1939, Remarks on the "letter classification" of the East Indian Tertiary, *De Mijningenieur in Nederlands Oost-Indië* **6** (7): 39–97.

van Vessem, E.J., 1978, Study of Lepidocyclinidae from South East Asia, particularly from Java and Borneo, *Utrecht Micropaleontological Bulletins* **19**: 1–163.

van der Vlerk, I.M., 1922, Studieën over Nummulinidae en Alveolinidae's Gravenhage-Moulton, 140 pp.

van der Vlerk, I.M., 1925, Het foraminiferengenus *Spiroclypeus* en zijn betekenis voor de stratigraphie van het Tertiair van den Indo-Australischen Archipel, *Verhandelingen van het Geologisch en Mijnbouwkundig Genootschap voor Nederland en Koloniën* **8**: 561–568.

van der Vlerk, I.M., 1928, Het genus *Lepidocycina* in het Indopacifische gebied. *Wetenschappelijke Mededeelingen, Dienst Mijnbouw Bandoeng* **8**: 7–86.

van der Vlerk, I.M., 1929, Groote foraminiferen van NO Borneo, *Wetenschappelijke Mededeelingen, Dienst Mijnbouw Bandoeng* **9**: 1–44.

van der Vlerk, I.M., 1951, Tabulation of determinations of larger foraminifera, in: Reinhard, M. and Wenk, E. (eds), *Geology of the Colony of North Borneo, Bulletin of the Geological Survey Department of the British territories in Borneo* **1**: 137–145.

van der Vlerk, I.M., 1955, Correlation of the Tertiary of the Far East and Europe, *Micropaleontology* **1**: 72–75.

van der Vlerk, I.M., 1968, Two methods of worldwide correlation, *Micropaleontology* **14**: 334–338.

van der Vlerk, I.M. and Postuma, J.A., 1967, Oligo-Miocene Lepidocyclinas and planktonic foraminifera from East Java and Madura, Indonesia, *Proceedings Konijnklijke Nederlandse Academie voor Wetenschappen* **B-70**: 391–398.

van der Vlerk, I.M. and Umbgrove, J.H.L., 1927, Tertiaire gidsforaminiferen uit Nederlandsch Oost-Indie, *Wetenschappelijke Mededeelingen, Dienst Mijnbouw Bandoeng* **6**: 1–31.

Verbeek, R.D.M. and Fennema, R., 1896, *Geologische Beschrijving van Java en Madoera*, J.G. Stemler Cz, Amsterdam.

Veron, J.E.N., 1995, *Corals in Space and Time, The Biogeography and Evolution of the Scleractinia*, University of New South Wales Press, Sydney.

Wang, L., 1994, Sea surface temperature history of the low latitude western Pacific during the last 5.3 million years, *Palaeogeography, Palaeoclimatology, Palaeoecology* **108**: 379–436.

Whittaker, J.E. and Hodgkinson, R.L., 1979, Foraminifera of the Togopi formation, eastern Sabah, Malaysia, *Bulletin of the British Museum (Natural History) Geology* **31**: 1–120.

Wilson, M.E.J., 1995, The Tonasa Limestone Formation, Sulawesi, Indonesia: Development of a Tertiary Carbonate Platform, Ph.D. thesis, Royal Holloway University, London.

Wilson, M.E.J., 2002, Cenozoic carbonates in Southeast Asia: implications for equatorial carbonate development, *Sedimentary Geology* **147**: 295–428.

Wilson, M.E.J. and Rosen, B.R., 1998, Implications of paucity of corals in the Paleogene of SE Asia: plate tectonics or Centre of Origin, in: Hall, R. and Holloway, J.D. (eds), *Biogeography and Geological Evolution of Southeast Asia*, Backhuys, Leiden, pp. 165–195.

Wonders, A.A.H. and Adams, C.G., 1991, The biostratigraphical and evolutionary significance of *Alveolinella praequoyii* sp. nov. from Papua New Guinea, *Bulletin of the British Museum (Natural History)* Geology **47**: 169–175.

Zachos, J., Stott, L.D., and Lohmann, K.D., 1994, Evolution of early Cenozoic marine temperature, *Paleoceanography* **9**: 353–387.

Chapter 7

# The Role of Spain in the Development of the Reef Brachiopod Faunas During the Carboniferous

COR F. WINKLER PRINS
Nationaal Natuurhistorisch Museum Naturalis, P.O. Box 9517, 2300 RA Leiden, The Netherlands, winkler@nnm.nl



**Abstract** After a short introduction on the reef development during the Late Palaeozoic, the tectono-stratigraphic history of the Cantabrian Mountains (northern Spain) during the Carboniferous is discussed, with an emphasis on the tectonically active Pennsylvanian (i.e., Late Carboniferous). The reef-bearing Valdeteja, San Emiliano, and Cuera formations are briefly described, and their brachiopod faunas are discussed with special emphasis on adaptations to a reef environment. The brachiopod faunas are compared with similar faunas from carbonate-platform deposits with reef structures of Mississippian (i.e., Early Carboniferous) age from Northwest Europe (the British Isles in particular), with Pennsylvanian-Permian faunas from the Alps, Urals, Spitsbergen and Arctic Canada, and with Permian reef faunas from Texas (USA). The Bashkirian brachiopod faunas of the Valdeteja Formation resemble the similarly aged Hare Fiord fauna from Arctic Canada most. This makes one wonder whether the connection between the Palaeotethys and Arctic Canada was through the Urals sea and Arctic, as generally believed, or whether there was another connection from the Cantabrian Mountains to the north along eastern North America.



*W. Renema (ed.), Biogeography, Time, and Place: Distributions, Barriers, and Islands,* 217–246

# 1. Introduction

After the Devonian crisis (Eder and Franke, 1982; Webb, 2002) reefs, and especially coral reefs, had become rather rare due to cooling associated with the shift from "greenhouse" to "icehouse" global climate. In the Mississippian, reefs were still reasonably common in Northwest Europe (e.g., Stubblefield, 1960; Bridges et al., 1995; Lees and Miller, 1995; Aretz, 2002). In the Pennsylvanian, further cooling and the influx of siliciclastic material prevented reef forming in most regions, such as Northwest Europe with its paralic deposits. According to Twenhofel (1950) the rarity of Pennsylvanian reefs over large parts of North America was due to muddy water. In the Pennsylvanian, reef structures developed in Texas (West, 1988) leading to the large Permian reef complexes. The Cantabrian Mountains of northwest Spain provide an exceptional example of Pennsylvanian reef mound development (Webb, 2002; Wahlman, 2002). During the entire Carboniferous, the biogenic structures on carbonate platforms were mainly formed by calcareous algae, bryozoans, pelmatozoans, and (chaetetid) sponges (Sommerville et al., 1996; Minwegen, 2001; Wahlman, 2002), although small coral reefs did occur (e.g., Rodríguez, 1996; Aretz and Herbig, 2003; Sano et al., 2004; García-Bellido and Rodríguez, 2005). Associated faunas, brachiopods in particular, sometimes show peculiarities, which may suggest an adaptation to reef environment (e.g., Mundy and Brunton, 1985; Brunton and Mundy, 1988). The presence of closely related or similar forms among the rich brachiopod faunas from the Permian reefs of Texas (Cooper and Grant, 1972–1977) support this interpretation.

In the present paper it is intended to show the importance of the Pennsylvanian (brachiopod) faunas from the Cantabrian Mountains for the survival of these faunas from the Mississippian into late Pennsylvanian and Permian. To explain the special position of these faunas, the geological history of the Cantabrian Mountains during the Devonian and Carboniferous is summarised with special emphasis on the Pennsylvanian, a period of tectonic activity in that area.

# 2. The Cantabrian Mountains

The Palaeozoic core of the Cantabrian Mountains in northwestern Spain (Cantabrian Zone of Lotze, 1945) represents an arcuate fold belt (Fig. 1), which consists of thrust slices and small nappes caused by diastrophic movements of various Carboniferous ages, resulting from a tightening of the arc. The further subdivision of the Cantabrian Zone, as proposed by Julivert (1967, 1971), reflects major faults in the area constituting a single palaeogeographical region. The thrust structures have been moulded around a foreland spur that became more and more restricted to the east as time went by (see Wagner and Martínez García, 1974; Wagner, 2004). This foreland massif was called the Cantabrian Block by Radig (1962). The general region of the Picos de Europa represents the more permanently stable area in which carbonate sedimentation was the norm throughout the Carboniferous. During the Early Palaeozoic a strongly subsident basinal area was situated in the West

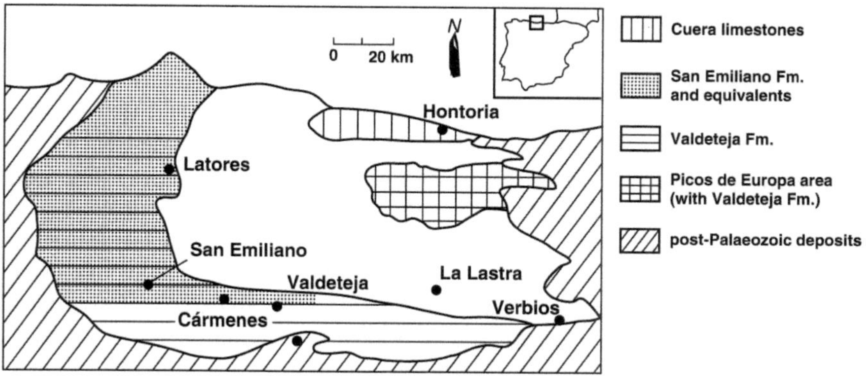


Fig. 1 Map of the Cantabrian Mountains, northwest Spain, showing the areas of deposition of the Valdeteja and San Emiliano formations, Cuera Limestones, and the Picos de Europa (based on data from Eichmüller and Seibert, 1984; Rodríguez Fernández, 1993; Sánchez de Posada et al., 1998; and Wagner and Winkler Prins, 1999). The Carboniferous formations crop out in thrust sheets showing near vertical successions ranging in age from Cambrian to Carboniferous.

Asturian–Leonese Zone (Lotze, 1945), receiving sediment from an enveloping hinterland to the west and south. It moved to the Cantabrian Zone during the Pennsylvanian. The Palaeozoic edifice is covered by an unconformable succession of highest Pennsylvanian (Stephanian C), Permian, Mesozoic, and Cenozoic sediments. A discussion of the different views on the complicated tectonics of the Cantabrian Mountains is beyond the scope of this paper.

The Devonian and Carboniferous deposits of the Cantabrian Zone can be divided into two sedimentary areas, the Asturo–Leonese and Palentian realms, which are separated by the Ruesga Fault, a line of major shortening due to the tightening of the arcuate fold belt (Wagner and Winkler Prins, 1999). I will practically confine myself to the Asturo–Leonese succession. During the Devonian, the southern and western part of the Asturo–Leonese realm subsided most, providing a complete succession with reef development in the Lower–Middle Devonian (e.g., Brouwer, 1964; Mohanti, 1972; Becker et al., 1979; Frankenfeld, 1982; Méndez-Bedia et al., 1994). Towards the north(east) a tectonic high developed and the formations wedge out in that direction. This is at least partly due to a phase of uplift during the Famennian (Comte, 1938; Parga, 1969; Reijers, 1973). The unconformable late Famennian – early Tournaisian Ermita Formation consists largely of sandstones derived from erosion of the underlying deposits.

## 2.1. Mississippian

After this phase of erosion, which lasted until the early Tournaisian, the whole area was leveled and covered by a quiet sea with an anaerobic mud floor (Vegamián Formation), surrounded by an area with limestone development (Baleas Formation;

see Wagner et al., 1971; Eichmüller and Seibert, 1984: Fig. 3). The Vegamián faunas are either pelagic (e.g., entomozoid ostracodes; Jordan and Bless, 1970) or consist of small invertebrates, concentrated at certain levels, that may represent material drifted in after a storm, perhaps attached to seaweed (cf. Amler and Winkler Prins, 1999; Winkler Prins and Martínez Chacón, 1999), with a connotation of quiet water below wave base (Winkler Prins, 1991; Martínez Chacón and Winkler Prins, 1993). Sediment supply diminished and practically stopped during the deposition of the succeeding Genicera Formation (also called Alba Formation), a succession of nodular to wavy bedded (griotte) limestones of greenish-grey and pink colours, usually enclosing a dark red chert layer (varying from black to white through bleaching), called the Lavandera Member (Wagner et al., 1971). This suggests that the hinterland area was peneplained and probably flooded most of the time. The Genicera Formation contains characteristically large ammonoids and is rich in conodonts. Radiolaria are also found (as in the phosphatic nodules of the Vegamián Formation) and foraminifera, which indicate a gradual deepening in upward succession (Balthasar and Amler, 2003). The faunas, including brachiopods, trilobites, corals, and molluscs are again indicative of a quiet water environment, but – oddly enough – large crinoids (belonging to the genus *Balearocrinus*) and a large brachiopod, i.e., typical *Martinia glabra*, do occur as well (Amler and Winkler Prins, 1999). The top of the Genicera Formation reaches the early Serpukhovian (Arnsbergian), when it is succeeded gradually by dark grey, laminated, fetid limestones of the Barcaliente Formation of Serpukhovian to earliest Bashkirian Age, representing turbidites in a starved basin (allodapic limestones of Reuther, 1977: 50). Subsidence increased, since during the Serpukhovian 50–200 m of limestone were deposited as compared to the less than 50 m for the combined Tournaisian and Viséan. The basin became shallower towards the top of the formation, where breccias (Porma Breccia of Reuther, 1977) and (replaced) gypsum crystals suggest sedimentation in a very shallow, probably intertidal sea (Winkler Prins, 1968: 59; Eichmüller, 1986). Also, the hinterland probably started to rise, as indicated by the development of siliciclastic turbidites replacing the lower part of the Barcaliente Formation in the extreme south (Olleros Formation), at the same place where the top part of the Genicera Formation is replaced by red fossiliferous shales (the Olaja beds; see Wagner et al., 1971: Fig. 6). These are the first signs of the strong tectonic movements that dominated the Cantabrian Mountains during the Pennsylvanian, when a cumulative thickness of some 16,000 m of flysch and molasse type sediments were deposited.

Fossils are extremely rare in the Barcaliente Formation, but conodonts have been found (Méndez and Menéndez Álvarez, 1985). A conodont fauna typical for the basal Bashkirian of Central Asia has been found at an atypical succession (Palentian facies) near La Lastra in northern Palencia (Nemyrovska, Personal communication 2005). At the La Lastra section no change in the sedimentation can be noticed at the mid-Carboniferous boundary, normally connected with an important eustatic lowering of the sea level. On the other hand, it is possible that the intertidal deposits with collapse breccias in the upper part of the Barcaliente Formation (spectacular examples of which are found in the type section in the Curueño Valley with blocks of 1 $m^3$ in a white spar matrix, see Winkler Prins, 1971) are linked to this event.

## 2.2. Pennsylvanian

During the larger part of Bashkirian times thick carbonate platform deposits were laid down around a central, more stable area. These platform deposits include reef limestones (Winkler Prins, 1968; Eichmüller and Seibert, 1984; Minwegen, 2001) of the Valdeteja Limestone Formation succession (675 m at the type section; Winkler Prins, 1968, 1971). The rather thin Barcaliente Formation and the Ricacabiello formation, a condensed succession of grey and purplish shales with manganese nodules, are found in the central part of the basin. These formations contain a fauna of "Culm" type indicating quiet, rather deep waters (Martínez Chacón et al., 1985). In late Bashkirian to early Moscovan times, siliciclastic turbidites grading into shales with limestone bands and occasional reefoid mounds (San Emiliano Formation and equivalents) rest upon or partially replace the Valdeteja Limestone in the west and south (Bowmann, 1985). This pattern of shallowing upwards cycles, ideally from turbidites through shallow marine and paralic deposits to coals and conglomerates (cf. Wagner and Winkler Prins, 2002), is repeated throughout the Moscovian, with the clastic sedimentation shifting from west to east (Bless and Winkler Prins, 1973) and a decreasing marine influence towards the top. Only in the extreme east, in the more stable Picos de Europa area, carbonate deposition was continuous into the Gzhelian (late Stephanian; see Villa and van Ginkel, 1999). The largely marine uppermost Moscovian (upper Myachkovsky) and Kasimovian deposits were laid down in an unconformable basin in eastern León and Palencia (the post-Leonian basin; see Wagner and Winkler Prins, 1985; Wagner et al., 2002). The upper Barruelian and Stephanian B was deposited in a subsequent unconformable basin covering the southern part of the Cantabrian Mountains (post-Asturian basin; see Wagner and Winkler Prins, 1985: 386). This basin, which is almost entirely marine in the area close to the foreland in eastern Asturias, is represented by generally non-marine coal-bearing successions in a string of coalfields following the arcuate fold belt in northern León, northeastern Palencia, and western Asturias.

The Valdeteja and San Emilano formations and the Cuera Limestones (Bashkirian–Upper Moscovian) are considered in some detail below, because their brachiopod faunas are the best-known faunas of reef affinity from the Carboniferous of the Cantabrian Mountains.

### *2.2.1. Valdeteja Formation*

Originally described as the Valdeteja Member of the Escapa Formation (Winkler Prins, 1968), the Valdeteja Limestone was raised to formation rank by Wagner et al. (1971) and its type section was redescribed (Winkler Prins, 1971). It typically consists of a thick (up to 700 m) succession of mainly light grey limestones containing algae, locally forming algal mounds and foraminifera and occasional fossiliferous bands with brachiopods, corals, bryozoans, conodonts, etc. These fossiliferous bands may be due to shallowing, which is presumably related to sea-level movements, which are the result of intermittent subsidence of the basin. The formation is exposed in nearly vertical successions in thrust sheets several kilometres apart; these successive sections do not allow a three-dimensional picture to be obtained. Both in East–West and North–South directions the formation can be traced for more than a hundred kilometres. The lack of a three-dimensional picture hampers the distinction of the reef structures, which are mainly algal mounds. Extensive dolomitisation and recrystallisation further complicate matters. For example, the importance of coral ghost structures in massive, recrystallised limestone beds cannot be assessed properly (Winkler Prins, 1968: 49). However, it appears obvious that no true coral reefs were present. The carbonate platform of the Valdeteja Formation formed an external rim around a basin, where the mud sedimentation did not keep up with subsidence, a situation comparable to the Viséan reefs in Derbyshire (Wolfenden, 1958). The Valdeteja Formation is of Bashkirian age ranging locally into the earliest Moscovian (Villa et al., 2001).

The massive limestones of the Valdeteja Formation, forming high mountains and deep gorges (hence the old names "caliza de montaña" and "calcaire des cañons"), are not inviting for fossil collecting and for a long time they were considered unfossiliferous (e.g., Martínez Díaz, 1969), though fossils have been found and a rich fauna from the Latores locality in Asturias has been listed by Delépine and Llopis Lladó (1956). The faunas of the Valdeteja Formation are linked to certain horizons (see Winkler Prins, 1968) or occur in pockets, probably related to reef structures (op. cit., loc. 10, north of Cármenes). The brachiopods are by far the most important part of the Valdeteja faunas. Among the microfaunas the foraminifera are most important for dating (Villa, 1982, 1989; Villa et al., 2001), but conodonts (van den Boogaard and Bless, 1985; Méndez and Menéndez Álvarez, 1985) and ostracodes (Bless and Sánchez de Posada, 1973; Sánchez de Posada, 1976; Becker, 1982) also occur (Sánchez de Posada et al., 1996). The latter are of Eifelian type, quite distinct from the Thuringian ones of quiet water deposits (Vegamián, Genicera and Ricacabiello formations). Bryozoans, porifera, anthozoa (de Groot, in Winkler Prins, 1971; Boll, 1985), crinoid ossicles, bivalves, and gastropods are occasionally found; trilobites (Gandl, 1987), ammonoids (Wagner-Gentis, in Martínez Chacón, 1979) and rostroconchs (Babin et al., 1999) are rare. Algae (Rácz, 1964; in Winkler Prins, 1968: Table 3; Eichmüller, 1985) play an important role as mud binders, forming *Donezella* mounds (Riding, 1978; Bowman, 1979). The first comprehensive list of brachiopods from the Valdeteja Formation was published by Delépine and Llopis Lladó (1956). Descriptions of the Productidina were given by Winkler Prins (1968), while Martínez Chacón (1977, 1979) described the Orthida, Productida, Orthotetida, and Rhynchonellida.

### *2.2.2. San Emiliano Formation*

The San Emiliano Formation was originally described by Brouwer and van Ginkel (1964) and formally introduced with a type section by Bowman (1979, 1982, 1985; see also Carballeira et al., 1985; Fernández, 1993). Bowman described the occurrence of *Donezella* mud mounds in its middle, La Majua, Member and associated oncolithic marls with well-preserved and varied brachiopod faunas. The San Emiliano Formation overlies, and partly replaces laterally, the Valdeteja Formation and is of late Bashkirian to early Moscovian age. Although the deltaic environmental setting is quite different from that of the earlier Valdeteja Formation, its faunas are rather similar but less specialised.

These mounds were also described from the Cármenes Syncline by Riding (1978), who warned that similar structures could be produced by diagenesis. The deposits described by him are, however, younger and attributed to the Lois-Ciguera Formation by Rácz (1964), who described the algae and mentioned the mounds (see also van Ginkel and Villa, 1996).

The lower, Pinos, Member is practically unfossiliferous. In the La Majua Member algae are found in the limestones (Rácz, 1964), occasionally forming mounds (Bowman, 1982). This member contains well-preserved brachiopod faunas in the marly layers, as well as corals (Kullmann and Rodriguez, 1986), gastropods, bivalves, rostroconchs (Babin et al., 1999), porifera (García-Bellido and Rigby, 2004), trilobites (Romano, 1971; Gandl, 1987), ostracodes (Fernández López and Sánchez de Posada, 1987), crinoids, and echinoids (Winkler Prins, 1968; Sánchez de Posada et al., 1996). Fusulinid foraminifera were helpful in dating the formation (van Ginkel, 1965; Bowman, 1982; van Ginkel and Villa, 1996); its conodont faunas have not yet been described. In the upper, Candemuela, Member the (brachiopod) faunas (Martínez Chacón, in Carballeira et al., 1985) are mainly found as moulds in mudstones. Macrofloral elements have been used for correlation with northwestern Europe (Wagner and Bowman, 1983). Brachiopods were first described by Winkler Prins (1968) and Martínez Chacón (1977,1978a, b, 1979); together they presented a summary of the brachiopod faunas (Martínez Chacón and Winkler Prins, 1986, 2000).

### 2.2.3. Cuera Limestones

The "Calizas del Cuera" (Cuera Limestones), an about 1,000 m thick limestone succession of Bashkirian and Moscovian ages, has been informally described by Navarro et al. (1986). It was deposited on a stable carbonate platform adjacent to the Picos de Europa area, but shows a variety of limestone facies with an occasional sandstone intercalation (op. cit.: Fig. 4). A sedimentological study of the type area (della Porta et al., 2004) showed the influence of tectonic subsidence, sea-level fluctuations and high carbonate accumulation rates. The occurrence of the brachiopod *Aseptella asturica* (Martínez Chacón and Winkler Prins, 1977) in the basal part, just above the Barcaliente Formation, suggests a rather quiet depositonal environment, showing some similarities with those of the San Emiliano Formation (cf. Martínez Chacón and Winkler Prins, 1993). It is mainly in the upper part (upper Moscovian) that *Donezella* mounds are found and the brachiopod fauna of that part (Martínez Chacón, 1990, 1991) compares better with that of the Valdeteja Formation.

The upper part of the Cuera Limestones, especially at the locality Hontoria, has rich faunas: bryozoans, corals (de Groot in Martínez Chacón, 1979; Rodriguez and Ramírez, 1987), bivalves, gastropods, crinoids, and ostracodes (Sánchez de Posada and Bless, 1999; Sánchez de Posada and Fohrer, 2001). Rich fusulinid faunas were found as well (Villa Otero, 1995), whereas conodonts are rare. Both below and above the limestone at Hontoria, miospores were found indicating a Westphalian D age (García Bartolomé et al., 2003). The brachiopods were described by Martínez Chacón (1975, 1977, 1979, 1990, 1991; Martínez Chacón and Bahamonde, in press; see also Sánchez de Posada et al., 1993).

## 3. Discussion of the Brachiopod Faunas from the Cantabrian Mountains

The analysis of the brachiopod faunas associated with these reefs is here confined to examples from the Bashkirian and Moscovian of Asturias and León, since these brachiopod faunas are currently the best known. The brachiopod faunas from the Cuera Limestones have been largely described (Martínez Chacón, 1990, 1991; Martínez Chacón and Bahamonde, in press), but complete descriptions of the other faunas, especially those from the Valdeteja Formation, are still outstanding. The information on the brachiopod faunas is summarised in Table 1. A short discussion of the brachiopod reef assemblages from the Valdeteja Formation and Cuera imestones was given by Martínez Chacón (in Sánchez de Posada et al., 2002).

The inarticulate brachiopods are rare in the reef faunas and are not considered here. The Rafinesquinidae, represented in the Viséan Cracoe reef environment of Yorkshire by *Leptagonia*, became extinct before the end of the Mississippian. Chonetidina are not specific for a reefoid environment with agitated water, and indeed more commonly associated with a muddy sea floor. The occasional specimens probably lived in sheltered areas. On the other hand, many of the Productidina and Strophalosiidina preferred agitated water (e.g., *Fluctuaria undata*; see Fig. 2 (6)) and some forms are considered characteristic of a reef environment (e.g., Mundy and Brunton, 1985; Brunton and Mundy, 1986, 1988; Brunton et al., 1994). Examples, also found in the Cantabrian Mountains, are *Productina*, *Eomarginifera*, *Heteralosia*, and possibly *Limbifera* (figured as *Institina*? sp. by Martínez Chacón and Winkler Prins, 1993; see Fig. 2 (2)); *Plicatiferina* had presumably a similar lifestyle as *Plicatifera* and thus may be considered another example (see specimen with attachment ring of Fig. 2 (9)). *Proboscidella proboscidea* was found in a reef limestone assigned to the Perapertú Formation near Verbios (Palencia), which is early Moscovian in age (Fig. 2 (7)). The "reef-building" Richthofenioidea (Flajs et al., 1996) are extremely rare in the Cantabrian Mountains; in fact, only one specimen of *Zalvera* sp. has been described so far from an upper Moscovian limestone in Palencia (Brunton, 1996; Fig. 2 (5)). *Parmephrix*, a characteristic element of the Viséan reefs in Derbyshire (Brunton et al., 1994), has only been found questionably in quiet water faunas of the Tournaisian black shales of the Vegamián Formation: *Parmephrix*? *aprathensis* (cf. Martínez Chacón and Winkler Prins, 1993). These specimens probably drifted in attached to goniatites as suggested by their ornamentation. A recent revision of the brachiopods from the Valdeteja Formation (see Martínez Chacón and Winkler Prins, 2006a) and notably of the locality of Latores (Martínez Chacón and Winkler Prins, 2006b) has shown the presence of rare specimens related to typical reef-related taxa from the Viséan of the British Isles (Mundy and Brunton, 1985; Brunton and Mundy, 1988), such as *Stipulina*? sp., *Institina*? sp., and *Retroplexus*? sp. (Fig. 2 (4, 5)). The Incisiini *gen. et sp. nov.* (the genus has been referred to with the nomen nudum "*Regrantia*", e.g., Martínez Chacón and Winkler Prins, 1993) has occasionally been found in the Valdeteja and San Emiliano formations and thus is the oldest known member of the tribe. It does not seem to be particularly adapted to a reef environment, lying anchored in the mud with its spines.

*Table 1.* *List of species occurring in the Cantabrian Mountains (Spain; A–D) and Hiare Fiord (Canada; E), and their occurrence in the Viséan of the British Isles (F–G). The Inarticulates are not included since they are not considered relevant. (From Winkler Prins, 1968: Table 3, which is continued after Table 4, 1983; Martínez Chacón, 1979, 1990, 1991; Martínez Chacón and Winkler Prins, 1986, 1993, 2000, 2006; Carter and Poletaev, 1998; Wolfenden, 1958; Brunton, 1984, 1987; Brunton and Tilsley, 1991.) A: Valdeteja Formation, mainly Bashkirian; B: Cuera Limestones, lower Moscovian (non-reefoid limestones); C: Cuera Lst, Moscovian (with reef structures); D: San Emiliano Formation, upper Bashkirian – lower Moscovian; E: Hiare Fjord, upper Bashkirian – lower Moscovian; F: Cracoean reefs in Yorkshire; Viséan; G: Ireland, Viséan; x = doubtful identification.*

| *Taxa* | *A* | *B* | *C* | *D* | *E* | *F* | *G* |
|---|---|---|---|---|---|---|---|
| *Tornquistia polita* (McCoy) | | | | cf. | | | |
| *Caenanoplia* sp. | x | | | | | | |
| *Globsochonetes waldenburgianus* (Paeckelmann) | | | | X | | | |
| *Globsochonetes waldschmidti* (Paeckelmann) | | | | X | | | |
| *Rugosochonetes acutus* (Demanet) | | | X | X | | | |
| *Rugosochonetes skipseyi* (Currie) | | X | X | X | | | |
| *Chonetinella crassiradiata* (Dunbar and Condra) | | | X | X | | | |
| *Chonetinella flemingi* (Norwood and Pratten) | X | | cf. | | | | |
| *Chonetinella jeffordsi* (Stevens) | | | | | cf. | | |
| *Neochonetes babianus* (Martínez Chacón and Winkler Prins) | | x | | X | | | |
| *Sokolskya* sp. | | | | x | x | | |
| *Quadrochonetes* sp. | | | | X | | | |
| *Productina pectinoides* (Phillips) | X | | | | | | |
| *Alitaria frechi* (Paeckelmann) | X | | sp. | | | | |
| *Alitaria nasuta* (Paeckelmann) | | | | X | | | |
| *Eomarginiferina* sp. | | | | | x | | |
| *Rugivestis pristina* (Carter and Poletaev) | sp. | | | | X | | |
| *Fimbrinia? borealis* (Carter and Poletaev) | | | | | X | | |
| *Quasiavonia aculeata* (Sowerby) | X | | | X | X | X | X |
| *Quasiavonia echidniformis* (Chao) | | | X | | | | |
| *Tuberculatella* sp. | | | X | | | | |
| Institiferini gen. and sp. nov. | | | X | | | | |
| *Krotovia granulosa* (Phillips) | X | | X | X | | | |

(continued)

*Table 1.* (continued)

| *Taxa* | *A* | *B* | *C* | *D* | *E* | *F* | *G* |
|---|---|---|---|---|---|---|---|
| *Krotovia lamellosa* (Brunton) | | | | | cf. | X | X |
| *Breileenia* sp. | X | X | | | | X | |
| *Desmoinesia* sp. | | | X | | | | |
| Incisiini gen. et sp. nov. | X | | | X | | | |
| *Hystriculina*? cf. *wabashensis* (Norwood and Pratten) | | | | | x | | |
| *Retimarginifera*? sp. | | | | | x | | |
| *Lazarevia stepanowensis* (Carter and Poletaev) | | | | | X | | |
| Semicostellini gen. nov. (aff. *Limbifera*) sp. nov. | | | X | | | | |
| *Maemia gelida* (Carter and Poletaev) | sp. | sp. | x | | | X | |
| *Admoskovia* sp. | | | X | | | | |
| *Bicarteria*? sp. | x | | x | | | | |
| *Duartea* sp. | X | | | | | | |
| *Inflatia* sp. | X | | X | | | | |
| *Latispinifera* cf. *chaykensis* (Lazarev) | X | | | X | | | |
| *Latispinifera* aff. *ivanovi* (Lapina) | X | | | X | | | |
| *Productus carbonarius* (de Koninck) | X | | | | | | |
| *Productus concinnus* (Sowerby) | X | | | | | | |
| *Kozlowskia bediae* (Martínez Chacón) | sp. | sp. | X | | | | |
| *Kozlowskia involuta* (Tschernyschew) | | X | | | | | |
| *Kozlowskia* cf. *pulchra* (Rotai) | | X | | | | | |
| *Kozlowskia splendens* (Norwood and Pratten) | | | | | x | | |
| *Eomarginifera setosa* (Phillips) | X | | | X | sp. | | x |
| *Eomarginifera minuta* (Muir-Wood) | X | | | | | | |
| *Eomarginifera praecursor* (Muir-Wood) | X | | | X | | | |
| *Antiquatonia costata* (Sowerby) | X | | X | | | | |
| *Antiquatonia hermosana* (Girty) | | | | | cf. | | |
| *Antiquatonia insculpta* (Muir-Wood) | X | | | | | | |
| *Tubaria genuina* (Kutorga)[a]) | cf. | | | | | | |
| *Kutorginella* cf. *mosquensis* (Lapina) | | | x | | | | |
| *Kutorginella stepanovi* (Ivanova[a]) | | | | | cf. | | |

| | | | | | | | |
|---|---|---|---|---|---|---|---|
| *Alexenia* sp. | X | | x | | | | |
| *Chaoiella bathycolpos* (Schellwien) | | | cf. | | | | |
| *Chaoiella gruenewaldti* (Krotow) | X | | | | | | |
| *Reticulatia americanus* (Dunbar and Condra)[b]) | | | | | x | | |
| *Buxtonia*? sp. | x | | | | | X | X |
| *Kochiproductus* sp. | x | | | | | X | |
| *Echinaria* cf. *knighti* (Dunbar and Condra) | X | | sp. | | x | | |
| *Echinoconchella elegans* (McCoy) | X | | | | | X | X |
| *Echinoconchella*? *venusta* (Thomas) | X | | | | | X | X |
| *Karavankina rakuszi* (Winkler Prins) | | cf. | X | | | | |
| *Karavankina wagneri* (Winkler Prins) | | | | X | | | |
| *Cubacula* sp. | | | X | | | | |
| *Pustula*? sp. | x | | | | x | X | X |
| *Linoproductus*? sp. | | | | | x | | |
| *Balakhonia continentalis* (Tornquist) | X | | | X | | | |
| *Fluctuaria undata* (Defrance) | X | | sp. | | cf. | X | X |
| *Marginovatia* sp. | X | | | | | | |
| *Cancrinella craigmarkensis* (Muir-Wood) | | | X | X | sp. | | |
| *Cancrinella retiformis* (Muir-Wood) | | | x | X | | | |
| *Globiella*? sp. | | | | x | | | |
| *Liraria paucispina* Carter and Poletaev | | | sp. | | X | | |
| *Ovatia laevicosta* (White) | X | | | X | | | |
| *Heteralosia* sp. | | | | X | | X | X |
| *Hontorialosia uniplicata* (Martínez Chacón) | | | X | | | | |
| *Plicatiferina sinecosta* (Martínez Chacón)[c]) | X | | x | X | | | |
| *Plicatiferina kalashnikovi* (Carter and Poletaev[c]) | | | | | | | |
| *Stipulina*? sp. | x | | | | | | |
| *Institina*? sp. | x | | | | | | |
| *Retroplexus*? sp. | x | | | | | X | X |
| *Rugicostella*? sp. | | | x | | | | |
| *Tapajotia* sp. | | | x | X | | | |
| *Diplanus posadai* (Martínez Chacón) | | | X | | | | |
| *Meekella* sp. | X | | | | | | |
| *Schuchertella*? *sajakensis asturica* (Martínez Chacón) | sp. | cf. | | | | | |

(continued)

*Table 1.* (continued)

| *Taxa* | *A* | *B* | *C* | *D* | *E* | *F* | *G* |
|---|---|---|---|---|---|---|---|
| *Streptorhynchus subpelargonatus* (sensu Demanet) | X | | | | | | |
| *Rhipidomella michelini* (Léveillé) | X | cf. | X | X | sp. | X | X |
| *Enteletes* sp. | | | sp. | | | | |
| *Orthotichia dorsistrigis* (Carter and Poletaev) | sp. | | sp. | | X | | |
| *Orthotichia* cf. *oklahomae* (Dunbar and Condra) | | | x | | | | |
| *Pugnax acuminatus* (Sowerby) | aff. | | sp. | | | X | X |
| *Pugnoides rosae* (Martínez Chacón) | x | | | | | | |
| *Stenoscisma winkleri* (Martínez Chacón) | | | X | X | sp. | | |
| *Psilocamara* sp. | X | | | | | | |
| *Careoseptum septentrionale* (Carter and Poletaev) | | | | | X | | |
| *Callaiapsida alcaldei* (Martínez Chacón) | X | | sp. | | | | |
| *Callaiapsida paucicostata* (Martínez Chacón) | | | | X | | | |
| *Lambdarina manifoldensis* (Brunton and Champion) | sp. | | | | | X | X |
| *Rhynchopora nikitini* (Tschernyschew) | X | | | | | | |
| *Trasgu minor* (Martínez Chacón) | X | | | | | | |
| *Septacamera* sp. | | | | | X | | |
| *Yanishewskiella globosa* (Martínez Chacón) | X | | | | | | |
| *Exlaminella insolita* (Carter and Poletaev) | | | | | X | | |
| *Cenorhynchia* sp. | | | | | X | | |
| *Phrenophoria*? sp. | | | | | X | | |
| *Hemileurus*? sp. | | | | | x | | |
| *Pontisia leonica* (Martínez Chacón) | | | | X | | | |
| *Antronaria annosa* (Carter and Poletaev) | | | | | X | | |
| *Elassonia? sverdrupensis* (Carter and Poletaev) | | | | | X | | |
| *Cleiothyridina* sp. | X | | X | X | | X | X |
| *Cardiothyris* sp. | X | | | | | | |
| *Composita ohioense* (Sturgeon and Hoare) | | | cf. | | | | |
| *Camarium nuperum* (Carter and Poletaev) | | | | | X | | |
| *Nucleospira aquilonaris* (Carter and Poletaev) | | | | | X | | |
| *Hustedia remota* (von Eichwald) | sp. | | cf. | sp. | sp. | | |
| *Crurithyris tschernyschewi* (Likharev) | | | | | cf. | | |

| | | | | | | |
|---|---|---|---|---|---|---|
| *Crurithyris urii* (Fleming) | cf. | | sp. | cf. | sp. | X |
| *Martinia glabra* (Sowerby) | X | | | cf. | | X |
| *Tiramnia* ex gr. *uralica* (Tschernyschew) | sp. | | X | | | |
| *Tiramnia walteri* (Carter and Poletaev) | | | x | | X | |
| *Tiramnia grunti* (Carter and Poletaev) | | | sp. | | X | |
| *Jilinmartinia*? cf. *sokolovi* (Tschernyschew) | | | | | x | |
| *Heteraria canadiensis* (Carter and Poletaev) | | | sp. | | X | |
| *Eomartiniopsis susanae* (Martínez Chacón) | | | | X | | |
| *Martiniopsis*? sp. | | | x | | | |
| *Donispirifer* sp. | | | X | | | |
| *Anthracospirifer* cf. *solenensis* (Poletaev) | x | | x | | sp. | |
| *Anthracospirifer occiduus* (Sadlick) | | | sp. | | cf. | |
| *Anthracothyrina llanisca* (Martínez Chacón) | x | x | X | | | |
| *Anthracothyrina pinica* (Martínez Chacón) | | | X | X | sp.? | |
| *Brachythyrina*? sp. | | | x | | | |
| *Elinoria ellesmerensis* (Carter and Poletaev) | | | sp. | | X | |
| *Skelidorygma asturica* (Martínez Chacón) | | | X | | | |
| *Alphachoristites (Prochoristites)* Sp. | x | | x | x | | |
| *Parachoristites tellevakensis* (Carter and Poletaev) | sp. | | sp. | | X | |
| *Trautscholdia*? ex gr. *Jigulensis* (Stuckenberg) | | | X | | sp. | |
| *Tangshanella*? sp. | | | | | x | |
| *Tegulispirifer*? ex gr. *dunbari* (King) | | | X | | | |
| *Tegulispirifer tegulatus* (Trautschold) | x | | X | X | | |
| *Gypospirifer* sp. | | | X | | X | |
| *Avisyrinx obsoleta* (Martínez Chacón) | | | X | | | |
| *Cantabriella schulzi* (Martínez Chacón) | sp. | x | sp. | X | sp. | |
| *Brachythyris* sp. | X | | sp. | | | |
| *Meristorygma arctica* (Carter) | sp. | | sp. | | X | |
| *Reticularia lineata* (Sowerby) | cf. | | | | | |
| *Kitakamithyris*? sp. | x | | | | | |
| *Phricodothyris asiatica* (Chao) | | | sp. | | X | |
| *Phricodothyris (Condrathyris) ovata* (Chao) | X | | | X | | |
| *Phricodothyris (Condrathyris) truyolsae* (Martínez Chacón) | | | sp. | X | | |

(continued)

*Table 1.* (Continued)

| *Taxa* | *A* | *B* | *C* | *D* | *E* | *F* | *G* |
|---|---|---|---|---|---|---|---|
| *Punctospirifer*? sp. | X | | | X | | X | X |
| *Spriferinella multispinosa* (Martínez Chacón) | | | X | | | | |
| *Altiplecus antiquus* (Martínez Chacón) | | | X | | | | |
| *Xestotrema*? sp. nov. | | | X | | | | |
| *Cranaena nassichuki* (Carter and Poletaev) | | | | | X | | |
| *Girtyella* sp. | X | | | | | | |
| *Dielasma vesiculare* (de Koninck) | | cf. | | | | | |
| *Beecheria itaitubensis* (Derby) | | | | | cf. | | |
| *Dielasmella*? sp. | x | | | | | | |
| *Pelaiella exigua* Martínez Chacón | | | X | | | | |
| *Notothyis (Ligatella) sarytchevae* (Martínez Chacón) | | | | X | | | |

[a]*The distinction between Kutorginella and Tubaria is not clear and both genera appear closely related, if not identical. The Spanish material (Fig. 2 (2)) is assigned with some reserve to Tubaria genuina because of its tube, transverse shape and large convex ears (see Brunton et al., 2000).*

[b]*It is doubtful if Reticulatia americanus really belongs to Reticulatia, it could rather belong to Latispinifera.*

[c]*Plicatiferina kalashnikovi is very similar to Plicatiferina sinecosta, their main difference being the presence of clasping spines and even an attachment ring in the latter species (Fig.2 (5)).*

Some Orthida, such as *Schizophoria*, are common in a reef environment but also occur in other deposits. The same holds true for some Orthotetidina, like *Diplanus* and *Streptorhynchus* with their large interareas, and for the Stenoscismatoidea with forms like *Stenoscisma* and *Callaiapsida*. The importance of differences in internal structures in many small, attached forms for their way of life and for their relations is difficult to judge. *Lambdarina* was a small rhynchonellid that lived sheltered in the reef (e.g., Brunton and Champion, 1974; Bassett and Bryant, 1993), possibly attached to algae or bryozoans. *Rhynchopora* also appears to be characteristic of reef environments in Pennsylvanian and Permian times. This genus may have been derived from the Viséan *Tretorhynchia*. Several Athyrididina species lived on the reefs but not exclusively so, since they were not particularly adapted to them. The same holds true for many representatives of the Spiriferida and Spiriferinida, but some forms were particularly well adapted to agitated water through their relatively large interareas (*Cantabriella*, *Avisyrinx*, *Altiplecus*) and large size (*Trautscholdia*), although *Cantabriella* and *Trautscholdia* are also found in mudstones. The Terebratulida have no apparent adaptations to life on a reef, other than their pedicle attachment, and the Spanish material has not been studied in sufficient detail to allow consideration here.

## 4. Relation with Other Areas

As an example of Viséan reefs from Northwest Europe the Craven reef belt of North Yorkshire (Table 2) is chosen for comparison, since the brachiopod faunas are well known through the summary by Wolfenden (1958) and the work of Brunton and others (e.g., Brunton and Mundy, 1988; Brunton and Tilsley, 1991; Brunton et al., 1994). Data from Ireland (Brunton, 1966, 1968, 1984, 1987; Harper and Jeffrey, 1996) only marginally modify the picture. They have many genera and species in common with the Valdeteja Formation, which at first glance make the impression of Viséan faunas. Only some Moscovian elements (e.g., *Chaoiella*, *Reticulatia*, and *Rhynchopora*) and the absence of some of the more typical Mississippian forms, such as *Leptagonia* and *Gigantoproductus*, suggest a younger, i.e., Bashkirian, age (Winkler Prins, 1968), which has been corroborated by fusulinid evidence (Villa, 1982).

Although it is less apparent when considering only the reef-related faunas, there is a close relationship between the Carboniferous brachiopod faunas from the Cantabrian Mountains and those from the Carboniferous-Permian of the Carnic Alps and the Karavanke Mountains in Slovenia with genera such as *Isogramma* and *Eolyttonia* (e.g., Ramovš, 1971; Winkler Prins, 1983; Martínez Chacón and Winkler Prins, 1993), and to a lesser extent also with the Bükk Mountains (Gulyás-Kis, 2004). Also other faunal elements, such as the corals (Rodriguez et al., 1986; Kullmann and Rodríguez, 1994) and ostracods (Sánchez de Posada and Fohrer, 2001), show a close relationship with the Carnic Alps. The connections extended along the Palaeotethys as far east as Thailand, though no

Fig. 2

typical reef forms were found there (e.g., *Comuquia*, *Tuberculatella*, Incisiini; cf. Winkler Prins, 1983).

The Urals were a less closely related area during the Carboniferous but the late Pennsylvanian and early Permian show some specialised forms belonging to genera known from earlier deposits in the Cantabrian Mountains, e.g., *Rugivestis kutorgae* (Tschernyschew, 1902) and *Avisyrinx*? *expansa* (Tschernyschew, 1902). The Spitsbergen faunas are similar to those from the Urals, but typical reef forms are seemingly absent. The Permian brachiopods from Greenland appear to be rather distinct, probably due to palaeoecological and climatic differences.

Surprisingly, the brachiopod fauna from the upper Bashkirian or lower Moscovian Hare Fiord Formation of Ellesmere Island (Canadian Arctic Archipelago) (Carter and Poletaev, 1998) proved quite similar to the Spanish faunas, particularly from the Valdeteja Formation. Excluding the terebratulids, Hare Fiord has 28 of its 42 genera in common with the Cantabrian Mountains, which is more than with the Urals or with Texas (see Table 2). The similarity is even greater when one considers such closely related forms, as *Tubaria* and *Kutorginella* (see Table 1; Fig. 2 (3)), *Pontisia* and *Antronaria* and *Eomartiniopsis* and *Heteraria*. The fact that they derived in part from bryozoan reef mounds (Carter and Poletaev, 1998: 106) is a partial explanation. One wonders, however, whether the connection with the Canadian Arctic was through the Arctic, Spitsbergen and the Urals, or whether there existed another seaway along eastern North America, as suggested by Bless and Winkler Prins (1972). Unfortunately, the Pennsylvanian brachiopod faunas from the Appalachians are practically undescribed.

Fig. 2 Some brachiopods from the Cantabrian Mountains characteristic of reef facies: (1) *Rugivestis* sp., pedicle valve (RGM 288956); from the Valdeteja Formation north of Cármenes (León; Winkler Prins, 1968, loc. 10); A: anterior view, ×4; B: lateral view, ×4; (2) *Limbifera*? sp. brachial valve (RGM 290805); from the upper part of the Valdeteja Formation at Latores (Asturias; loc. WP101); dorsal view, ×1.6; (3) *Tubaria* cf. *genuina* (Kutorga, 1844), brachial valve (RGM 288953); from the local equivalent of the Valdeteja Formation at 1 km East of Santa María de Nava (Palencia; loc. Wa21); A: dorsal view, ×1; B: anterior view, ×1; (4) *Institina*? sp., pedicle valve (RGM 288957); from the upper part of the Valdeteja Formation at Latores (Asturias; loc. WP101); lateral view, ×3; (5) *Institina*? sp., brachial valve (RGM 288958); from the upper part of the Valdeteja Formation at Latores (Asturias; loc. WP101); dorsal view, ×3.5; (6) *Fluctuaria undata* (Defrance, 1826), pedicle valve (RGM 142545); from the Valdeteja Formation north of Vegacervera (León; loc. Wa925); posterior view, ×1; (7) *Proboscidella proboscidea* (de Verneuil, 1840) pedicle valve (RGM 142805); from a reef limestone of the Perapertú Formation westnorthwest of Verbios (Palencia; loc. dG511); A: lateral view, ×2; B: posterior view, ×2; (8) *Plicatiferina sinecosta* (Martínez Chacón, 1979) (RGM 290597); brachial valve and fragment of pedicle valve with spines from its type locality, the upper part of the Valdeteja Formation at Latores (Asturias); ventral view, ×2; (9) *Plicatiferina sinecosta* (Martínez Chacón, 1979) (RGM 288955); pedicle valve from its type locality, the upper part of the Valdeteja Formation at Latores (Asturias); umbonal view, note attachment ring, ×12.

Table 2. *List of genera occurring in the faunas from the Cantabrian Mountains described above (Spain: CM), Hiare Fiord (HF), and in the Viséan of the British Isles (C); and their occurrence in other areas; D: Alps (mainly Permian); E: Urals (Carboniferous-Permian); F: Spitsbergen (Pennsylvanian-Permian); G: Texas (Permian); x = doubtful occurrence (counted as half). The Inarticulates and Terebratulida are not included since they are not considered relevant. Due to a lack of modern revisions are the numbers in columns D–F minima. For literature see Table 1 and the References.*

| *Genus* | *CM* | *HF* | *C* | *D* | *E* | *F* | *G* |
|---|---|---|---|---|---|---|---|
| *Leptagonia* | | | X | | | | |
| *Tornquistia* | X | | | | | X | |
| *Caenanoplia* | x | | | | | | |
| *Globosochonetes* | X | | X | | | | |
| *Rugosochonetes* | X | | X | | | | |
| *Chonetinella* | X | X | | | X | X | X |
| *Neochonetes* | X | | | | X | X | X |
| *Sokolskya* | x | x | | | | | |
| *Megachonetes* | | | X | | | | |
| *Plicochonetes* | x | | X | | | | |
| *Quadrochonetes* | x | | | | | | X |
| *Productina* | X | | X | | | | |
| *Argentiproductus* | | | X | | | | |
| *Alitaria* | X | | X | | | | |
| *Eomarginiferina* | | x | X | | | | |
| *Rugivestis* | X | X | | | X | | x |
| *Overtonia* | | | X | | | | |
| *Fimbrinia* | | x | X | | X | | X |
| *Avonia* | | | X | | X | | |
| *Quasiavonia* | X | | X | x | | | |
| *Tuberculatella* | X | | | | | | |
| *Institifera* | x | | X | | | | |
| *Krotovia* | X | X | X | X | X | X | |
| Breileenia | X | | X | | | | |
| *Desmoinesia* | X | | | | | | |
| Incisiini *gen. nov.* | X | | | | | | |
| *Hystriculina* | | x | | x | | | |
| *Retimarginifera* | | x | | | | | |
| *Plicatifera* | | | X | | | | |
| *Lazarevia* | | X | | | | | |
| *Semicostella* | | | x | | X | | |
| *Acanthoplecta* | | | X | | | | |
| *Admodorugosus* | | | X | | | | |
| *Geniculifera* | | | X | | | | |
| *Carringtonia* | | | X | | | | |
| *Cinctifera* | | | X | | | | |
| *Limbifera* | x | | X | | | | |
| *Maemia* | X | X | | | | | |
| *Admoskovia* | X | | | | | | |
| *Bicarteria* | x | | | | | X | |
| *Duuartea* | X | | | | | x | |
| *Inflatia* | X | | X | | | | |
| *Latispinifera* | X | x | | | | X | |
| *Tenaspinus* | | | X | | | | |
| *Productus* | X | | X | | X | | |
| *Kozlowskia* | X | x | | x | X | x | X |
| *Eomarginifera* | X | X | X | | | | |
| *Antiquatonia* | X | X | X | | X | | X |

Table 2. (Continued)

| *Genus* | *CM* | *HF* | *C* | *D* | *E* | *F* | *G* |
|---|---|---|---|---|---|---|---|
| *Tubaria* | X | x | | | X | X | x |
| *Kutorginella* | X | X | | | X | | |
| *Alexenia* | X | | | | X | | |
| *Dictyoclostus* | | | X | | | x | |
| *Chaoiella* | X | | | x | X | X | |
| *Pugilis* | x | | X | | | | |
| *Reticulatia* | | x | | | X | X | X |
| *Buxtonia* | x | | X | | X | X | |
| *Kochiproductus* | x | | X | | X | | X |
| *Marginicinctus* | | | X | | | | |
| *Echinoconchus* | | | X | | | X | |
| *Echinaria* | X | x | | | X | x | X |
| *Echinoconchella* | X | | X | | | X | |
| *Karavankina* | X | | | X | X | | |
| *Cubacula* | X | | | | X | | |
| *Pustula* | x | x | X | | | X | |
| *Stegacanthia* | | | X | | | | |
| *Linoproductus* | x | x | x | x | X | X | X |
| *Balakhonia* | X | | X | | x | | |
| *Fluctuaria* | X | X | X | | X | | |
| *Marginovatia* | X | | | | | | |
| *Cancrinella* | X | X | | x | X | X | X |
| *Globiella?* | x | | | | | | |
| *Linoprotonia* | | | X | | | | |
| *Liraria* | X | X | | | | | |
| *Ovatia* | X | | X | | | X | |
| *Undaria* | | | X | | | | |
| *Vitiliproductus* | | | X | | | | |
| *Gigantoproductus* | | | X | | X | | |
| *Semiplanus* | | | X | | | | |
| *Latiproductus* | | | X | | | | |
| *Semiplanella* | | | X | X | | | |
| *Striatifera* | | | X | | X | x | X |
| *Proboscidella* | X | | X | | | | |
| *Heteralosia* | X | | X | | | | X |
| *Dasyalosia* | | | X | | | X | |
| *Crossalosia* | | | X | | | | |
| *Hontorialosia* | X | | | | | | |
| *Pamephrix* | | | X | | | | |
| *Semenewia* | | | X | | | | |
| *Plicatiferina* | X | X | | | X | | |
| *Stipulina* | x | | X | | | | |
| *Institina* | x | | X | | | | |
| *Retroplexus* | x | | X | | | | |
| *Rugicostella* | x | | X | | | | |
| *Sinuatella* | | | X | | | X | |
| *Apsocalyma* | | | X | | | | |
| *Brochocarina* | | | X | | | | |
| *Tapajotia* | X | | | | | | |
| *Meekella* | X | | | | X | X | X |
| *Schellwienella* | | | X | | | X | |
| *Diplanus* | X | | | | | | X |
| *Schuchertella* | x | | X | | | | x |
| *Serratocrista* | | | X | | | | |
| *Streptorhynchus* | X | | X | X | X | X | |

(continued)

Table 2. (continued)

| *Genus* | *CM* | *HF* | *C* | *D* | *E* | *F* | *G* |
|---|---|---|---|---|---|---|---|
| *Rhipidomella* | X | X | X | x | X | X | X |
| *Enteletes* | X | | | X | X | | X |
| *Schizophoria* | x | | X | | X | | |
| *Aulacophoria* | | | X | | X | | |
| *Pocockia* | | | X | | | | |
| *Orthotichia* | X | X | | | X | X | X |
| *Pugnax* | X | | X | | | | |
| *Pleuropugnoides* | | | X | | | | |
| *Propriopugnus* | | | X | | | | |
| *Pugnoides* | X | | X | | | | |
| *Stenoscisma* | X | X | X | x | X | | X |
| *Coledium* | | | x | X | | | |
| *Psilocamara* | X | | | | | | X |
| *Careoseptum* | | X | | | | | |
| *Callaiapsida* | X | | | | | | |
| *Lambdarina* | X | | X | | | | |
| *Rhynchopora* | X | | | x | X | X | X |
| *Tretorhynchia* | | | X | | | | |
| *Trasgu* | X | | | | | | |
| *Septacamera* | | X | | | | | |
| *Yanishewskiella* | X | | | | X | | |
| *Exlaminella* | | X | | | | | |
| *Cenorhynchia* | | X | | | | | |
| *Phrenophoria* | | x | | | | | |
| *Hemileurus* | | x | | | | | |
| *Pontisia* | X | | | | | | X |
| *Antronaria* | | X | | | | | |
| *Elassonia* | | x | | | | | |
| *Actinoconchus* | | | X | | | | |
| *Athyris* | | | X | | | | |
| *Lamellosathyris* | | | X | | | | |
| *Cleiothyridina* | X | | X | x | X | X | X |
| *Cardiothyris* | X | | | | | | |
| *Composita* | X | | X | | X | X | X |
| *Camarium* | | X | | | | | |
| *Nucleospira* | | X | X | | | | X |
| *Hustedia* | X | X | X | X | X | X | X |
| *Plectospira* | | | x | | | | |
| *Crurithyris* | X | X | X | | X | X | X |
| *Martinia* | X | | X | X | X | X | X |
| *Tiramnia* | X | X | | | X | | |
| *Jilinmartinia* | | x | | | X | | |
| *Heteraria* | X | x | | | | | |
| *Merospirifer* | | | X | | | | |
| *Eomartiniopsis* | X | | | | | | |
| *Martiniopsis* | x | | | | X | | |
| *Spirifer* | | | X | | | | |
| *Donispirifer* | x | | | | X | | |
| *Anthracospirifer* | X | x | | | | | |
| *Podtsheremia* | | | X | | | | |
| *Angiospirifer* | | | X | | | | |
| *Anthracothyrina* | X | x | X | | | | |
| *Brachythyrina* | x | x | | X | X | X | |

Table 2. (continued)

| *Genus* | *CM* | *HF* | *C* | *D* | *E* | *F* | *G* |
|---|---|---|---|---|---|---|---|
| *Elinoria* | | X | | | | | |
| *Choristites* | x | x | | | X | X | |
| *Alphachoristites (Prochoristites)* | X | | | | X | | |
| *Parachoristites* | X | X | | | X | | |
| *Trautscholdia* | x | x | | X | X | | X |
| *Tangshanella* | | x | | | | | |
| *Tegulispirifer* | X | | | X | X | X | |
| *Gypospirifer* | X | x | | | | | |
| *Avisyrinx* | X | | | | X | | |
| *Cantabriella* | X | X | | x | | | |
| *Fusella* | | | X | | | | |
| *Brachythyris* | X | | X | x | X | | |
| *Meristorygma* | X | X | | | | | |
| *Skelidorygma* | X | | X | | | | |
| *Tylothyris* | | | X | | | | |
| *Reticularia* | x | | X | | x | X | |
| *Georgethyris* | | | X | | | | |
| *Kitakamithyris* | x | | | | | | |
| *Phricodothyris* | X | X | x | | x | | X |
| *Phricodothyris (Condrathyris)* | X | | | | | | |
| *Cyrtina* | | | X | | | | |
| *Davidsonina* | | | X | | X | | |
| *Syringothyris* | | | X | | | | |
| *Asyrinxia* | | | X | | | | |
| *Punctospirifer* | x | | X | | X | X | |
| *Altiplecus* | X | | | | | | X |
| *Spriferellina* | X | | X | | | | X |
| *Crenispirifer* | X | | | | | | |
| *Xestotrema* | x | | | | | | |
| *Minithyra* | | | X | | | | |
| Total number of genera | 97 | 42 | 106 | 17 | 57 | 37 | 34 |
| Total genera in common with CM | | 28 | 39 | 14 | 43 | 27 | 28 |
| Total genera in common with HF | 28 | | 13 | 7 | 21 | 12 | 14 |

Although the connections with the Pennsylvanian faunas from the American midcontinent were rather poor (see Winkler Prins, 1983), there are definite links with the Permian reef faunas from Texas (e.g., *Heteralosia*, *Diplanus*, *Altiplecus*) and Oregon (*Rugivestis*; Fig. 2 (1)), which could have been through the Canadian Arctic. The presence of several taxa in Texas that may have been derived from northwestern Europe (cf. Mundy and Brunton, 1985) and which have no direct relatives known from the Canadian Arctic (nor from the Cantabrian Mountains) is puzzling, but the excellent preservation with their spines attached in specimens from Britain and Texas make a direct comparison with the less well-preserved Hare Fiord and Spanish material difficult. Furthermore, these specialised forms are rather rare and could have been missed (collecting has been less intensive in Spain and Ellesmere Island than in the British Isles and Texas).

## 5. Conclusions

The Bashkirian–Moscovian basin with reefoid deposits in the Cantabrian Mountains has apparently played an important role in providing a niche for the survival of reef-related brachiopod faunas during the Pennsylvanian, bridging the gap between the Mississippian brachiopods from reefoid deposits in northern Europe and the Permian faunas from reefs in the Urals and Texas. Surprisingly, they have their counterpart on Ellesmere Island in the Canadian Arctic, another refuge for these brachiopods during the Bashkirian–Moscovian.

Also, the Cantabrian Mountains provided the first examples of some specialised brachiopods, such as the Incisiini (*Regrantia* nomen nudum), *Avisirinx* and *Rugivestis*, the latter genus being also present on Ellesmere Island.

## Acknowledgements

The author gratefully acknowledges the constructive criticism of Dr. R.H. Wagner and of the referees Drs. H.C.H. Brunton and M.L. Martínez Chacón, which helped to improve the manuscript significantly. Figure 1 was expertly drawn by E.J. Bosch and J. Leloux is thanked for taking some of the photographs of Figure 2.

## References

Amler, M.R.W. and Winkler Prins, C.F., 1999, Lower Carboniferous bivalves from the Cantabrian Mts (Spain), *Scripta Geologica* **120**: 1–45.

Aretz, M., 2002, Habitatanalyse und Riffbildungspotential kolonialer rugoser Korallen im Unterkarbon (Mississippium) von Westeuropa, *Kölner Forum für Geologie und Paläontologie* **10**: 1–155.

Aretz, M., and Herbig, H.G., 2003, Coral-rich bioconstructions in the Viséan (Late Mississippian) of Southern Wales (Gower Peninsula, UK), *Facies* **49**: 221–242.

Babin, C., García-Alcalde, J., Gutiérrez-Marco, J.C., and Martínez Chacón, M.L., 1999, Conocardiacea (Mollusca, Rostroconchia) du Dévonien et du Carbonifère d'Espagne, *Revue de Paléobiologie* **18**: 173–186.

Balthasar, U., and Amler, M.R.W., 2003, Silicified Foraminifera from the Lower Carboniferous Genicera Formation (Cantabrian Mountains, Northern Spain) and their palaeoenvironmental interpretation, *Courier Forschungsinstitut Senckenberg* **242**: 1–19.

Bassett, M.G. and Bryant, C., 1993, The micromorphic rhynchonelloidean brachiopod *Lambdarina* from the type Dinantian, *Journal of Paleontology* **67**: 518–527.

Becker, G., 1982, Fazies-anzeigende Ostracoden-Vergesellschaftungen aus dem frühen Oberkarbon des Kantabrischen Gebirges (N-Spanien), *Neues Jahrbuch für Geologie und Paläontologie, Abhandlungen* **164**: 307–338.

Becker, G., Frankenfeld, H., and Schulze, R., 1979, Neue Daten zum Riffsterben im Oberdevon des Kantabrischen Gebirges (N-Spanien), *Clausthaler Geologische Abhandlungen* **30**: 19–33.

Bless, M.J.M. and Sánchez de Posada, L.C., 1973, Sobre la aparición de Ostrácodos nectónicos en la Cordillera Cantábrica, *Breviora Geológica Astúrica* **17**: 25–30.

Bless, M.J.M. and Winkler Prins, C.F., 1972, Paleoecology and paleogeography of the Aegir Marine Band and its equivalents in North-western Europe, *Compte Rendu Septième Congrès International de Stratigraphie et de Géologie du Carbonifère*, Krefeld, 1971, **1**: 231–239.

Bless, M.J.M. and Winkler Prins, C.F., 1973, Paleoecology of Upper Carboniferous strata in Asturias (N. Spain), *Compte Rendu Septième Congrès International de Stratigraphie et de Géologie du Carbonifère*, Krefeld, 1971, **2**: 129–137.

Boll, F.Ch., 1985, Rugose Korallen der Flachwasser-Fazies im Oberkarbon des Kantabrischen Gebirges (Nordspanien), *Palaeontographica, Abteilung A* **190**: 1–81.

Boogaard, M. van den and Bless, M.J.M., 1985, Some conodont faunas from the Aegiranum Marine Band, *Proceedings Koninklijke Nederlandse Akademie van Wetenschappen, Serie B* **88**: 133–154.

Bowman, M.B.J., 1979, The depositional environments of a limestone unit from the San Emiliano Formation (Namurian/Westphalian), Cantabrian Mts., NW Spain, *Sedimentary Geology* **24**: 25–43.

Bowman, M.B.J., 1982, The stratigraphy of the San Emiliano Formation and its relationship to other Namurian/Westphalian A sequences in the Cantabrian Mts., NW Spain, *Trabajos de Geología* **12**: 23–35.

Bowman, M.B.J., 1985, The sedimentology and palaeogeographic setting of late Namurian-Westphalian A basin-fill successions in the San Emiliano and Cármenes areas of NW León, Cantabrian Mts, NW Spain, in: Lemos de Sousa, M.J. and Wagner, R.H. (eds), *Papers on the Carboniferous of the Iberian Peninsula (Sedimentology, Stratigraphy, Palaeontology, Tectonics and Geochronology). Anais da Faculdade de Ciências, Universidade do Porto*, **64** (Suppl.): 117–169.

Bridges, P.H., Gutteridge, P., and Pickard, N.A.H., 1995, The environmental setting of early Carboniferous mud-mounds, in: Monty, C.L.V., Bosence, D.W.J., Bridges, P.H., and Pratt, B.R. (eds), *Carbonate Mud-Mounds Their Origin and Evolution. International Association of Sedimentologists Special Publication* **23**: 171–190.

Brouwer, A., 1964, Devonian biostromes and bioherms of the southern Cantabrian Mountains, northwestern Spain, in: Straaten, L.M.J.U. van (ed.), *Deltaic and Shallow Marine Deposits. Developments in Sedimentology* **1**: 48–53.

Brouwer, A. and Ginkel, A.C. van, 1964, La succesion carbonifère dans la partie méridionale des Montagnes Cantabriques, *Compte Rendu V Congrès International de Stratigraphie et de Géologie du Carbonifère*, Paris, 1962, **1**: 307–319.

Brunton, C.H.C., 1966, Silicified brachiopods from the Visean of County Fermanagh, I. *Bulletin of the British Museum (Natural History), Geology Series* **12**: 175–243.

Brunton, C.H.C., 1968, Silicified brachiopods from the Visean of County Fermanagh, II, *Bulletin of the British Museum (Natural History), Geology Series*, **16**: 1–70.

Brunton, C.H.C., 1984, Silicified brachiopods from the Viséan of County Fermanagh, Ireland (III). Rhynchonellids, Spiriferids and Terebratulids, *Bulletin of the British Museum (Natural History), Geology Series* **38**: 27–130.

Brunton, C.H.C., 1985, Some Carboniferous brachiopod distributions, *Compte Rendu Dixième Congrès International de Stratigraphie et de Géologie du Carbonifère*, Madrid, 1983, **4**: 75–81.

Brunton, C.H.C., 1987, The palaeoecology of brachiopods, and other faunas, of Lower Carboniferous (Asbian) limestones in West Fermanagh, *Irish Journal of Earth Sciences* **8**: 97–112.

Brunton, C.H.C., 1996, A new protorichthofenoid brachiopod (Productida) from the Upper Carboniferous of the Urals, Russia, *Bulletin of the Natural History Museum London (Geology)* **51**: 51–59.

Brunton, C.H.C. and Champion, C., 1974, A Lower Carboniferous brachiopod fauna from the Manifold Valley Staffordshire, *Palaeontology* **17**: 811–840.

Brunton, C.H.C., Lazarev, S.S., Grant, R.E., and Jin Yu-gan, 2000, Productidina, in: Williams, A. et al. *Treatise on Invertebrate Paleontology. Part H. Brachiopoda Revised, Volume 3: Linguliformea, Craniiformea, and Rhynchonelliformea (part)*. The Geological Society of America and The University of Kansas, Boulder, CO and Lawrence, KA, pp. 424–609.

Brunton, C.H.C. and Mundy, D.J.C., 1985, Growth and shell shape in Productacean Brachiopods, *Bulletin of the British Museum of Natural History (Geology)* **38**: 273–281.

Brunton, C.H.C. and Mundy, D.J.C., 1986, Some Dinantian chonopectinid productaceans (Brachiopoda) from the British Isles, *Proceedings of the Yorkshire Geological Society* **46**: 1–10.

Brunton, C.H.C. and Mundy, D.J.C., 1988, Strophalosiacean and aulostegacean productoids (Brachiopoda) from the Craven Reef Belt (late Viséan) of North Yorkshire, *Proceedings of the Yorkshire Geological Society* **47**: 55–88.

Brunton, C.H.C. and Mundy, D.J.C., 1994, A new productoid brachiopod from the late Visean Craven Reef Belt of North Yorkshire, *Proceedings of the Yorkshire Geological Society* **50**: 119–123.

Brunton, C.H.C., Mundy, D.J.C., and Lazarev, S.S., 1994, Productellid and Plicatiferid (Productoid) Brachiopods from the Lower Carboniferous of the Craven Reef Belt, North Yorkshire, *Bulletin of the British Museum of Natural History (Geology)* **49**: 99–119.

Brunton, C.H.C. and Tilsley, J.W., 1991, A checklist of brachiopods from Treak Cliff, Derbyshire, with reference to other Dinantian (Lower Carboniferous) localities, *Proceedings of the Yorkshire Geological Society* **48**: 287–295.

Carballeira, J., Corrales, I., Valladares, I., Naval, A., Ruiz, F., Lorenzo, S., Martínez Chacón, M.L., Méndez, C., Sánchez de Posada, L.C., and Truyols, J., 1985, Aportaciones al conocimiento de la estratigrafía de la Formación San Emiliano (Carbonífero, Cordillera Cantábrica) en su área-tipo, *Compte Rendu Dixième Congrès International de Stratigraphie et de Géologie du Carbonifère*, Madrid, 1983, **1**: 345–362.

Carter, J.L. and Poletaev, V.I., 1998, Atokan (Late Bashkirian or Early Moscovian) brachiopods from the Hare Fiord Formation of Ellesmere Island, Canadian Arctic Archipelago, *Annals of the Carnegie Museum* **67**: 105–180.

Comte, P., 1938, La transgression du Famennien supérieur dans la Cordillère Cantabrique, *Compte Rendu de l'Académie des Sciences* **206**: 1741–1743.

Cooper, G.A. and Grant, R.E., 1972–1977, Permian Brachiopods of West Texas, I–VI., *Smithsonian Contributions to Paleobiology* **14**: 1–231; **15**: 233–793; **19**: 795–1921; **21**: 1923–2607; **24**: 2609–3159; **32**: 3161–3370.

Defrance, M.J.L., 1826, Article *Productus, Dictionaire des Sciences Naturelles*, Paris, **43**: 349–355.

Delépine, G. and Llopis Lladó, N., 1956, Nouvelle faune Carbonifère à Latores (Asturies, Espagne), *Comptes Rendus sommaires des Séances de la Société Géologique de France* 1956, pp. 106–108.

Eder, W. and Franke, W., 1982, Death of Devonian reefs, in: Kullmann, J., Schönenberg, R., and Wiedmann, J. (eds), *Subsidenz-Entwicklung im Kantabrischen Variszikum und an passiven Kontinentalrändern der Kreide. Teil I Variszikum. Neues Jahrbuch für Geologie und Paläontologie, Abhandlungen* **163**: 241–243.

Eichmüller, K., 1985, Die Valdeteja Formation: Aufbau und Geschichte einer oberkarbonischen Karbonatplattform (Kantabrisches Gebirge, Nordspanien), *Facies* **13**: 45–154.

Eichmüller, K., 1986, Some Upper Carboniferous (Namurian; Westphalian) Lithostratigraphic units in Northern Spain, results and implications of an environmental interpretation, *Boletín del Instituto Geológico y Minero de España* **97**: 590–607.

Eichmüller, K. and Seibert, P., 1984, Faziesentwicklung zwischen Tournai und Westfal D im Kantabrischen Gebirge (NW-Spanien), *Zeitschrift der Deutschen Geologischen Gesellschaft* **135**: 163–191.

Fernández, L.P., 1993, La Formación San Emiliano (Carbonífero de la Zona Cantábrica, NO de España): estratigrafía y extensión lateral. Algunas implicaciones paleogeográficas, *Trabajos de Geología* **19**: 97–122.

Fernández López, M.G., and Sánchez de Posada, L.C., 1987, Hollinomorpha y Kirkbyocopina (Ostracoda) del Bashkiriense del N. de León (Cordillera Cantábrica, NW. de España), *Revista Española de Micropaleontología* **19**: 465–485.

Flajs, G., Hüssner, H., Fenninger, A., and Hubmann, B., 1996, Upper Permian Richthofeniid buildups of Chios Island (Aegean Sea) – preliminary report, *Jahrbuch der Geologischen Bundesanstalt* **139**: 21–28.

Frankenfeld, H., 1982, Das Ende der devonischen Riff-Fazies im nordspanischen Variszikum, *Neues Jahrbuch für Geologie und Paläontologie, Abhandlungen*, **163** (2): 238–241.

Gandl, J., 1987, Trilobiten aus dem höheren Namur und tieferen Westfal, *Abhandlungen der Senckenbergischen Naturforschenden Gesellschaft* **543**: 1–79.

García Bartolomé, A., Rodríguez, R.M., and Sánchez de Posada, L.C., 2003, Miosporas de los niveles superiores de la sucesión de la Playa de la Huelga (Carbonífero del E de Asturias, España), *Revista Española de Micropaleontología* **35**: 29–42.

García-Bellido, D.C. and Rigby, K., 2004, Devonian and Carboniferous sponges from Spain. *Journal of Paleontology* **78**: 431–455.

García-Bellido, D.C. and Rodríguez, S., 2004, Palaeobiogeographical relationships of poriferan and coral assemblages during the late Carboniferous and the closure of the western Palaeotethys Sea-Panthalassan Ocean connection, *Palaeogeography, Palaeoclimatology, Palaeoecology* **219**: 321–331.

Ginkel, A.C. van, 1965, Spanish Carboniferous fusulinids and their significance for correlation purposes, *Leidse Geologische Mededelingen* **34**: 175–225.

Ginkel, A.C. van, and Villa, E., 1996, Palaeontological data of the San Emiliano Formation (Cantabrian Mountains, Spain) and their significance in the Carboniferous chronostratigraphy, *Geobios* **29**: 149–170.

Gulyás-Kis, Cs., 2004, Paleobiogeographic analysis of Moscovian (Pennsylvanian) brachiopods Productida, Orthotetida, Orthida, Rhynchonellida from Bükk Mts, Hungary, *Hantkeniana* **4**: 17–35.

Harper, D.A.T. and Jeffrey, A.L., 1996, Mid-Dinantian brachiopod biofacies from western Ireland, *Geological Society Special Publication* **107**: 427–436.

Jordan, H. and Bless, M.J.M., 1970, Nota preliminar sobre los ostrácodos de la Formación Vegamián, *Breviora Geológica Astúrica* **4**: 37–44.

Julivert, M., 1967, La ventana del Río Monasterio y la terminación meridional del Manto del Ponga, *Trabajos de Geología* **1**: 59–67.

Julivert, M., 1971, Décollement tectonics in the Hercynian cordillera of Northwest Spain, *American Journal of Science* **270**: 1–29.

Kullmann, J. and Rodríguez, S., 1986, Hornförmige Einzelkorallen (Rugosa) aus früh oberkarbonischen Flachwasser-Sedimenten im Kantabrische Gebirge (Nordspanien), *Neues Jahrbuch für Geologie und Paläontologie, Monatshefte* **1986**: 293–306.

Kullmann, J. and Rodríguez, S., 1994, Biostratigraphic range and biogeographic relationships of the undissepimented solitary corals from the Picos de Europa Formation (Moscovian, Cantabrian Mountains, NW Spain), *Courier Forschungsinstitut Senckenberg* **172**: 15–22.

Kutorga, S.S., 1844, Zweiter Beitrag zur Paläontologie Russlands, *Verhandlungen der Russisch-kaiserlichen Mineralogischen Gesellschaft* **1844**: 62–104.

Lees, A. and Miller, J., 1995, Waulsortian banks, in: Monty, C.L.V., Bosence, D.W.J., Bridges, P.H., and Pratt, B.R. (eds), *Carbonate Mud-Mounds Their Origin and Evolution. International Association of Sedimentologists Special Publication* **23**: 191–271.

Lotze, F., 1945, Zur Gliederung der Varisziden der Iberischen Meseta, *Geotektonische Forschungen* **6**: 1–12

Martínez Chacón, M.L., 1975, *Avisyrinx* n. gen. (Syringothyrididae, Brachiopoda) del Carbonífero de Asturias (España), *Breviora Geológica Astúrica* **19**: 33–40.

Martínez Chacón, M.L., 1977, New Carboniferous stenoscismatacean brachiopods from Oviedo and León, Spain, *Palaeontology* **20**: 209–223.

Martínez Chacón, M.L., 1978a, Nuevos braquiópodos de la Formación San Emiliano (Cordillera Cantábrica, NO. de España), *Breviora Geológica Astúrica* **22**: 17–32.

Martínez Chacón, M.L., 1978b, Syringothyridacea (Brachiopoda) del Carbonífero de la Cordillera Cantábrica (N de España), *Trabajos de Geología* **10**: 317–330.

Martínez Chacón, M.L., 1979, *Braquiópodos carboníferos de la Cordillera Cantábrica (Orthida, Strophomenida y Rhynchonellida). Memorias del Instituto Geológico y Minero de España* **96**: 291 pp.

Martínez Chacón, M.L., 1988, Braquiópodos del Carbonífero Medio del corte de la Playa de La Huelga (Asturias, NW de España), *Geogaceta* **5**: 13–16.

Martínez Chacón, M.L., 1990, Braquiópodos carboníferos de la costa E de Asturias (España). I: Orthida, Strophomenida, Rhynchonellida y Athyridida, *Revista Española de Paleontología* **5**: 91–110.

Martínez Chacón, M.L., 1991, Braquiópodos carboníferos de la costa E de Asturias (España). II: Spiriferida y Terebratulida, *Revista Española de Paleontología* **6**: 59–88.

Martínez Chacón, M.L. and Bahamonde, J.R., in press, Brachiopods in the steep slope of a Pennsylvanian carbonate platform (Sierra del Cuera, Asturias, N. Spain), *Proceedings V International Brachiopod Congress*, Copenhagen, 2005.

Martínez Chacón, M.L., Menéndez Álvarez, J.R., Sánchez de Posada, L.C., and Truyols, J., 1985, Aportaciones al conocimiento de la Formación Ricacabiello (Carbonífero de la Zona Cantábrica, N de España) y su contenido paleontológico, *Trabajos de Geología* **15**: 53–65.

Martínez Chacón, M.L. and Winkler Prins, C.F., 1977, A Namurian brachiopod fauna from Meré (Province of Oviedo, Spain), *Scripta Geologica* **39**: 1–67.

Martínez Chacón, M.L. and Winkler Prins, C.F., 1986, The brachiopod fauna of the San Emiliano Formation (Cantabrian Mountains, NW Spain) and its connection with other areas, *Compte Rendu Neuvième Congrès International de Stratigraphie et de Géologie du Carbonifère*, Washington and Champaign-Urbana, 1979, **5**: 233–244.

Martínez Chacón, M.L. and Winkler Prins, C.F., 1993, Carboniferous brachiopods and the palaeogeographic position of the Iberian Peninsula, *Compte Rendu XII Congrès International de Stratigraphie et de Géologie du Carbonifère et Permien*, Buenos Aires, 1991, **1**: 573–580.

Martínez Chacón, M.L. and Winkler Prins, C.F., 2000, New Rugosochonetidae (Brachiopoda) from the upper Bashkirian and Moscovian of the Cantabrian Mountains (N Spain), *Revista Española de Paleontología* **15**: 219–232.

Martínez Chacón, M.L. and Winkler Prins, C.F., 2006a, Braquiópodos de la Formación Valdeteja (Pensilvánico, Cordillera Cantábrica, N de España), in: Fernández-Martínez (ed.), *XXII Jornadas de la Sociedad Española de Paleontología y simposios de los proyectos PICG 493, 503, 499, y 467, Libro de Resúmenes*: 145–147.

Martínez Chacón, M.L. and Winkler Prins, C.F., 2006b (in press), Medio siglo de estudio de los Braquiópodos del yacimiento de Latores, *I Congreso de Estudios Asturianos*, Oviedo, 2006.

Martínez Díaz, C., 1969, Carbonífero marino de la Zona de Riosa (Asturias, España), *Revista Española de Micropaleontología* **1**: 59–80.

Méndez, C.A. and Menéndez Álvarez, J.R., 1985, Conodontos carboníferos de las regiones del Manto del Ponga y Picos de Europa (Oriente de Asturias, N de España), *Compte Rendu Dixième Congrès International de Stratigraphie et de Géologie du Carbonifère*, Madrid, 1983, **1**: 71–82.

Méndez-Bedia, I., Soto, F., and Fernández-Martínez, E., 1994, Devonian reef types in the Cantabrian Mountains (NW Spain) and their faunal composition, *Courier Forschungsinstitut Senckenberg* **172**: 161–184.

Minwegen, E., 2001, Die Biokonstruktionen im Pennsylvanium des Kantabrischen Gebirges (Nordspanien), *Kölner Forum für Geologie und Paläontologie* **9**: 1–139.

Mohanti, M., 1972, The Portilla Formation (Middle Devonian) of the Alba Syncline, Cantabrian Mountains, Prov. León, northwestern Spain: carbonate facies and rhynchonellid palaeontology, *Leidse Geologische Mededelingen* **48**: 135–205.

Mundy, D.J.C. and Brunton, C.H., 1985, Morphological Similarities in Some British Dinantian and Texas Permian Reef Brachiopods, *Compte Rendu Neuvième Congrès International de Stratigraphie et de Géologie du Carbonifère*, Washington and Champaign-Urbana, 1979, **5**: 225–232.

Navarro, D., Leyva, F., and Villa, E., 1986, Cambios laterales de facies en el Carbonífero del Oriente de Asturias (Cordillera Cantábrica, Norte de España), *Trabajos de Geología* **16**: 87–102.

Parga, J.R., 1969, Consideraciones sobre la erosión Fameniense en la Cordillera Cantábrica, *Breviora Geológica Astúrica* **13**: 46–48.

Porta, G. della, Kenter, J.A.M., and Bahamonde, J.R., 2004, Depositional facies and stratal geometry of an Upper Carboniferous prograding and aggrading high-relief carbonate platform (Cantabrian Mountains, N Spain), *Sedimentology* **51**: 267–295.

Rácz, L., 1964, Carboniferous calcareous algae and their associations in the San Emiliano and Lois-Ciguera formations (Prov. León, NW Spain), *Leidse Geologische Mededelingen* **31**: 1–112.

Radig, F., 1962, Ordovizium/Silurium und die Frage prävariszischer Faltungen in Nordspanien, *Geologische Rundschau* **52**: 346–397.

Ramovš, A., 1971, Connections of the Upper Carboniferous brachiopod faunas from the Carnic Alps and Karavanke Mountains with those of the Cantabrian Mountains (Spain), in: R.H. Wagner (ed.), *The Carboniferous of Northwest Spain, 2. Trabajos de Geología* **4**: 373–377.

Reuther, C.-D., 1977, *Das Namur im südlichen Kantabrischen Gebirge (Nordspanien); Krustenbewegungen und Faziesdifferenzierung im Übergang Geosynklinale – Orogen, Clausthaler Geologische Abhandlungen* **28**: 1–122.

Reijers, T.J.A., 1973, Hinge movements influencing deposition during the Upper Devonian in the Esla area of the Cantabrian Mountains, Spain, *Leidse Geologische Mededelingen* **53**: 13–21.

Riding, H., 1978, *Donezella* bioherms in the Carboniferous of the southern Cantabrian Mountains, Spain, *Bulletin du Centre de Recherches d'Exploration Petrolifère Elf-Aquitaine* **3**: 787–794.

Rodríguez, S., 1996, Development of coral reef-facies during the Viséan at Los Santos de Maimona, SW Spain, in: Strogen, P., Somerville, I.D., and Jones, G.L. (eds), *Recent Advances in Lower Carboniferous Geology. Geological Society Special Publication* **107**: 145–152.

Rodríguez, S. and Ramírez, C., 1987, Los siringopóridos de la sección de la Playa de la Huelga (Carbonífero, Asturias, Noroeste de España), *Boletín de la Real Sociedad Española de Historia Natural, sección Geológica* **83**: 57–82.

Rodríguez, S., Sando, W.J. and Kullmann, J., 1986, Utility of corals for biostratigraphic and zoogeographic analyses of the Carboniferous in the Cantabrian Mountains, Northern Spain, *Trabajos de Geología* **16**: 37–60.

Rodríguez Fernández, L.R., 1993, Tectonosedimentary evolution of a Carboniferous foreland basin related with arcuated fold-thrust belt. The example of NW Iberian Variscan belt, *Compte Rendu XII Congrès International de Stratigraphie et de Géologie du Carbonifère et Permien*, Buenos Aires, 1991, **1**: 435–445.

Romano, M., 1971, A new Proetid trilobite from the Lower Westphalian of North-West Spain. in: Wagner, R.H. (ed.), *The Carboniferous of Northwest Spain, 2. Trabajos de Geología* **4**: 379–383.

Sánchez de Posada, L., 1976, Quelques remarques au sujet de la répartition des faunes d'Ostracodes carbonifères dans la chaîne Cantabrique, *Annales de la Société Géologique du Nord* **96**: 407–412.

Sánchez de Posada, L.C. and Bless, M.J.M., 1999, Hollinéllidos del Moscoviense Superior (Carbinífero) de la Playa de la Huelga (Zona Cantábrica, N de España): fauna y bioestratigrafía, *Revista Española de Paleontología* número extraordinario: 29–41.

Sánchez de Posada, L.C. and Fohrer, B., 2001, Kirkbyoid ostracodes (Upper Carboniferous) from the Cantabrian Mountains (Spain) and Carnic Alps (Austria and Italy), *Journal of Paleontology* **75**: 972–984.

Sánchez de Posada, L.C., García-Lopez, S., and Villa, E., 1998, The Paleozoic of the Cantabrian Zone, in: Lamolda, M.A. (ed.), *Libro Guía. 24° Coloquio Europeo de Micropaleontología*, Bilbao, Spain, pp. 11–22.

Sánchez de Posada, L.C., Martínez Chacón, M.L., Méndez, C.A., Menéndez Álvarez, J.R., and Villa, E., 1993, El Carbonífero de las regiones de Picos de Europa y Manto del Ponga (Zona Cantábrica, N de España): fauna y bioestratigrafía, *Revista Española de Paleontología* número extraordinario: 89–108.

Sánchez de Posada, L.C., Martínez Chacón, M.L., Méndez, C.A., Menéndez Álvarez, J.R., Río, L.M., Rodríguez, S., Truyols, J., and Villa, E., 1996, El Carbonífero marino del ámbito Astur-Leonés (Zona Cantábrica): síntesis paleontológica, *Revista Española de Paleontología* número extraordinario: 82–96.

Sánchez de Posada, L.C., Martínez Chacón, M.L., Méndez, C.A., and Villa, E., 2002, Rasgos paleontológicos del Carbonífero marino cantábrico, in: Gámez Vintaned, J.A. and Liñan, E. (eds), *Memorias de las VII Jornadas Aragonesas de Paleontología. La era paleozóica. El desarrollo de la vida marina. Homenaje al Prof. Jaime Truyols*. Institución "Fernando el Católico", Zaragoza, Spain, pp. 191–222.

Sano, H., Fujii, S., and Matsuura, F., 2004, Response of Carboniferous-Permian mid-oceanic seamount-capping buildup to global cooling and sea-level change: Akiyoshi, Japan, *Palaeogeography, Palaeoclimatology, Palaeoecology* **213**: 187–206.

Sommerville, I.D., Strogen, P., Jones, G.Ll., and Somerville, H.E.A., 1996. Late Viséan buildups at Kingscourt, Ireland: possible precursors for Upper Carboniferous bioherms, *Geological Society Special Publication* **107**: 127–144.

Stubblefield, C.J., 1960, Sessile marine organisms and their significance in pre-Mesozoic strata, *Quarterly Journal of the Geological Society of London* **116**: 219–238.

Tschernyschew, Th., 1902, *Die obercarbonische Brachiopoden des Ural und des Timan. Mémoires du Comité Géologique* **16** (2): 749 pp.

Twenhofel, W.H., 1950, Coral and other organic reefs in geologic column, *Bulletin of the American Association of Petroleum Geologists* **34**: 182–202.

Verneuil, E. de, 1840, Sur quelques espèces intéressantes de brachiopodes des terrains anciens, *Bulletin de la Société Géologique de France* **11**: 257–262.

Villa, E., 1982, Observaciones sobre la edad de la Formación Valdeteja (Carbonífero de la Cordillera Cantábrica) en su area-tipo, *Revista Española de Micropaleontología* **14**: 63–72.

Villa, E., 1989, Foraminíferos Bashkirienses de la Sierra del Sueve (Cuenca Carbonífera Central, Zona Cantábrica, NW de España), *Revista Española de Paleontología* **4**: 81–89.

Villa Otero, E., 1995, *Fusulináceos Carboníferos del Este de Asturias (N de España). Biostratigraphie du Paléozoïque* **13**: 1–261.

Villa, E. and Ginkel, A.C. van, 1999, First record of Gzhelian fusulinaceans from the Carboniferous of northern Spain, *Revista Española de Paleontología* número extraordinario: 205–216.

Villa, E., Sánchez de Posada, L.C., Martínez Chacón, M.L., and Stavros, Ch., 2001, Foraminifera and biostratigraphy of the Valdeteja Formation stratotype (Carboniferous, Cantabrian Zone, NW Spain), *Facies* **45**: 59–86.

Wagner, R.H., 2004, The Iberian Massif: a Carboniferous assembly, *Journal of Iberian Geology* **30**: 93–108.

Wagner, R.H. and Bowman, M.B.J., 1983, The position of the Bashkirian/Moscovian boundary in West European chronostratigraphy, *Newsletter on Stratigraphy* **12**: 132–161.

Wagner, R.H. and Martínez García, E., 1974, The relation between geosynclinal folding and foreland movements in Northwest Spain, *Studia Geológica* **7**: 131–158.

Wagner, R.H., Sánchez de Posada, L.C., Martínez Chacón, M.L., Fernández, L.P., Villa, E., and Winkler Prins, C.F., 2002, The Asturian Stage: a preliminary proposal for a replacement of the Westphalian D. *Canadian Society of Petroleum Geologists, Memoir* **19**: 832–850.

Wagner, R.H. and Winkler Prins, C.F., 1985, The Cantabrian and Barruelian stratotypes: a summary of basin development and biostratigraphic information, in: Lemos de Sousa, M.J. and Wagner, R.H. (eds), *Papers on the Carboniferous of the Iberian Peninsula. Anais da Faculdade de Ciências, Universidade do Porto*, **64** (Suppl.): 359–410.

Wagner, R.H. and Winkler Prins, C.F., 1999, Carboniferous stratigraphy of the Sierra del Brezo in northern Palencia (Cantabrian Mts, NW Spain): evidence of major uplifts, *Trabajos de Geología* **21**: 385–403.

Wagner, R.H. and Winkler Prins, C.F., 2002, Tectonics vs. cyclothems: Carboniferous sedimentation in the Cantabrian Mountains, Spain, *Canadian Society of Petroleum Geologists, Memoir* **19**: 228–238.

Wagner, R.H., Winkler Prins, C.F., and Riding, R.E., 1971, Lithostratigraphical units of the lower part of the Carboniferous in northern León, Spain (with a Note on some goniatite faunas by C.H.T. Wagner-Gentis), in: R.H. Wagner (ed.), *The Carboniferous of Northwest Spain*, 2. *Trabajos de Geología* **4**: 603–633.

Wahlman, G.P., 2002, Upper Carboniferous-Lower Permian (Bashkirian-Kungurian) mounds and reefs, in: Kiessling, W., Flügel, E., and Golonka, J. (eds), *Phanerozoic Reef Patterns, SEPM Special Publication* **72**: 271–338.

Webb, G.E., 2002, Latest Devonian and Early Carboniferous reefs: depressed reef building after the Middle Paleozoic collapse, *In*: Kiessling, W., Flügel, E., and Golonka, J. (eds), *Phanerozoic Reef Patterns, SEPM Special Publication* **72**: 239–269.

West, R.R., 1988, Temporal changes in Carboniferous reef mound communities, *Palaios* **3**: 152–169.

Winkler Prins, C.F., 1968, Carboniferous Productidina and Chonetidina of the Cantabrian Mountains (NW Spain): systematics, stratigraphy and palaeoecology, *Leidse Geologische Mededelingen* **43**: 41–126.

Winkler Prins, C.F., 1971, The road section East of Valdeteja with its continuation along the Arroyo de Barcaliente (Curueño River Valley, León), in: Wagner, R.H. (ed.), *The Carboniferous of Northwest Spain, 2. Trabajos de Geología* **4**: 677–684.

Winkler Prins, C.F., 1983, A general review of the Carboniferous brachiopods from the Cantabrian Mountains (North Spain), in: Lemos de Sousa, M.J. (ed.), *Contributions to the Carboniferous Geology and palaeontology of the Iberian Peninsula*, Faculdade de Ciências, Universidade do Porto, Porto, Portugal, pp. 69–91.

Winkler Prins, C.F., 1991, Brachiopod distributions and the palaeogeographic reconstructions for the Carboniferous, *Compte Rendu XIe International Congress on Carboniferous Stratigraphy and Geology*, Beijing, 1987, **2**: 382–390.

Winkler Prins, C.F. and Martínez Chacón, M.L., 1999, The brachiopod fauna of the Lower Carboniferous Vegamián Formation (Cantabrian Mts, Spain); part 1: Introduction, Linguliformea, *Revista Española de Micropaleontología* número extraordinario: 173–183.

Wolfenden, E.B., 1958, Paleoecology of the Carboniferous reef complex and shelf limestones in northwest Derbyshire, England, *Bulletin of the Geological Society of America* **69**: 871–898.

Chapter 8

# Contrasting Patterns and Mechanisms of Extinction During the Eocene–Oligocene Transition in Jamaica

STEPHEN K. DONOVAN[1], ROGER W. PORTELL[2], AND DARYL P. DOMNING[3]

[1]Nationaal Natuurhistorisch Museum Naturalis, P.O. Box 9517, 2300 RA Leiden, The Netherlands, Donovan@naturalis.nl

[2]Florida Museum of Natural History, P.O. Box 117800, University of Florida, Gainesville, FL 32611, portell@flmnh.ufl.edu

[3]Department of Anatomy, Howard University, 520 W Street NW, Washington, DC 20059, ddomning@howard.edu



**Abstract** The Eocene–Oligocene transition was an important period of global extinction. Jamaica is used as an example of how local influences in both the terrestrial and shallow marine realms contributed to the global pattern. In the early Middle Eocene, Jamaica had a coastal terrestrial fauna that included a rhinoceros, amphibious prorastomid sirenians, an archontan, eusuchian crocodiles, and an anolid? lizard. Before the late Middle Eocene, the island was completely submerged, drowning non-amphibious members of this fauna. This extinction was apparently driven by local changes of relative sea level.



*W. Renema (ed.), Biogeography, Time, and Place: Distributions, Barriers, and Islands,* 247–273

In the marine realm, the change from a mixed siliciclastic/carbonate (Yellow Limestone Group) to carbonate depositional system (White Limestone Group) during the Early–Middle Eocene was thought to have produced a reduction in diversity of at the echinoids, but, until the end of the Eocene, the White Limestone sea was probably populated by typically Yellow Limestone species. By the Late Oligocene, members of this Paleogene assemblage had been largely replaced by immigrants. Local effects related in part to environmental change probably produced a reduction in diversity; subsequent faunal turnover, during a period of global extinction, but not related to major changes of sedimentary substrate, were presumably largely driven by global climatic changes.

## 1. Introduction

Although smaller in effect than any one of the "Big Five" mass extinctions (Late Ordovician, Late Devonian, end Permian, Late Triassic, and end Cretaceous; Donovan, 1989), the Eocene–Oligocene transition is recognized as an important global event, comprised of several separate and successive extinctions that together combined to produce a marked reduction in organic diversity (Prothero and Berggren, 1992; Prothero, 1994). Although most probably driven primarily by climatic effects, other local influences contributed in combination with global changes in producing these extinctions. The present paper considers how asynchronous Eocene extinctions affected the terrestrial and marine biota of one limited area, the Antillean island of Jamaica. In turn, this is used to illustrate how local influences in both the terrestrial and shallow marine realms contributed to the global pattern, but also how they differed from it.

In the early Middle Eocene, Jamaica had a coastal terrestrial fauna that included a primitive rhinocerotoid (*Hyrachyus* sp.), amphibious prorastomid sirenians, an archontan (either primate or plesiadapiform), eusuchian crocodiles, and an anolid? lizard, amongst others (Portell et al., 2001). Coeval terrestrial plants show a moderate diversity (Graham, 1993). However, before the late Middle Eocene, the island had become completely submerged, driving the non-amphibious members of this unique fauna to extinction. As far as can be ascertained, this extinction was entirely driven by local changes of relative sea level.

In the marine realm, the change from a mixed siliciclastic/carbonate (Yellow Limestone Group) to carbonate depositional system (White Limestone Group) during the Early–Middle Eocene has hitherto been considered to have produced a reduction in diversity of at least some groups, such as the echinoids (Donovan, 1995, 2001). The late Middle Eocene White Limestone sea was populated by a fauna of reduced diversity, albeit of typical Yellow Limestone echinoid species. By the Late Oligocene, members of this Eocene assemblage had been largely replaced by taxa that had migrated from elsewhere and which persist in the shallow marine environments of the Caribbean to the present day. Thus, it was considered that local effects related in part to environmental change produced a reduction in diversity; subsequent changes, during a period of global extinction, but not related to major changes of sedimentary substrate, were presumably largely driven by global climatic changes. The present contribution is adapted from Donovan (1995) and Portell et al. (2001), revised and including the latest relevant published information on the study area.

## 2. Tectonics and Palaeogeography

Draper (1987, 1998) proposed a model which divided the geological and tectonic evolution of Jamaica into four distinct phases (see also Robinson, 1994); Pindell (1994) discussed the plate tectonic context and the changing position of Jamaica relative to other land masses (Fig. 1).

During the Early Cretaceous to Early Cenozoic, Jamaica formed part of the greater Antillean Island Arc; volcanism migrated from the central to the eastern part of the island in the Late Cretaceous (Phase 1). The island was largely emergent in the latest Cretaceous to Early Eocene (Phase 2), a time of intrusion and rifting

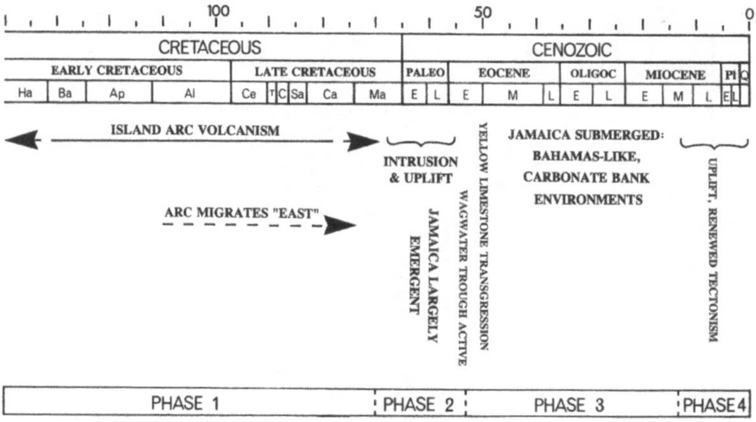


Fig. 1 The geological evolution of Jamaica, based on the narrative account of Draper (1987, 1998; after Donovan et al., 2002, fig. 24.2). For most of the Cretaceous, Jamaica formed part of an island arc that migrated in position from the central to the eastern part of the island in the Late Cretaceous (Phase 1). The island was largely emergent in the latest Cretaceous-earliest Eocene (Phase 2), a time of intrusion and graben formation. In the Eocene, the island became a submerged carbonate bank similar to the Bahamas at the present day (Phase 3). The island was again uplifted *c.* 10 Ma and has remained tectonically active to the present day (Phase 4). Al = Albian; Ba = Barremian; C = Coniacian; Ca = Campanian; Ce = Cenomanian; E = Early; Ha = Hauterivian; L = Late; M = Middle; Ma = Maastrichtian; Oligoc = Oligocene; Paleo = Paleocene; Pl = Pliocene; Q = Quaternary; Sa = Santonian; T = Turonian. Ages in million years ago at top.

(see below) (Fig. 2). Later in the Eocene, the island became a submerged carbonate bank, perhaps similar to the Bahamas Bank at the present day (Phase 3). This phase of near-continuous limestone deposition in shallow to deeper water settings persisted throughout the mid-Cenozoic. The island was again uplifted *c.* 10 Ma and has remained tectonically active throughout the Late Cenozoic (Phase 4).

The Early–Middle Eocene was a time of marine transgression, which flooded western Jamaica (= early Phase 3, Fig. 3a). Volcanism continued, but was waning. Deposition of the Yellow Limestone Group commenced in the west (Robinson, 1988a, Fig. 3), spreading eastwards in the early Middle Eocene. Deep-water lithofacies were developed off the north coast and in the Wagwater graben, which flanks

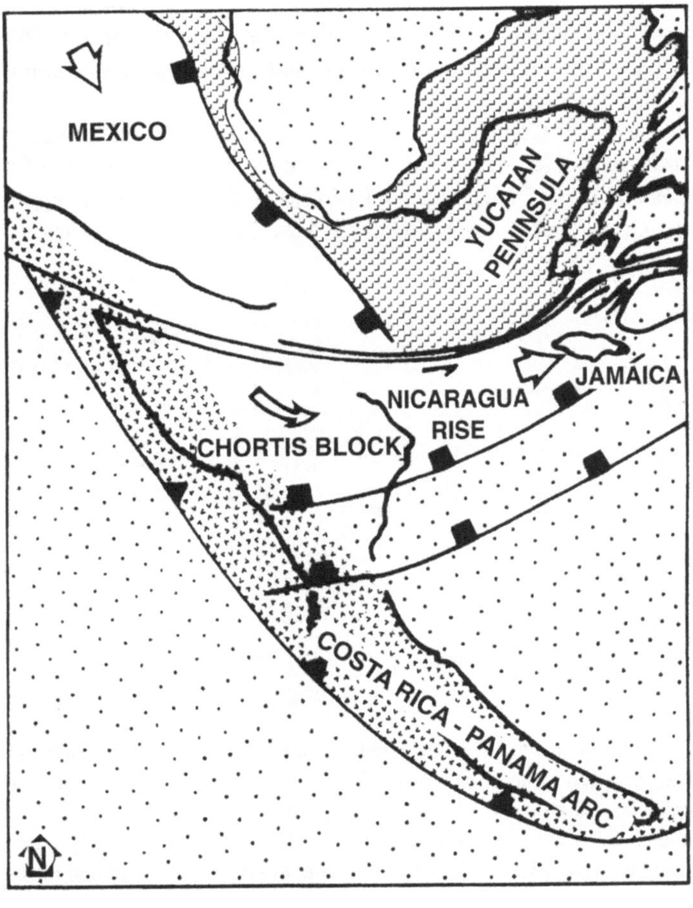


Fig. 2 Reconstruction of the Central American region during the late Early–early Middle Eocene (after Portell et al., 2001, Fig. 1; redrawn after Robinson, 1988b, Fig. 6; and Pindell, 1994, Fig. 2.6k). The probable direction of migration of *Hyrachyus* is indicated by the arrows. White = land areas; v v v = active volcanic arc; ⌋⌋⌋ = shallow water marine; stipple = deep water marine; filled triangles = subduction zone; ▫ = major thrust faults.

the southwestern Blue Mountain block in eastern Jamaica (Eva and McFarlane, 1985, Figs. 5, 6; Mitchell, 2004, Fig. 10) (Fig. 3).

## 3. Marine Environment

Fossil echinoids occur in Jamaican rocks from the Lower Cretaceous (Aptian) to Pleistocene (Sangamonian) and can be divided into four main faunal associations, in the Lower Cretaceous, Upper Cretaceous, Paleocene(?) to Eocene, and post-Eocene. While the Lower Cretaceous echinoids remain poorly known (hemicidarids; Draper et al., 1998), those of the Upper Cretaceous (mainly Campanian to Maastrichtian) are well documented and show some similarities with the coeval Cuban fauna (Donovan, 1993a). Presumably this Late Cretaceous fauna was adversely affected by the mass extinction at the end of the Mesozoic. Jamaican (indeed, Caribbean) Paleocene echinoids are rare (Donovan et al., 2005), so that any meaningful comparison of faunas across the Cretaceous/Cenozoic boundary is not, as yet, possible. However, echinoids are abundant in the succeeding Eocene series, both in Jamaica (Donovan, 1993b) and throughout the Caribbean (Jackson,

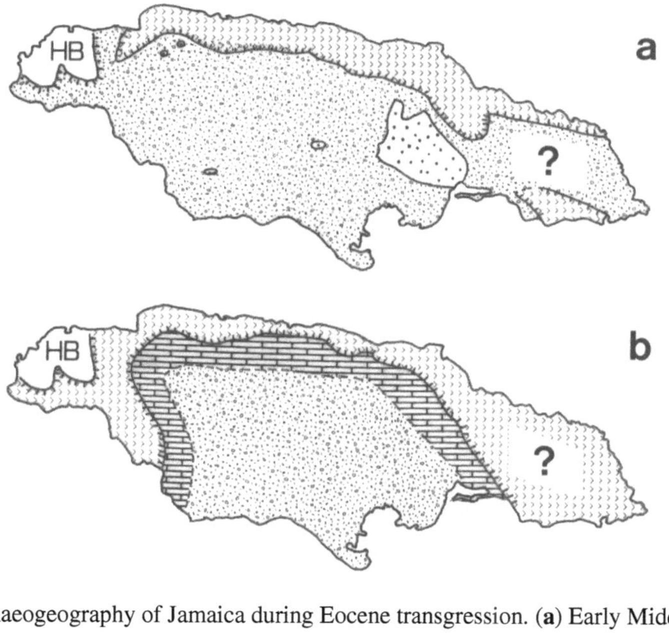


Fig. 3 Palaeogeography of Jamaica during Eocene transgression. (**a**) Early Middle Eocene (redrawn after Eva and McFarlane, 1985, Fig. 6), shallow water high, and low-energy environments not differentiated. (**b**) Late Middle Eocene (based on Eva and McFarlane, 1985, Fig. 7; Mitchell, 2004, Fig. 10, top). > > > = deeper water; gravel pattern = shallow water, low energy; brick pattern = shallow water, high energy; coarse stipple = land; ? = Blue Mountain region; HB = Hanover Block (possibly, in part, emergent).

1922; McKinney et al., 1992). This fauna is very different from those known from the Upper Cretaceous.

Between the Middle Eocene and Late Oligocene a faunal turnover occurred, related to the pattern of extinctions around the Eocene/Oligocene boundary (Prothero and Berggren, 1992; Prothero, 1994) (Fig. 4). By the Late Oligocene the Caribbean echinoid fauna showed many similarities to that of the present day and may be considered the precursor of the modern fauna of the region. This review of the evidence for extinction in the marine environment considers the transition from the Paleocene?/Eocene echinoid fauna to that of the Late Oligocene and after. This report builds on previous syntheses for the Caribbean (such as McKinney et al., 1992), and particularly using the data presented for Jamaica by Dixon (1995), Dixon and Donovan (1994, 1998a), and Donovan (1994a, 1995, 2001).

The Jamaican rock record includes fossiliferous strata from all series of the Cenozoic, commonly represented by a variety of limestone or limestone and siliciclastic sedimentary facies. The biostratigraphy of Jamaica is probably the best known amongst the Greater Antillean islands (Wright and Robinson, 1993). Although certain other larger Caribbean islands have a comparable rock record, this may be less well-constrained biostratigraphically and/or may lack the

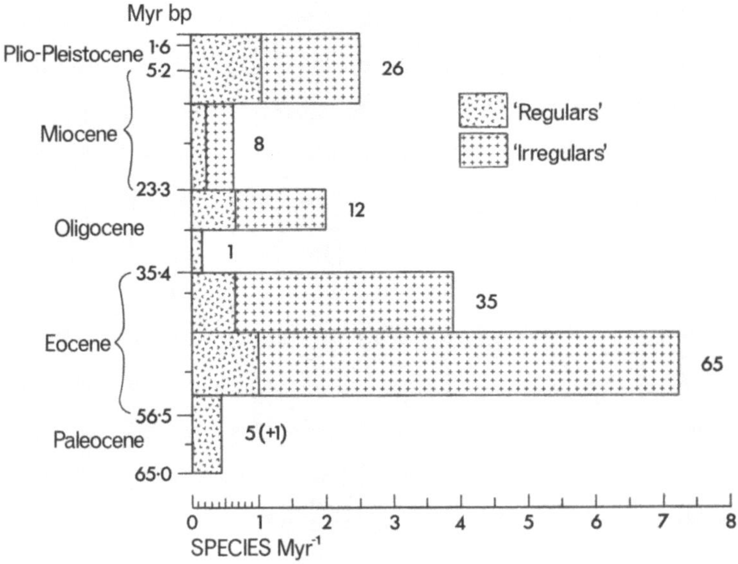


Fig. 4 Temporal distribution of fossil echinoids of Jamaica, including unpublished records (after Donovan, 2001, Fig. 2). The Eocene record is divided into three unequal parts: lowermost (= turbidites of the Richmond Formation); mid-Lower to mid-Middle Yellow Limestone Group; and overlying White Limestone Group (Troy, Swanswick and Somerset formations). Timescale after Harland et al. (1990). See Cooper (2004) for a critique of the use of species per million years as a metric of mean standing diversity.

macroinvertebrates that occur so commonly at certain horizons in Jamaica. For these various reasons, the Jamaican fossil macroinvertebrate fauna is an important sampling point for the fossil record of the Greater Antilles.

Jamaica has one of the best documented fossil echinoid faunas of the Greater Antilles region, particularly from the Eocene. Monographic studies (Arnold and Clark, 1927, 1934) concentrated on the mid-Lower to mid-Middle Eocene Yellow Limestone Group, an impure, poorly to well-lithified unit including limestones, evaporites, lignites, and siliciclastic rock units. It represents a succession of marginal marine-to-marine rocks, divided into a number of formations of contrasting lithologies (Robinson, 1988a, Fig. 3) that were deposited as Jamaica sank below sea level. Quartzo-feldspathic sandstones with limestone lenses occur close to the palaeoshoreline, with impure limestones occurring further offshore. Eva and McFarlane (1985: 213) noted: "The sedimentary environments present [in the early Middle Eocene] were particularly diverse, because the seas and landmasses contributed contrasting types of sediment, resulting in many mixed carbonate/clastic shorelines [in eastern Jamaica]." In western Jamaica at this time more distal, carbonate-rich sediments were being deposited in the absence of exposed land masses (Robinson, 1988a, Fig. 3). Variations in the distributions of larger benthic foraminifera in the shallow water rocks of the Yellow Limestone Group (Chapelton Formation) indicate the existence of numerous micropalaeoenvironments (Eva, 1977; Robinson, 1988a; Robinson and Mitchell, 1999), in both space and time. The Yellow Limestone Group is composed of two principal units, the Chapelton Formation *sensu lato* (including those former members now regarded as formations in their own right; Robinson and Mitchell, 1999) and Font Hill Formation, deposited in shallow and deeper water marine environments, respectively. Echinoids are particularly well known from the Chapelton Formation *sensu lato* (over 50 nominal species; Donovan, 1993b, Fig. 4 herein), although the distribution of taxa between various lithofacies remains incompletely understood (see Fig. 5).

Decrease in the proportion of siliciclastic impurities through the sequence of the Yellow Limestone Group reflects the progressive submergence of the land masses. In the late Middle Eocene, continued erosion and submergence led to the complete disappearance of exposed land areas (Eva and McFarlane, 1985; Mitchell, 2004). The onset of White Limestone deposition was thus diachronous and occurred earlier in the west (Hose and Versey, 1957). Total submergence led to sedimentation dominated by pure, well-lithified limestones of the White Limestone Group, perhaps totalling about 2.75 km in thickness (Robinson, 1994: 121), which persisted until the Middle–Late Miocene, that is, *c.* 30 My. Although Perfit and Williams (1989: 70) speculated that Jamaica may have been, in part, subaerially exposed during the mid-Cenozoic, available evidence supports near-continuous deposition of the White Limestone Group (for further discussion, see Domning et al., 1997: 638), apart from exposure in low-energy lagoonal settings (Mitchell, 2004).

The extension of the area of pure limestone deposition resulted in the loss of siliciclastic and mixed carbonate/siliciclastic sedimentary environments. In this totally marine setting pure carbonates were accumulated in deeper water, high-energy shelf edge, and low-energy shallow bank (lagoonal) environments (Eva and

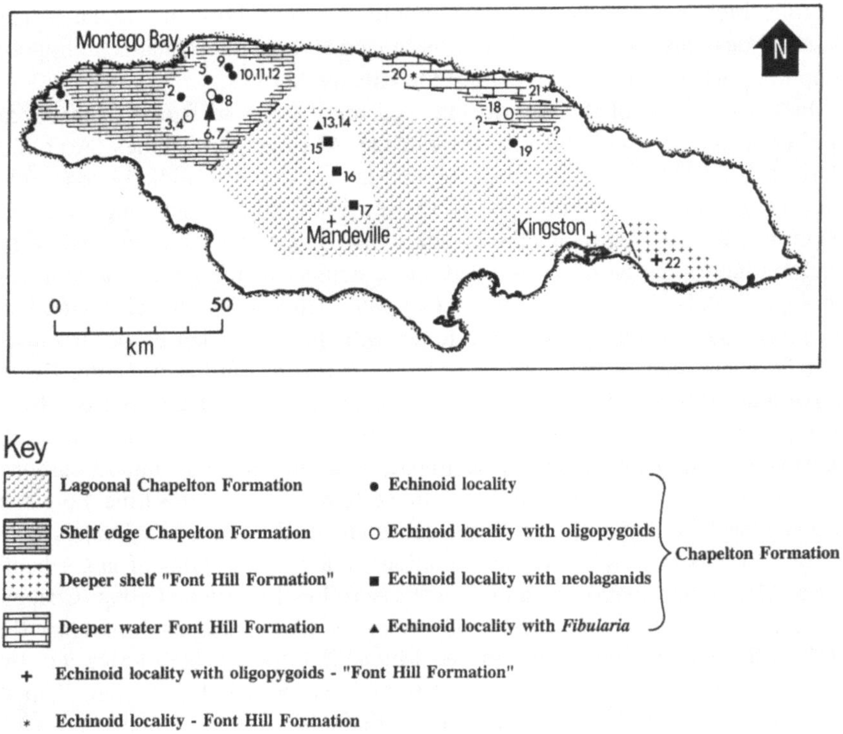


Fig. 5 Geographical distribution of echinoid localities of the Yellow Limestone Group in Jamaica (after Miller and Donovan, 1996, Fig. 4, but without the north–south elongation apparent in the original publication), based on time-averaged data sets of unequal sizes that were not necessarily precisely coeval. Localities (parishes in parentheses): (1) Abingdon (Hanover); (2) Montpelier; (3) Seven Rivers; (4) Cambridge; (5) John's Hall; (6) Spring Mount; (7) Springfield/Welcome Hall; (8) Point; (9) Glasgow; (10) Leyden; (11) Ginger Valley; (12) Somerton (all parish of St. James); (13) Albert Town; (14) Freemans Hall; (15) Wait-a-Bit Cave (all parish of Trelawny); (16) south of Christiana (Manchester); (17) Peace River (Clarendon); (18) Lucky Hill and Gayle (St. Mary); (19) Guy's Hill (St. Catherine); (20) south of Mt. Zion (St. Ann); (21) Ceran Hill (St. Mary); (22) Easington and Yallahs River valley (St. Thomas). For discussion of the formations of the Yellow Limestone Group, see Robinson (1994). Note that oligopygoids also occur at locality 20 and *Fibularia jacksoni* is only known from locality 13.

McFarlane, 1985, Fig. 3b). Echinoids are present in the Eocene formations of the White Limestone Group, albeit at a reduced absolute diversity when compared to the Yellow Limestone Group (about 35 compared to 65 species, Fig. 4). These White Limestone faunas have received much less study than those of the Yellow Limestone Group; none of the specimens documented by Arnold and Clark (1927, 1934) seems to have come from the Eocene of the White Limestone Group.

The pattern of extinctions from the Middle Eocene to the middle of the Oligocene is well documented (Prothero, 1989, 1994; Prothero and Berggren,

1992), with most explanations for these events relying on climatic change (McKinney and Oyen, 1989) or, less probably, extraterrestrial influences (Hut et al., 1987). A recent analysis of the pattern of occurrence of echinoids in Jamaica during the Paleogene (Donovan, 2001) recognized a low diversity of species in the Paleocene, followed by a Cenozoic peak in the Eocene and a subsequent decrease in the Oligocene. Until the mid-1990s, the only moderately diverse, stratigraphically well-constrained echinoid faunas of Oligocene age in the Antilles were from Puerto Rico (Gordon, 1963) and Antigua (Poddubiuk and Rose, 1985). The lack of described Oligocene echinoids from Jamaica supported the theory of a decline in diversity after the Eocene (Kier, 1977; McKinney et al., 1992). However, it was suspected that it was due at least in part to incomplete collection from the Oligocene formations of the island, deposited in carbonate bank and shelf edge environments, respectively. A programme to collect echinoids from the Oligocene of Jamaica was initiated during 1991 by Donovan and the late Dixon (1995; Dixon and Donovan, 1994, 1998a).

## 3.1. Eocene Echinoids

### *3.1.1. Yellow Limestone Group*

Figure 5 reproduces the preliminary palaeoenvironmental map of Jamaica during Yellow Limestone deposition, based on the evidence of the fossil echinoid fauna (after Miller and Donovan, 1996, Fig. 5; compare with Fig. 3 herein). Some broad trends in distribution are interpreted as being indicative of palaeoenvironmental signals. Oligopygoids mainly occurred in shelf edge environments (also true in Georgia, but not Florida; B.D. Carter, written communication); clypeasteroids (neolaganids and *Fibularia jacksoni* Hawkins, 1927) were commonest in lower energy "lagoonal" environments; and spatangoids and cassiduloids (apart from *Eurhodia*) were better represented in shelf edge environments. In the Chapelton Formation the clypeasteroid *Tarphypygus* occurs in association with the oligopygoid *Haimea* at Seven Rivers and Spring Mount, while oligopygoids are not found in association with neolaganids and *F. jacksoni*. Localities with the former association are considered to be shelf edge, while the latter are "lagoonal". Spatangoids and cassiduloids are most diverse in shelf edge settings, although *Eurhodia matleyi* (Hawkins, 1927) is found in deeper water shelf, shelf edge, and "lagoonal" environments (localities 6, 14, 15, 18, 22). The "lagoonal" sites are in the centre of the stable Clarendon Block, with shelf edge localities towards the edge of the block. Oligopygoids occurred in deeper water environments (localities 20, 22) of the Font Hill Formation, perhaps broadly comparable to those that contained *Fibularia* and small neolaganids during White Limestone deposition.

### *3.1.2. Troy Formation*

The Troy Formation (mid to upper Middle or lower Upper Eocene) was deposited in a low-energy lagoonal palaeoenvironment and locally yields abundant echinoids of limited diversity. The fossiliferous part of the Troy Formation *sensu* Mitchell (2004;

including Claremont Formation of previous usage) is a pale grey, poorly fossiliferous micritic limestone (ibid.: 14 *et seq.*), locally with a rich molluscan–foraminifer fauna (Robinson, 1988a, 1994). It includes at least 11 species of echinoids (Donovan, 2004: Table 1); clypeasteroids are locally abundant. Most of the known species from the Troy Formation were first documented by McFarlane (1974, 1977), and include the clypeasteroids *F. jacksoni*, *Cubanaster* cf. *acunai* (Lambert and Sánchez Roig *in* Sánchez Roig, 1926) and *Wythella* sp., and cassiduloids *E. matleyi* and *Eurhodia* cf. *rugosa* (Ravenel, 1848) (Donovan, 2004). The thesis collections of McFarlane (1974) were made in northern central Jamaica. The collections of the University of the West Indies Geology Museum also include an indeterminate regular (non-cidaroid) echinoid and the oligopygoid *Haimea* sp. from the same area. At Burt's Run, parish of St. Ann, between Browns Town and Bamboo, in McFarlane's study area, a modest abundance of echinoids includes the neolaganid *Cubanaster* cf. *acunai* and the spatangoid *Eupatagus* cf. *antillarum* (Cotteau, 1875; Donovan and Rowe, 2000). Donovan (1994b) described poorly preserved neolaganid clypeasteroids as cf. *Durhamella* cf. *floridana* (Twitchell *in* Clark and Twitchell, 1915) from the Somerset Formation, but this stratigraphic assignment was erroneous, the specimens actually coming from the underlying Troy Formation (Mitchell, 2004). An undescribed, locally abundant neolaganid is known from Green Acres, near Spanish Town, west of Kingston (Dixon and Donovan, 1998b).

*3.1.3. Skibo Limestone*

Donovan et al. (1991) described a moderately diverse fauna of small tests and fragments of larger specimens from the Skibo Limestone, which conformably overlies the Font Hill Formation, Yellow Limestone Group, and is in turn overlain by the deeper water limestones of the White Limestone Group (Robinson and Jiang, 1990). This unit is exposed at Content, parish of Portland, northeast Jamaica. It has yielded the cidaroid *Eucidaris* sp., regular echinoid *incerti ordinis* (both based on radioles only), *Fibularia*? sp. and two other species of clypeasteroid, and various indeterminate fragments. Its stratigraphic position is upper Middle Eocene (Robinson *in* Donovan et al., 1991: 2).

*3.1.4. Swanswick Formation*

The echinoid fauna of the Swanswick Formation (mid to upper Middle Eocene), a foraminiferal, bioclastic limestone (Robinson, 1988a, 1994; Mitchell, 2004), consists of at least 22 species (Donovan et al., 1989; Donovan, 1994a, 2004; Rowe and Stemann, 1999; Donovan and Rowe, 2000), including those left in open nomenclature. The most fossiliferous echinoid locality known from the Eocene of the White Limestone Group is at Beecher Town, near Ocho Rios, parish of St. Ann (Donovan et al., 1989; Donovan and Gordon, 1989; Donovan, 1994a), which has contributed over 700 specimens. Preservation is only moderately good at best, but complete tests of smaller irregular echinoids are usually identifiable. Clypeasteroids are noticeably absent at this locality, apart from one small specimen doubtfully placed in this group. In contrast, *Haimea ovumserpentis* (Guppy, 1866), a member of the sister group, Oligopygoida (Kier, 1967; Mooi, 1990; Rowe and

Stemann, 1999), is particularly common, associated with spatangoids, cidaroids, and rarer cassiduloids. To the Beecher Town fauna can be added three species of irregular echinoid from the Dry Harbour Mountains (McFarlane, 1974, 1977).

*3.1.5. Somerset Formation*

The Somerset Formation (lower Upper Eocene?; Mitchell, 2004) has only yielded one species of echinoid, the cassiduloid *E. matleyi*, which was recorded by McFarlane (1974, 1977). The specimen is lost and the locality that yielded this species remains unknown.

## 3.2. Oligocene Echinoids

In his revision of the lithostratigraphy of the White Limestone Group of Jamaica, Mitchell (2004) subsumed both the Browns Town and Walderston formations into the Moneague Formation. In his Appendix 2, he suggested use of "*Lepidocyclina*-dominated biofacies of the Moneague Formation" as a replacement term for the Browns Town Formation (or Brown's Town Limestones therein) and "miliolid-dominated biofacies of the Moneague Formation" for the Walderston Formation/ Limestones. These terms are unwieldy, although they have been used as suggested elsewhere (see, e.g., Donovan, 2004). Herein, we refer to these former formations as Moneague Formation (ex-Browns Town Formation) and (ex-Walderston Formation), respectively.

*3.2.1. Moneague Formation (ex-Walderston Formation)*

A partial test of the camarodont *Gagaria*? sp. is known from the Lower Oligocene Walderston Formation, south of Walderston, parish of Manchester (Donovan, 1996). Clypeasteroids occur in the same unit (H. L. Dixon, Personal Communication, 1996).

*3.2.2. Moneague Formation (ex-Browns Town Formation)*

About 12 distinct echinoid species have now been recognized from the Upper Oligocene (Robinson, 2004) of the type area of the former Browns Town Formation (Dixon, 1995; Dixon and Donovan, 1998a). This diversity is comparable with that reported from the Antigua Formation of Antigua, the largest echinoid fauna documented from the (Upper) Oligocene of the Caribbean plate, comprising 14 species (Poddubiuk and Rose, 1985). Similarly, eleven species of echinoid are recorded from the Lower and Upper Oligocene of Puerto Rico, but spread between three stratigraphic units; the San Sebastian Formation (six species), the Lares Limestone (five species), and the Juana Díaz Formation (two species) (Gordon, 1963; Larue, 1994).

The most common echinoids in the Brown's Town Formation belong to the genus *Clypeaster*. This genus does not occur in the Caribbean fossil record until the Oligocene (Poddubiuk, 1985) and it is well known for its high preservation potential. *Clypeaster oxybaphon* Jackson is present in all Caribbean, Upper Oligocene faunas (Fig. 6), as (probably) is *C. batheri* (Lambert, 1915). *Clypeaster* cf. *julii*

(Roman, 1952), is somewhat rarer. In Jamaica, spatangoids, other irregulars and disarticulated plates of cidaroids are relatively rare, whereas cidaroids are abundant and spatangoids are common to abundant in the Antigua Formation.

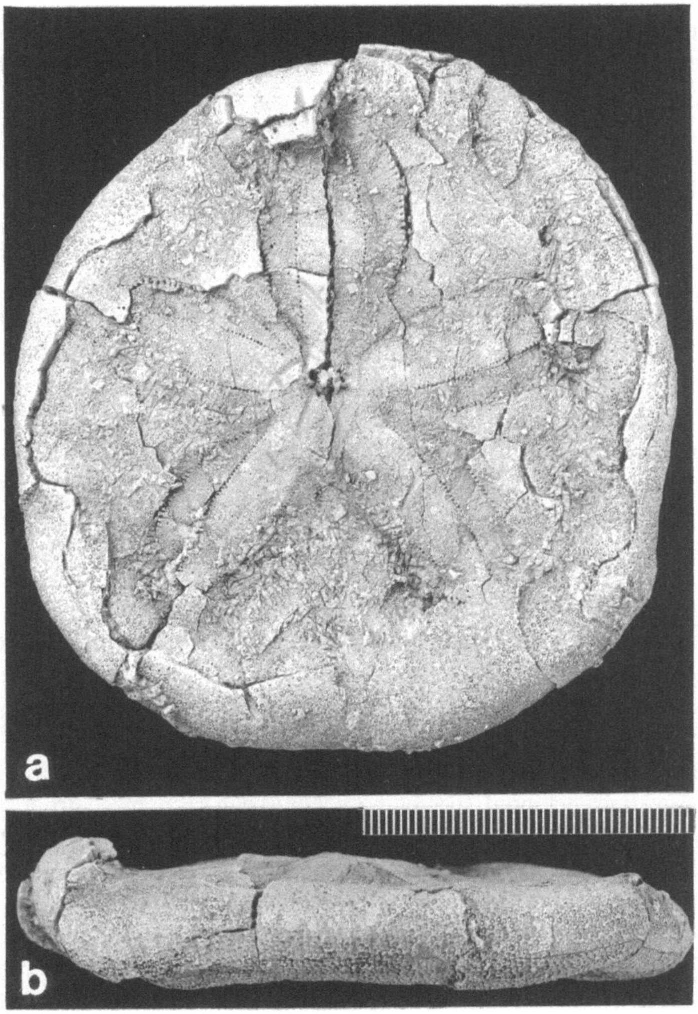


Fig. 6 *Clypeaster oxybaphon* (Jackson, 1922), Moneague Formation (ex-Browns Town Formation), parish of St. Ann, Jamaica, The Natural History Museum, London (BMNH), EE 5690 (after Donovan, 2004, pl. 2, Figs. 4, 3, respectively). Queen of the Oligocene sea floor, the most obvious new echinoid taxon in the post-Eocene of the Caribbean was *Clypeaster*, generally large in size and with a high preservation potential. (a) Apical view. (b) Left lateral view (anterior to left). Scale bar in millimeters. Photographs by Phil Crabb, Photographic Unit, The Natural History Museum, London (BMNH).

## 4. Terrestrial Environment

Tertiary terrestrial tetrapods are very rare in the fossil record of the Antillean region. Those findings that have been made mainly occur in the Greater Antilles. These uncommon fossils are particularly important for testing theories of island biogeography and Caribbean tectonics, both of which have been heavily debated for many years. Recent significant finds have included a fascinating array of mammals from the Oligo–Miocene of Cuba, the Dominican Republic, and Puerto Rico (see, e.g., MacPhee and Iturralde-Vinent, 1994, 1995a, b, 2000; MacPhee and Grimaldi, 1996; MacPhee et al., 2003; Dávalos, 2004). These fossils are notable for their antiquity and their diversity, and include sloths, platyrrhine monkeys, rodents, and insectivores.

Williams (1989: 24) proposed a twofold division of Antillean biogeographic history into Stages I and II. Terrestrial faunas of Stage II are those which colonized the Antillean islands post-inundation, that is, after the islands rose above sea level for the last time. These faunas would obviously include those flightless mammals that have inhabited the Greater Antilles in the Pleistocene and Holocene, consisting solely of ground sloths, monkeys, rodents, and insectivores (Williams, 1989: Table 1), while lacking survivors of the pre-inundation terrestrial fauna (=Stage I). The known Oligo-Miocene taxa of the Greater Antilles represent a post-inundation, Stage II assemblage. As such, they potentially support a vicariant model for the origin of this mammal fauna (see, e.g., discussion in MacPhee and Wyss, 1990: 3–7), at least for these islands, although data are admittedly sparse. However, and more probably, their depauperate nature is indicative of limited over water dispersal (Hedges, 2001). While at least some of these taxa may also have been present in pre-inundation, Stage I assemblages, they would have been extirpated by inundation and would have had to repopulate the islands following final emergence. Thus, a true Stage I terrestrial mammal in the Greater Antilles would have to belong to a taxon that does not occur in the Oligo–Miocene to Quaternary biota and predates an undoubted "final" major period of inundation.

The first Stage I mammal to be recognized from the Greater Antilles was collected from western Jamaica by Roger W. Portell (Domning et al., 1997). This specimen is a right dentary and dp3–m3 from the rhinocerotoid perissodactyl *Hyrachyus* sp. Domning and Clark (1993) recorded few Tertiary tetrapods from Jamaica, comprising the sirenian *Prorastomus sirenoides* (Owen, 1855), from the Eocene Stettin and Guys Hill formations, the crocodilian *Charactosuchus*? *kugleri* (Berg, 1969), and unidentifiable fragments of turtle from the Guys Hill Formation. Domning (1999) recorded the first Oligocene sea cow remains from the Moneague Formation (ex-Browns Town Formation). Therefore, in addition to the Seven Rivers material discussed here, the depauperate Jamaican Tertiary tetrapod record comprises Eocene and Oligocene marine taxa. Only in the Pleistocene are terrestrial tetrapods common in the fossil record of Jamaica (Morgan, 1993; MacPhee, 1997).

## 4.1. Seven Rivers

The succession that includes the Seven Rivers vertebrate site (Fig. 5, locality 3) forms part of the Guys Hill Formation, Yellow Limestone Group, either upper Lower or lower Middle Eocene. Diagnostic features of this formation include the dominance of sandstones and associated siliciclastic sedimentary rocks, and locally abundant oysters, *Carolia*, and carbonized plant remains (Robinson, 1988a). Sedimentological and palaeontological evidence favours an estuarine/deltaic environment of deposition for the Seven Rivers locality.

The exposed section is ~7 m thick, consisting of a sequence of mudrocks, siltstones and fine- to medium-grained sandstones. Some units are gypsiferous or contain calcareous nodules. The associated fossil biota contains plant debris, a limited diversity of marine invertebrates and moderately abundant bones. The large collection of vertebrate remains, mainly tetrapod, now known from this site is the result of perhaps two man years of collecting.

The Yellow Limestone Group in western Jamaica yielded the unique type specimen of *P. sirenoides* (Owen, 1855), consisting of only the skull, mandible, and atlas vertebra, which is the world's oldest and most primitive sirenian (Savage et al., 1994). Other, roughly coeval Jamaican localities have produced additional remains of primitive sirenians (Donovan et al., 1990), but provided only minimal evidence of the postcranial skeleton of the animal. The Seven Rivers site has yielded abundant postcranial and cranial specimens since 1994 (Domning et al., 1995; Portell et al., 2001; Domning, 2001). These remains formed the type series for a new prorastomid sirenian, *Pezosiren portelli* (Domning, 2001); a second, new and undescribed taxon of the same family is present. Both are close in morphology to *Prorastomus* and give a clear picture of the sort of creatures that were the prorastomid sea cows. Perhaps remarkably, these fossils confirm the accuracy of the speculative reconstruction by the late Professor R.J.G. Savage (Dixon et al., 1988: 229), that was based only on the type cranial material of *Prorastomus*. This reconstruction depicts a pig-sized, barrel-chested animal with four stout legs, stubby toes, and a long and muscular, but slender, tail that was not much modified as a swimming organ (Fig. 7).

The prorastomids had well-developed legs and could support their bodies on land, but it is clear that they spent most of their time in the water. Their aquatic habit is indicated by their enlarged and retracted nasal openings (which facilitated breathing at the water's surface) and their typically sirenian thick, dense ribs (which provided ballast for submerging). They evidently swam by forcefully extending the spinal column, kicking backward and upward with the hind feet used simultaneously, and, perhaps, deriving some additional thrust from the tail. This is the swimming style hypothesized for early whales such as *Ambulocetus* (Thewissen et al., 1994) and is also observed in modern otters.

Among the hundreds of sirenian bones collected from Seven Rivers, a single right dentary and dp3–m3 of a further large mammal is attributable to the rhinocerotoid perissodactyl *Hyrachyus* (Fig. 8), close to *H. affinis* (Marsh, 1871). This specimen was described in detail by Domning et al. (1997), where its affinities and

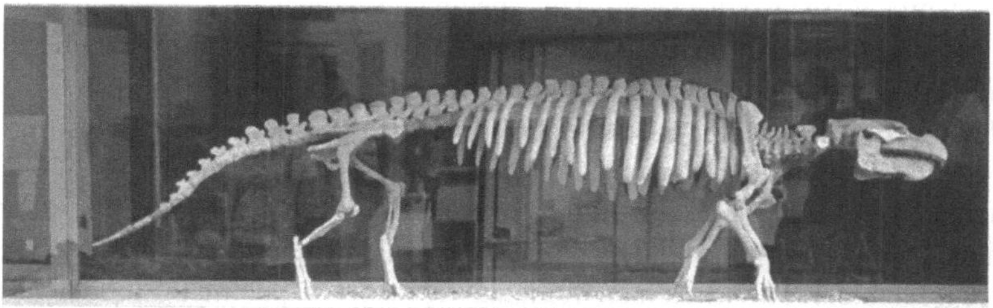

Fig. 7 Mounted cast of the Eocene quadrupedal sirenian *Pezosiren portelli* Domning, from the Guys Hill Formation, Yellow Limestone Group of Seven Rivers, parish of St. James, western Jamaica (locality 3 in Fig. 5), on display in the Geology Museum, University of the West Indies, Mona (UWIGM). Photograph by Shakira Kahn, UWIGM.

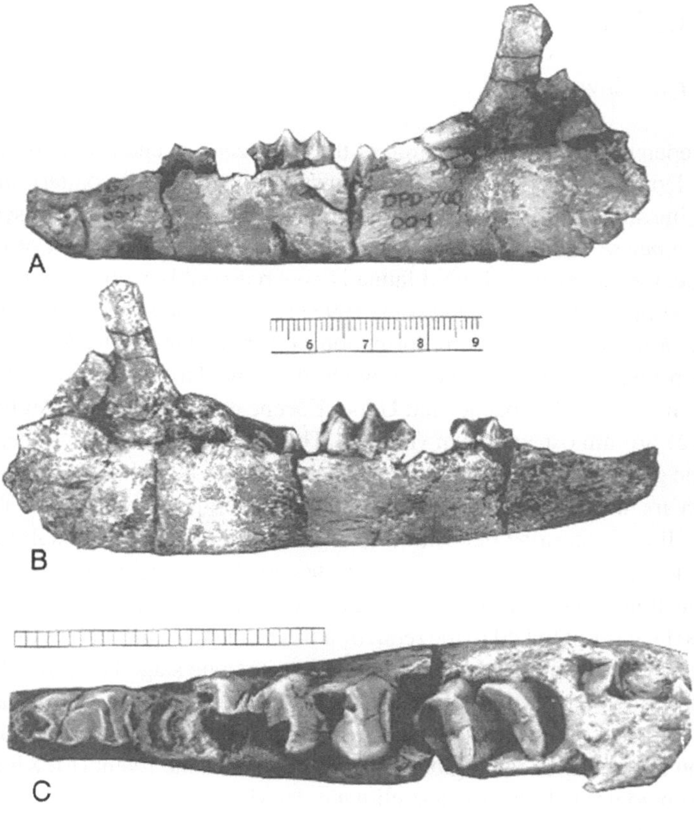


Fig. 8 Right dentary and dp3–m3 of *Hyrachyus* sp. from the upper Lower or lower Middle Eocene of Seven Rivers, parish of St. James, western Jamaica, Nationaal Natuurhistorisch Museum Naturalis, Smithsonian Institution (USNM) 489191 (after Domning et al., 1997, fig. 1; Portell et al., 2001, fig. 4.). (**A**) Medial view. (**B**) Lateral view. (**C**) Occlusal view. Scale bars in cm (A, B) and mm (C).

relationships were also discussed. This specimen is considered to be *c.* 12 My older than the next oldest Antillean mammal, an Early Oligocene sloth from Puerto Rico (MacPhee and Iturralde-Vinent, 1995b).

The Seven Rivers site has also yielded fossils of fishes, crocodilians, turtles, a lizard, and a therian? mammal. The fishes include sharks and rays. Crocodilians are eusuchians and may represent *Charactosuchus*? *kugleri* Berg, otherwise described from the Dump Limestone lenticle of the Guys Hill Formation, parish of Manchester, western Jamaica. The chelonian remains represent a pelomedusoid pleurodiran (side-necked) turtle. The lizard is an iguanian and possibly an anoloid (Family Polychrotidae; Pregill, 1999). The therian? is of uncertain affinity and is only known from a broken petrosal bone, "Possibly but doubtfully primate" (MacPhee and Iturralde Vinent, 2000: 148; see also MacPhee et al., 2003: Table 1).

## 5. Discussion

### 5.1. Marine Environment

The palaeoenvironmental distribution of the Jamaican Eocene echinoids was delineated by Donovan (1994a, 1995), and shows a broadly similar pattern in both the Yellow Limestone and White Limestone groups (Fig. 5); it will not be discussed further herein *per se*. Formerly, it was considered that, with the exception of the Lower Oligocene, the Jamaican echinoid fauna is now reasonably well known throughout the Eocene and Oligocene (Donovan, 1995) (Fig. 4). However, reexamination of known taxa in the Eocene, using formation ranges recognized by Mitchell (2004), makes it probable that all of the echinoids from the Troy and Swanswick formations are upper Middle Eocene, and Upper Eocene echinoids (Somerset Formation; see above) are almost unknown (Fig. 9). There is thus an Upper Eocene–Lower Oligocene gap and the greatest decline in echinoid diversity, in terms of species per million years, now appears to be at the end of the Middle Eocene, rather than at the end of Yellow Limestone deposition (=mid Middle Eocene), although the decline in absolute number of known species is nonetheless large at the Yellow/White Limestone transition. That is, despite previous statements to the contrary (Donovan, 1995, 2001), it is not linked to the reduction of sedimentary lithofacies that followed the end of Yellow Limestone deposition and occurs one stage later, at or before the Bartonian/Priabonian (Middle/Upper Eocene) boundary. This pattern is at variance with the evidence of individual genera, which appear to have suffered their greatest extinction in the Caribbean region at the Terminal Eocene Event (McKinney et al., 1992), coincident with the greatest climatic deterioration. It is relevant to note that Berggren and Prothero (1992) considered the major Eocene–Oligocene extinction event to have occurred at the Middle–Late Eocene transition.

The Upper Eocene–Lower Oligocene gap in Jamaica is almost certainly exaggerated and most probably indicates that palaeontologists need to look harder for echinoids in this interval in Jamaica. For comparison, Oyen and Portell (2001)

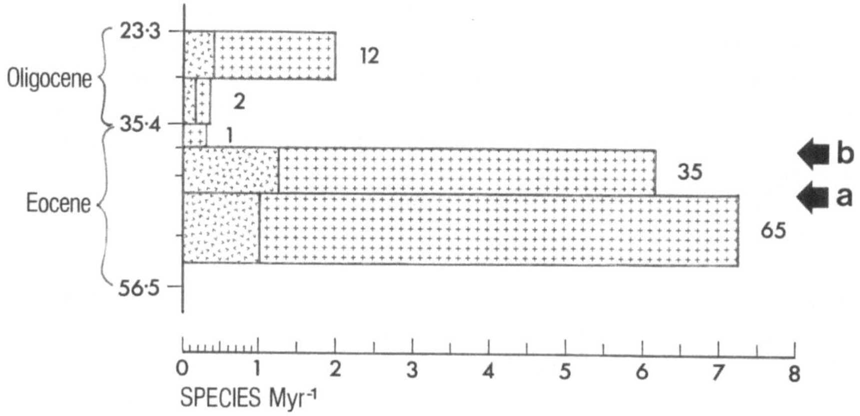


Fig. 9 Revised temporal distribution of Eocene–Oligocene echinoids of Jamaica, assuming all known echinoids of the Eocene White Limestone Group, apart from the Somerset Formation, are upper Middle Eocene (contrast with same interval in Fig. 4). (a) The approximate position of the Yellow Limestone/White Limestone transition in the mid Middle Eocene. This boundary is diachronous and (a) has been arbitrarily positioned midway between the Lower/Middle (Ypresian/Lutetian) and Middle/Upper (Bartonian/Priabonian) boundaries. This also coincides with the presumed time of demise of the Jamaican Stage I terrestrial biota through habitat loss. (b) The Bartonian/Priabonian boundary, which marks the presumed major extinction of the Jamaican Eocene echinoid fauna. Key as in Figure 4. Numbers at ends of bars indicate the total number of species recognized from startigraphic interval. Timescale after Harland et al. (1990).

noted 40 Eocene and only 11 Oligocene echinoid species from Florida. Previously, McKinney et al. (1992) and Carter (1987) documented an almost complete absence of echinoids in the Lower Oligocene of the US Gulf Coastal Plain (Heller and Bryan, 1992), sandwiched between a peak of diversity in the Middle to Upper Eocene and a moderate recovery in the Upper Oligocene. However, Carter (2003), using a data set with improved stratigraphic control and new information concerning Lower Oligocene echinoids, demonstrated an uppermost Upper Eocene peak in diversity, declining into the Lower Oligocene and maintained in the Upper Oligocene.

That extinction events in the latter part of the Eocene had a marked influence on the composition of the Jamaican fauna is not doubted. Many of the species known from the Eocene Swanswick Formation are relicts from the Yellow Limestone Group fauna; other later Eocene echinoid faunas of Jamaica include at least some comparable taxa (Donovan, 1993a). In contrast, the Upper Oligocene fauna of the island is dominated numerically by immigrant *Clypeaster* and includes few taxa that are apparently directly descended from Jamaican Eocene species, although Kier (1984) suggested a close relationship between Oligocene *Eupatagus hildae* (Hawkins, 1927), and Eocene *E. clevei* (Cotteau, 1875), and *Agassizia* and *Echinolampas* are present in both intervals.

The importance of this transition is that it saw a changeover, at least in the echinoids (and, by inference, possibly in a number of other groups of marine invertebrates), from the Paleocene(?) to Eocene fauna, typified by the echinoids of the Yellow Limestone Group, to the precursors of the modern Caribbean echinoid fauna. The modern shallow water echinoid fauna of the region was largely established by the Oligocene (Donovan and Veale, 1996) and was derived only partly from the more diverse Eocene fauna. Additions to this fauna since the Paleogene have mainly involved the appearance of new genera of sand dollars. Where known from sufficiently well-preserved specimens, morphologically similar, congeneric species of many of the extant taxa are recognizable back into the latest Paleogene. Well-documented examples include the irregular echinoids *Echinoneus*, *Brissus*, and *Moira* (Donovan and Veale, 1996).

The dynamics of echinoid evolution during the Eocene–Oligocene transition in Jamaica thus appear to be as follows. It is apparent that the Yellow Limestone fauna contributed a greater proportion of residual species to the shelf edge fauna of the Eocene White Limestone than to lagoonal environments (Donovan, 1994a, 1995). The lagoonal fauna was dominated by clypeasteroids, some of which were Yellow Limestone relicts, but others were migrants. Although the details of the succeeding Late Eocene to Early Oligocene fauna are largely unknown, an indeterminate clypeasteroid that occurs in lagoonal limestones of this age (Moneague Formation, ex-Walderston Formation) is perhaps suggestive. By the Late Oligocene the shelf edge fauna was dominated by clypeasteroids (particularly *Clypeaster*), rather than oligopygoids, although these were not derived from Eocene, lagoonal environments of the Antilles, but by migration from the Mediterranean (Poddubiuk, 1985). This Late Oligocene, shelf edge fauna included many taxa typical of, and ancestral to, the modern Caribbean echinoid fauna.

## 5.2. Terrestrial Environment

Donnelly (1988: 15–16) noted that "the Caribbean has remained one of the most controversial areas in the world for geologic reconstructions". The complex geological history of the Caribbean region has been explained by a plethora of published tectonic evolutionary models. These models have relied on a variety of interpretations of the many sources of data available, both plate tectonic (such as Donnelly, 1988, and references therein) and otherwise (Morris et al., 1990, and references therein). Herein, we favour the interpretation of Pindell and co-workers (see, e.g., Pindell and Barrett, 1990; Pindell, 1994), which has received wide acceptance, at least in its broad details. It also permits the most reasonable explanation of the occurrence of *Hyrachyus* in Jamaica, which we regard as strong and independent supporting evidence for the essential correctness of an important aspect of Pindell's model (Fig. 2).

The crust of the Caribbean Plate is one of the largest Phanerozoic oceanic plateau basalt provinces in the world (Donnelly, 1994). Available evidence supports an origin for this plate in the Pacific (Pindell, 1990); to grossly simplify a complex process (see Pindell, 1994, for detail), the plate was pushed between North and

South America, resulting in the subduction of "Proto-Caribbean Plate". Perhaps the most compelling evidence for such an origin of the Caribbean Plate is that the best date we have for the opening of the Caribbean seaway is late Middle Jurassic, about 165 Ma. However, obducted sea floor sedimentary rocks in southwest Puerto Rico contain microfossils that are 30 My older, indicating that the Caribbean Plate was formed before the opening of the seaway (Montgomery et al., 1994).

During the Middle–Late Cretaceous, an island arc, including the terranes that constitute the Greater Antilles, formed at the leading edge of the Caribbean Plate as the "Proto-Caribbean Plate" was subducted beneath it. This island arc system was disrupted and fragmented from the Late Cretaceous onwards, as the Caribbean Plate was transported eastwards between the North and South American Plates. From the Paleocene to the Early Eocene, at least the western portion of Jamaica, at the northeastern edge of the Nicaragua Rise (itself a topographic extension of northern Central America), was subaerially exposed and situated adjacent to the Yucatan Peninsula, from which it was separated by a transform fault (Pindell, 1994, Figs. 2.6j, Figs. 1, 2 herein) (=Phase 2 of Draper, 1987; see above). This subaerial exposure is interpreted as being due to the Yucatan block preventing simple east-northeast movement of the Chortis Block – Nicaragua Rise – Jamaica, resulting in uplift and block faulting of the Nicaragua Rise. As was recognized by Donnelly (1988: 28), this formed a terrestrial connection between Central America and Jamaica via the Nicaragua Rise. This connection was to be severed by the Middle Eocene following submergence of the Nicaragua Rise. Only after it had moved past this "obstruction" formed by the Yucatan block did Jamaica and the Nicaragua Rise enter a period of tectonic quiescence, and become a Bahamas Bank-like carbonate platform (=Phase 3 of Draper, 1987) (Fig. 3). However, during Phase 2, migration of the North American terrestrial biota would have been facilitated by the continuous land mass of North America–Mexican Arc–Chortis Block–Nicaragua Rise–Jamaica, as indicated by the trail of arrows in Fig. 2.

*Hyrachyus* sp. and the therian? from the Seven Rivers site pre-date the Phase 3 inundation of Jamaica, and are thus part of the Stage I terrestrial mammal fauna of the island, based on the evidence of both taxonomic assignment and geology. Except for the iguanian lizard, which may have been dispersed by over-water transport (Censky et al., 1998), all other vertebrates so far discovered from Seven Rivers are obligate aquatic or, at best, amphibious organisms, the latter group including *C.*? *kugleri* and prorastomid sirenians. *Hyrachyus* is well known from the Eocene of Eurasia and North America (Radinsky, 1969). Such a large terrestrial mammal (the size of a large dog) may potentially have been dispersed across broad water barriers, so it remains debatable if its presence in Jamaica is supporting evidence, or not, for an ancient land bridge connection with North America (Fig. 2). Although MacPhee and Wyss (1990: 3) correctly considered that the "land-bridge argument … suffers from an acute lack of supporting geological fact" for explaining the evolution of the Stage II fauna, it must now be considered a plausible explanation for at least part of the Stage I fauna, that of Jamaica. It may be relevant to note that the non-chiropteran, Stage II mammals of Jamaica are uniformly small and of limited taxonomic diversity, with no taxa approaching the size of *Hyrachyus* sp. (Morgan, 1993).

## 6. Conclusions

1. The presence of terrestrial tetrapods at the Seven Rivers site in western Jamaica provides a unique glimpse into an Antillean Stage I association. This fauna would have been extirpated at or by the mid Middle Eocene (Lutetian/Bartonian boundary), when the inundation of the Jamaican land mass would have caused the demise of the island's terrestrial biota. Although this was formerly considered to be approximately coincident with the main Eocene extinction of the marine biota (such as Donovan, 1994a, 1995), this is now less certain.
2. The main drop in echinoid diversity (in species per million years) in the Jamaican Paleogene fossil record most probably occurs at or near the Bartonian/Priabonian (Middle/Upper Eocene) boundary, within rocks of the White Limestone Group. This is at variance with the earlier idea (Donovan, 1995, 2001) that it was associated with facies changes at the transition from Yellow to White Limestone deposition in the Middle Eocene (at or near the Lutetian/Bartonian boundary, that is, one stage earlier. However, the known change from 65 to 35 known species across the Yellow to White Limestone transition (Fig. 9) is nonetheless a marked change.
3. The "similarity" of echinoid diversity (in species per million years) between the Lower and upper Middle Eocene (Fig. 9) indicates that the White Limestone environmental complex was able to support as wide a variety of echinoids as the Yellow Limestone Group, despite the undoubted reduction in the variety of sedimentary palaeoenvironments. Although this example is parochial, the Jamaican echinoids formed an important part of the data base of the Caribbean Paleogene fauna (McKinney et al., 1992), although this is not now considered to be a "global" pattern to which it can usefully be compared (Carter, 2003).
4. The importance of this transition is that it saw the emergence of a Caribbean echinoid fauna comparable in taxonomic composition to that seen at the present day. The Eocene–Oligocene extinctions led to the demise of many taxa that now appear to be archaic. By the Late Oligocene many species with a decidedly modern aspect were present in the Antillean region. In contrast, repopulation of the terrestrial biota had to wait until subaerial exposure of the island in the Middle to Late Miocene.

## Acknowledgements

This research was supported by the First Shell Distinguished Research Fellowship in Science to Stephen K. Donovan, National Geographic Society grants #5116-93,

5327-94, and 5562-95, the Department of Geology at University of the West Indies, Mona (UWI), and Barbara and Reed Toomey, all of which we are delighted to acknowledge. We thank the many collaborators who made vital contributions to our Jamaican fieldwork, including the late Trina MacGillivray, the late Hal Dixon, Eamon Doyle, Trevor Jackson, Simon Mitchell, Ted Robinson, Deborah-Ann Rowe and Tom Stemann (all currently or formerly UWI), Brian Beatty, Kevin Schindler, Barbara and Reed Toomey, George Hecht and the late Craig Oyen (all currently or formerly Florida Museum of Natural History, Gainesville), and Ross MacPhee and Clare Fleming (both American Museum of Natural History, New York). We thank Phil Crabb, Photographic Unit, BMNH, and Shakira Khan, UWI, for photographing specimens depicted in Figs. 6 and 7, respectively. Our reviewers, Professors Edward Robinson (UWI), Donald A. McFarlane (W.M. Keck Science Center, Claremont, California) and Burchard D. Carter (Georgia Southwestern College, Americus), are thanked for their numerous and perspicuous comments. This contribution is dedicated to the late Mr. William F. Schickler, who introduced Stephen K. Donovan to the echinoids of the Swanswick Formation that occurred in the flowerbeds in his garden. This is University of Florida Contribution to Paleobiology 573.

## References

Arnold, B.W., and Clark, H.L., 1927, Jamaican fossil echini, *Memoirs of the Museum of Comparative Zoology, Harvard* **50**: 1–75.

Arnold, B.W., and Clark, H.L., 1934, Some additional fossil echini from Jamaica, *Memoirs of the Museum of Comparative Zoology, Harvard* **54**: 139–156.

Berg, D.M., 1969, *Charactosuchus kugleri*, eine neue Krokodilart aus dem Eozän von Jamaica, *Eclogae Geologicae Helvetiae* **62**: 731–735.

Berggren, W.A., and Prothero, D.R., 1992, Eocene–Oligocene climatic and biotic evolution: an overview, in: Prothero, D.R. and Berggren, W.A. (eds), *Eocene–Oligocene Climatic and Biotic Evolution*, Princeton University Press, Princeton, NJ, pp. 1–28.

Carter, B.D., 1987, Paleogene echinoid distributions in the Atlantic and Gulf Plains, *Palaios* **2**: 390–404.

Carter, B.D., 2003, Diversity patterns in Eocene and Oligocene echinoids of the southeastern United States, in: Prothero, D.R., Ivany, L.C., and Nesbitt, E.A. (eds), *From Greenhouse to Icehouse: The Marine Eocene–Oligocene Transition*, Columbia University Press, New York, pp. 354–365.

Censky, E.J., Hodge, K., and Dudley, J., 1998, Over-water dispersal of lizards due to hurricanes, *Nature* **395**: 556.

Clark, W.B., and Twitchell, M.W., 1915, The Mesozoic and Cenozoic Echinodermata of the United States, *U.S. Geological Survey Monograph* **54**: 341 pp.

Cooper, R.A., 2004, Measures of diversity, in: *The Great Ordovician Biodiversification Event* (B.D. Webby, F. Paris, M.L. Droser and I.G. Percival, eds.), Columbia University Press, New York, pp. 52–57.

Cotteau, G.H., 1875, Description des échinides Tertiaires des îles St. Barthélemy et Anguilla, *Kungliga Svenska Vetenskaps-Akademiens Handlingar* **13** (6): 1–48.

Dávalos, L.M., 2004, Phylogeny and biogeography of Caribbean mammals, *Biological Journal of the Linnean Society* **81**: 373–394.
Dixon, D., Cox, B., Savage, R.J.G., and Gardiner, B., 1988, *MacMillan Illustrated Encyclopedia of Dinosaurs and Prehistoric Animals*, MacMillan, London, 312 pp.
Dixon, H.L., 1995, *Upper Oligocene Echinoids of Jamaica*, Unpublished M.Phil. thesis, University of the West Indies, Mona, v + 115 pp.
Dixon, H.L., and Donovan, S.K., 1994, Local extinction patterns and the decline of the Jamaican Paleogene echinoid fauna, *Palaios* **9**: 506–511.
Dixon, H.L., and Donovan, S.K., 1998a, Oligocene echinoids of Jamaica, *Tertiary Research* **18**: 95–124.
Dixon, H.L., and Donovan, S.K., 1998b, The occurrence and distribution of neolaganid clypeasteroid echinoids in Jamaica, *Contributions to Geology, UWI, Mona* **3**: 77.
Domning, D.P., 1999, Appendix 1. Oligocene Sirenia of the Caribbean Region, in: H.L. Dixon, and S.K. Donovan, with appendices by D.P. Domning, and P.D. Taylor, Report of a field meeting to the area around Browns Town, parish of St. Ann, north-central Jamaica, 21st February, 1998, *Journal of the Geological Society of Jamaica* **33** (for 1998): 24–30.
Domning, D.P., 2001, The earliest known fully quadrupedal sirenian, *Nature* **413**: 625–627.
Domning, D.P., and Clark, J.M., 1993, Jamaican Tertiary marine Vertebrata, in: *Biostratigraphy of Jamaica* (R.M. Wright and E. Robinson, eds.), *Geological Society of America Memoir* **182**: 413–415.
Domning, D.P., Donovan, S.K., Dixon, H.L., Portell, R.W., and Schindler, K.S., 1995, The world's most primitive seacow: a new sirenian site in western Jamaica, *Geological Society of America Abstracts with Programs* **27** (6):A386.
Domning, D.P., Emry, R.J., Portell, R.W., Donovan, S.K., and Schindler, K.S., 1997, Oldest West Indian land mammal: rhinocerotoid ungulate from the Eocene of Jamaica, *Journal of Vertebrate Paleontology* **17**: 638–641.
Donnelly, T.W., 1988, Geologic constraints on Caribbean biogeography, in: Liebherr, J.K. (ed.), *Zoogeography of Caribbean Insects*, Comstock Publishing, Cornell University Press, Ithaca, NY, 15–37.
Donnelly, T.W., 1994, The Caribbean sea floor, in: *Caribbean Geology: An Introduction* (S.K. Donovan and T.A. Jackson, eds.), University of the West Indies Publishers' Association, Kingston, pp. 41–61.
Donovan, S.K. (ed.), 1989, *Mass Extinctions: Processes and Evidence*, Belhaven Press, London, xiv + 266 pp.
Donovan, S.K., 1993a, Jamaican Cretaceous Echinoidea, in: *Biostratigraphy of Jamaica* (R.M. Wright and E. Robinson, eds.), *Geological Society of America Memoir* **182**: 93–103.
Donovan, S.K., 1993b, Jamaican Cenozoic Echinoidea, in: *Biostratigraphy of Jamaica* (R.M. Wright and E. Robinson, eds.), *Geological Society of America Memoir* **182**: 371–412.
Donovan, S.K., 1994a, Middle and Upper Eocene echinoids of Jamaica, in: *Echinoderms through Time: Proceedings of the Eighth International Echinoderm Conference, Dijon, 6th-10th September, 1993* (B. David, A. Guille, J.-P. Firal, and M. Roux, eds.), Balkema, Rotterdam, pp. 629–635.
Donovan, S.K., 1994b, Echinoids of the Upper Eocene Somerset Formation of Jamaica, *Tertiary Research* **15**: 43–48.
Donovan, S.K., 1995, Evolution of the Jamaican echinoid fauna during the Eocene–Oligocene extinction crisis, *Jamaican Journal of Science and Technology* **5** (for 1994): 49–62.
Donovan, S.K., 1996, A regular echinoid from the Walderston Formation (Lower Oligocene) of Jamaica, *Caribbean Journal of Science* **32**: 78–82.

Donovan, S.K., 2001, Evolution of Caribbean echinoderms during the Cenozoic: moving towards a complete picture using all of the fossils, *Palaeogeography, Palaeoclimatology, Palaeoecology* **166**: 177–192.

Donovan, S.K., 2004, Echinoderms of the mid Cainozoic White Limestone Group of Jamaica, *Cainozoic Research* **3** (for 2003): 143–156.

Donovan, S.K., Domning, D.P., Garcia, F.A., and Dixon, H.L., 1990, A bone bed in the Eocene of Jamaica, *Journal of Paleontology* **64**: 660–662.

Donovan, S.K., and Gordon, C.M., 1989, Report of a field meeting to selected localities in St Andrew and St Ann, 25 February 1989, *Journal of the Geological Society of Jamaica* **26**: 51–54.

Donovan, S.K., Gordon, C.M., Schickler, W.F., and Dixon, H.L., 1989, An Eocene age for an outcrop of the 'Montpelier Formation' at Beecher Town, St Ann, Jamaica, using echinoids for correlation, *Journal of the Geological Society of Jamaica* **26**: 5–9.

Donovan, S.K., Jackson, T.A., and Miller, D.J., 2002, Fieldwork: a key component in teaching geology and geomorphology at the University of the West Indies, Jamaica, in: Jackson, T.A. (ed.), *Caribbean Geology: Into the 3rd Millennium: Transactions of the 15th Caribbean Geological Conference, 29th June to 2nd July, 1998, Kingston, Jamaica*, The Press, University of the West Indies, Mona, pp. 269–279.

Donovan, S.K., Renema, W., and Pickerill, R.K. 2005, The ichnofossil *Scolicia prisca* de Quatrefages from the Paleogene of eastern Jamaica and fossil echinoids of the Richmond Formation, *Caribbean Journal of Science* **41**: 876–881.

Donovan, S.K., and Rowe, D.-A.C., 2000, Spatangoid echinoids from the Eocene of Jamaica, *Journal of Paleontology* **74**: 654–661.

Donovan, S.K., Scott, A.D., and Veltkamp, C.J., 1991, A late Middle Eocene echinoid fauna from Portland, northeastern Jamaica, *Journal of the Geological Society of Jamaica* **28**: 1–8.

Donovan, S.K., and Veale, C., 1996, The irregular echinoids *Echinoneus* Leske and *Brissus* Gray in the Cenozoic of the Antillean region, *Journal of Paleontology* **70**: 632–640.

Draper, G., 1987, A revised tectonic model for the evolution of Jamaica, in: Ahmad, R. (ed.), *Proceedings of a Workshop on the Status of Jamaican Geology, Journal of the Geological Society of Jamaica, Special Issue* **10**: 151–169.

Draper, G., 1998, Geological and tectonic evolution of Jamaica, *Contributions to Geology, UWI, Mona* **3**: 3–9.

Draper, G., Donovan, S.K., and Lewis, J.F., 1998, Cretaceous echinoid fossils from central Hispaniola and Jamaica, in: Ali, W., Paul, A., and V. Young On (eds), *Transactions of the 3rd Geological Conference of the Geological Society of Trinidad and Tobago and the 14th Caribbean Geological Conference (in two volumes)*, **1**: 83.

Eva, A.N., 1977, The palaeoecology and sedimentology of Middle Eocene larger Foraminifera in Jamaica, *Maritime Sediments, Special Publication* **1**: 467–475.

Eva, A., and McFarlane, N., 1985, Tertiary to Early Quaternary carbonate facies relationships in Jamaica, in: *Transactions of the 4th Latin American Geological Congress, Port-of-Spain, Trinidad, 7th–15th July, 1979*, **1**: 210–219.

Gordon, W.A., 1963, Middle Tertiary echinoids of Puerto Rico, *Journal of Paleontology* **37**: 628–642.

Graham, A., 1993, Contribution toward a Tertiary palynostratigraphy for Jamaica: the status of Tertiary paleobotanical studies in northern Latin America and preliminary analysis of the Guys Hill Member (Chapelton Formation, Middle Eocene) of Jamaica, in: Wright, R.M. and E. Robinson (eds), *Biostratigraphy of Jamaica, Geological Society of America Memoir* **182**: 443–461.

Guppy, R.J.L., 1866, On Tertiary echinoderms from the West Indies, *Quarterly Journal of the Geological Society of London* **22**: 297–301.

Harland, W.B., Armstrong, R.L., Cox, A.V., Craig, L.E., Smith, A.G., and Smith, D.G., 1990, *A Geologic Time Scale 1989*, Cambridge University Press, Cambridge, 263 pp.

Hawkins, H.L., 1927, Descriptions of new species of Cainozoic Echinoidea from Jamaica, *Memoirs of the Museum of Comparative Zoology, Harvard* **50**: 76–84.

Hedges, S.B., 2001, Biogeography of the West Indies: an overview, in: Woods, C.A. and Sergile, F.E. (eds), *Biogeography of the West Indies: Patterns and Perspectives*, 2nd edition, CRC Press, Boca Raton, FL, pp. 15–33.

Heller, J.A., and Bryan, J.R., 1992, First Oligocene occurrence of the irregular echinoid *Macropneustes mortoni* (Conrad), *Journal of Paleontology* **66**: 345–346.

Hose, H.R., and Versey, H.R., 1957, Palaeontological and lithological divisions of the Lower Tertiary limestones of Jamaica, *Colonial Geology and Mineral Resources* **6** (for 1956): 19–39.

Hut, P., Alvarez, W., Elder, W.P., Hansen, T., Kauffman, E.G., Keller, G., Shoemaker, E.M., and Weissman, P.R., 1987, Comet showers as a cause of mass extinctions, *Nature* **329**: 118–125.

Jackson, R.T., 1922, Fossil echini of the West Indies, *Carnegie Institution of Washington Publication*, **306**: 103 pp.

Kier, P.M., 1967, Revision of the oligopygoid echinoids, *Smithsonian Miscellaneous Collections* **152** (2): 147 pp.

Kier, P.M., 1977, The poor fossil record of the regular echinoid, *Paleobiology* **3**: 168–174.

Kier, P.M., 1984, Fossil spatangoid echinoids of Cuba, *Smithsonian Contributions to Paleobiology* **55**: 336 pp.

Lambert, J.M., 1915, Échinides néogènes des Antilles Anglaises, *Mémoires de la Société Académique d'Agriculture, des Sciences, Artes et Belles Lettres du Département de l'Aube* **79**: 17–33.

Larue, D.K., 1994, Puerto Rico and the Virgin Islands, in: Donovan, S.K. and T.A. Jackson (eds), *Caribbean Geology: An Introduction*, University of the West Indies Publishers Association, Kingston, pp. 151–165.

MacPhee, R.D.E., 1997, Vertebrate paleontology of Jamaican caves, in: Fincham, A. (ed.), *Jamaica Underground: The Caves, Sinkholes and Underground Rivers of the Island*, 2nd edition, The Press, University of the West Indies, Mona, pp. 47–55.

MacPhee, R.D.E. and Grimaldi, D.A., 1996, Mammal bones in Dominican amber, *Nature* **380**: 489–490.

MacPhee, R.D.E. and Iturralde-Vinent, M.A., 1994, First Tertiary land mammal from Greater Antilles: an Early Miocene sloth (Xenarthra, Megalonychidae) from Cuba, *American Museum Novitates* **3094**: 13 pp.

MacPhee, R.D.E. and Iturralde-Vinent, M.A., 1995a, Earliest monkey from Greater Antilles, *Journal of Human Evolution* **28**: 197–200.

MacPhee, R.D.E. and Iturralde-Vinent, M.A., 1995b, Origin of Greater Antillean land mammal fauna, 1: New Tertiary fossils from Cuba and Puerto Rico, *American Museum Novitates* **3141**: 30 pp.

MacPhee, R.D.E. and Iturralde-Vinent, M.A., 2000, A short history of Greater Antillean land mammals: biogeography, paleogeography, radiations, and extinctions, *Tropics* **10**: 145–154.

MacPhee, R.D.E., Iturralde-Vinent, M.A., and Gaffney, E.S., 2003, Domo de Zaza, an Early Miocene vertebrate locality in south-central Cuba, with notes on the tectonic evolution of Puerto Rico and the Mona Passage, *American Museum Novitates* **3394**: 42 pp.

MacPhee, R.D.E. and Wyss, A.R., 1990, Oligo-Miocene vertebrates from Puerto Rico, with a catalog of localities, *American Museum Novitates* **2965**: 45 pp.

Marsh, O.C., 1871, Notice of some fossil mammals from the Tertiary formation, *American Journal of Science and Arts* (series 3) **11** (8): 36–37.

McFarlane, N., 1974, *The Geology of the Dry Harbour Mountains, St. Ann, Jamaica*, Unpublished M.Sc. thesis, University of the West Indies, Mona, 87 pp.

McFarlane, N., 1977, Some Eocene and Oligocene faunas from central Jamaican, in: *Memoria Segundo Congreso Latinoamericano de Geologia, Caracas, Venezuela, 11 al 16 de Noviembre de 1973* **3**: 1393–1411.

McKinney, M.L., McNamara, K.J., Carter, B.D., and Donovan, S.K., 1992, Evolution of Paleogene echinoids: a global and regional view, in: Prothero, D.R. and Berggren, W.A. (eds), *Eocene–Oligocene Climatic and Biotic Evolution*, Princeton University Press, Princeton, NJ, pp. 349–367.

McKinney, M.L., and Oyen, C.W., 1989, Causation and nonrandomness in biological and geological time series: temperature as a proximal cause of extinction and diversity, *Palaios* **4**: 3–15.

Miller, D.J., and Donovan, S.K., 1996, Geomorphology, stratigraphy and palaeontology of Wait-a-Bit Cave, central Jamaica, *Tertiary Research* **17** (for 1995): 33–49.

Mitchell, S.F., 2004, Lithostratigraphy and palaeogeography of the White Limestone Group, *Cainozoic Research* **3** (for 2003): 5–29.

Montgomery, H., Pessagno, E.A., Jr., and Pindell, J.L., 1994, A 195 Ma errane in a 165 Ma sea: Pacific origin of the Caribbean Plate, *GSA Today* **4** (1): 1, 3–6.

Mooi, R., 1990, Paedomorphosis, Aristotle's lantern, and the origin of the sand dollars (Echinodermata: Clypeasteroida), *Paleobiology* **16**: 25–48.

Morgan, G.S., 1993, Quaternary land vertebrates of Jamaica, in: Wright, R.M. and E. Robinson (eds), *Biostratigraphy of Jamaica, Geological Society of America Memoir* **182**: 417–442.

Morris, A.E.L., Taner, I., Meyerhoff, H.A., and Meyerhoff, A.A., 1990, Tectonic evolution of the Caribbean region; alternative hypothesis, in: Dengo, G. and Case, J.E. (eds), *The Geology of North America, Volume H, The Caribbean Region*, Geological Society of America, Boulder, CO, pp. 433–457.

Owen, R., 1855, On the fossil skull of a mammal (*Prorastomus sirenoides*, Owen), from the island of Jamaica, *Quarterly Journal of the Geological Society* **11**: 541–543.

Oyen, C.W., and Portell, R.W., 2001, Diversity patterns and biostratigraphy of Cenozoic echinoderms from Florida, *Palaeogeography, Palaeoclimatology, Palaeoecology*, **166**: 193–218.

Perfit, M.R., and Williams, E.E., 1989, Geological constraints and biological retrodictions in the evolution of the Caribbean Sea and its islands, in: Woods, C.A. (ed.), *Biogeography of the West Indies: Past, Present and Future*, Sandhill Crane Press, Gainesville, FL, pp. 47–102.

Pindell, J.L., 1990, Geological arguments suggesting a Pacific origin for the Caribbean Plate, in: Larue, D.K. and Draper, G. (eds), *Transactions of the 12th Caribbean Geological Conference, St. Croix, Virgin Islands, 7th–11th August, 1989*, Miami Geological Society, Miami, FL, pp. 1–4.

Pindell, J.L., 1994, Evolution of the Gulf of Mexico and the Caribbean, in: Donovan, S.K. and Jackson, T.A. (eds), *Caribbean Geology: An Introduction*, University of the West Indies Publishers' Association, Kingston, pp. 13–39.

Pindell, J.L., and Barrett, S.F., 1990, Geological evolution of the Caribbean region; a plate-tectonic perspective, in: Dengo, G. and Case, J.E. (eds), *The Geology of North America,*

*Volume H, The Caribbean Region*, Geological Society of America, Boulder, CO, pp. 405–432.

Poddubiuk, R.H., 1985, Evolution and adaptation in some Caribbean Oligo-Miocene *Clypeasters*, in: Keegan, B.F. and O'Connor, B.D.S. (eds), *Echinodermata: Proceedings of the 5th International Echinoderm Conference, Galway, 24–29 September 1984*, Balkema, Rotterdam, pp. 75–80.

Poddubiuk, R.H., and Rose, E.P.F., 1985, Relationships between mid-Tertiary echinoid faunas from the central Mediterranean and eastern Caribbean and their palaeobiogeographic significance, *Annales Géologiques des Pays Hélleniques* **32** (for 1984): 115–127.

Portell, R.W., Donovan, S.K., and Domning, D.P., 2001, Early Tertiary vertebrate fossils from Seven Rivers, parish of St. James, Jamaica, and their biogeographic implications, in: Woods, C.A. and Sergile, F.E. (eds), *Biogeography of the West Indies: Patterns and Perspectives*, 2nd edition, CRC Press, Boca Raton, FL, pp. 191–200.

Pregill, G.K., 1999, Eocene lizard from Jamaica, *Herpetologica* **55**: 157–161.

Prothero, D.R., 1989, Stepwise extinctions and climatic decline during the later Eocene and Oligocene, in: Donovan, S.K. (ed.), *Mass Extinctions: Processes and Evidence*, Belhaven Press, London, pp. 217–234.

Prothero, D.R., 1994, *The Eocene–Oligocene Transition: Paradise Lost*, Columbia University Press, New York, xviii + 291 pp.

Prothero, D.R. and Berggren, W.A. (eds), 1992, *Eocene–Oligocene Climatic and Biotic Evolution*, Princeton University Press, New Jersey, xiv + 568 pp.

Radinsky, L.B., 1969, The early evolution of the Perissodactyla, *Evolution* **23**: 308–328.

Ravenel, E., 1848, *Echinidae, Recent and fossil, of South Carolina*, Burgeo and James, Charleston, South Carolina, 4 pp.

Robinson, E., 1988a, Late Cretaceous and Early Tertiary sedimentary rocks of the Central Inlier, Jamaica, *Journal of the Geological Society of Jamaica* **24** (for 1987): 49–67.

Robinson, E., 1988b, Early Tertiary larger foraminifera and platform carbonates of the northern Caribbean, in: Barker, L. (ed.), *Transactions of the 11th Caribbean Geological Conference, Dover Beach, Barbados, July 20–26, 1986*, Energy and Natural Resources Division, National Petroleum Corporation, Barbados, pp. 5–12.

Robinson, E., 1994, Jamaica, in: Donovan, S.K. and Jackson, T.A. (eds), *Caribbean Geology: An Introduction*, University of the West Indies Publishers' Association, Kingston, pp. 111–127.

Robinson, E., 2004, Zoning the White Limestone Group of Jamaica using larger foraminiferal genera: a review and proposal, *Cainozoic Research*, **3** (for 2003): 39–75.

Robinson, E., and Jiang, M.-J., 1990, Calcareous nannofossils from western Portland and the ages and significance of the Richmond and Mooretown formations, *Journal of the Geological Society of Jamaica*, **27**: 17–25.

Robinson, E., and Mitchell, S.F., 1999, Upper Cretaceous to Oligocene stratigraphy in Jamaica, *Contributions to Geology, UWI, Mona* **4**: 1–47.

Roman, J., 1952, Sur les structures internes des Clypéastres, *Bulletin de la Société Géologique de France* (séries 6) **2**: 403–416.

Rowe, D.-A.C., and Stemann, T.A., 1999, Defining species boundaries in Eocene oligopygoid echinoids: a case study from the Swanswick Formation, Jamaica, *Geological Society of America Abstracts with Programs* **31** (7): A-470.

Sánchez Roig, M., 1926, Contribucion a la paleontologia Cubana: los equinodermos fôsiles de Cuba, *Boletin de Minas* **10**: 179 pp.

Savage, R.J.G., Domning, D.P., and Thewissen, J.G.M., 1994, Fossil Sirenia of the West Atlantic and Caribbean region. V. The most primitive known sirenian, *Prorastomus sirenoides* Owen, 1855, *Journal of Vertebrate Paleontology* **14**: 427–449.

Thewissen, J.G.M., Hussain, S.T., and Arif, M., 1994, Fossil evidence for the origin of aquatic locomotion in archaeocete whales, *Science* **263**: 210–212.

Williams, E.E., 1989, Old problems and new opportunities in West Indian biogeography, in: Woods, C.A. (ed.), *Biogeography of the West Indies: Past, Present and Future*, Sandhill Crane Press, Gainesville, FL, pp. 1–46.

Wright, R.M., and Robinson, E. (eds), 1993, *Biostratigraphy of Jamaica, Geological Society of America Memoir*, **182**: xi + 492 pp.

Chapter 9

# Long-Lived Lake Molluscs as Island Faunas: A Bivalve Perspective

FRANK P. WESSELINGH
Nationaal Natuurhistorisch Museum Naturalis, P.O. Box 9517, 2300 RA Leiden, The Netherlands and Biology Department, University of Turku, SF 20014 Turku, Finland, wesselingh@naturalis.nnm.nl



**Abstract** The evolutionary biology of long-lived lake bivalves, a group that has received comparatively little attention compared to, for example, gastropods and ostracods, is reviewed. Bivalve faunas of different (fossil and extant) long-lived lakes are characterised, and evolutionary aspects, such as the paucity of radiations of common cosmopolitan freshwater groups like sphaeriid and corbiculid clams are addressed. Special attention is given to the corbulid radiations in Miocene Lake Pebas of Western Amazonia. The ability to evolve morphological and ecological characteristics in long-lived lake biota that exceed the range of variation of their relatives in "ordinary" (non-long-lived lake) environments is discussed and termed



*W. Renema (ed.), Biogeography, Time, and Place: Distributions, Barriers, and Islands*, 275–314

"supralimital evolution". Several examples are discussed that show that such evolution is facilitated by the availability of empty biotopes after ecological crises in ecosystems at the onset of long-lived lake stages that are stable on ecological time scales. Implications for uniformitarian applications of long-lived lake biota are discussed.

## 1. Introduction

Long-lived lakes are considered to be (aquatic) islands of diversity and endemicity in continental interiors. Their faunas are almost completely isolated from freshwater faunas of surrounding (and inflowing) rivers, streams, and lakes, or in the case of saline lakes from seas and oceans. In many clades, extensive radiations have occurred, resulting in relatively high species numbers and a bewildering array of morphologies. Because it is assumed that long-lived lake faunas evolved *in situ* (within the lake), these organisms have been considered good model groups for the study of evolution. Only recently the notion of molluscan *in situ* evolution has been challenged (Hausdorf et al., 2003 for Baikalian rissooid gastropods; Wilson et al., 2004 for Tanganyikan cerithoidean gastropods; Anderson et al., 2006 for Lake Pebas pachydontine bivalves). Despite a pre-lake history for some groups (mainly at genus and family levels), *in situ* evolution appears to be the norm for most groups, especially at the species level. Well-studied groups from long-lived lakes that have yielded much insight into evolutionary processes include cichlid fish, gastropod molluscs, and gammariid and ostracod crustaceans. Only molluscs (gastropods and bivalves) and ostracods produce shells that fossilise easily, and have been studied widely from fossil long-lived lakes. Gastropods have received most attention because they are a very distinct and visible part of the faunas of these lakes and relatively easy to collect and study. Bivalve radiations are much less common. Substantial bivalve radiations occurred in only 7 of the approximately 20 known mollusc-bearing fossil and extant long-lived lakes. In recent years bivalve evolution in these lakes has received more attention (Müller and Magyar, 1992; Danukalova, 1996; Korniushin et al., 2000; Nevesskaja et al., 2001; Glaubrecht et al., 2003; Harzhauser and Mandic, 2004; Wesselingh, 2006b). In the lakes where they undergo radiations, bivalves can dominate mollusc assemblages in abundance, but in species numbers they are usually (much) less diverse than gastropods. Cosmopolitan freshwater groups, such as sphaeriids, corbiculids, and unionoids are comparatively rare elements of long-lived lake faunas. This paper reviews bivalve evolution in long-lived lakes. It covers both extant and fossil examples, and explores the ecological context of such radiations. In the first part the corbulid radiations of Miocene Lake Pebas (western Amazonia) are treated. The central issue to be addressed is the sheer dominance of Corbulidae in the Pebas system. The various bivalve radiations of other long-lived lakes are discussed in the second part. Finally, general aspects of bivalve evolution in long-lived lakes are addressed.

## 2. Corbulid Radiations in Miocene Lake Pebas (Western Amazonia)

### 2.1. Lake Pebas: an Introduction

Lake Pebas was a huge (>1 million $km^2$), shallow system of lakes and wetlands, that straddled the equator in western Amazonia between *c.* 9 and 18 Ma (Wesselingh et al., 2002). The lake experienced similar tropical rainforest-like climates as the region does today (Kaandorp et al., 2005, 2006). In the lake, profuse radiations of endemic ostracods (Whatley et al., 1996; Muñoz-Torres et al., 1998; 2006), and gastropods and bivalves (Nuttall, 1990; Wesselingh et al., 2002; Wesselingh, 2006a, b) occurred. In species numbers the cochliopid snails dominate the Pebas fauna (~84 species of a total of ~160: Wesselingh, 2006a), but in abundance the corbulid bivalves dominate (~67%). The bivalve fauna of the Pebas Formation is mainly composed of endemic corbulids, known as the Pachydontinae. Unionoids (pearly freshwater mussels) occur in low numbers with few species (often dominating marginal lacustrine to fluvial settings). Only three unionoid species (*Diplodon indianensis* and *D. longulus* which may belong to a single lineage, and *Anodontites batesi*) were adapted to living in the lake. Two species of dreissenid bivalves lived in the system, where they form the second-most abundant group (typically making up 1% or so of samples). Corbiculid and sphaeriid clams are very rare in the Pebas fauna, and only represented by one and two species respectively, which appear to be non-endemic fluvial species. Unionoids, dreissenids, and corbiculids are common in Miocene fluvial faunas in the adjacent Andean regions, and of slightly younger faunas from the Amazon region.

The pachydontine corbulids are represented by 6 genera and 23 species in the Pebas system (Nuttall, 1990; Wesselingh, 2006a, b), 15 of which belong to *Pachydon*. Recently, three of the six genera (*Pachydon*, *Ostomya*, and *Anticorbula*) have been found in Paleocene deposits of the USA (Anderson et al., 2006), but it appears that the Pebasian *Pachydon* species (with *Exallocorbula*) form a single or two endemic diversification events within the Pebas system. Corbulids are a common constituent of marginal marine and marine soft bottom communities worldwide. They are typically represented by few co-occurring species, with unspectacular, small shells. Various species are opportunists, and are capable of dealing with unfavourable conditions such as dysoxic settings. In contrast, the Pebasian corbulids form a remarkable morphological diverse assemblage (Figs. 1–7, 9–10). The dominance and variety of pachydontine bivalves has been attributed to isolated brackish (anomalohaline) conditions (Nuttall, 1990), which appeared to be corroborated by ostracod data (Whatley et al., 1998). The brackish nature of the Pebas system was supported by other data that also implied marine connections of the system (e.g., mangrove pollen (Hoorn, 1993); estuarine fish (Monsch, 1998); tidal sedimentary structures (Räsänen et al., 1998); and brackish ichnofossil assemblages (Gingras et al., 2002)). An open (marginal marine) system however does not comply with predominant endemic ostracods and mollusc faunas (typically over 90% in abundance (Muñoz Torres et al., 1998; Wesselingh et al.,

2002)). Furthermore, a brackish nature conflicts with the presence of unionoids and the absence of typical marginal marine taxa such as arc shells, oysters, mussels, and mangrove cerithioidean snails. Strontium and stable oxygen data point to freshwater settings in the Pebas system as well (Vonhof et al., 1998, 2003, Wesselingh et al., 2002). If the system were a truly freshwater system, why are unionoids and corbiculids so scarce? On the other hand, if the system were brackish, why then are cosmopolitan marginal marine groups such as arc shells, oysters, mussels, and mangrove cerithoideans lacking (that were present in South American coastal regions at the time)? In order to understand the dominance of the pachydontine bivalves in the Pebas system, their living modes need to be understood. Below, nine species present in the middle and upper part of the Pebas Formation (*Crassoretitriletes* and *Grimsdalea* zones of Hoorn (1993): Middle to early Late Miocene) are shortly diagnosed, followed by an interpretation of morphology in terms of adaptation to specific ecological conditions. Nuttall (1990) and Wesselingh (2006a) provide the systematic background for this discussion.

## 2.2. Pachydontine Species

*Pachydon obliquus* (Fig. 1a–i) has some unusual shell characteristics for a corbulid. It is a globose small to medium-sized *Pachydon* (length (L) 8–17 mm) that has strong inaequivalve and inaequilateral shells, with highly incurved umbones (Fig. 1e–g). The hinge is extremely robust. The pallial line is located far from the shell's margin, although a deep cavity exists behind the hinge plate, apparently to compensate for the loss of living space near the shell edge (Nuttall, 1990). The shell is markedly thickened near the umbones and at the anterior margin. The anterior adductor scars are deeply embedded and very prominent. The shell's rim outside the pallial line is much thicker than inside, which appears to be corroded. A pallial sinus is very shallow or lacking. Together with the massive and complex hinge, a furrow in the right valve (RV) in which the left valve (LV) fits, forms a system that tightly interlocks the valves. Estimates from growth lines (that are often poorly and irregularly developed) suggest ages between 4 and 10 years for adult specimens. *Pachydon obliquus* is the most common species of the Pebas fauna, except for fluviolacustrine settings it was present in all lacustrine biotopes (Wesselingh et al., 2002). The species is dominant in organic-rich clay intervals presumably representing dysoxic fluid mud depositional settings. The development of thick shells seems counterintuitive for a species living in such unsolidified environments. The shell's point of maximum density, located near the posterior margin towards the umbo, provides a clue. It suggests that the umbones anchored the shell into the substrate. Valve thickness, valve overlap, deep attachment of the mantle margins onto the shell and the general globose nature of these shells are features that might have deterred predation (Fig. 1h,i).

A small species is *P. trigonalis* (Fig. 2a–h) whose shell length typically ranges between 6 and 10 mm. This triangular species has a solid shell with a truncate posterior margin. A broad postero-ventral ridge leans extra strength to the shell. The pallial sinus is short, indicating this species to be a shallow burrower.

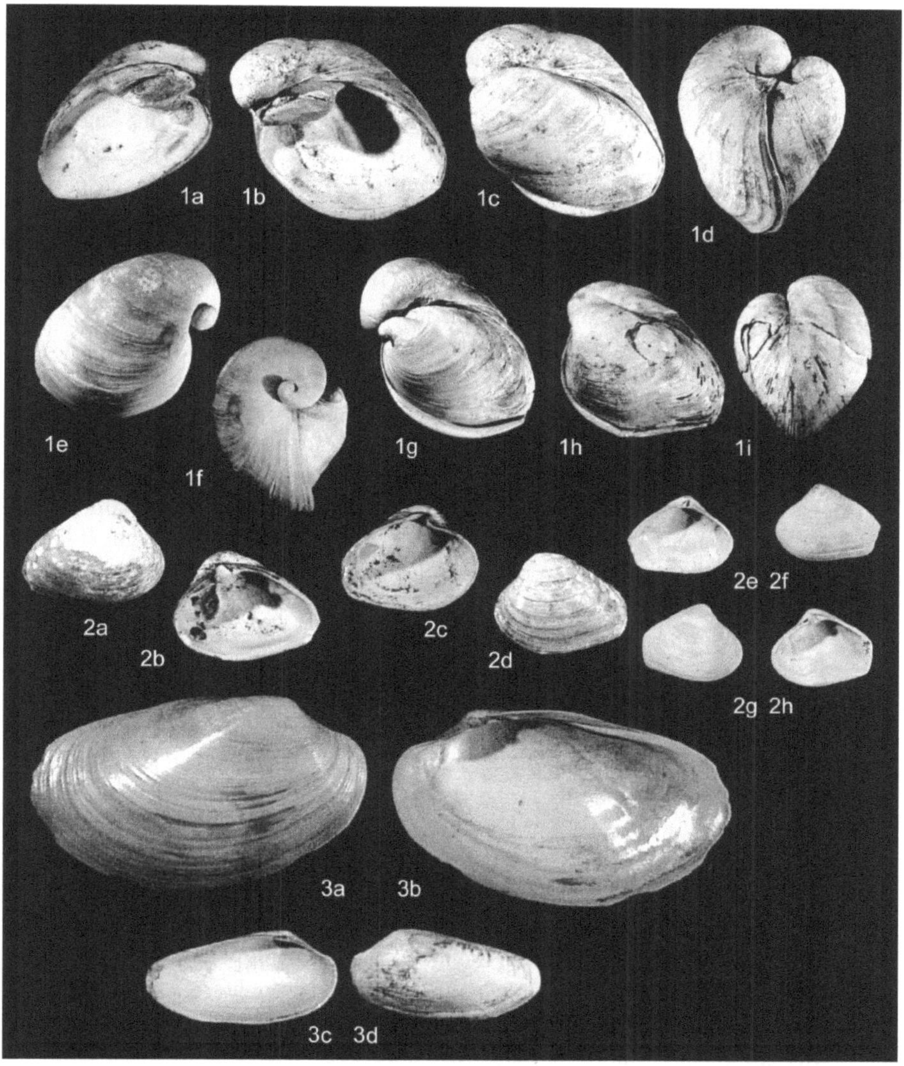


Figs. 1–3 *Pachydon* from the Miocene Pebas Formation of western Amazonia. Figure 1. *Pachydon obliquus*. Figures 1a–d: RGM 456192; Santa Rosa de Pichana, Loreto, Peru; L(ength) 16 mm. Figures 1e–g: RGM 456193; Puerto Almendras, Loreto, Peru; L 13 mm. This specimen is strongly coiled (1.5 revolution). Figures 1h, i: RGM 456194; Iquitos, Puerto Ganso-Azul, Loreto, Peru; L 10 mm. The specimen survived a cracking-type predator attack, and continued growth after. If the specimen were flatter, the cracking effort probably would have succeeded. Figure 2: *Pachydon trigonalis*. Figures 2a–d: RGM 456195; Puerto Nariño, Amazonas, Colombia; L RV 9.5 mm. Specimen from marginal (riverine influenced) lacustrine assemblage. Figures 2e–h: RGM 456196; Los Chorros, Amazonas, Colombia; L RV 7 mm. Specimen from lacustrine assemblage, notably smaller and thinner than specimens from marginal lacustrine settings. Figure 3: *Pachydon amazonensis*. Figures 3a, b: RGM 456197; Santa Elena, Loreto, Peru; L 14.5 mm. Large and thin-shelled specimen. Figures 3c, d: RGM 456198; Santo Tomas Amazonas, Loreto, Peru. L 6.5 mm.

*P. trigonalis* replaces *P. obliquus* within the upper part (in the upper *Grimsdalea* zone of Hoorn, 1993) of the Pebas Formation as the dominant species. The turnover coincides with an interval of marine influence. *P. trigonalis* occupied most parts of the ecosystem, ranging from fluvially influenced marginal lacustrine settings to fluid mud bottoms of the lake proper. In lacustrine clayish intervals, specimens tend to be small and thin-shelled (L adult ~6–7 mm): in more coarse-grained marginal lacustrine settings with fluvial influences, shells are larger (L ~10 mm), thicker, and more inflated (compare Figs. 2a–d and 2e–h). Suggested ages from growth line counts average 5.5 (standard deviation (SD) 1.4; $n = 31$) and 5.6 (SD 1.4; $n = 20$) years in a "lacustrine" and a "fluviolacustrine" population respectively. These data indicate that smaller size in the former is the result of slower growth, not of shorter life spans.

*Pachydon amazonensis* (Fig. 3a–d) is a medium-sized (L 7–20 mm), elongate, oval shaped *Pachydon* species. It is aequivalve, comparatively thin-shelled and often markedly polished. From growth lines counts, ages 4–9 years are estimated. Large variations in shell outline and size exist. Smaller shells can be either elongate ovate or elongate tapering, with a slightly truncate posterior margin (Fig. 3c, d). The umbo is located between one fifth and one eight from the anterior margin. Smaller ovate-shelled populations tend to have rather convex shells (in cross section), specimens from populations with very large shells are comparatively flatter. The cardinal tooth is rather small in comparison with other *Pachydon* species. The adductor scars and the pallial line are very shallow. The pallial sinus is truncate, suggesting shallow burrowing behaviour. A furrow may develop in the RV just inside the ventral edge, marking the location of the LV margin. The pedal retractor scars are small, but well defined and deeply embedded. On the interior surface outside the pallial line, vague and broad radial striae occur, best developed on the posterior margin. In many specimens the largest height is attained in the posterior half, thereby counteracting the umbonal regions as the centre of gravity. From the shell itself no information about living position can be interpreted, other than shallow burrowing behaviour. Given the abundance of shell predators in the Pebas Formation (see also below), the thin shell is a remarkable feature, especially when taking into account that this species only rarely shows signs of (failed) predation. One explanation could be that every attack resulted into the destruction of the whole shell. Another possibility is shown by the Chinese unionoid *Soleinaia oleivora* (Savazzi and Peiyi, 1992) that combines inefficient burrowing and a distasteful sulphur-laden animal with a shell that forms a rather poor defence against predation (like the shell of *P. amazonensis*).

The largest of the pebasian corbulids is *P. erectus* (Fig. 4a–e) that attains sizes of L 55 mm, H 50 mm. This species has a solid but thin shell for its size. It is a tumid, aequivalve, almost triangular species, with a stout anterior ridge that bounds a wide flat to concave anterior platform on which paired specimens can balance. The posterior margin of paired specimens contains a slit-like gape with reinforced margins. The shells are somewhat thicker on the anterior side, where the adductor scars and pallial line are more deeply embedded. The pallial sinus is lacking or developed as a very irregular and unimpressive indentation, suggesting shallow burial. Oxygen isotope profiles along growth increments (Kaandorp et al., 2006) indicated an age of 5 for an intermediate-sized specimen (L ~30 mm). The largest specimen (L 55 mm) possessed approximately 14 rings that can be interpreted as years; smaller species (L 25–30 mm) have 5–7 possible year rings, which agrees with the observed isotope age-estimate. Many specimens did not yield growth lines regular enough to be interpreted as year-rings. Shell density distribution indicates an anterior side down orientation. The broad, concave anterior margin may have enhanced an epifaunal or semi-infaunal living mode on soft sediments, ensuring that the species did not sink, leaving the posterior margins with the siphons closely enough near or at the sediment surface (iceberg strategy; Thayer, 1975).

*P. telliniformis* strikingly resembles in outline a number of *Tellina* species (Fig. 5a–e). The intermediate-sized (L 16 mm) flat and rather thick-shelled species is somewhat inaequivalve. The RV is slightly more convex and larger than the LV, but almost no valve overlap exists apart from the ventral most point. The shell has a posterior margin that is slightly rostrate. The postero-ventral margin of the LV is slightly concave, giving that valve a rostrate outline. The posterior dorsal margin is quite robust. The commissural plane is wavy and the posterior margin is bent (Fig. 5e). From the posterior adductor, the pallial line runs without a sinus towards the central part of the ventral margin leaving a substantial area posteriorly outside the mantle edge attachment. The posterior commissural plane is bend towards the RV, very similar as in various deposit-feeding western Atlantic tellinid species that burrow until they lay horizontal to the sediment surface (Stanley, 1970). Although it is likely that *P. telliniformis* was a (sub)-horizontal burrower, nothing is known about its feeding ecology. A subhorizontal living mode produced a snowshoe effect (Thayer, 1975), enhancing its stability in unstable sediments. Specimens from marginal lacustrine sandy depositional environments tend to be somewhat larger, thicker and slightly more inflate then those from lacustrine muddy environments. Estimated ages for this species are between 6 and 7 years (based on growth line counts).

*P. ledaeformis* (Fig. 6a–h) is an aequivalve species that has very inaequilateral, elongate rostrate shells, resembling elongate crassatelloid shells. The species is small to intermediate-sized (L 8–10 mm) and flat. The umbo is located not far from the robust and rounded anterior margin. The posterior end of the shell is tapering. A well-defined ridge that runs from the umbo to the postero-ventral boundary delimits a rostrum-like posterior slope. Shells from fine-grained lacustrine environments (Fig. 6a–d) tend to be thinner than those from more coarse grained (sandy) nearshore environments (Fig. 6e–h), but no obvious differences in size were seen. Dentition is not very strong. The pallial sinus is truncate. Growth line counts suggest ages between 4 and 6 years. There is no indication of the living position based on the shell, other than that the rostrum may have served as a protection for siphons, possibly indicating a beak-down position. The truncate pallial sinus suggests a shallow burial depth.

*Exallocorbula dispar* (Fig. 7a–j) is an intermediately sized pachydontine (L 14–25 mm) with a remarkable outline. It has a strongly convex bilobed RV and a concave-flat LV. The shell has a rounded anterior and a rostrate posterior. The RV has a strong and well-defined overlap with the LV along the posterior two thirds

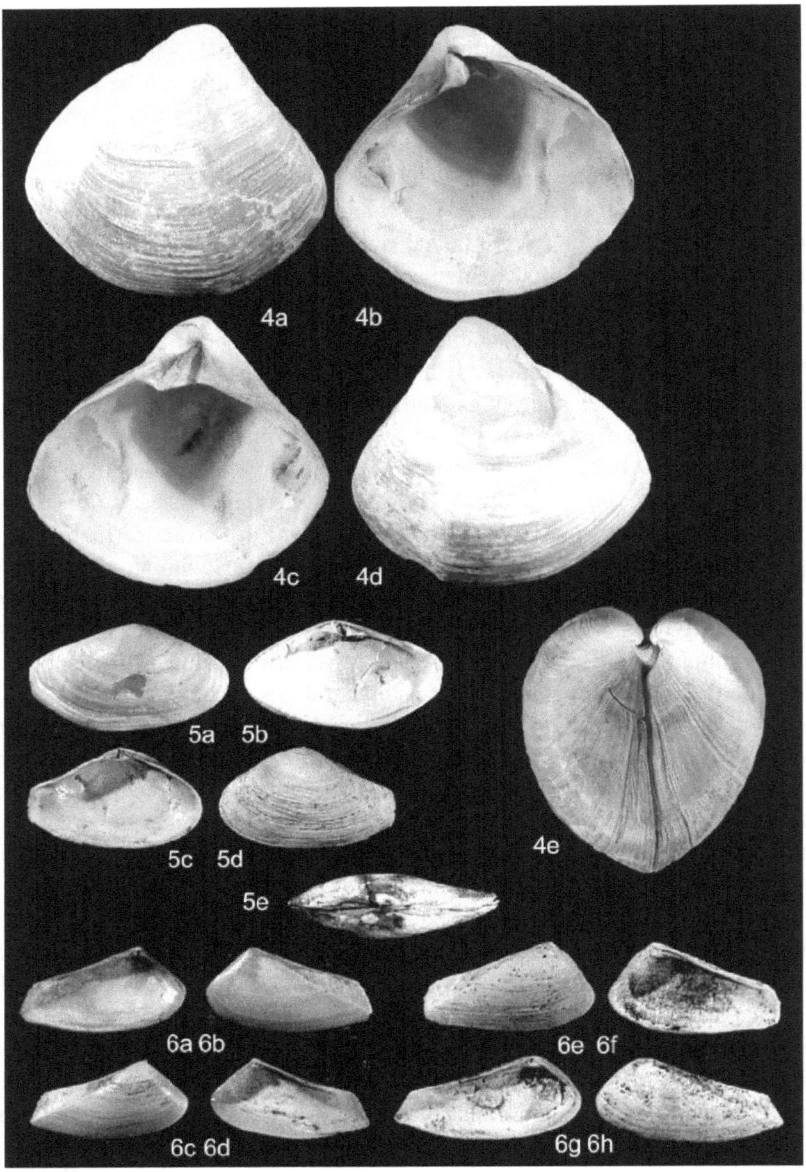


Figs. 4–6 *Pachydon* from the Miocene Pebas Formation of western Amazonia. **Figures 4a-e.** *Pachydon erectus*. RGM 456199. Santa Rosa de Pichana, Peru. L 50 mm. Figure 4e shows the concave, wide posterior platform that presumably enhanced stability in semi-fluid substrates. **Figures 5a-e**. *Pachydon telliniformis*. RGM 456200. Nuevo Horizonte, Loreto, Peru. L RV (5a, b) 13 mm. Notice the bend posterior tip in figure 5e. **Figure 6**. *Pachydon ledaeformis*. **Figures 6a-d**. RGM 456201. Santa Elena, Loreto, Peru. L RV (6a, b) 8 mm. **Figures 6e-h**. RGM 456202. Puerto Caiman–Caqueta, Amazonas, Colombia. L LV (6g, h) 9 mm.

of the ventral margin (Fig. 7i, j). Typical adult shells reached ages between 5 and 6 years (based on growth line counts). The shells are usually not extremely thick, but in some populations extremely thickened shells occur next to normal shells.

These heavily calcified shells occur in random size classes, so apparently they are not a geriatric feature. These shells may provide an important clue about the living mode of *Exallocorbula*, and possibly of other pachydontines as well. The LV of shells illustrated in Fig. 7c–f show a threefold scarring typical of many of these heavily calcified specimens. The posterior adductor is more deeply embedded than the anterior adductor, and both yield a typical brain-like surface structure. It is, however, a circular to elongate kidney shaped third scar that is located at the center of the shell just above the pallial line that is most striking. This third scar can be very large (almost one third of the shell L) and bears very well developed radial striae. Its position varies somewhat in different shells. The scar bears a striking resemblance to scars seen in marine lucinid bivalves (Taylor and Glover, 2000). In this latter group the anterior adductor is greatly enlarged (located in the same area as the third scar of *Exallocorbula*, the latter being detached from the mantle line) and very similarly striated (Fig. 8). The anterior adductor muscle facilitates chemosymbiosis in the lucinid bivalves in a variety of ways (Taylor and Glover, 2000). In several of the lucinids, a linear impression from the scar towards the apex occurs in the shells interior, which is an impression of an enlarged pallial blood vessel that transports large quantities of blood between the heart and the bacteria-laden gills. Very similar imprints were seen in some of the studied *Exallocorbula* specimens. This is no certain proof of a chemosymbiotic lifestyle, but the combination of features (including the capability of *Exallocorbula* to live in dysoxic mud) makes it a feasible explanation. The recurring central location of the third scar makes an origin as a repair mark from predatory attack of infestation unlikely. *Exallocorbula* occurs in most shallow lacustrine to offshore lacustrine facies in the Pebas Formation, though it is more common in the latter. In coarser grained substrates it is larger and thicker than specimens from fine-grained substrates. The concavo-convex architecture of the shells resembles a ship form, also seen in for example Jurassic gryphaeid bivalves (Savazzi, 1999). In the latter, the form is an adaptation to floating in unstable mud.

Unusual in Pebasian Pachydontines is the very large anterior margin of *Ostomya myiformis* (Fig. 9a–c). This intermediately sized (L 23 mm) flat shell has a slightly trapezoid outline, with a broadly rounded anterior margin. The posterior ventral margin is slightly elevated compared to the anterior dorsal margin. On the exterior surface, the shells are riddled with rather irregular concentric growth lines overlain by irregular oblique ridges that indicate the presence of a very thick former periostracum. This is a rare species that is typically found in non-agitated lacustrine bottom assemblages (Wesselingh et al., 2002). No age could be estimated from the material available. The pallial line is broad and truncate posteriorly. Despite being flat, the shell is robust, and it is thickened in the posterior region, as shown by more deeply embedded adductor scars. The posterior adductor scar has strongly calcified striae. No paired specimens have been found, but the loose valves indicate that the commissural plane was more or less flat as well as the presence of a very distinct anterior gape. In the LV, a ridge and furrow are developed near the ventral margin, presumably to fix the RV. I am uncertain of the living position of this species. The combination of aequivalve shells with the point of maximum density in the posterior half suggests a (semi) vertical living position, but the flatness of the shell would make it cut like a knife into soft substrates.

*Concentricavalva concentrica* (Fig. 10a–g) combines a massive flat hinge platform with a paper-thin concentrically ribbed shell, superficially resembling Mesozoic *Posidonia* (i.e., *Bositra*) species. This intermediately sized species (L 21 mm) has several unusual characteristics for a pachydontine shell: the shell is extremely thin, the hinge plate is very broad, the cardinal teeth are completely reduced and broad lateral grooves exist. A prominent lanceolate resilifer (an attachment structure for the internal ligament) that is embedded within the hinge platform dominates the hinges of LV and RV alike. Above the resilifer, a deep (RV) or shallow (LV) depression is present that bounds a prominent nymph above. This nymph (an attachment structure for the external ligament) is rather strong and auriculate in the LV; it is thin and wrinkled in the RV. Apart for more or less regular concentric folds, the shell's exterior is in places corrugate, and bears irregular/oblique wrinkles (especially prominent on the margins) that represent a formerly thick periostracum. Paired specimens seen *in situ* (but not collected) contained almost completely sealed margins, although it cannot be out ruled that post-depositional compaction is in part responsible. The shell has a subquadrate outline, with the pointed, triangular umbo located at approximately one third from the anterior margin. No estimate of the life span of individual specimens could be made on the basis of the available material. I am uncertain as to the living position and characteristics of this species. Its overall resemblance with jurassic *Posidonia* (*Bositra*), which was a benthic inhabitant of low-energetic oxygen-stressed sea bottoms (Oschmann, 1994), may provide some comparison for the possible living conditions of *C. concentrica*. The very thin shell would not have been a match to any predator. The very flat nature of the outer half of this shell (when paired) may have proved efficient in surviving possible breakage, but more likely *C. concentrica* lived in oxygen-stressed fluid muddy environments that were hostile to potential predators. The very flat nature of this species may also have enhanced oxygen absorption (Oschmann, 1994). The comparatively robust hinge-plate of *C. concentrica* is unlike that of *Posidonia* (*Bositra*) and may have counteracted a floating living position.

## 2.3. Possible Ecological Adaptations

What can explain the sheer abundance of pachydontine bivalves in Lake Pebas? If the system was brackish, why are common marginal marine taxa lacking? On the

Figs. 7–10 (continued) Eocene (Lutetian), Falun de Bois Gouët; L 59 mm. Note enlarged and striate posterior adductor scar. Figure 9: *Ostomya myiformis*. Figures 9a, b: RGM 456206; Beiruth, Loreto, Peru; L 23 mm. Figure 9c: RGM 456207; Los Chorros, Amazonas, Colombia; L 21 mm. Figure 10: *Concentricavalva concentrica*. Figures 10a–c: RGM 456208, fragment RV; Nuevo Horizonte, Loreto, Peru; L. 8 mm. Figures 10d–f: fragment LV. RGM 456209; Same locality; L 10 mm. Figure 10 g: Damaged interior of LV RGM 456210; Los Chorros, Amazonas, Colombia; L 21 mm.

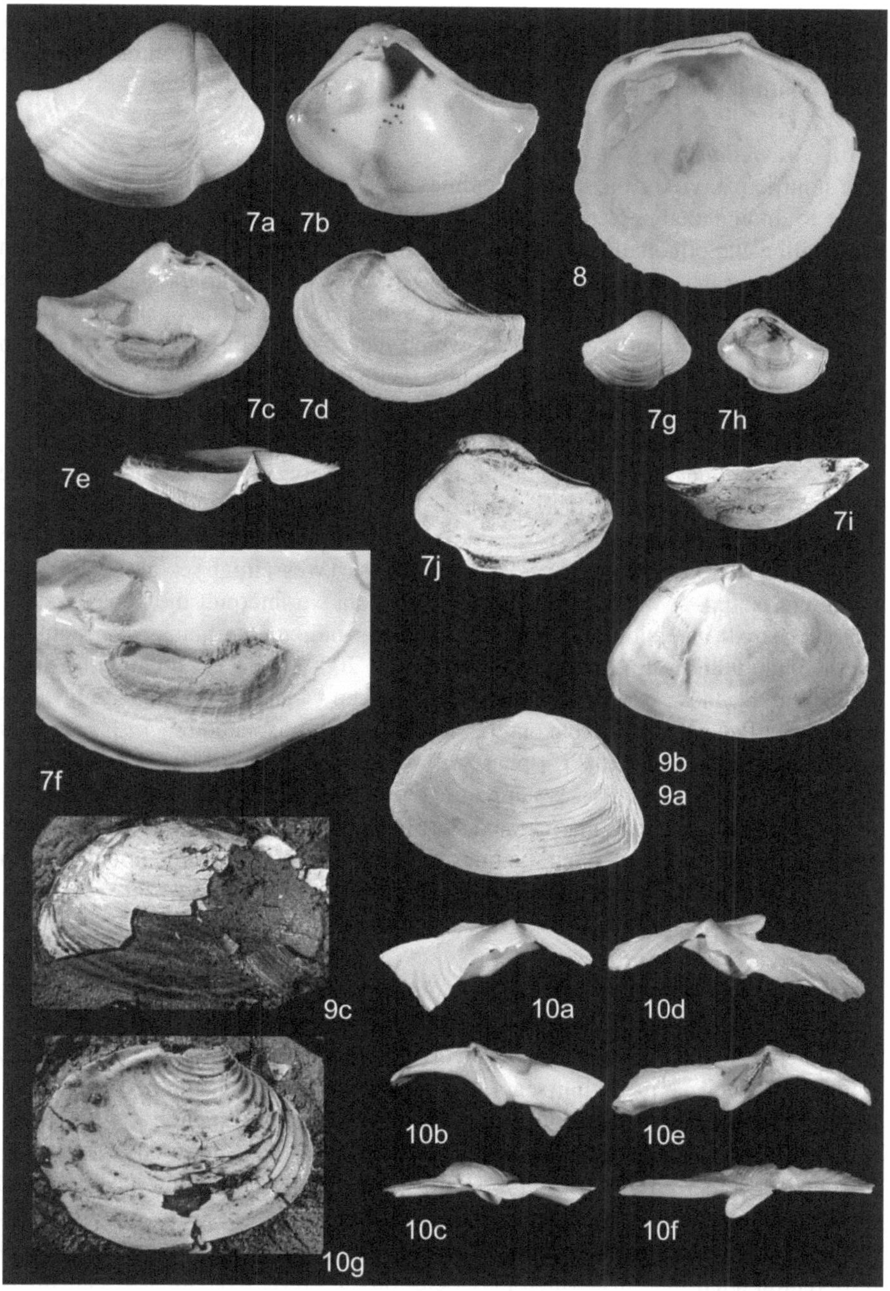


Figs. 7–10 Pachydontinae from the Miocene Pebas Formation. Figure 7: *Exallocorbula dispar*. Figures 7a–h: RGM 456203; Nuevo Horizonte, Loreto, Peru; L LV in Figures 7a, b 14 mm. Note a third filamentous scar in the interior of specimens in Figs 7c, f, and h. These specimens have thickened shells. Figures 7 i, j: RGM 456204; Santa Sofia, Amazonas, Colombia; L 13 mm. Figure 8: *Lucina menardi*; RGM 456205; Le Bois Gouët, Loire Atlantique, France.

other hand, if the system was fresh, why are corbiculids, sphaeriids (and to a lesser extent unionoids) so rare? Below, four possible reasons for the dominance of the Pachydontinae are explored.

*2.3.1. Anomalohaline Conditions*

The conflicting views on salinity regimes in the Pebas system might appear to point to anomalohaline settings. Ichnofossil assemblages, for example, indicated mesohaline and oligohaline conditions as well as water stratification to be common in the Pebas system (Gingras et al., 2002). However, as a group the Pachydontinae appear very well adapted to freshwater settings, despite the generally (marginal) marine nature of the family to which they belong. In the North American Paleogene, pachydontine clams co-occurred with strictly freshwater viviparoid and unionoid species (Anderson et al., 2006). The single extant pachydontine species, *Anticorbula fluviatilis*, is known from the upper reaches of the Guyana estuaries, possibly experiencing freshwater-mesohaline salinity variations, but the same species has been recorded from Amazon floodplain lakes in central Brazil, possibly even up to Peruvian Amazonia, at least 2,500 km and possibly even 4,000 km away from marine influence. Based on faunal (Wesselingh et al., 2002) and strontium isotope (Vonhof et al., 1998, 2003) data we interpret the Pebas system, apart for small incursion levels actually yielding oligohaline faunas (Vermeij and Wesselingh, 2002; van Aartsen and Wesselingh, 2000) to be essentially a freshwater system. Negative $\delta^{18}O$ values (typically between –4 and –8‰) found in bivalves throughout Pebasian strata rule out the possibility of an isolated brackish sea under arid climatic settings like the Caspian Sea (Wesselingh et al., 2002; Kaandorp et al., 2005). The predominant freshwater nature of the Pebas system explains the absence of widespread marginal marine taxa such as oysters, mussels, and mangrove cerithioideans, but leaves open the question of the Pachydontinae dominance over ordinary freshwater mussels such as unionoids and corbiculids.

*2.3.2. High Predation Pressure*

Several morphological characteristics of the Pachydontinae are interpretable as adaptations against very high (more or less marine-like) predation intensity. Lake Pebas abounded with molluscivoran predators as shown by their fossil remains, as well as predation scars and common sharply edged shell fragments. Major molluscivoran groups identified from the Pebas Formation are sciaeniid, myliobatiform, and serrasalmine fish (Monsch, 1998) and decapod crustaceans. Presumably other molluscivoran predators, such as waterfowl, crocodiles and otters were present as well. Thick shells commonly observed in the Pebasian pachydontines can deter predation. Several *Pachydon* species have strong postero-ventral shell margins, which may deter predation (*P. trigonalis, P. cuneatus*). The posterodorsal keel seen in several species (e.g., *P. carinatus, P. ledaeformis*) contribute to stronger shells. However, in several species (*P. obliquus, P. carinatus*) selective shell-thickening was found at the anterior margin, the area most difficult to reach for predators. These thickenings cannot be explained as antipredatory adaptations. Predation scars are often observed on the posterior parts of pachydontine shells, indicating a predominant umbo-down living position. The extreme convex nature of *P. obliquus* may also have helped deter cracking (Fig. 1h, i). Other potential antipredatory traits in pachydontine bivalves include (1) thickened ventral margins (common), (2) valve overlap (common), (3) the development of a rostrum to protect siphons (seen in varying extents in several species, with the most extreme case *P. ledaeformis*), and (4) the vacation of the interior marginal areas as indicated by the deep location of the pallial line in most pachydontines. However, these features may have evolved for reasons other than as a response to intense predation. Finally, a suit of adaptations for hermetic sealing (complex interlocking hinge, inaequivalve condition with LV fitting into a groove in RV and the seal being improved by internal outcrop of the main conchiolin layer) may also have been beneficial in deterring recognition from predators.

### *2.3.3. Adaptation to Living in Soft Substrate*

Corbulids have very short siphons reflecting a shallow burrowing life style. A non-streamlined, inflated morphology typical of some species limits burrowing capacities. As a result, these species are highly prone to predation and dislodgement by currents and bioturbation (Lewy and Samtleben, 1979). Much of the Pebas pachydontine corbulids are found in organic-rich clayish silts and clays that during deposition presumably formed fluid bottoms. Several works have dealt with adaptations of bivalves and other organisms to living in soft substrates (Stanley, 1970; Thayer, 1975; Savazzi, 1999). Various characters seen in the pachydontine bivalves indicate them to be very well adapted to fluid substrates, despite their in general large and heavy shells. The selective thickening of the anterior margin seen in, e.g., *P. obliquus* and *P. carinatus* stabilized these shells in the soft substrate umbo-down ensuring that the siphons were close to the surface. A broad concave anterior platform on which *P. erectus* could balance provided a floating surface that may have circumvented sinking. Reduction of shell size in muddy lacustrine bottoms (e.g., *P. trigonalis, P. obliquus*) enhanced floating capacity (Stanley, 1970). Age estimates from *P. trigonalis* populations suggests that smaller size was accomplished by slower growth rates. Oblique/lateral burrowing in *P. telliniformis* created a snowshoe effect (Thayer, 1975). The gryphaeid cup form of *E. dispar* also possibly circumvented sinking. Large valve overlap along the ventral–anterior half of *Exallocorbula* may have served as a protection against mud entering the mantel cavity when the shell was opened. Extreme thinning and flattening of shells did reduce effective weight in *C. concentrica*, although the massive hinge plate is not well understood. The presence but low abundance of epibyssate *Mytilopsis* shows that hard substrates (possibly in the form of dead shells) were available in the Pebas system, but as a whole the Pachydontinae appear to have developed a wide array of strategies to cope with fluid substrate types.

### *2.3.4. Coping with Dysoxia*

The prevalence of blue organic-rich clays, as well as the common preservation of organic tissue such as periostracum, points to common dysoxia or anoxia within the bottoms of Lake Pebas. Several marine corbulids are known to tolerate episodic low oxygen levels (Lewy and Samtleben, 1979). These species seal their valves and await favourable conditions, or can lower their metabolic rates, diminishing oxygen consumption. Several morphological characteristics in Pebasian pachydontines indicate the possibility of very tight closure, including the furrow in the RV margin to receive the LV in various pachydontine species and the complex interlocking hinge structures (clearly developed in, e.g., *P. obliquus*). Possible cessation or lowering of metabolic rates might explain the generally rather irregular growth line successions, commonly seen in these bivalves (and often seriously complicating age estimates). *E. dispar* and possibly *P. amazonensis* may have developed a chemosymbiotic lifestyle. A similar chemosymbiotic lifestyle has been indicated for *Congeria* species in Miocene long-lived lake Pannon (Harzhauser and Mandic, 2004), and is suspected in Neogene unionoids from East African lakes (van Damme, personal communication, 2005). Possibly (facultative) chemosymbiosis was more widespread in the pachydontinae. Several of the species discussed above have a general morphology that indicate inefficient burrowing behaviour. This implies that *Pachydon* in general lived in little agitated environments, both physical and biological. The presence of a very thin oxygenated zone at best in the bottom does explain a paucity of deeper burrowing benthos that could dislodge these shells.

In summary, the lack of marginal marine bivalves in the Pebas is explained by the predominant freshwater nature of the system. The abundance of Pachydontinae and paucity of ordinary freshwater groups is explained by the adaptation of the former to oxygen stress and soft (fluid) substrates and possibly a superior adaptation to high predation pressure. Corbiculid clams, widespread in Miocene South American rivers, are notably poor in dealing with lowered oxygen levels, explaining their paucity in the Pebas fauna. Several unionoids can tolerate lowered oxygen settings. Savazzi and Peiyi (1992) provide examples of morphological adaptations to dysoxia, as well as likely chemosymbiosis in some Chinese unionoids. Possibly a combination with very high predation pressure and soft substrates prevented them from becoming very abundant. Sphaeriids might have floated in liquid mud, as a result of their small size, but are very easy targets for predators. The predominance of Pachydontinae in the Pebas system is therefore likely the result of their successful ability to deal with soft substrates, oxygen stress and possibly high predation intensities at the same time.

## 3. An Overview of Long-Lived Lake Bivalve Radiations

### 3.1. Introduction

Below, fossil and extant long-lived lakes that harbour bivalve radiations are discussed. The extent of radiations varies immensely from few species-poor unionoid flocks in Lakes Biwa and Turkana to tens to hundreds of endemic cardiid and dreissenid species in Lake Pannon.

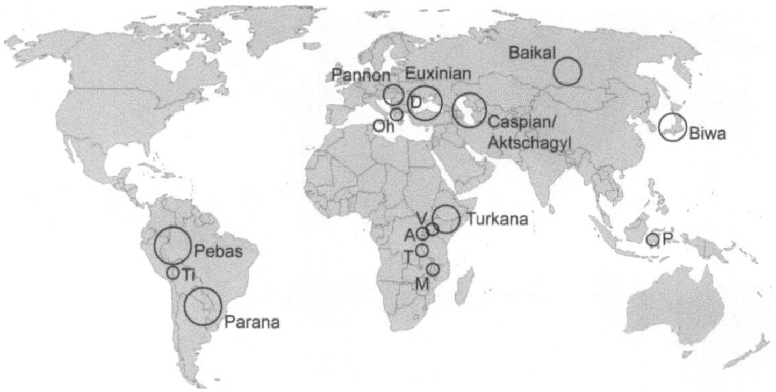


Fig. 11 Long-lived lake systems discussed below. A = Albert-Edward, D = Dacian, Oh = Ohrid, P = Pozo/Malili, T = Tanganyika, Ti = Titicaca, V = Victoria.

## 3.2. Lake Parana (Late Permian, Southern Brazil, Northern Argentina, Uruguay, and Paraguay)

Late Permian Lake Parana (*c.* 248–256 Ma) is the oldest known molluscan-bearing long-lived lake. The lake and its bivalve fauna have been reviewed by Runnegar and Newell (1971) and have recently gained renewed systematic and ecological interest (see, e.g., Simões et al., 1997, 1998, 2000; Mello, 1999). Lake Parana was located in the southern interior of Pangea. During the Permian, an epicontinental sea with connections to the Tethys Ocean occupied the Parana Basin. The basal (large-scale transgressive) part of the Parana Basin fill (the Tubarao Group) is made up of marine formations with diverse marine biota. At the maximum flooding stage or just after, the basin became isolated (Simões et al., 1997). During this stage, extensive organic clays of the Iratai Formation were deposited, indicating widespread anoxic conditions in the basin. Only in marginal areas, where oolitic carbonates were deposited, fauna could survive. It is likely that this period represented an ecological crisis, where many taxa went extinct, and others went through population bottlenecks, although such bottlenecks occurred also in earlier times. With the end of the Iratai times, a large scale regression set in, deposits of which are grouped into the Pasa Dois Group. It is this Pasa Dois interval, which is indicated as Lake Parana here. Smaller scale transgressions did occur in the lake, that existed less then 8 Ma and ended with the establishment of terrestrial (fluvial, eolian) conditions. Lake Parana was large (>1.5 million $km^2$), underwent various climatic changes (humid to arid), and harboured diverse (and periodically variable) salinity regimes (hypersaline, brackish, and freshwater). The lake yielded a diverse (taxonomically as well as morphologically) endemic bivalve fauna (Figs. 12–14). Interestingly, it lacked endemic gastropods, a unique feature among long-lived lakes. Permian freshwater gastropods were, however, comparatively rare and undifferentiated (unlike Cenozoic faunas).

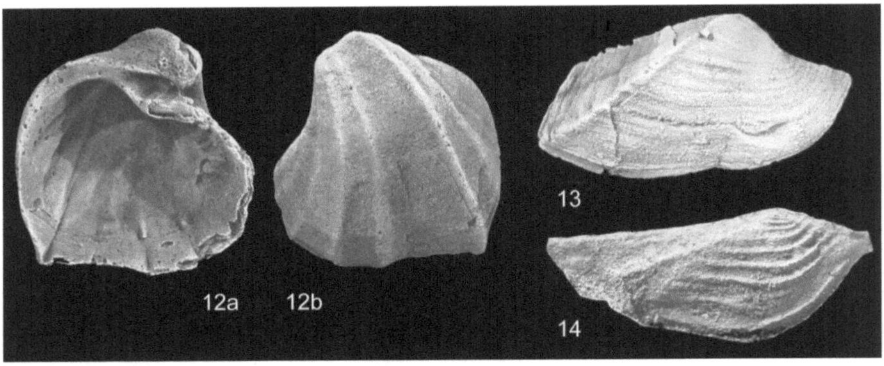


Figs. 12–14 Permian bivalves from the Estrada Nova Formation of the Parana basin. (Figures reproduced with permission from the author and editors of the *Bulletin of the American Museum of Natural History*.) Figure 12: *Ferrazia cardinalis*; Morro Azul farm, Rio Claro, Sao Paulo, Brazil; L 26 mm (Runnegar and Newell, 1971, Figs 11a, b). Interior (12a) and exterior (12b) views. This species is thick-shelled (unusual for a Permian bivalve) and has strongly developed ribs. It superficially resembles some of the Paratethyan cardiids (see Fig. 20). Figure 13: *Jacquesia elongata*; Manoel Pereira Primo property, Sao Paulo, Brazil. L 41 mm (Runnegar and Newell, 1971, Fig. 16h). One of the more elongate Paranan mesodesmiids. Figure 14: *Leinzia similes*. Road cut, 109.7 km from Prudentópolis to Guarapuana, Paraná, Brazil; L 18 mm (Runnegar and Newell, 1971, Figure 24j). Runnegar and Newell (1971) suggested the anterior notch to be an adaptation to living in dysoxic mud.

The different papers about the Paranian fossils are not clear on the exact number of species present (owing to the poor preservation of some earlier described taxa, as well as the unpublished nature of some important systematic studies), thus numbers below are indications only. Megadesmidae (~22 species) and Veneroidea (~8 species) dominate the Parana fauna. The Megadesmidae contain endemic species attributed to the Megadesminae (at least 3 species) and to the subfamily Plesiocyprinellinae (at least 13 species: Simões et al., 1997, 1998; Mello, 1999). The latter subfamily evolved within Lake Parana, and was endemic to it. Other taxa have uncertain affinities or are relatively rare, and include a mytiliform species. The total fauna comprises approximately 39 species.

Only a single immigrant (*Kidokia* cf. *stocklei*) has been reported in the Parana fauna, as well as two taxa appearing in the final stages with a potential wider freshwater/terrestrial distribution (Simões et al., 1998). The bivalve fauna is almost entirely composed of infaunal, burrowing filter feeding taxa. Epibyssate taxa are few (Simões et al., 1998). According to these authors bivalve speciation was allopatric, and the result of environmental restriction and interruption of larval dispersion. The authors are not specific whether geographic isolation was from the marine realm or within the lake.

Simões et al. (1998) described four lacustrine assemblages. Among the burrowers, the morphology of the elongate *Leinzia*, with a marked shell-extension at its posterior margin (Fig. 14), as well as its *in situ* occurrence in fine-grained organic-rich sediments suggests adaptations to dysoxic settings. The four assemblages were deposited in subtidal to intertidal settings, exposed to varying salinity regimes. Species appeared part of the time spatially restricted to areas within the lake, but at least one transgressive event caused bivalve species to occupy the whole lake. The uppermost lake assemblage represents a decline in species numbers. This is followed by a fifth assemblage, that represents the final basin fill phase, containing fluvial/terrestrial taxa and possibly two endemic megadesmiid taxa that adapted to fluvial settings (Simões et al., 1998). The Parana bivalve fauna is unusual that it comprises many thick-shelled species that formed extensive "modern" type of shell beds, resembling post-Paleozoic rather then Paleozoic ecosystems (Simões et al., 2000).

### 3.3. Lake Pannon (Miocene, Central Europe)

During the Middle Miocene (between *c.* 11.5 and 13 Ma), a Tethyan relic sea, the Sarmatian Sea, covered large tracks of central–eastern Europe and western Asia. This anomalohaline sea broke into (semi-)isolated basins; first at *c.* 11.5 Ma Lake Pannon became isolated, located in the western margin of this Paratethyan system. At *c.* 6 Ma the Euxinian (more or less covering the present-day Black Sea region) and Caspian basins to the east had developed (Popov et al., 2006). These basins underwent stages of lake formation, exhumation and periods of marine influence. Sometimes basins were connected, and faunal exchange was possible. These basins are commonly referred to as the Paratethyan system. In order to discuss the bivalve evolution in "long-lived lake episodes" in these basins, they are treated as four long-lived lakes, namely the Pannonian, Euxinian, Aktschagylian, and Caspian lakes (see Müller et al., 1999; Nevesskaja et al., 2001; Grigorovich et al., 2003 and Popov et al., 2006, for indications of further potential long-lived lake and semi-isolated marine stages in the history of these Paratethyan basins).

Lake Pannon formed *c.* 11.5 Ma in the western margin of this Paratethyan complex, and persisted well into the Pliocene, *c.* 4 Ma. It became isolated from the Sarmatian Sea as the result of uplift of the Carpathian-Dinariid mountains, and developed in a back-arc basin setting. The lake covered most of Hungary and the margins of all neighboring countries, with a maximum area of some 280,000 $km^2$ (Magyar et al., 1999) and attaining depths over 1000 m. Density stratification and associated deepwater anoxia occurred but was not common. For most of its duration, the lake was mesohaline, but freshened in the latest stages of its history. It was located in a climate belt with alternating warm temperate humid and arid climates. Several molluscan lineages radiated within the lake including hydrobioid, viviparoid, melanopsid, planorbid, and limnaeid snails as well as dreissenid, unionoid, and cardiid bivalves. At about the same time, various small and isolated lakes developed in the Dinariid region to the west, that housed poorly documented and understood radiations of dreissenid bivalves (M. Harzhauser, personal communication, 2005).

The evolution of dreissenid bivalves in Lake Pannon has been dealt with recently by Harzhauser and Mandic (2004). In Lake Pannon, the family is represented by five genera covering tens of species (Müller et al., 1999 list 135 species, but point to extensive synonymy). Almost all species are endemic to the lake. Dreissenids are byssate filter feeders, capable of colonizing hard substrate in freshwater and coastal environments. They have a planktotrophic juvenile stage. Colonization is typically swift (opportunistic), as is seen, for example, by the rapid conquest of North American freshwater ecosystems by two co-occurring *Dreissena* species introduced with ballast waters from Europe since 1988 (Therriault et al., 2004). In contrast, Pannonian dreissenids yield a large number of co-occurring species and an unusual large range of morphologies. Both opportunistic colonizers as well as highly specialized, possibly long-living K-strategist lived in Lake Pannon. For example, *Sinucongeria primiformis* was found by Harzhauser and Mandic (2004) to be an "extreme" opportunistic colonizer of soft grounds in periods when prevailing dysoxia gave way to some oxygenation, related to shifts in the epilimnion–hypolimnion boundary within the lake as well as episodic explosions of algal and zooplankton blooms, their staple food. Thus, these shells formed temporal boom-and-bust populations. They can be considered opportunists that evolved in highly specialized settings. On the other hand, the large and thick-shelled *Congeria subglobosa* (Fig. 15) that lived in muddy oxygen-deficient environments avoided by all other molluscs was found by these authors to have a chemosymbiotic lifestyle. It was characterized as a highly adapted K-strategist. A chemosymbiotic living mode is unknown in "ordinary" dreissenids (e.g., other than Pannonian *Congeria* species). Other autecological traits, such as infaunal living mode in *Dreissenomya* (Fig. 16) and *Sinucongeria* species, are also beyond the usual range of living modes in dreissenids. *Sinucongeria* species evolved specialized shallow infaunal living modes from their epibiotic r-strategist founder species, *S. primiformis* (Harzhauser and Mandic, 2004).

The second bivalve group with profuse radiations resulting into species flocks in Lake Pannon is the lymnocardiine cockles. Müller et al. (1999) list 225 species, grouped in 13 genera, as mainly endemic to the lake, or shared with the subsequent Euxinian lakes to the east (see below). As these authors noted, cleaning out the extensive synonymy will certainly lower these numbers. The origin of the Pannonian cardiid radiations is uncertain. They presumably originated from diminutive *Cerastoderma* stocks that developed in the preceding brackish Sarmatian Sea, and survived through bottleneck populations during an ecological crisis at the transition of the Sarmatian Sea to Lake Pannon (Müller et al., 1999; Nevesskaja et al., 2001). The Pannonian radiations resulted into a spectacular range of morphologies and life habits of these cockles. Closely related *Cerastoderma* today is a common inhabitant of coastal regions and saline lakes in Europe and adjacent North Africa and Asia. Only two species are currently recognized (*C. edule* and *C. glaucum*) that have very simple morphologies compared to Pannonian (and Euxinian) relatives. The Pannonian cockles developed a wide array of outlines, sizes, globosity, thickness, ribbing-types, hinge teeth and siphonal apertures (Müller et al., 1999, see also Figs. 17–20). These cockles inhabited oligohaline coastal zones where they could be subject to extensive freshwater influence, as well as the stable mesohaline depths of the lake. For example, smoothened and reduced ribs allowed cockles living in the lakes shallows for improved ploughing, a trait considered by Müller and Magyar (1992) to be beneficial for efficient burrowing in such environments (Fig. 17). A striking contrasting example was cited by Müller et al. (1999) in the form of *Lymnocardium* (*Budmania*) species that lived in fluid mud at greater depths (Fig. 18). This taxon developed high sail-like ribs in order to stabilize itself in such conditions (Savazzi and Sälgeback, 2004). The deep muddy bottoms were also inhabited by *Paradacna* species that had very thin shells with reduced hinges that minimize drag (Fig. 19). Most of the evolutionary change of the Pannonian cardiids was gradual, as is shown for example by the extensive documentation of Müller and Magyar (1992). Pannonian cardiids dispersed into the Euxinian lakes to the east between 6 and 7 Ma, to contribute there to further cardiid radiations (see, e.g., Nevesskaja et al., 2001 and below). Near the Mio-Pliocene boundary (*c.* 5 Ma) the cardiids went extinct in Lake Pannon, to be replaced by a freshwater lake fauna of the *Paludina* stage (Müller et al., 1999).

The third Pannonian bivalve group to undergo radiations is the Unionoidea. In the main "Pannonian" stage of Lake Pannon these clams formed a subordinate group, mostly restricted to the lake margins and inflowing rivers. At least two species of *Unio* (one of which contains two chronostratigraphic forms) were common in littoral sediments in association with cardiids and other open-lake forms. This led Müller et al. (1999) to the conclusion that these species were possibly adapted to elevated salinities, which would be a unique feature for this cosmopolitan group. However, Harzhauser (personal communication, 2005) indicated that unionoid species are most probably transported from lower parts of streams into the lake. Unionoid radiations became profuse in the final freshwater stage of Lake Pannon, the Early Pliocene *Paludina* stage (located mostly in Serbia and eastern Croatia). The evolution of these latter unionoid radiations has received scant attention in literature; their extent awaits study.

## 3.4. Euxinian Lakes (Late Miocene–Pliocene, Eastern Europe)

The Euxinian lakes are a composite term for various lake stages that developed in and around the present-day Black Sea Basin between the Late Miocene and Late Pliocene (*c.* 2–6 Ma, Snel et al., 2001, 2006; Popov et al., 2006). Extensive literature about molluscs from these basins is available, but this is mostly descriptive, with the exception of the recent review of cardiid evolution by Nevesskaja et al. (2001). Below, the faunal successions of the western part of the Euxinian complex, the Dacian basin in Rumania, are summarized (following Wenz, 1942 and Pana et al., 1981). Although the history of the Euxinian complex dates back well into the Miocene, the long-lived lake stages emerged in the Late Miocene only. Until recently the timing of the major Euxinian stages has been very uncertain. Snel et al. (2001, 2006) established the date of the transition of the restricted marine Maeotian phase to the brackish long-lived Lake Pontian phase at 6.15 Ma. A duration of 0.85 Ma has been estimated by these authors for the Pontian. In the western Dacian part, the Pontian was preceded by a continental stage. Immigrant Pannonian bivalve taxa became established, and continued diversification (Müller et al.,1999; Nevesskaja

Figs. 15–21 Some highly adapted bivalves from the Mio-Pliocene Pannonian and Euxinian lakes. Figure 15: *Congeria subglobosa*, RGM 456211; Baden, Austria; H 71 mm. This very large and unusual dreissenid presumably adopted a chemosymbiontic living mode. Figure 16: *Dreissenomya* spec. RGM 53820; Vertsch, Crimea, Ukraine; L 47 mm. The ovate, flat outline minimizes drag during burrowing. Burrowing behaviour is unusual in Dreissenidae. Figure 17: Evolution in the Miocene *Lymnocardium decorum – Prosodacnomya* lineage from lake Pannon. Figure 17a: *Prosodacnomya dainellii*; Kötcse, úrilak, Hungary; L 22 mm. The central ribs have become fused in this species (Müller and Magyar, 1991, Reproduced with permission).

et al., 2001). The Pontian phase was succeeded by a second brackish long-lived lake phase, the Dacian. This phase lasted between 5.3 and either 4.25 or 4.6 Ma, encompassing 0.7 or 1 My. The Dacian fauna formed a continuation of the Pontian fauna. The Dacian phase was succeeded by a third, freshwater, lake phase, the Rumanian, that lasted from 4.25/4.6 Ma to the beginning of the Quaternary (*c.* 2 Ma, but estimates of the upper boundary are uncertain), that yielded a diverse freshwater fauna with unionoid bivalves and viviparoid snails.

The origin and migrational pathways of Pannonian and Euxinian cardiids is the subject of discussion (see Popov et al., 2006). Several paleontologists (e.g., Nevesskaja et al., 2001 and references therein) believe that several of the Euxinian cardiid genera evolved from marine ancestors in the Aegean Basin before establishing in the Euxinian regions. These authors also claim that several of the Euxinian taxa migrated into Lake Pannon and gave rise to small radiations in the lake's latest stages. Alternatively, I. Magyar and P. Müller (personal communication, 2005) consider Lake Pannon as the single cradle of Euxinian cardiids. They claim that all the ancestors of Euxinian taxa evolved within Lake Pannon, and that given new, improved age estimates of Euxinian successions (Stel et al., 2001), alternative migration patterns can be ruled out. The origin and migration pathways between Pannonian, Euxinian and possibly Aegean cardiids are, therefore, not resolved.

Between *c.* 6.15 Ma and 4.25/4.6 Ma cardiids and dreissenids dominated the bivalve fauna of the Dacian subbasin, giving rise to a diversity of forms and species somewhat similar to that seen in Lake Pannon. In the central and eastern Euxinian basins (not discussed herein) these cardiid/dreissenid radiations persisted until *c.* 2 Ma (Nevesskaja et al., 2001). Up to six contemporaneous dreissenid and 18 cardiid species have been mentioned by Wenz (1942) and Pana et al. (1981) from the western Euxinian (Dacian) Basin. In the Early Pliocene (*c.* 4.25/4.6 Ma), these two groups went into decline and a second radiation wave, comprising freshwater taxa such as viviparoid snails and pearly freshwater mussels (Unionoidea), set

Figs. 15–21 (continued) Figure 4e, from authors and editors.) Figure 17b: *Lymnocardium decorum decorum*; Kötcse, Csillagó clay pit, Hungary; L 21 mm (Müller and Magyar, 1991, Fig 3f). These two species belong to a single lineage, with *Prosodacnomya* being the stratigraphic younger member. Figures 18a, b: *Lymnocardium (Budmania) semseyi*. RGM 456212; Tirol (Königsgnad), Rumania; H 55 mm. The high sail like ribs (that are hollow) stabilized this large shell in the soft substrate. Figure 19: *Paradacna abichi*. RGM 456213; "Schela" Arbanasi, Buzeu, Rumania; Length lower margin picture 54 mm. This very thin-shelled and flat species is found in deeper off-shore mud with other thin-shelled molluscs. The reduction of shell thickness and flat outline of these shells greatly reduced the chance of sinking into the unstable mud. Figure 20: *Prosodacna neumayri*. RGM 456214; Coca, Jud Buzeu, Rumania; H 35 mm. An extremely thick-shelled, coiled, and heavily ribbed cardiid. The maximum shell density is located near the posterior margin. Figure 21: *Psilunio bielzi*. RGM 456215; Bukovatz, Rumania; H 39 mm. This unionoid is small, very thick and inflated, almost resembling a corbiculid. The point of maximum density is located at the postero-dorsal margin. The thick shell possibly stabilized the animal into soft substrates.

in. Unionoid faunas reached maximum diversity (22 species according to Wenz (1942); 30 species according to Pana et al., 1981) in the western Euxinian basins during the Late Pliocene, and declined before the establishment of continental settings at the onset of the Quaternary (*c.* 2 Ma). The geological record from the region is very good, but no morphometric documentation of evolutionary change, similar to those from the Lake Pannon cardiids (Müller and Magyar, 1992) exists, although bivalve morphological diversity is great (Fig. 20).

Extensive dreissenid/cardiid radiations and unionoid radiations did not co-occur (Pana et al., 1981). The overall succession of dreissenid/cardiid radiations followed by unionoid diversification is strikingly similar to the radiation in Lake Pannon, though the latter occurred *c.* 1 My earlier. The western Euxinian unionoid faunas can be divided into two groups. The first group contains various genera, each represented by a single or at most two contemporaneous species. Many of these taxa, such as *Sinanodonta* and *Pseudohyropisis*, appear to be relics of Tertiary (sub)tropical Eurasian freshwater unionoid faunas. The second group consists of genera that underwent diversifications, represented by *Rugunio* and *Cuneopsidea* (both with approximately five contemporary endemic species: Pana et al., 1981). The exact species numbers are uncertain, as ecophenotypic variation has so far not been addressed using ontogenetic-morphometric analyses. The prominent existence of rather small, very globose, circular and very thick-shelled unionoids is a remarkable feature of these faunas, although similar globose forms are known from fluvial systems in the southeastern USA. Some of the Euxinian unionoids resemble *Corbicula* in outline (Fig. 21).

In the Early Pliocene the Dacian Basin became separated from the central and eastern Euxinian Basins. By the early Late Pliocene, the lacustrine dreissenid and cardiid taxa had disappeared in the Dacian Basin, but their evolution (including new profuse cardiid radiation waves) continued in more easterly Euxinian Basins into the Quaternary, and from the Early Quaternary on in the Caspian Basin. Dreissenid and cardiid species living in the lemans (estuarine regions) of the present-day Black Sea are descendants from this later (Pliocene-Quaternary) Caspian Sea fauna. This modern fauna is indicated as "Ponto-Caspian", and will be treated below under the Caspian Sea.

## 3.5. Lake Aktschagyl (Pliocene, Eastern Europe, and Western Asia)

During the Late Pliocene (between *c.* 1.8 and 3.3 Ma), Lake Aktschagyl covered the Caspian Sea Basin and its surroundings, reaching more then twice the size of the present-day Caspian Sea. The lake was fresh along the northern margins, and attained salinities up to approximately 18‰ in the southern parts (Danukalova, 1996). Connections with the Euxinian Basins to the west existed initially as well as during maximum high stands (Popov et al., 2006), when the fauna migrated from Lake Aktschagyl towards the Euxinian region. Aktschagylian deposits are found in the entire lower Volga region, as well as in the Aral Sea, Caspian Sea, and Caucasus regions. Lake Aktschagyl harboured a spectacular radiation of cardiids and mactrids. These faunas are not part of the modern Caspian faunas (e.g., Nevesskaja et al., 2001). Therefore, Lake Aktschagyl is treated as a long-lived lake on its own. Data discussed below are mostly from Danukalova (1996).

The assumed Mediterranean cockle *Cerastoderma dombra* (Fig. 22) gave rise to a profuse cardiid radiation, containing five endemic genera (next to *Cerastoderma*) with 30 contemporaneous endemic species. A similar profuse radiation originated from an ordinary looking mactroid, *Aktschagylia subcaspia*, resulting in another 13 endemic species assigned to four genera (Danukalova, 1996). These two radiations are classic examples of species flocks: each is monophyletic and contains many closely related species that are endemic to the lake. Most of the species occurred in the southern half of the lake, apparently restricted from the northern half by lower salinities. The morphological variation in both groups is staggering. For example for the cardiids, some species are alate and have gaping shells and others have hypertrophied hinges. There is an overall increase in size, but different size classes are present. Some shells develop very robust ribs and others are remarkably smooth.

Most endemic lymnocardiid species were infaunal suspension feeders with well-developed siphons. Some adopted an epifaunal living mode (Nevesskaja et al., 2001). Alate forms with thin-walled shells (*Andrusovicardium* and *Avicardium*)

Figs. 22–26 Some Pliocene Aktschagylian bivalves, all from Aktschagylian Suite deposits. (From Danukalova, 1996, Reproduced with permission from the author). Figure 22: *Cerastoderma dombra vogdti*; Naphtalan, Azerbaijan; L 27 mm. This species closely resembles the basal species of the Aktschagylian radiations, and is almost indistinguishable from modern widespread *Cerastoderma* species (Danukalova, 1996; pl. 5, Fig. 15). Figure 23: *Miricardium alexinum*; Danata, Turkmenistan; L 11 mm. This species resembles modern *Parvicardium* species (Danukalova, 1996; pl. 11, Fig. 13). Figure 24: *Cerastoderma sanani*; Danata, Turkmenistan; L 36 mm. This flat, thin-shelled species with fused ribs bears some resemblance to the modern rapid/deep burrowing Caspian *Adacna laeviuscula* (see Fig. 32) (Danukalova, 1996; pl. 6, Fig. 17). Figure 25: *Aktschagylocardium uspenskaiae*; Bozdag, Azerbaijan; L 24 mm. A marked elongate lanceolateform outline, otherwise unknown from the Lymnocardiidae. The outline resembles that of some marine pterioid species, possibly indicating an attached living mode for this species (Danukalova, 1996; pl. 17, Fig. 3). Figure 26: *Avicardium transcaspicum*; Danata, Turkmenistan; L 44 mm. The flat modioliform outline of this species strikingly resembles the outline of the totally unrelated Aktchagylian mactroid *Avimactra preaviculoides*. This species may have been a recliner, living on fluid bottoms, and possibly capable of swimming by valve flapping (Danukalova, 1996; pl. 13, Fig. 11).

probably floated on liquid muddy sediments. Several of these endemics have been interpreted as adapted to lowered oxygen levels. During the Late Pliocene, only *c.* 1 My after their origination, most endemics went extinct and *C. dombra* was widespread again for a few hundred thousand years.

The total fauna from Lake Aktschagyl comprised approximately 49 bivalve species and only 23 gastropod species (Danukalova, 1996), many of which were endemic to the lake. At *c.* 1.8 Ma, the rapid cardiid/mactroid extinction coincided with immigration of dreissenid bivalves from the Euxinian. Danukalova (1996) mentioned six taxa that are conspecific with or closely related to modern Caspian taxa.

## 3.6. Caspian Sea (Quaternary, Southeastern Europe, and Northwestern Asia)

The Caspian Sea is an enclosed saline lake located on the boundary of southeastern Europe and northwestern Asia. It measures approximately 374,000 $km^2$, and reaches a maximum depth of 1,025 m. The lake level (currently approximately 27 m below sealevel) is the result of a balance between run-off and direct precipitation, and evaporation. Lake-level fluctuations are large (for example a 3 m sea level rise occurred between 1978 and 1992), and not related to eustatic sealevel changes. The northern Caspian Sea is a shallow platform (depths typically less then 10 m), where freshwater input from the Volga and other rivers depress salinities that range from freshwater in the delta regions to approximately 10‰ off shore. The main body of the Caspian Sea contains two deep basins that experience stable mesohaline salinities (approximately 12–14‰). Hypersalinity occurs in coastal lagoons, freshening occurs locally near river deltas at the northern, but also at the western and southern margins of the lake. The Caspian Sea fauna is dominated numerically by (mainly endemic) cardiids (three genera, approximately 12 species) and dreissenids (one genus, possibly four species), as well as hydrobioid snails. The taxonomy of Caspian molluscs is very unsatisfactory resolved. Taxon boundaries are uncertain for many species and even some genera. Apart from immigrants, the northern Caspian Sea faunas are dominated by the *Monodacna*-complex (whose species numbers and boundaries are totally unclear) and *Dreissena polymorpha*. The central–southern parts of the Caspian Sea are dominated by species of the currently endemic genus *Didacna* and saline forms and species of *Dreissena* (*D. rostriformis, D. polymorpha elata*, and *D. caspia*). *Didacna profundicola* has been documented from water depths down to 870 m (Tarasov, 1996). *Adacna* and *Hypanis* species are found all over the Caspian Sea.

The modern Caspian Sea originated at *c.* 1.8 Ma, and underwent profound expansions and contractions through its history. Nevesskaja et al. (2001) proposed that the Caspian cardiids originated from Aktschagylian cardiid stocks (with *Didacna* making a bypass through the Euxinian region before establishing in the Caspian Sea), whereas Grigorovich et al. (2003) cite evidence for a Miocene Pannonian–Euxinian origin of this group. In the Apsheronian interval (Early Pleistocene) a substantial

Figs. 27–32 Some Holocene and recent cardiids from the Caspian Sea. All specimens were collected at the beach of Turali, Dagestan, Russia. Figure 27: *Cerastoderma glaucum*; RGM 456216; L 22 mm. Holocene or Late Pleistocene immigrant from the Black Sea region, now very common in coastal regions of the entire Caspian Sea. Figure 28: *Didacna praetrigonoides*; RGM 456217; L 55 mm. The largest species of the Caspian *Didacna* flock. Figure 29: *Didacna baeri*; RGM 456218; L 37 mm. Robust, thick-shelled species, typical of coastal environments with high levels of physical disturbance and high predation intensities. Predation scars are common on shells of this species. Figure 30: *Monodacna caspia* s.l. RGM 456219; L 24 mm. The taxonomic status of the group is unclear. It may comprise a single highly variable species, or various closely resembling species. The group is most abundant in the northern oligohaline-mesohaline regions of the Caspian Sea. It has closely related or conspecific relatives in the Black Sea lemans. Figure 31: *Adacna laeviuscula*. RGM 456220; L valve 44 mm, L. pair 36 mm. Very thin but rather flexible shell with reduced hinge and ribbing, large gapes at the anterior margin and the posterodorsal margin, and a deep pallial sinus. Nevesskaja et al., 2001 suggest these traits to represent a deep burrowing living mode. This species, however, has been found in Holocene deposits reflecting agitated near-shore environments near Turali, suggesting a (complementary) rapid burrowing behavior. Figure 32: *Adacna plicata*; RGM 456221; L 34 mm. This species has a very closely related sister species in the Black Sea. The strengthened posterior ribs enhanced burrowing.

cardiid radiation occurred, giving rise to eigth genera, and tens of species. This rather sudden burst was followed by a gradual decrease of the fauna resulting in an extinction event in the late Early Pleistocene. During the Middle Pleistocene two more modern Caspian taxa were added to the Apsheronian survivors. These eventually evolved (mainly through anagenetic change) into the present-day Caspian Sea cardiid species.

During the larger part of the Quaternary, the Caspian Sea was an isolated water body, but from the late Middle Pleistocene on, episodic connections with the Black Sea existed (Popov, 1983; Nevesskaja, 1965), allowing for migration of Caspian taxa into the Black Sea Basin and vice versa. As a result, Caspian endemics are found in discrete intervals in the Quaternary fossil sequences of the Black Sea Basin (Nevesskaja, 1965). Several of these immigrants went extinct there; others evolved into sister species that now live in coastal areas (lemans) of the Black Sea. *C. glaucum* also may have become established in the Caspian Sea during the latest Pleistocene or early Holocene through this connection the other way around. A single *Adacna* species lives (or lived until recently) in the (desiccating) Aral Sea. The combined Caspian Sea, Aral Sea. and Black Sea leman fauna is referred to the Ponto-Caspian fauna. The sister species pairs of the Caspian and Black Sea (e.g., *Hypanis plicata* vs. *H. relicta*, *Monodacna caspia*-complex vs *M. angusticostata* and *Adacna laeviuscula* vs. *A. luciae*) are morphologically almost indistinguishable. This is also seen with similar crustacean species-pairs, that furthermore show very little genetic divergence (Grigorovich et al., 2003), in agreement with a very recent (Late Pleistocene) connection.

The latest overflow period has been dated between *c.* 12 and 16ka, based on $^{14}C$ and U/Th analyses (Tschepalyga, 2003; Arslanov et al., 2002). During this Khvalynian stage the Caspian Sea became overfilled by meltwater-laden Russian rivers. As a result, overflow was routed through the Manych depression towards the Black Sea. The overflow threshold is over 100m above the present-day water level (Tschepalyga, 2003). Such high-volume variations have resulted in strong variable salinities in the Caspian Sea, possibly accounting for large salinity tolerances seen in many of the modern Caspian taxa.

Recent molecular phylogenetic work (Therriault et al., 2004) has shed light on other possible speciation mechanisms. These authors discovered that *D. rostriformis* (a Caspian Sea species typically living in salinities of 12–14‰) and *D. bugensis* (a Black Sea basin species living in salinities between 0 and 3‰) are so closely related that they should be considered forms of a single species. *D. rostriformis* forma *bugensis* has recently expanded into the Volga system, and the authors expressed their concerns about a possible subsequent interbreeding of these two forms in the Caspian Sea. However, despite the very small molecular (and morphological) divergence of these two forms, their respective salinity tolerances have diverged immensely, the *rostriformis* form is restricted to mesohaline settings and the *bugensis* form inhabits the freshwater to oligohaline waters. Almost no dreissenids live in the transition zone between oligohaline and mesohaline settings in the northern Caspian Sea, which is dominated by the immigrant *Mytilaster minimus* (Kosarev and Yablonskaya, 1994). The latter appears to out compete the dreissenids and therefore may form an extra barrier for contact between mixture of the two *Dreissena* forms. It may therefore well be that we witness an allopatric speciation event that through re-immigration is leading to the addition of a species in the Caspian Sea. Reid and Orlova (2002) argued that the long history of Ponto-Caspian dreissenids in successive regimes with large salinity variations has resulted in increased ecological (salinity) tolerances of modern dreissenid species, facilitating their establishment through introduction in other parts of the world. In the 20th century several immigrants were added to the Caspian fauna through shipping or deliberate introduction. These immigrants, among them the Mediterranean bivalves *M. minimus* and *Abra ovata*, now numerically dominate large zones in the Caspian Sea. Little is known about their effects on the endemic faunas of the Caspian Sea, but possibly the endemic *D. caspia* and *D. polymorpha elata* have disappeared (Kosarev and Yablonskaya, 1994).

## 3.7 Lake Baikal (Late Oligocene-Extant, Siberia)

Lake Baikal is a classic rift-lake. It is an elongated, deep (1637 m) lake with a surface area of approximately 31,500 km$^2$. The Baikal rift zone has been in place since *c.* 70 Ma. Lakes have existed within the rift *c.* 60 My, and the current Lake Baikal (that may have been divided in several lakes during its history) developed *c.* 28 Ma (Sherbakov, 1999). With global cooling the lake ecosystem experienced a dramatic shift from a relatively warm and stratified lake to the modern cold, deep and fully oxygenated lake between 2 and 3 Ma (Sherbakov, 1999). Consequently, deep-water taxa could only evolve after this transition.

Lake Baikal contains one endemic unionoid species, and 30 sphaeriid species, 14 of which are endemic (Slugina et al., 1994; Slugina, 1995). The sphaeriids are assigned to nine genera, several of which contain small flocks. The most speciose genus is *Euglesia*, which comprises seven species, divided over two subgenera. Baikal sphaeriid diversity is highest in the littoral zones (down to 5 m), but three species have been reported from as much as 60 m water depth (Slugina et al., 1994). The sphaeriids inhabit a variety of substrate types. Very little is known about the evolutionary biology of these Baikalian clams. Several Baikalian endemic sphaeriids have closely related widespread sister species outside the lake (Sitnikova, 1994 mentioned five endemic/non-endemic species pairs). The existence of various species pairs strongly indicates independent penetration and *in situ* evolution of common Siberian sphaeriids into Lake Baikal. Four species are shared between Lake Baikal and Mongolian Lake Hövsgöl (Slugina, 2000), some 250 km away.

I am not aware of molecular studies on the Baikal bivalves, but evidence from other groups shows that the diversity of Lake Baikal comprise old (30–50 Ma: amphipods, pulmonate gastropods, and tubelarians) and young (Neogene and Quaternary: baicaliid gastropods and sculpine fish) species flocks (Sherbakov, 1999; Hausdorf et al., 2003). Baikalian endemic sphaeriids are considered to be “young” endemics (Sitnikova, 1994), that diverged in Lake Baikal after the establishment of modern cold and oxygenated conditions.

## 3.8. Lake Biwa (Pliocene-Extant, Central Japan)

Lake Biwa, located in a tectonic depression in central Japan, has a history of *c.* 5 My. It is 104 m deep, and measures 674 km$^2$. Both the fauna of the modern lake, as well as its fossil faunas have been investigated (Matsuoka, 1987; Nishino and Watanabe, 2000). The present-day bivalve fauna of Lake Biwa contains eight endemic unionoid (sub)species and an endemic corbiculid (*C. sandai*). The history of Lake Biwa can be subdivided into six stages (Matsuoka, 1987). The lake originated approximately 50 km southeast of the present-day location and migrated northward in each successive stage (Nishino and Watanabe, 2000). The northern deeper part of the present-day lake came into existence 430 ka. Each of the paleolake phases ended with a major extinction, mainly related to climatic deterioration and/or partial exhumation. For the successive lake stages molluscan assemblages representing lagoon/swamp, stream, stream/lake margin, littoral, and profundal lake settings were described, providing an ecological framework for the documentation of evolution (Matsuoka, 1987). In general the fauna developed from a typical (sub)tropical to temperate east Asian fauna, with increasing endemism. The lake has been freshwater throughout its history.

Unionoid species numbers remained constant throughout the lake's history (between 15 and 18 species: Matsuoka, 1987). The modern unionoids do not contain species-flocks, the group is non-speciose. In earlier stages possible small flocks (of up to four species) did occur. The presence of many of the modern endemic unionoid taxa in the fossil deposits of the lake, together with likely predecessors, indicate the *in situ* evolution of most of these species. Today, only the gastropod *Sulcospira* (*Biwamelania*) forms an endemic mollusc species flock, comprising 15 species. In total, 55% of the molluscan species of Lake Biwa (excluding recent immigrants) are endemic (Nishino and Watanabe, 2000). The systematic status of some of the bivalves, as well as the endemic status of various groups, are still subject of some uncertainty. Fossils of two species of currently endemic unionoid species of Lake Biwa (*Hyriopsis schlegeri* and *Inversiunio hiraseus*) have been found in fossil series beyond Lake Biwa, indicating these species to be relics, for which the lake served as a "museum" (Nishino and Watanabe, 2000).

The phylogenetic status of the endemic corbiculid, *C. sandai*, has been addressed recently by Nishino and Watanabe (2000) and Glaubrecht et al. (2003). The former authors quoted a differentiation from the widespread *C. leana c.* 900 ka, based on isozyme divergence estimates. The record of *C. sandai* from modern Lake Tai Hu (central China) needs investigation, in order to reject (or confirm) its conspecifity with the Biwa corbiculid. Molecular phylogenetic work by Glaubrecht et al. (2003) indicates a surprisingly close relationship with two endemic corbiculid species from the Malili lake system of Sulawesi. The direct development of young from benthic egg masses of *C. sandai* is unique among the Corbiculidae that are ovoviviparous (Glaubrecht et al., 2003).

### 3.9. Lake Turkana (Pliocene-Quaternary, Kenya)

Today, Lake Turkana is an alkaline lake, located in the eastern limb of the African rift system. The lake contains a reduced gastropod fauna composed of widespread East

African freshwater species (Brown, 1992). Comparison with other African lake faunas led Brown to the conclusion that the current high salinity/alkalinity is at the maximum that freshwater mollusc faunas can endure. The modern depauperate fauna is in strong contrast with the fossil record of the lake. In total, 42 aquatic mollusc species have been identified from Pliocene and Quaternary (*c.* 0.5–4 Ma) lake deposits. In 1981 a documentation of molluscan evolution from Pliocene and Quaternary deposits of Lake Turkana by P.G. Williamson provided a starting point for lengthy discussions (Williamson, 1981; see Brown, 1992 for references of the successive discussions). Williamson's documentations and especially interpretations have been fiercely contested and most of his interpretations have not gained wide acceptance. However, his original documentation provides us with insight of bivalve evolution in a long-lived lake. Lake Turkana was most of the time inhabited by seven unionoid lineages, one *Corbicula* lineage and two sphaeriid lineages. All these were widespread east African taxa. Turkana bivalve evolution can be summarized in three episodes:

1. Strong and concerted anagenetic change in Williamson's Suregei interval (now part of the lower Burgi member of the Lokridede complex, *c.* 2.2 +/−0.3 Ma), also seen in the gastropod faunas, followed by extinction. All six examined bivalve lineages were affected. Morphological variation increased, and overall morphology shifted away from the parental morphologies. Novel forms evolved outside the phenotypic range of the widely distributed ancestor species, and morphological variability of these novel forms declined, suggesting morphological stabilization. A major lake level drop occurred during the Suregei interval that finally resulted in extinction of these new forms. Thereafter, ancestral species reinvaded the lake from their riverine habitats.
2. Gradual appearance of new species, possibly evolving from and living next to (widespread) freshwater inhabitants of Turkana Lake during the lower Koobi Fora interval (now dated at *c.* 1.89–1.95 Ma: Brown and Feibel, 1986). Evolutionary changes occurred in two unionoid and one sphaeriid lineage. Newly evolved forms persisted until a regressive event, when they went extinct. Ancestral species either survived this event, or went extinct but re-established from surrounding riverine populations.
3. An episode of strongly diminished diversity in the Guomde stage (now attributed to the Chari member, dated between 0.7 and 1.3 Ma). Only four mollusc species were present, including two unionoids. These two are strongly altered morphologically from their assumed predecessors with whom they show no morphological overlap. This stage was followed by extinction and the remigration of ordinary freshwater taxa from surrounding rivers (comprising of the same lineages that are found throughout the lake's history). The Guomde interval is interpreted as coinciding with a major lacustrine regression and by inference increased alkalinity, followed by extinction of the fauna.

As suggested by Brown (1992) who compared the fossil Turkana faunas with modern African lake faunas, much of the morphological variation in the short evolutionary intervals is outside the domain of the living species; even those living in chemically stressed circumstances. He, therefore, concurred with Williamson that the unusual morphs must have been manifestations of considerable genetic change (and are not mere ecophenotypic forms as suggested by various other authors). Stable oxygen isotope studies from these same sequences led Abell (1982) to conclude that the lake basin experienced drastic changes between wet and arid climate phases in the lake's catchment, resulting into strong shifts in the lake chemistry (salinity and alkalinity). Mollusc extinction was found to be related to strongly elevated salinities that were shown to have occurred repeatedly from his isotope studies in the Pliocene-Pleistocene Turkana sequence. Lake Turkana was, therefore, not a single long-lived lake, but instead a series of successive lake stages in which normal, stressed and uninhabitable conditions occurred under which the faunas waxed and waned. Detailed restudy of these intervals might teach us still a great deal more of molluscan evolution, especially of the apparent very short time intervals in which massive faunal change can occur.

### 3.10. Other Long-lived Lakes

Other fossil and modern long-lived lakes yield non-diversified endemic bivalve faunas. These include Lake Titicaca (Peru–Bolivia: sphaeriid bivalves), Lake Ohrid (Macedonia–Albania: sphaeriids and a dreissenid bivalve, the latter species shared with nearby Lake Prespa), Lake Poso and the Malili Lake system (Sulawesi, Indonesia: corbiculids among which a highly unusual attached species in Lake Poso: Bogan and Bouchet, 1995), Lake Victoria (Uganda, Kenya and Tanzania: unionoid and sphaeriid bivalves), Pliocene Lake Edward-Albert (central eastern Africa: unionoid bivalves) and Lake Tanganyika and Lake Malawi (eastern Africa: unionoid bivalves). Whenever studied, the lacustrine endemic species have closely related sister species from widespread species living in other aquatic biota around the lake (Korniushin et al., 2000). Oddly enough, the young Lake Victoria (*c.* 12.4 ka) harbours approximately ten endemic (sub) species of bivalves (Daget, 1998). A similar number of endemic (sub-) species is found in the ancient Lake Tanganyika (9–12 My), although numbers may alter in the light of ongoing studies (Scholz and Glaubrecht, 2004).

## 4. Discussion

Apart from radiations, high bivalve diversity and endemicity in long-lived lakes can result from survival of species with a previously wider distribution, as is shown by two extant endemic unionoid species in Lake Biwa with fossil records beyond the lake (Nishino and Watanabe, 2000) and subtropical unionoid relic taxa in the Pliocene Dacian faunas. Also, the evolution of endemics that diverged from widespread sister species without radiating further in lakes is a common contributor to bivalve diversity. Examples include the presence of species pairs of lake endemics and their widespread sister species in Baikalian, Ohrid and Titicaca sphaeriids, the Ohrid/Prespa dreissenid species and Lake Biwa and Turkana unionoids. The study of reproductive biology of the corbiculid clams in southeast Asian lakes Biwa, Poso and Malili sheds light on the mechanisms behind such speciation (Glaubrecht et al., 2003). In these corbiculid species, which generally have a widespread distribution and flexible living and reproduction modes (typically r-strategists), a shift towards K-reproductive strategies has been documented in the closely related endemic species (Glaubrecht et al., 2003). These authors concluded that the combined isolation from parental populations and the stability of the lacustrine ecosystems (relative to fluvial ecosystems) allowed for this shift. However, they show that long-lived lakes are not the sole environments that promote such speciation through isolation, as endemic species with similar shifts towards K-reproductive strategies also developed in isolated river systems on Sulawesi. Neither is their isolation total, as is shown by the occurrence of four Baikalian species in Lake Hövsgöl and the shared *Dreissena* species between Lake Ohrid and Prespa. Long-distance dispersal vectors such as birds may have contributed to the (low-intensity) dispersal of species that originated as lake endemics. Also episodic connections (possibly as overflows) have allowed for the dispersal of originally endemic lake faunas, as is shown in the Pannonian-Euxinian and Ponto-Caspian Basins.

Profuse endemic molluscan radiations are also known from a number of river systems. These include mostly gastropods: cerithoideans in the Coosa River and adjacent river systems (southern USA), rissooids in the Mekong River (Laos, Thailand) and cerithoideans in various African and southeast Asian river systems. Unionoid bivalve radiations are known from North American and Chinese river systems. The combination of trophic specialization (Glaubrecht and Köhler, 2003) and the mosaic nature of suitable substrates fuels the continuous separation of communities of gastropod species, restricting gene-flow and enhancing speciation. Such mechanisms are difficult to apply to fluvial unionoid bivalves. At this point I can only speculate that high species numbers and endemicity rates in fluvial unionoids must reflect a long river basin history with drainage reorganizations and plenty of relatively isolated rivers where speciation could occur. In general, endemic species are not that densely packed (syntopic) in rivers as they are in long-lived lakes.

Cosmopolitan freshwater bivalve groups diversify very rarely in long-lived lakes, and yield rather modest radiations only (unionoids in the Pannonian, Euxinian, Biwa and East African lakes, sphaeriids in Lake Baikal). It is rather remarkable that these cosmopolitan bivalve groups, that are very successful in occupying almost any freshwater habitat, are rather poorly represented in long-lived lake faunas. In the case of unionoids, their complex reproductive strategies that include (very often a very host-specific) parasitic stage on fish, may limit diversification potential in long-lived lakes. Short life spans seen in sphaeriids, as well as their small size, which can make them prey of choice for many predators, may be explanations for the paucity of their radiations. Sphaeriids and unionoids also appear to be highly intolerant to variable salinity (and possibly alkalinity) regimes, which may also explain their paucity in long-lived lake environments that often can develop (episodic) anomalous aquatic chemical regimes. Corbiculids also are restricted to (warm-) temperate and tropical settings. They favour well-oxygenated waters, and appear to be intolerant of low oxygen conditions. Such settings can be very widespread in tropical long-lived lakes, as a result of water stratification or of high organic decomposition rates (Thayer, 1975; Stanley, 1970). Up to two co-occurring endemic corbiculid species, however, have been found in Lake Poso (Bouchet, 1995; Bogan and Bouchet, 1998) and in the Malili lake system (Glaubrecht et al., 2003) of Sulawesi.

Extensive bivalve radiations predominantly apply to groups of marine descent, although the status of Dreissenidae as being a marine or freshwater group is debatable. Extensive freshwater unionoid radiations have been found only in the Euxinian lakes, and possibly in the Paleogene Lake Zaysan, Kazakhstan (G. Danukalova, Personal Communication, 2005) and unnamed Saharan Cretaceous lakes (D. van Damme, Personal Communication, 2005). Given the dominance of bivalve groups of marine ancestry in long-lived lakes it appears odd that not more marine gastropod groups (other then the marine and freshwater Rissooidea and Cerithoidea) have developed such profuse radiations. Endemic radiations of marine gastropod groups have existed, in (semi-) enclosed anomalohaline seas like the Neogene Sarmatian and Maeotian seas of central–eastern Europe. Here, marine snail groups such as Trochidae, developed (modest) endemic radiations. This point shows that the separation of water bodies between long-lived lakes and semi-enclosed anomalohaline seas is somewhat artificial.

From the fossil record several cases have been documented where radiations occurred in or after episodes of ecological crisis. These include widespread anoxia and salinity stress at the base of the Parana radiations, salinity stress at the base of the Pannonian and Aktschagylian radiations and the reorganization of a stratified warm-temperate to fully oxygenated cold lake in Baikal with the onset of severe glaciations. Environmental stress provides survivors with empty ecological niches allowing for rapid expansion and diversification (Nevesskaja et al., 2001). Subsequent diversification rates are lower (although stratigraphic control over most fossil long-lived lakes is not precise enough to determine exact rates), and often involve the occupation of very specific niches and life styles (see below). Secondary opportunists also may evolve that can deal with highly specialized opportunistic settings in long-lived lakes, such as short periods of oxygenation in otherwise dysoxic settings in Lake Pannon (exploited by *S. primiformis*) and adaptation to strong salinity variations in the Ponto-Caspian *Monodacna* and *Dreissena* species. The continuous presence of a complex, but on ecological time scales stable, set of biotopes adds to the opportunity to develop divergent life styles and increased diversity, as is shown by parallel evolutionary adaptations to specific environments in unrelated radiations. For example, fusion of central ribs in Pannonian cardiids, that enhances ploughing behavior beneficial in agitated near-shore zones (Müller and Magyar, 1992) is also seen in independent radiations of the Euxinian, Aktchagylian and Caspian cardiids. Similar parallel developments are seen in the evolution of chemosymbiosis, arising independently in Pannonian dreissenids

(Harzhauser and Mandic, 2004) and possibly in Pebasian corbulids. Another example is the adaptation to fluid bottom conditions seen in Pebasian corbulids, Pannonian dreissenids and cardiids and Aktchagylian cardiids. Repeated evolution of epifaunal suspension feeders lying on the substrate with beak downwards, having an anteriorly thickened shell, and hypertrophied anterior teeth occurred in five independent cardiid lineages in four different lakes and marginal seas (Nevesskaja et al., 2001). These authors also cite the repeated evolution of deep burrowing taxa with thin shells, deep sinuses and reduced ribbing. These examples clearly indicate that adaptation to stable and complex ecosystems provided by long-lived lakes is an important feature facilitating diversification.

Long-lived lake bivalves show repeatedly the evolution of ecological tolerances and morphological characteristics that exceed the range of variation of their families in ordinary (non long-lived) environments. These include expanded tolerance of salinity regimes (Pannonian, Euxinian, Aktschagylian and Caspian cardiids, Aktschagylian mactrids), expansion into fluid substrates (Aktschagylian and Pannonian cardiids), settlement of extreme depths (Pannonian cardiids and dreissenids, the Caspian cardiid *Didacna profundicola*) or low oxygen conditions (Pebasian *Exallocorbula*, Parana *Leinzia* and Pannonian *Congeria*). Nevesskaja et al. (2001: 172) cite the description "specialization beyond the limit" for this feature. Specialization beyond the limit suggests that such adaptations are not beneficial beyond these highly specialized settings, but increased salinity tolerances developed in Ponto-Caspian dreissenids together with their reproductive mode instead has made them very invasive (Reid and Orlova, 2002). Instead, the term "supralimital evolution", introduced by Myers (1960: 327) for the same phenomenon in cyprinoid fishes in Lake Lanao (Philippines) should be used. Evolution beyond usual tolerances and characters in long-lived lake taxa implies that reconstructing paleoecological conditions using strict uniformitarian principles with these taxa is hazardous. More general reservations about the application of uniformitarianism in the geological record in general have been put forward by several authors, including Bottjer et al. (1995), Vermeij and Dudley (1985) and Vermeij (1987). These authors stress the need for the simultaneous use of independent faunal, geochemical and sedimentary criteria to reconstruct past ecological settings. For example, applying uniformitarianism to the (endemic) corbulids led Nuttall (1990) and others to the interpretation of predominantly brackish settings in Miocene Lake Pebas, where the presence of unionoids and the absence of typical oligo- and mesohaline mollusc taxa as well as strontium and stable oxygen isotope analyses clearly indicated predominant freshwater conditions (Vonhof et al., 1998; 2003; Wesselingh et al., 2002). Similar uncertainties also exist over the paleosalinities in Lake Pannon (see e.g., Mátyás et al., 1996).

Long-lived lake bivalves are less diverse than the co-occurring gastropods (with the exception of Permian Lake Parana, where gastropods were lacking altogether and Pliocene Lake Aktschagyl). In only 7 of the approximately 20 known (fossil and extant) long-lived lakes substantial bivalve radiations occur, where gastropod radiations occur in 19 out of 20 long-lived lakes. Gastropod populations are usually smaller than bivalve populations, and often have more specific habitat requirements (such as common substrate dependency in cerithoidean snails). Smaller populations are more prone to speciation by isolation, but also more prone to extinction (Martens, 1997). Bivalves are rather immobile after settling, and thus must cope with more environmental variation, resulting in larger ecological tolerances compared to gastropods. Bivalve species appear to be in general more widely distributed in long-lived lakes. The wider geographic and ecological distribution is matched with increased stratigraphic longevity. Species longevity in diverse taxa in marine and terrestrial environments has been shown to correlate positively with the geographic distribution (Kammer et al., 1997 and references therein). This may explain lower diversity and propensity of bivalves to radiate relative to gastropods in long-lived lakes. Limited larval dispersal of ostracod groups that evolved brooding habits is correlated with high species numbers in Lake Tanganyika (Cohen and Johnston, 1987, but see Johnston and Cohen, 1987). Extreme genetic isolation as a result of poor dispersal may cause rapid speciation (and extinction) owing to the vagaries of such small populations. No indications exist at present that these brooding habits are present in long-lived lake bivalves (with the exception of the southeast Asian corbiculids, Glaubrecht et al., 2003), whereas it is very common in long-lived lake gastropods (Cohen and Johnston, 1987; Michel, 1994).

Major faunal turnover events in long-lived lakes may partially be attributed to climate change. For example, episodic extinctions, as well as intervals of aberrant faunal development have been linked to aridification resulting into increased alkalinity and salinity in Lake Turkana (Abell, 1982; Williamson, 1981). Freshening of the final stages of Lake Pannon and western Euxinian (Dacian) lakes coincided with the establishment of more humid climates during the Pliocene in southeast Europe. The remaining Euxinian lakes and Lake Aktschagyl terminated with the onset of severe northern hemisphere glaciations *c.* 1.8 Ma, at the same time the modern Caspian Sea developed. Finally, the onset of glaciations also caused a major reorganization of Lake Baikal, from a temperate stratified lake to a fully mixed cold lake. Such large-scale ecosystem reorganizations have influenced the evolution of the biota, as seen by the timing of a young diversification event in a number of Baikalian groups contemporary with the ecosystem reorganization *c.* 2 Ma (Sherbakov, 1999; Martens, 1997).

Much remains to be learned about the evolution of long-lived lake bivalve faunas. However, an absolute prerequisite is a combined rigorous taxonomic, stratigraphic, and ecologic treatment of these faunas, many of which have been poorly documented. Especially good opportunities exist in the Ponto-Caspian region, where molecular and ecological data of living taxa can be compared with the fossil record. The fossil record in itself yields vast quantities of ecological data, as well as indications of morphological change and insights into temporal scales of evolution. Mining these data sets needs the application of very different disciplines, from ecology to sedimentology and geochemistry. For many of the lakes such combined approaches have not (yet) been made, with the exception of Lake Biwa, but if they are made they will yield much more insight into the ecological context of speciation.

## 5. Conclusions

Relative high diversity and endemism in long-lived lake bivalves is a result of three factors: (1) the survival of relic species with a formerly widespread distribution, (2) isolation from widespread sister species and a shift to K-reproductive strategies without further diversification, and (3) *in situ* diversification. Diversifications are often accelerated at times of ecological stress that remove competitors and predators, and result in the establishment of complex and stable set of biotopes (on a ecological time scale) to which taxa can adapt. Parallel adaptations to specific ecological conditions in unrelated lakes and bivalve groups show that ecology is an important facilitating feature behind diversification. Morphological diversity and ecological characteristics are often beyond the ranges seen in closely related groups in ordinary, non-long-lived, environments. This is termed supralimital evolution, which can complicate the reconstruction of past systems using strict uniformitarian methods. Long-lived lake bivalves are a very good model-group for the study of evolution, but in order to fulfil their potential more combined rigorous taxonomic, ecologic, and stratigraphic documentation will be necessary.

## Acknowledgements

Many people have helped me throughout the years in my understanding and study of long-lived lake faunas. They include Dirk van Damme (Gent University, Belgium), Matthias Glaubrecht (MNH, Berlin, Germany), Javier Guerrero (UN Bogota, Colombia), Carina Hoorn (UvA, Amsterdam, The Netherlands), Salle Kroonenberg (TU Delft, The Netherlands), Tom Meijer (Naturalis, Leiden, The Netherlands), Ellinor Michel (BMNH, London, UK), Pal Müller (Geological Institute of Hungary, Budapest), Patrick Nuttall (BMNH, London, UK), Matti Räsänen (UTU, Turku, Finland), Willem Renema (Naturalis, Leiden, The Netherlands), Lidia Romero Pittman (INGEMMET, Lima, Peru), Jukka Salo (UTU, Turku, Finland), Gustavo Sarmiento (UN Bogota, Colombia), Alexander Svytoch (MSU, Moscow, Russia) and Tamara Yanina (MSU, Moscow, Russia). The section on Pachydontinae has been published in modified form elsewhere and I thank Peter Skelton (Open University, Milton Keynes, U.K.) and Arie Janssen (Xewkija, Malta) for their suggestions that improved that manuscript (and the section on Pachydontinae herein). I thank Gusel Danukalova (ANBR, Ufa, Russia) for her help with the section on Lake Aktschagyl. Jan Snel (UU, Utrecht, The Netherlands) provided me with unpublished data on the age of Dacian deposits. Jose Joordens (VU, Amsterdam, The Netherlands) provided me with an update of the Turkana stratigraphy including age estimates. I am indebted to Pal Müller, Imre Magyar (MOL, Budapest, Hungary), Gusel Danukalova, and Professor B. Runnegar (AMNH, Washington, DC) as well as the managing editors of the *Bulletin of American Museum of Natural History*, Acta Palaeontologica Polonica and Trudy Paleontologischeski Instituta Rossia Nauka for permission to reproduce figures. Suggestions of Geerat Vermeij (UC, Davis, USA), Laurie Anderson (University of Louisiana, Baton Rouge, USA), and Mathias Harzhauser (NHM, Vienna, Austria) were very helpful in improving the manuscript. Finally, I am deeply indebted to Imre Magyar (MOL, Budapest, Hungary) who helped me out with several issues concerning the Paratethys for this review.

## References

Abell, P.I., 1982, Palaeoclimates at Lake Turkana, Kenya, from oxygen isotope ratios of gastropod shells, *Nature* **297**: 321–323.

Anderson, L.C., Hartman, J.H., and Wesselingh, F.P., 2006, Close evolutionary affinities between freshwater corbulid bivalves from the Neogene of Western Amazonia and Paleogene of the Northern Great Plains, USA, *Journal of South American Earth Sciences* **21**: 28–48.

Arslanov, K.A., Tertychny, N.I., Kuznetsov, Y.Y., Chernov, S.B., Lokshin, N.V., Gerasimova, S.A., Maksimov F.E., and Dodonov, A.E., 2002, $^{230}$Th/U and $^{14}$C dating of mollusc shells from the coasts of the Caspian, Barents, White and Black Seas, *Geochronometria* **21**: 49–56.

Bogan, A. and Bouchet, P., 1998, Cementation in the freshwater bivalve family Corbiculidae (Mollusca, Bivalvia): a new genus and species from Lake Poso, Indonesia, *Hydrobiologia* **389**: 131–139.

Bottjer, D.A., Campbell, K.A., Schubert, J.K., and Droser, M.L., 1995, Palaeoecological models, non-uniformitarianism, and tracking the changing ecology of the past, *Geological Society Special Publication* **83**: 7–26.

Bouchet, P., 1995, A major new molluscan radiation discovered in the ancient lakes of Sulawesi, in: Guerra, A., Rolan, E., and Rocha, F. (eds), *Abstracts 12th International Malacological Congress, Vigo, 1995*. CSIC, Vigo, pp. 14–15.

Brown, D.S., 1992, Interpreting the Turkana fossil gastropods: evidence from the living snails, *Proceedings ninth International Malacological Congress*, pp. 69–75.

Brown, F.H. and Feibel, C.S, 1986, Revision of lithostratigraphic nomenclature in the Koobi Fora region, Kenya, *Journal Geological Society, London* **143**: 297–310.

Cohen, A.S. and Johnston, M.R., 1987, Speciation in brooding and poorly dispersing lacustrine organisms, *Palaios* **2**: 426–435.

Daget, J., 1998, *Catalogue Raisonné des Mollusques bivalves d'eau douce Africains*. Backhuys Publishers, Leiden.

Danukalova, G., 1996, Bivalvian molluscs and stratigraphy of the Aktchagilian, *Trudy Palaeontogischeskih Instituta Rossia Akademyia Nauka* **265**: 1–132 (in Russian).

Gingras, M.K., Räsänen, M.E., Pemberton, S.G., and Romero, L.P., 2002, Ichnology and sedimentology reveal depositional characteristics of bay-margin parasequences in the Miocene Amazonian foreland basin, *Journal of Sedimentary Research* **72**: 871–883.

Glaubrecht, M., von Rintelen, T., and Korniushin, A.V., 2003, Towards a systematic revision of brooding freshwater Corbiculidae in southeast Asia (Bivalvia, Veneroida): on shell morphology, anatomy and molecular phylogenetics of endemic taxa from islands in Indonesia, *Malacologica* **45**: 1–40.

Glaubrecht, M. and Köhler F., 2003, Radiating in a river: systematics, molecular genetics and morphological differentiation of viviparous freshwater gastropods endemic to the Kaek River, central Thailand (Cerithioidea, Pachychilidae), *Biological Journal of the Linnaean Society* **82**: 275–311.

Grigorovich, I.A., Therriaul, T.W., and MacIsaac, H.J., 2003, History of aquatic invertebrate invasions in the Caspian Sea, *Biological Invasions* **5**: 103–115.

Harzhauser, M. and Mandic, O., 2004, The muddy bottom of Lake Pannon – a challenge for dreissenid settlement (Late Miocene; Bivalvia), *Palaeogeography, Palaeoclimatology, Palaeoecology* **204**: 331–352.

Hausdorf, B., Röpstorf, P., and Riedel, F., 2003, Relationships and origin of endemic Lake Baikal gastropods (Caenogastropoda: Rissooidea) based on mitochondrial DNA sequences, *Molecular Phylogenetics and Evolution* **26**: 435–443.

Hoorn, C., 1993, Marine incursions and the influence of Andean tectonics on the Miocene depositional history of northwestern Amazonia: results of a palynostratigraphic study, *Palaeogeography, Palaeoclimatology, Palaeoecology* **112**: 187–238.

Johnston, M.R. and Cohen, A.S., 1987, Morphological divergence in endemic gastropods from Lake Tanganyika: implications for models of species flock formation, *Palaios* **2**: 413–425.

Kaandorp, R., Wesselingh, F., and Vonhof, H., 2006, Ecological implications from stable isotope records of Miocene western Amazonian bivalves, *South American Journal of Earth Sciences* **21**: 54–74.

Kaandorp, R.J.G., Vonhof, H.B., Wesselingh, F.P., Romero Pittman, L., Kroon, D., and van Hinte, J.E., 2005, Seasonal Amazonian rainfall variation in the Miocene climate optimum, *Palaeogeography, Palaeoclimatology, Palaeoecology* **221**: 1–6.

Kammer, T.W., Baumiller, T.K., and Ausich, W.I., 1997, Species longevity as a function of niche breadth: evidence from fossil crinoids, *Geology* **25**: 219–222.

Korniushin, A.V., Krstanovski, Z., and Kostoski, G., 2000, Anatomical evidence of close affinity between two endemic species of *Pisidium* (Bivalvia, Sphaeriidae) from some ancient lakes and the widely distributed taxa, *Journal of Zoological, Systematical and Evolutionairy Research* **38**: 81–86.

Kosarev, A.N. and Yablonskaya, E.A., 1994, *The Caspian Sea*. Academic Publishers, The Hague.

Lewy, Z. and Samtleben, C., 1979, Functional morphology and palaeontological significance of the conchiolin layers in corbulid pelecypods, *Lethaia* **12**: 341–351.

Magyar, I., Geary, D.H., and Müller, P., 1999, Paleogeographic evolution of the Late Miocene Lake Pannon in Central Europe, *Palaeogeography, Palaeoclimatology, Palaeoecology* **147**: 151–167.

Martens, K., 1997, Speciation in ancient lakes, *Trends in Ecology and Evolution* **12**: 177–182.

Matsuoka, K., 1987, Malacofaunal succession in Pliocene to Pleistocene non-marine sediments in the Omi and Ueno basins, central Japan, *Journal Earth Sciences Nagoya University* **35**: 23–115.

Mátyás, J., Burns, S.J., Müller, P., and Magyar, I., 1996, What can stable isotopes say about salinity? An example from the Late Miocene Pannonian Lake, *Palaios* **11**: 31–39.

Mello, L.H.C., 1999, *Analise cladistica dos bivalves do Grupo Passa Dois (Neopermiano), Bacia do Parana: implicaçãos taxonomicas, evolutivas e paleobiogeograficas*. Unpublished MSc thesis, Instituto de Geociências, Universidade de Sao Paulo, Sao Paulo. http://www.sbgeo.org.br/580.html for summary

Myers, G.S., 1960, The endemic fish fauna of Lake Lanao and the evolution of higher taxonomic categories, *Evolution* **14**: 323–333.

Michel, E., 1994, Why snails radiate: a review of gastropod evolution in long-lived lakes, both recent and fossil, *Archiv für Hydrobiologie, Beiheft Ergebnisse Limnologie* **44**: 285–317.

Monsch, K.A., 1998, Miocene fish faunas from the northwestern Amazonia basin (Colombia, Peru, Brazil) with evidence of marine incursions, *Palaeogeography, Palaeoclimatology, Palaeoecology* **143**: 31–50.

Müller, P., Geary, D.H., and Magyar, I., 1999, The endemic molluscs of Late Miocene Lake Pannon: their origin, evolution and family-level taxonomy, *Lethaia* **32**: 47–60.

Müller, P. and Magyar, I., 1992, Continuous record of the evolution of lacustrine cardiid bivalves in the Late Miocene Pannonian Lake, *Acta Palaeontologica Polonica* **36**: 353–372.

Muñoz Torres, F., Whatley, R., and van Harten, D., 1998, The endemic non-marine Miocene ostracod fauna of the Upper Amazon Basin, *Revista Española de Micropaleontologia* **30**: 89–105.

Muñoz Torres, F., Whatley, R., and van Harten, D., 2006, Miocene ostracod (Crustacea) biostratigraphy of the upper Amazon Basin and evolution of the genus *Cyprideis*, *Journal of South American Earth Sciences* **21**: 75–86.

Nevesskaja, L.A., 1965, Bivalve molluscs of the Late Quaternary of the Black Sea, their classification and ecology, *Trudy Paleontological Institute USSR* **105**: 1–391.

Nevesskaja, L.A., Paramonova, N.P., and Popov, S.V., 2001, History of the Lymnocardiinae (Bivalvia, Cardiidae), *Paleontological Journal* **35**: s147–s217.

Nishino, M. and Watanabe, N.C., 2000, Evolution and endemism in Lake Biwa, with special reference to its gastropod mollusc fauna, *Advances in Ecological Research* **31**: 151–180.

Nuttall, C.P., 1990, A review of the Tertiary non-marine molluscan faunas of the Pebasian and other inland basins of north-western South America, *Bulletin of the British Museum of Natural History* (Geology Series) **45**: 165–371.

Oschmann, W., 1994, Adaptive pathways of benthic organisms in marine oxygen-controlled environments, *Neues Jahrbuch Geologie und Paläontologie, Abhandlungen* **191**: 393–444.

Pana, I., Enache, C., and Andreescu, I., 1981, *Fauna de molluste a depozitelor cu ligniti din Oltenia*. Institutul de Cercetari, inginerie technologica si proiectari miniere pentru lignit, Craiova.

Popov, G.I., 1983, *Pleistocene Black Sea-Caspian Sea straits*. Academy of Sciences of the USSR Nauka, Moscow.

Popov, S.V., Shcherba, I.G., Ilyina, L.B., Nevesskaya, L.A., Paramonova, N.P., Khondkarian, S.O., and Magyar, I., 2006, Late Miocene to Pliocene palaeogeography of the Paratethys and its relation to the Mediterranean, *Palaeogeography, Palaeoclimatology, Palaeoecology* **238**: 91–106.

Räsänen, M.E., Linna, A., Irion, G., Rebata Hernani, L., Vargas Huaman, R., and Wesselingh, F., 1998, Geologia y geoformas de la zona de Iquitos, *Anales Turkuensis* AII **114**: 59–137.

Reid, D.F. and Orlova, M.I., 2002, Geological and evolutionary underpinnings for the success of Ponto-Caspian species invasions in the Baltic Sea and North American Great Lakes, *Canadian Journal of Fisheries and Aquatic Sciences* **59**: 1144–1158.

Runnegar, B. and Newell, N.D., 1971, Caspian-like relict molluscan fauna in the South American Permian, *Bulletin of the American Museum of Natural History* **146**: 1–66.

Savazzi, E., 1999, Soft-bottom dwellers and the leaning tower of Pisa: adaptive exploitations of unstable positions, in: Savazzi, E. (ed.), *Functional Morphology of the Invertebrate Skeleton*, Wiley, London, pp. 123–128.

Savazzi, E. and Peiyi, Y., 1992, Some morphological adaptations in freshwater bivalves, *Lethaia* **25**: 195–209.

Savazzi, E. and Sälgeback, J., 2004, A comparison of morphological adaptations in the cardiid bivalves *Cardium* and *Budmania*. *Paleontological Research* **8**: 221–239.

Scholz, H. and Glaubrecht, M., 2004, Evaluating limnic diversity: towards a revision of the unionoid bivalve *Coelatura* Conrad, 1853 in the Great Lakes of East Africa and adjacent drainage systems (Mollusca, Bivalvia, Unionidae), *Mitteillungen Museum fürNaturkunde, Berlin, Zoologische Reihe* **80**: 89–121.

Sherbakov, D.Y., 1999, Molecular phylogenetic studies on the origin of biodiversity in Lake Baikal, *Trends in Ecology and Evolution* **14**: 92–95.

Simões, M.G, Kowalewski, M., Freitas Torello, F. de, Ghilardi, R.P., and Cruz de Mello, L.H., 2000, Early onset of modern-style shell beds in the Permian sequences of the Parana Basin: implications for the Phanerozoic trend in bioclastic accumulations, *Revista Brasilera de Geociencias* **30**: 495–499.

Simões, M.G., Marques, A.C., Mello, L.H.C., and Anelli, L.E., 1997, Phylogenetic analysis of the genera of the extinct family Megadesmiidae (Pelecypoda, Anomalodesmata), with remarks on its palaeoecology and taxonomy, *Journal of Comparative Biology* **2**: 75–90.

Simões, M.G., Rocha-Campos, A.C., and Anelli, L.E., 1998, Paleoecology and evolution of Permian pelecypod assemblages (Parana Basin) from Brazil, in: Johnston, P.A. and Haggart J.W. (eds), *Bivalves, an Eon of Evolution. Paleobiological Studies honouring Norman D. Newell*, Calgary, University of Calgary Press, pp. 443–452.

Sitnikova, T.Y., 1994, Recent views on the history and diversity of the Baikalian malacofauna, *Archiv für Hydrobiologie, Beiheft Ergebnisse Limnologie* **44**: 319–326.

Slugina, Z.V., 1995, Systematic and distribution of bivalvian molluscs of lake Baikal, in: Guerra, A., Rolan, E., and Rocha, F. (eds), *Abstracts 12th International Malacological Congress*, Vigo, 1995, CSIC, pp. 50–51.

Slugina, Z.V., 2000, Comparative analysis of bivalves from lakes Baikal and Hövsgöl, *Ruthenica* **11**: 37–41.

Slugina, Z.V., Starobogatov, Y.I., and Korniushin, A.V., 1994, Bivalves (Bivalvia) of Lake Baikal, *Ruthenica* **4**: 111–146.

Snel, E., Marunteanu, M., and Meulenkamp, J.E., 2001, Late Miocene-Early Pliocene marine connections between Atlantic/Mediterranean and the Paratethys, in: Agusti, J., and Oms, O. (eds), *Abstracts of the 2nd EEDEN Plenary Workshop*, Sabadell, p. 69.

Snel, E., Marunteanu, M., Macalet, R., Meulenkamp, J.E., and van Vugt, N., 2006, Late Miocene to Early Pliocene chronostratigraphic framework for the Dacic Basin, Romania, *Palaeogeography, Palaeoclimatology, Palaeoecology* **238**: 107–124.

Stanley, S.M., 1970, Relation of shell form to life habits of the Bivalvia (Mollusca), *Geological Society of America, Memoir* **125**: 1–296.

Tarasov, A.G., 1996, Deep-water Caspian benthic fauna, 1. Genesis and vertical zonality, *Zoologichesky Zhurnal* **75**: 1763–1775.

Taylor, J.D. and Glover, E.A., 2000, Functional anatomy, chemosymbiosis and evolution of Lucinidae, *Geological Society of London, Special Publications* **177**: 207–225.

Tchepalyga, A., 2003, Late glacial great flood in the Black Sea and Caspian Sea, *GSA Abstracts with Programs* **35**: 460.

Thayer, C.W., 1975, Morphological adaptations of benthic invertebrates to soft substrata, *Journal of Marine Research* **33**: 177–189.

Therriault, T.W, Docker, M.F., Orlova, M.I., Heath, D.D., and MacIsaac, H.J., 2004, Molecular resolution of the family Dreissenidae (Mollusca, Bivalvia) with emphasis on Ponto-Caspian species, including first report of *Mytilopsis leucophaeata* in the Black Sea basin, *Molecular Phylogenetics and Evolution* **30**: 479–489.

van Aartsen, J.J. and Wesselingh, F.P., 2000, New *Odostomia* species (Gastropoda, Heterobranchia, Pyramidellidae) from the Miocene Pebas formation of Western Amazonia (Peru, Colombia), *Basteria* **64**: 163–168.

Vermeij, G.J., 1987, *Evolution and escalation: an ecological history of life,* Princeton University Press, Princeton..

Vermeij, G.J. and Dudley, E.C., 1985, Distributions of adaptations: a comparison between functional shell morphology of freshwater and marine pelecypods, in: Trueman, E.R. (ed.), *Biology of the Mollusca, Vol. 10, Evolution*, Academic Press, London, pp. 461–478.

Vermeij, G.J. and Wesselingh, F.P., 2002, Neogastropod molluscs from the Miocene of Western Amazonia, with comments on marine to freshwater transitions in mollusks, *Journal of Paleontology* **76**: 265–270.

Vonhof, H.B., Wesselingh, F.P., and Ganssen, G.M., 1998, Reconstruction of the Miocene western Amazonian aquatic system using molluscan isotope signatures, *Palaeogeography, Palaeoclimatology, Palaeoecology* **141**: 85–93.

Vonhof, H.B., Wesselingh, F.P., Kaandorp, R.J.G., Davies, G.R., van Hinte, J.E., Guerrero, J., Räsänen, M., Romero-Pitmann, L., and Ranzi, A., 2003, Paleogeography of Miocene Western Amazonia: isotopic composition of molluscan shells constrains the influence of marine incursions, *GSA Bulletin* **115**: 983–993.

Wenz, W., 1942, Die Mollusken des Pliozäns der rumänischen Erdöl-Gebiete, *Senckenbergiana* **24**: 1–293.

Wesselingh, F.P., Cadée, G.C., and Renema, W., 1999, Flying high: on the airborne dispersal of aquatic organisms as illustrated by the distribution histories of the gastropod genera *Tryonia* and *Planorbarius*, *Geologie en Mijnbouw* **78**: 165–174.

Wesselingh, F.P., Räsänen, M.E., Irion, G., Vonhof, H.B., Kaandorp, R., Renema, W., Romero-Pittman, L., and Gingras, M., 2002, Lake Pebas: a palaeo-ecological reconstruction of a Miocene long-lived lake complex in Western Amazonia, *Cainozoic Research* **1**: 35–81.

Wesselingh, F.P., 2006a, Molluscs from the Miocene Pebas Formation of Peruvian and Colombian Amazonia, *Scripta Geologica* **133**: 19–290.

Wesselingh, F.P., 2006b, Evolutionary ecology of the Pachydontinae (Bivalvia, Corbulidae) in the Pebas lake/wetland system (Miocene, western Amazonia), *Scripta Geologica* **133**: 395–417.

Whatley, R., Muñoz-Torres, F., and van Harten, D., 1998, Des ostracods d'un lac sale Néogène àl'ouest du basin Amazonien. *Proceedings of the Third European Symposium on Ostracoda* (Bierville).

Williamson, P.G., 1981, Palaeontological documentation of speciation in Cenozoic molluscs from Turkana Basin, *Nature* **293**: 437–443.

Wilson, A.B., Glaubrecht, M., and Meyer A., 2004, Ancient lakes as evolutionairy reservoirs: evidence from the thalassoid gastropods of Lake Tanganyika. *Proceedings of the Royal Society London, Biological Sciences* **271**: 529–536.

Chapter 10

# Patterns in Insular Evolution of Mammals: A Key to Island Palaeogeography

JOHN DE VOS, LARS W. VAN DEN HOEK OSTENDE, AND GERT D. VAN DEN BERGH
Nationaal Natuurhistorisch Museum Naturalis, P.O. Box 9517, 2300 RA Leiden, The Netherlands, hoek@naturalis.nl, vos@naturalis.nl



**Abstract** The clearest examples of dwarfism and gigantism on islands are found in the fossil record. They form part of unbalanced faunas, which attest that only a few non-volant mammals were able to reach the island. Pygmy elephants and giant rats evolved in the isolation of these insular environments. Thus, the telltale signs of an insular fauna can be used to deduce the island's palaeogeography. The faunas from the Gargano (Italy), a region presently forming part of the mainland, contain various giant rodents and a giant insectivore indicating that the Gargano was an island during the Mio–Pliocene. On the other hand, the Miocene and Pliocene faunas from the present-day Greek islands are balanced, indicating that they were connected with the continent at that time. The Pleistocene faunas from the same islands, however, are unbalanced, showing they lived in an isolated insular environment, and thus the faunas bear witness of the timing of tectonic processes. The same patterns as in the Mediterranean can also be found on the Indonesian islands. The islands of the Sunda shelf, which during glacials are connected to the mainland, have balanced faunas. In Wallacea we find the familiar pattern of dwarfed large mammals



*W. Renema (ed.), Biogeography, Time, and Place: Distributions, Barriers, and Islands*, 315–345

and giant murids in unbalanced faunas. The island of Flores even yielded remains of *Homo floresiensis*, a small hominin that shows that Man as well could be subject to the pattern of insular evolution.

## 1. Introduction

Islands have been of vital importance in shaping the ideas of evolution formulated by Darwin and Wallace. On his voyage around the world on board of the "Beagle", Darwin visited the Galapagos Islands. Here he discovered a fauna that, although showing signs of resemblance with the fauna of Patagonia, yet had peculiarities of its own. Moreover, he observed that species differed from island to island, even if less than 100 km apart. The remark of Mr. Lawson, vice-governor of the Archipelago, that he could tell immediately from which island a particular tortoise came, made him realize that every island had its own evolutionary history. Similarly, Wallace, while travelling among the islands of the East Indies, developed his ideas of evolution, independently of Darwin.

Darwin and Wallace were by no means the only naturalists, or even the first, to recognize pecularities in island faunas. The first descriptions of the fauna from the Pacific islands date back to the 18th century, when naturalists Johann Forster and Joseph Banks accompanied James Cook on his exploration voyages. Ever since Darwin and Wallace, islands have played a dominant role in our thinking on evolution, biogeography and ecology. This is understandable, as was pointed out by Lomolino et al. (2006: 469): "Islands and other insular habitats, such as mountaintops, springs, lakes, and caves, are ideal subjects for natural experiments. They are well defined, relatively simple, isolated, and numerous – often occurring in archipelagos of tens or hundreds of islands."

The study of island faunas and floras got a major boost in the 1960s, when MacArthur and Wilson (1963, 1967) published their equilibrium theory of island biogeography. They considered the species richness of islands, which as a rule is lower than on the mainland, as a result of a balance between ongoing colonization and extinctions. Thus, they introduced a dynamic view on island communities, where earlier workers had assumed a more static history. Until then, low species richness was considered the result of limited resources, and after initial settlement by early colonizers the niches were filled, leaving no room for change. This is sometimes referred to as the static theory of islands (Dexter, 1978). The opposing view of MacArthur and Wilson inspired a wide range of articles on island biogeography (see Lomolino et al., 2006 and references therein).

Also in the 1960s, Foster (1964) introduced the concept of dwarfism and gigantism for insular large and small mammals, respectively. The general pattern he noted became later known as the "Island Rule" (Van Valen, 1973). Foster first recognized this pattern in extant mammals and its general applicability was stressed by Lomolino (2005). Meiri et al. (2006), however, refuted the conclusions of Lomolino. Whereas neontologists debate the general character of the island rule, it has become widely accepted among palaeontologists. This is not surprising, since the clearest example of dwarfism (pygmy elephants and hippo's) and gigantism (giant rats and dormice) can be found in the fossil record. The presence of these typical island mammals on the islands of the Mediterranean was already known through, for instance, the work of Dorothy Bate, who travelled the various islands in search of fossils (Bate, 1904, 1905, 1907, 1909). It is no coincidence that the best examples of insular evolution are within the realm of palaeontology. Most of these animals became extinct during the Holocene as their ecosystems were disturbed. In the Mediterranean these extinctions seem to have occurred when Neolithic Man introduced alien species, including domesticated cattle and rats (e.g., Reese, 1996). In other parts of the world such extinctions took place in historical times, such as for the moas on New Zealand and the dodo on Mauritius. Other possible causes for extinctions are the climate change at the end of the Pleistocene, or, in specific cases, volcanic eruptions, as Morwood et al. (2004) suggested for Flores.

The patterns governing these fossil island faunas were described by Paul Sondaar, who first worked on the Mediterranean insular faunas (Sondaar and Boekschoten, 1967; Sondaar, 1976; Sondaar et al., 1986) and later continued studying the island faunas of the Indonesian Archipelago (Sondaar et al., 1994, van den Bergh et al., 2001). Sondaar (e.g., in Dermitzakis and Sondaar, 1978) explored to what extend the classical dispersal methods for mammals listed by Simpson (1940) could apply to the insular faunas of the Mediterranean. These possible dispersal methods are: (1) A corridor in which faunal interchange from one region to another is possible; (2) Filter dispersal: spread is probable for some organisms, but definitely improbable for others; (3) The pendel route: a route that is easily crossed vice versa between regions by some mammals, but an insurmountable barrier for others; and (4) A sweepstake: spread is impossible for most and very improbable for some organisms, but does occur accidentally.

The patterns found on islands that displayed clear dwarfism or gigantism suggested that chance dispersal, the sweepstake route, was the most important way for colonizing these islands (Sondaar, 1976). In particular, the highly unbalanced nature of the faunas suggested the fauna was nearly isolated. The same groups of mammals appear over and over again as colonizers of islands. Proboscideans, which are known to be prolific swimmers, hippopotami, which have a semiaquatic life style and float more readily because of their fat, and artiodactyls (mostly deer), which gain buoyancy because of the gasses in their gut. Small mammals may have reached islands floating on driftwood. Mammal groups with more limited swimming capabilities, such as carnivores and perissodactyls, are invariably missing. This suggests that islands showing such selective pattern lay at the fringes of the reach of non-volant mammals. The isolation also gave room for speciation as evidenced by the various dwarf and giant species, and in some cases to adaptive radiations, such as found for the deer of Crete (de Vos, 2000).

Even though the fossil record has yielded the clearest examples for dwarfism and gigantism, their role in theories on island biogeography seems limited. There appears to be a gap between the work done by palaeontologists and ecologists, and the first attempts to bridge that gap are only now being made (e.g., Raia and Meiri, 2006).

A possible explanation of the apparent differences lies in the isolation that allowed the evolution of dwarf elephants or giant rats on these islands. Since the islands could be reached by chance dispersal only, the dynamic equilibrium of ongoing colonizations and extinctions that plays a pivotal role in MacArthur and Wilson's theory is lacking. It is doubtful whether the faunas of islands which are nearer to the mainland and do have a certain degree of exchange would be recognizable as insular environments in the fossil record. But some oceanic islands yielded because of their isolated position typical faunas, which are totally different from those found on the nearest mainland.

Reversing the argument we can state that the presence of an unbalanced mammal fauna in the fossil record, particularly when some of its members display gigantism or dwarfism, points to an isolated insular environment. If, on the other hand, on a present-day island a balanced fauna is found, it indicates that in the past the island was connected to, or at least in the vicinity of, the mainland. In this paper we discuss several examples of the way in which fossil mammal faunas can be used to explore the palaeogeography of islands, focussing on the regions we know from personal experience.

## 2. Gargano, Island Faunas on the Present Mainland

The Mediterranean is an area of intense tectonic activity, leading to dramatic changes in the palaeogeography throughout the Cainozoic. A large submergence zone, as a result of the northward movement of Africa sets the stage. One of the more active orogenetic zones is found in present-day Italy. Mammal faunas from that region show that during the Tertiary, islands emerged and submerged again. One such island fauna is found on Sardinia, where an unbalanced small mammal fauna with endemic species was recovered at Oschiri Road Cut (de Bruijn and Rümke, 1974). The most important Italian island faunas, however, were discovered in the 1970s on the mainland, when M. Freudenthal studied fossils from fissure fillings on the Gargano peninsula.

Basically the Gargano consists of a block of uplifted Mesozoic limestone, which is exploited as "marble". This block is connected to the mainland by a low area, the so-called Foggia Graben or Tavoliere plain. The limestone of Gargano shows extensive superficial karst development, and the various quarries in the western area contain numerous fissure fillings, many of which are fossiliferous. Soon it became apparent that the assemblages found represent an island environment. The faunas are unbalanced, lacking perissodactyles, proboscideans, and carnivores, with the exception of the otter *Paralutra garganensis* (Willemsen, 1983). The artiodactyls are represented only by one aberrant form, the deer-like *Hoplitomeryx* (Leinders, 1984). Another indication that we are dealing with island faunas is the presence of various giant forms among the micro mammals, for example the murid *Mikrotia* (Freudenthal, 1976, 2006), the dormouse *Stertomys* (Daams and Freudenthal, 1985), and the erinaceid *Deinogalerix* (Freudenthal, 1972; Butler, 1980).

The importance of the Gargano lies in the occurrence of insular faunas from different ages from a single palaeo-island. Usually only one or a few sites are known from a certain island, but in Gargano *c.* 75 fissure fills were sampled, which seem to represent sequential time slices. Unfortunately, we can only reconstruct the time sequence based on the fossil contents of the different fissure fillings. A biostratigraphy of the Gargano faunas was made by Freudenthal (1976), who based his sequence primarily on the stage of evolution in the various *Mikrotia* lineages. Thus, the development of the fauna and its components can be reconstructed through time.

As to the age of the fauna, an upper limit is set by the marine sedimentary rock covering, at least in some places, the top of the quarries and fissures. Unfortunately, opinions about the age of these marine sediments vary. D'Allessandro et al. (1979) considered them to be Tortonian, whereas Valleri (1984) described a marine sequence of Lower and Upper Pliocene, in an area where d'Allesandro et al. (1979) had recognized Upper Miocene sedimentary rocks. The faunas set the lower limit themselves, but here too, there is little agreement. Freudenthal (1985) saw in the presence of *Cricetulodon* in Biancone, one of the oldest fissure fillings, an indication that the fauna must have entered the area in the Turolian. The presence of an *Apodemus*-like murid in the older faunas he considered in line with such an age. de Giuli et al. (1985; 1986a, b; 1987a, b) preferred a Messinian or even post-Messinian age for the colonization of the island. However, colonization after the Miocene does not seem very plausible, given the absence of arvicolids.

Thus, the correlation of the insular faunas from Gargano to the continental faunas of that period remains uncertain. Of course, one of the problems with correlating island faunas to the mainland is that the insular forms have changed so much that it is difficult to pinpoint the mainland ancestor. In the case of Gargano the typical examples of island evolution such as *Hoplitomeryx* and *Deinogalerix* have been studied most extensively. Answers to the origin of the fauna may, however, lie with the relatively unaltered micro mammals, some of which are currently under study.

In Gargano a second major phase of superficial karst development occurred during the Early Pleistocene, affecting the Neogene sedimentary rocks covering the Mesozoic block and the fissure fillings it contains. Cave deposits intercalated in these Neogene sediments have yielded a Villafranchian fauna with, a.o., *Mammuthus meridionalis*, *Equus*, *Stephanorhinus* cf. *etruscus*, *Homotherium latidens*, *Panthera gombagzoegensis*, *Pachycrocuta brevirostris*, *Apodemus*, and *Allophaiomys* (Freudenthal, 1971; Abbazzi et al., 1996). This is the typical Early Pleistocene mammal fauna found elsewhere on the continent. Therefore, we can conclude that at that time Gargano was already connected to the mainland.

The island faunas from Gargano are not the only Tertiary insular faunas from the Mediterranean. The islands of Mallorca and Menorca (Balears, Spain) have also yielded insular faunas, the oldest of which dates back to the Pliocene. There are some remarkable similarities to Gargano. For instance, the faunas are also characterized by the presence of an aberrant artiodactyle, here the caprine *Myotragus* (Bover and Alcover, 2003). Like in Gargano, there are fissure fillings of various ages, allowing us to follow the changes in the fauna over a respectable period of time. Unlike

Gargano, this faunal evolution continued right up to the beginning of the Holocene. The island faunas disappeared as Man entered the scene (Bover and Alcover, 2003), just like they did on other Mediterranean Islands, e.g., in the Greek archipelago.

## 3. The Greek Isles, a Developing Archipelago

The Greek islands make up only a minor part of the total surface of Greece. Nevertheless, many of the mammal faunas found in that country have been excavated on the islands, and together they give an almost continuous record of 20 My of faunal evolution.

### 3.1. Tertiary Faunas

The oldest small mammal fauna from the Greek islands (Fig. 1) was retrieved from a lignite mine near Aliveri on the island of Evia. The site yielded an MN (=Mammal Neogene Zone) 4 smaller mammal fauna (de Bruijn et al., 1980; van der Meulen and de Bruijn, 1982; Klein Hofmeijer and de Bruijn, 1985; Doukas, 1986; Lopez-Martinez, 1986; Alvarez-Sierra et al., 1987), and a felid (Schmidt-Kittler, 1983). Palaeogeographic reconstructions of the Early Miocene of the area suggest that Aliveri was situated on the same landmass as present-day Anatolia (de Bruijn and Saraç, 1991, Fig. 1). However, Anatolia and Europe formed two clearly separated bioprovinces at the time (van den Hoek Ostende, 2001a). The presence of Eomyidae and, e.g., the insectivore *Heterosorex* shows Aliveri has a fauna similar to the ones found in Central and Western Europe, and unlike the ones found in Anatolia.

The oldest larger mammals from the Greek archipelago are found on the island of Psara, Lesbos and Chios. From Psara an upper right M1 is known from *Prodeinotherium bavaricum* (Besenecker and Symeonidis, 1974). The same species is also known from Lesbos from St. Gavathas (Koufos et al., 2003). Based on K/Ar dating of the overlying volcanic rocks the authors placed the Gavathas specimen in the upper part of MN 3, suggesting that it is the oldest deinothere in Europe. In 2004 small bones from an artiodactyle were found at Gavathas (de Vos, unpublished data).

*P. bavaricum* is also known from Chios. This species is found south of Cape St. Helena, between Thymniana and Keramia, together with an Aragonian fauna (MN 5 till MN 7/8, maybe MN 9; Kondopulou et al., 1993; Lehmann and Tobien, 1995) consisting of the rodent *Megapedetes aegaeus*, the bovids *Hypsodontus* cf. *gaopense* and *Tethytragus* cf. *koehlerae*, the tragulid *Dorcatherium* sp., *Lophocyon paraskevaidisi*, *Sanitherium slagintweiti*, *Listriodon* sp., aff. *Euprox furcatus*, *Georgiomeryx georgalasi*, *Choerolophodon chioticus* (Paraskevaidis, 1940; Thenius, 1956; Melentis and Tobien, 1968; Tobien, 1969, 1980; Sen, 1977; Rothausen, 1977; Lehman and Tobien, 1995; Koufos et al., 1995; de Bonis et al., 1997a, b, 1998).

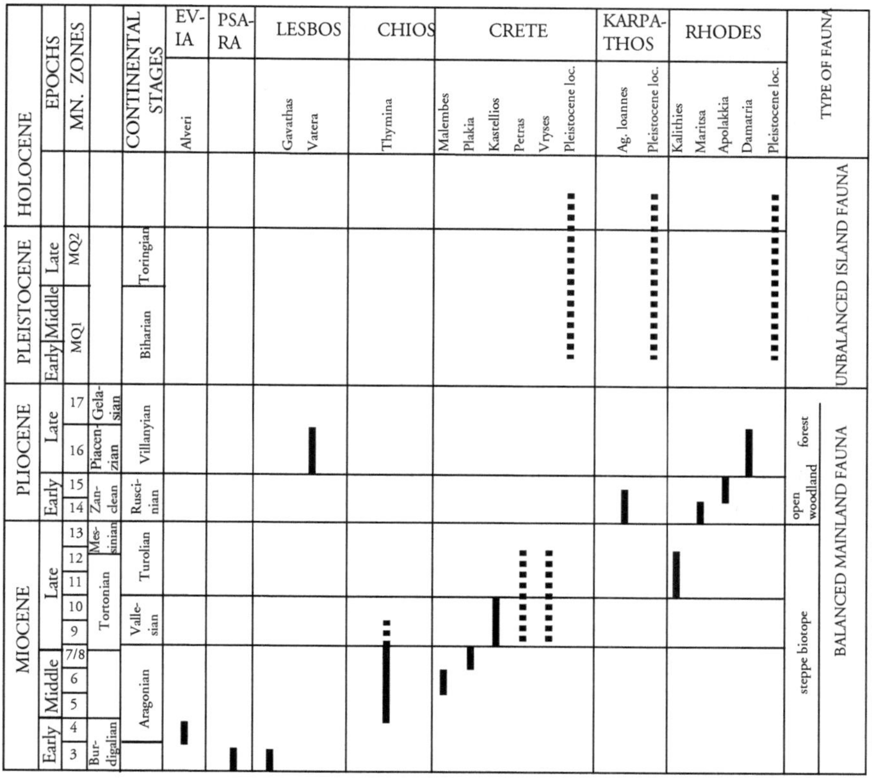


Fig. 1 A survey of the mammal localities of Evia, Psara, Lesbos, Chios, Crete, Karpathos, and Rhodes, and their supposed ages (altered after Sondaar et al., 1986).

Most Miocene faunas from the Greek isles have been found on Crete. At Malembes a faunule was found with cf. *Prohyrax hendeyi*, *Dorcatherium naui* and a bovid. The fauna was placed in MN 6, Middle Aragonian (van der Made, 1996). At Plakia a smaller mammal fauna was found consisting of *Spermophilinus* cf. *bredai*, *Blackia*? sp., *Forsythia*? sp., *Democricetodon* aff. *cretensis*, *Cotimys* sp., and *Glirudinus* sp. (de Bruijn and Meulenkamp, 1972). van der Made (1996) added the pig cf. *Propotamochoerus palaeochoerus* to the faunal list and assumed a Late Aragonian age for the fauna (MN 7/8).

Somewhat younger are the sites at Kastellios Hill, which are placed in the Vallesian (MN 9-10, van der Made, 1996). The fauna consists of *Cricetulodon* cf. *sabadellensis*, *Progonomys woelferi*, *P. cathalai*, *Spermophilinus bredai*, *Muscardinus* cf. *crusafonti*, *Schizogalerix* sp., *Hipparion* sp., *Taucanamo?/ Yunnanochoerus* sp., cf. *Pliocervus pentelici*, and unidentified remains of a bovid and a carnivore (van der Made, 1996).

Apart from these faunas there are some isolated finds, like *Microstonyx* cf. *major* at Petras (van der Made, 1996), and a mastodon at Vryses (Benda et al., 1968).

The latter site also yielded the ochotonid *Prolagus* sp. (de Bruijn in: van der Made, 1996) and is considered to be of Late Miocene age (Benda et al., 1968).

The sparse finds from Vryses are the youngest of the Miocene mammals on Crete. During the Late Miocene tectonic changes led to the fragmentation of the Cretan area, transforming it into an archipelago during the Late Tortonian/Middle Pliocene (Benda et al., 1974; Drooger and Meulenkamp, 1973). The whole island of Crete became subject to submergence from the Early Tortonian onwards and became a shallow sea with islands and shoals (Meulenkamp, 1971). Crete was divided into at least four islands during the Pliocene (Sondaar and Boekschoten, 1967). The absence of terrestrial mammals from this period may indicate that the area was too small to support stable populations (Sondaar et al., 1986) or that the fossilization potential was so small that they just not have been found. Crete re-emerged as an island during the Late Pliocene (Benda et al., 1974).

Although, there are no mammal faunas known from the Neogene of Crete, there are Late Miocene–Pliocene faunas from Rhodes. The oldest fauna has been found in Kalithies (Sondaar et al., 1986), and consists of the typical elements from Early or Middle Turolian (MN11/12) from the Greek mainland, the so-called Pikermi fauna (Sen et al., 1978), including the three-toed horse *Hipparion* sp., the hyaenid *Ictitherium orbingyi* and the sabre-toothed cat *Machairodus aphanistus*. A similar fauna was found on Samos (Bernor et al., 1996).

At Maritsa on Rhodes a large assemblage of small mammals was collected, which included sixteen rodents species, two insectivores and one lagomorph (de Bruijn et al., 1970). This fauna is considered to be of an Early Pliocene age (MN 14). From Apolakkia an MN 15 fauna was described by van de Weerd et al. (1982). The Apolakkia fauna contains among others *Castor fiber*, *Hipparion* aff. *crassum*, *Cervus rhenanus*, and *Mimomys occitanus*. Theodorou et al. (2000) described *Anancus arvernensis* from this site.

Additionally, Karpathos has yielded a Pliocene faunule. At Ag. Ioannis a faunal association with *Muscardinus* and *Kowalskia* was found. This faunule is placed in MN 14/first part MN 15 (van de Weerd et al., 1982).

Doukas and Athanassiou (2003), in their review of the Plio/Pleistocene Proboscidea from Greece, mention the Late Pliocene/Early Pleistocene continental proboscideans *A. arvernensis* and *M. meridionalis* from Kos. The latter species was also found on Evia and Kythera. The best-known faunas from the Greek isles from that period are found in different sites near Vatera on Lesbos (de Vos et al., 2002). Here the typical Villafranchian fauna, also known from Balkan sites such as Volax, Sesklo, Dafnero, Gerakarou, and Pyrgos in Greece (van der Meulen and Van Kolfschoten, 1986; Koufos et al., 1991; Athanassiou, 1996; Kostopoulos, 1996; Koufos and Kostopoulos, 1997), Slivnitsa in Bulgaria (Spassov, 1998) and Valea Gràunceanului and Fîntîna lui Mitilan in Roumania (Radulescu and Samson, 1991), is found. The presence of both *Anancus* and *Mammuthus* suggests a Late Pliocene age for the Vatera sites. Since *Equus* is already present at the sites, they must be younger than the first occurrence of this genus in Eurasia (2.5–2.7 My).

None of the Tertiary faunas from the Greek islands show any signs of endemism. The faunal elements from the various sites are the same as found in coeval localities on the mainland. Therefore, Sondaar et al. (1986) concluded that the various islands studied at the time were still connected to the mainland during the Tertiary. This certainly also holds true for Samos, which was studied later. The close resemblance of the Vatera assemblages to the mainland faunas, and the presence of carnivores such as the canid *Nyctereutes* and a sabre-toothed cat, clearly shows that Samos was still connected to the mainland during the Late Pliocene.

## 3.2. Pleistocene Faunas of the Aegean Islands

Pleistocene Crete is a classical example of an island that was colonized by sweepstake dispersal. From the time of colonization until the Holocene it had an unbalanced endemic island fauna. Crete got its present configuration in the late Early Pleistocene (Sondaar et al., 1986). The Pleistocene island fauna only contains cervids (de Vos, 1996a), elephants (Mol et al., 1996), hippos (Spaan, 1996), murids (Mayhew, 1996), shrews (Reumer and Payne, 1986; Reumer, 1996), birds (Weesie 1988), and reptiles (Brinkerink, 1996) (Fig. 2). Perissodactyles and carnivores, with the exception of an otter (Willemsen, 1996), are lacking.

Mayhew (1996) distinguished five biozones based on the endemic Pleistocene murid species of Crete. These species belong to two genera, *Kritimys* and *Mus*. The endemic murid *Kritimys* was larger than the brown rat, *Rattus norvegicus* (Mayhew, 1996), while the species of *Mus* are of small size. If we look at the ungulates, there is one faunal turnover. The faunal assemblages can be summarized as follows (Mayhew, 1996):

### *3.2.1.* Mammuthus creticus – Hippopotamus creutzburgi *Fauna, or the Kritimys Zone*

The oldest Pleistocene land vertebrates are from the locality Siteia 1. Besides *Kritimys* aff. *kiridus* (Mayhew, 1996), a rib of *Hippopotamus creutzburgi* was found (Spaan, 1996). At Cape Maleka 1, the dwarf elephant *M. creticus* is associated with *Kritimys* (Mayhew, 1996). The evolutionary stage of *Kritimys* indicates that this site is somewhat younger in age than Siteia 1, but it is still placed in the Early Pleistocene. The probable ancestors of *Kritimys* and the shrew *Crocidura zimmermanni*, found from the *K. catreus* subzone onwards, are of a Late Pliocene/Early Pleistocene mainland stock (Mayhew, 1996; Reumer, 1996). *H. antiquus* is considered to be the ancestor of the dwarf *H. creutzburgi*, which had a more unguligrade stance than the mainland species (Spaan, 1996). *M. creticus* probably descended from *M. meridionalis* (Mol et al., 1996). Apparently the Early Pleistocene *H. antiquus* and *M. meridionalis* dispersed by sweepstake route to Crete and adapted to the island environment by becoming dwarfed.

### *3.2.2.* Elephas creutzburgi – Candiacervus *Fauna, or the Mus Zone*

The earliest occurrence of *Mus* on Crete is at the Stavros Micro site. The earliest find of the cervid *Candiacervus* is in Charoumbes 2, the first co-occurrence of *Elephas* and *Candiacervus* in Charoumbes 3. The younger site (Gerani 5) has an

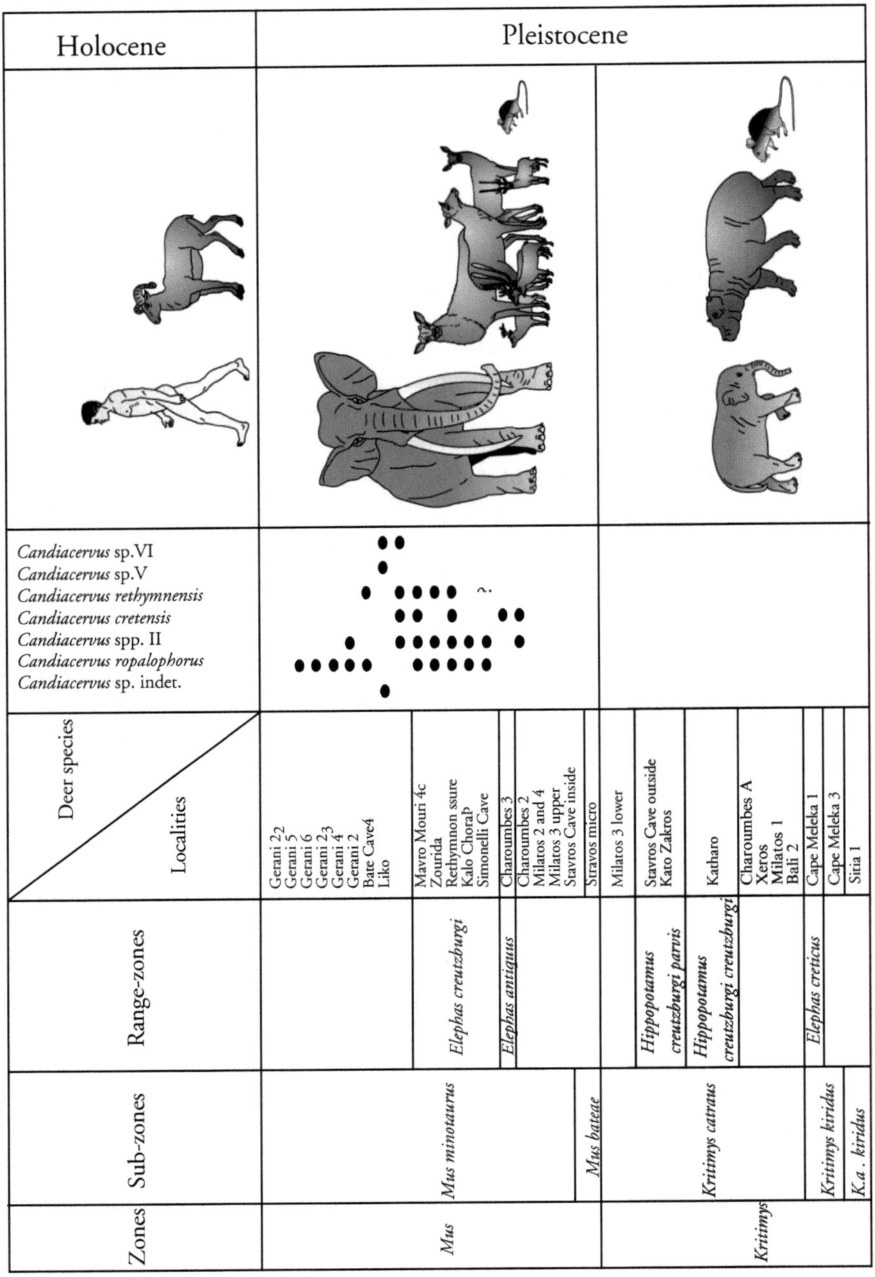


Fig. 2 Biostratigraphy and faunal turnovers on Crete during the Pleistocene (after de Vos, 1996).

absolute Electro Spin Resonance (ESR) age of 127,000 years ± 20% (Reese, 1996). The youngest locality is Gerani 2, which is AAR (Amino Acid Racemization) dated at 47,000 years ± 20% (Reese, 1996). Early Neolithic deposits cover the Pleistocene deposits in this cave.

de Vos (1984a) recognized six sizes in the endemic genus *Candiacervus*, which can be explained as an adaptive radiation of the ancestral stock (de Vos, 2000). Since there is still considerable overlap, particularly in the postcranial elements, no true species can be recognized. Nevertheless, species names are sometime attributed to extremes of the size spectrum (e.g., Capasso Barbato and Petronio, 1986). The adaptive radiation probably resulted from sympatric speciation (de Vos, 1996a). The large *Elephas* from Crete is a little smaller than the continental *Elephas antiquus*. Its taxonomic status is not clear and continues to be under discussion. Dermitzakis and Sondaar (1978) classified their material as *Elephas* cf. *antiquus*. Other authors have considered it to be a species on its own like *E. creutzburgi* by Kuss (1965) or *E. chaniensis* by Symeonides et al. (2001). Poulakakis et al. (2002) took an intermediate position by considering the Cretan elephant as a subspecies of the continental form (*Elephas antiquus creutzburgi*). The otter *Lutrogale cretensis* is the only mammalian carnivore in the Cretan fauna (Willemsen, 1996). It shows an adaptation to terrestrial life and was feeding on fish, crustaceans and small land vertebrates (Willemsen, 1996).

The genus *Mus* probably arrived in the early Middle Pleistocene (Mayhew, 1996). The first occurrence of *E. antiquus* in Europe is also in the early Middle Pleistocene, *c.* 700 ka. So, the faunal turnover from the *M. creticus – H. creutzburgi* fauna to the *E. creutzburgi – Candiacervus* fauna was no earlier than the transition from Early to Middle Pleistocene.

Due to the many localities of different ages, Crete can be considered as a case history for colonization, island adaptation of ungulates and rodents, and extinction of island endemics during the Holocene. It was, however, certainly not the only Greek island with a Pleistocene insular fauna. Doukas and Athanassiou (2003) give an overview from the islands of the Aegean with unbalanced endemic island faunas, mostly consisting of solely (dwarf) proboscideans. Apart from Crete these are Rhodes, Tilos, Dilos, Astypalaea, Seriphos, Milos, Naxos, Paros, and Kytnos. On Karpathos and Kassos, probably one single island at the time, also endemic cervids are found (Sondaar et al., 1996).

Examples of insular evolution, besides the Greek archipelago, can also be found on other Mediterranean islands. The Balears were mentioned above, with their *Myotragus* faunas that survived up to the end of the Pleistocene. Sicily is famed for having the smallest of all dwarf elephants, *E. falconeri*. Other than that, the giant dormouse *Leithia* was found here. On Cyprus, remains have been found of a dwarfed hippopotamus and elephant.

The study of the Mediterranean islands, which for a large part was the work of the late Paul Sondaar at Utrecht University, thus showed a consistent pattern occurring over and over again. Sondaar (1977) proposed a model for these islands. Colonization by sweepstake dispersal resulted in unbalanced endemic island faunas. At the time, dwarfism among large insular mammals was considered a token of degeneration. Sondaar (1977) showed that dwarfism, as well as gigantism among small mammals, was part of an adaptive pattern.

Although the consistency of the pattern could be demonstrated by comparing different Mediterranean islands, a weakness was that the island faunas of a single region were compared, often based on the same ancestral species and under comparable paleo-climatic conditions. The need was felt to compare the Mediterranean islands with another archipelago. The Indonesian islands offered a promising perspective, particularly since collections were already available in Naturalis.

## 4. Southeast Asia

Naturalis holds the Dubois collection, with fossils from sites on Sumatra and Java, extensively described in taxonomical studies by Hooijer (e.g., 1947, 1955). In the 1980s they formed the basis of a new stratigraphic and palaeoecological interpretation of the hominid bearing deposits of Java (de Vos et al., 1982). The other research areas were the islands of Wallacea. Here too, a lot of taxonomical work had already been published by Hooijer (e.g., 1949, 1953a, b, c, 1957a, 1964, 1969, 1972a, b) on material stored in Naturalis. These collections formed the basis for research of the islands of the Sunda Shelf and the islands of Wallacea.

### 4.1. Sunda Shelf

#### *4.1.1. Early/Middle Pleistocene of Java*

Based on work of de Vos et al. (1982), two faunal turnovers can be recognized on Java. Comparison of them with mainland faunas can be used to deduce dispersal routes.

*The Satir Fauna*

The oldest proboscidean on Java, the mastodon *Sinomastodon bumiajuensis*, originates from the so-called Satir Fauna from the late Early Pleistocene. The age, considered to be 1.5 Ma, is based on the faunal similarity between the Satir fauna from the type locality and the lower part of the Sangiran Formation, which has been fission track dated in the Sangiran area (Suzuki et al., 1985). The only other species in the fauna are a hippo (*Hexaprotodon simplex*), cervids and a giant tortoise *Geochelone*. In this fauna no fossil hominids have been found. The pollen spectrum suggests that the environment had been swampy.

*Sinomastodon* and *Hexaprotodon* are insufficiently well studied to know to what degree they are endemic. The unbalanced character of the fauna, with the same families as found on the isles of the Mediterranean and the presence of a giant tortoise, points to island conditions. As *Hexaprotodon* is an element of the Siwalik and Burma fauna, a Siva-Malayan origin and migration route is plausible.

In conclusion, there are indications that Java was an island during the late Early Pleistocene, and was colonized by hippos, cervids and proboscideans by sweepstake dispersal.

*Stegodon-Homo erectus fauna*

The sites Trinil, Kedung Brubus and Ngandong are attributed to the *Stegodon-H. erectus* fauna association. This association clearly shows affinities with the faunal association from the Indian Subcontinent (the Siwaliks) and Burma. However, species diversity is much lower. For example, horses and camels never reached Java. Five species and one genus occur both in the Siwaliks faunas and at Java, viz., *Hexaprotodon sivalensis, P. brevirostris, Caprolagus cf. sivalensis, H. ultimum [? = H. latidens; Galobart et al., 2003], Nestoritherium cf. sivalense, and Megantereon megantereon.*

Three species from this Javan fauna are closely related to Siwalik species, viz., *Stegodon trigonocephalus* with *Stegodon ganesa, E. husudrindicus* with *E. hysudricus*, and *Duboisia santeng* with the Boselaphini. de Vos (1996b) postulated that the *Stegodon-H. erectus* fauna association originated from the Siwaliks and reached Java via the so-called Siva-Malayan Route. The absence of camels and horses, but the presence of three carnivore species (the *Pachycrocuta, Homotherium,* and *Megantereon*) indicates that colonization took place via filter dispersal.

*4.1.2. Late Pleistocene of Java*

At the end of the Middle Pleistocene the species of the *Stegodon-H. erectus* fauna association became extinct. A faunal turnover took place and a new fauna migrated into the Indonesian Archipelago, the *Pongo-H. sapiens* fauna to which the Punung site is attributed (de Vos, 1996b). In this fauna we find the Indian elephant (*E. maximus*), orang-utan (*Pongo pygmaeus*), gibbon (*Hylobates syndactylus*), pig-tailed macaque (*Macaca nemestrina*) and Malayan bear (*Ursus malayanus*), all species which are still extant on the continent or in other places of the Indonesian Archipelago, but are no longer found on Java (Badoux, 1959). The large number of orang-utan and the presence of other primates indicate a humid tropical rainforest environment. Recently, a Punung-like fauna was discovered at Gunung Dawung (Storm et al., 2005). The faunal association from the Gunung Dawung site has been OSL (Optical Stimulated Luminescence) dated at 128 ka (Westaway et al., in preparation), corresponding with the onset of the last interglacial.

A younger faunule from the terminal phase of the last interglacial (radiocarbon dated at 35 ka) originates from the site Cipeundeuy, containing *E. maximus, Rhinoceros sondaicus, Muntiacus* sp., cervids, and *Bubalus* sp. (van den Bergh, 1999).

*Late Pleistocene of Sumatra*

Before searching in Java for the "missing link", Dubois (1889) excavated in the Padang Highlands, central Sumatra, during 1889–1890. The bulk of the material originated from three caves, viz., the Lida Ajer Cave near Pajakombo, the Sibrambang Cave and the Djambu Cave near Tapisello, all containing more or less the same fauna. Hooijer (1947, 1948, 1955) described the fossils of these caves and de Vos (1983) gave their faunal lists. The Sumatran cave faunas are characterized by an abundance of *Pongo*, suids, *Hylobates, Acanthion* and *Rhinoceros* teeth. Further, there are *E. maximus* molars and teeth of *P. tigris*. According to de Vos (1983), the Sumatran cave fauna suggests the presence of a tropical rainforest.

A right upper central incisor, which is semi-shovel shaped, and a left upper molar, both from Lida Ajer cave, were identified as *H. sapiens* by Hooijer (1948). The Cave material of Djambu is dated at 60–70ka, while the Lida Ajer material gives a date of 80 ka, all by AAR.

#### *4.1.3. Late Pleistocene of Borneo*

Lydekker (1885) described a tooth of a Mastodon (*Stegolophodon latidens*) from Borneo. Whether or not this small-sized proboscidean is an island form is not yet clear. Medway (1979) provided an overview of the fauna in the Niah Cave, which is dated as 50 ka, to Recent. The list is composed of both Pleistocene and Holocene species found in Niah. Next to domesticated mammals (like *Canis familiaris* in the upper 15–30cm level and *Sus scrofa* dom. in the 0–60cm level), *Pongo pygmaeus* in the levels of 260–270cm and *Hylobates* in the levels of 150–180cm are also present. The last two species indicate a tropical rainforest. The fauna is balanced.

Similar faunas as in the Sumatran caves, Borneo, and Java (Punung) with orang-utan are also found in fossil sites on the Asian continent, like Vietnam (Lang Trang Cave), Cambodia, (Phnom Loang) and China (de Vos, 1984b; Vu the Long et al., 1996)

Since the faunas of Punung, Borneo, and Sumatra are balanced and represent a tropical rainforest, which cannot cross a water barrier, we may assume a land connection with the mainland. A lowering of the sea level between 126 and 81 ka (Storm, 2001), but slightly earlier according to the new OSL dating evidence from Gunung Dawung (Westaway et al., in preparation), connected Sumatra, Java, and Borneo (the Sunda shelf) to the continent (Fig. 3), allowing corridor dispersal of the mainland fauna unto the Sunda Shelf (de Vos et al., 1999; de Vos and Vu the Long, 2001). Among the species that entered the region at this stage was *H. sapiens* (Storm et al., 2005).

### 4.2. Wallacea

The many islands and seas between the Sunda and Sahul shelves have been given the name Wallacea. In contrast to the islands of the Sunda Shelf, these islands were not connected to the mainland at times of low sea levels. Here we find unbalanced endemic island faunas. Every island had its own evolutionary history, although like in the Mediterranean a general pattern can be recognised.

#### *4.2.1. Sulawesi*

The first dwarf proboscidean remains from Southwest Sulawesi (Fig. 4) were described by Hooijer (1949) as *Archidiskodon celebensis*. Van Heekeren had recovered the fossil material near the village of Sompoh west of the Walanae River in the Sengkang district. Later collecting at several localities in the same area yielded abundant fossil material that permitted Hooijer (1953a, 1954a, 1955, 1972a) to assign the particular characteristics to this dwarf proboscidean, which was later reclassified as *E. celebensis*.

Fig. 3 Geographic position of the sites in southeast Asia mentioned in the text.

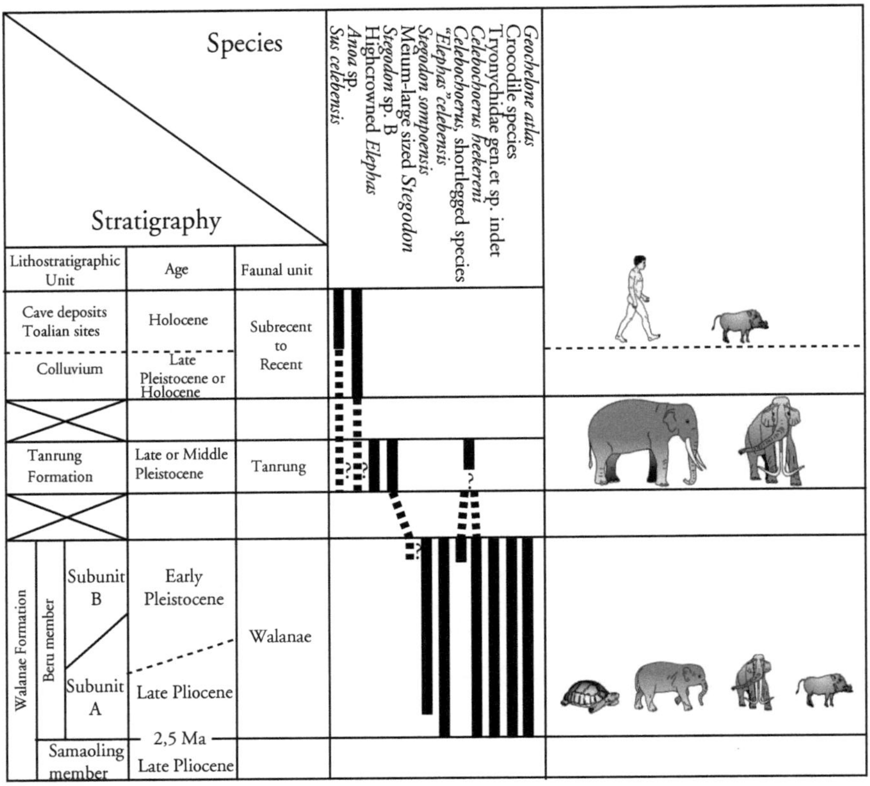


Fig. 4 Biostratigraphy and faunal turnovers on Sulawesi during the Late Pliocene–Pleistocene (after van den Bergh et al., 2001).

In addition to *E. celebensis*, Hooijer (1953a) announced the presence of *Stegodon*. Initially he doubted whether the fragmentary material should be attributed to a pygmy or a normal-sized *Stegodon*. In 1964, after additional fossil *Stegodon* material had been collected, he concluded that all the *Stegodon* material described so far belonged to a dwarf species, which he named *S. sompoensis*. Hooijer (1972b) described some *Stegodon* molar fragments collected in the previous years, which he attributed to *S.* cf. *trigonocephalus*, based on their supposedly large size. He concluded that there must have been both a large- and small-sized *Stegodon* in the fossil fauna. Hooijer (1948b, 1954) also described remains of an endemic suid from Southwest Sulawesi, *Celebochoerus heekereni*, and a giant tortoise (Hooijer 1948c, 1954b).

Stratigraphic data of the material described by Hooijer are lacking. Fieldwork during 1990–1994 clarified the stratigraphic sequence (van den Bergh et al., 2001; van den Bergh, 1999). In the Late Pliocene sedimentary rocks of *c.* 2.5 Ma a pigmy *Stegodon* (*Stegodon sompoensis*) and a pigmy *Elephas* (*"Elephas" celebensis*) are present in association with *Celebochoerus heekereni* and giant tortoise. During the Early Pleistocene, a large-sized *Stegodon*, represented by the few large-sized molar fragments and postcranial fragments, migrated into south Sulawesi. This might either be *Stegodon trigonocephalus* from Java or another large-sized *Stegodon* from the Philippines or the Asian mainland. By the middle or late Pleistocene, both pygmy proboscideans had become extinct, while the large-sized *Stegodon* continued or, alternatively, a new immigration of a large *Stegodon* took place besides a new immigration of a highly advanced *Elephas* species.

### *4.2.2. Flores*

The discovery of stone artefacts in association with remains of a large *Stegodon*, *S. trigonocephalus florensis*, and large murids, *Hooijeromys nusatenggara* at the localities Mata Menge and Boa Leza in west Central Flores (Fig. 5) was reported by Maringer and Verhoeven (1970). The same fauna had also been recovered from the locality Ola Bula, though from the latter locality no *in situ* artefacts were reported.

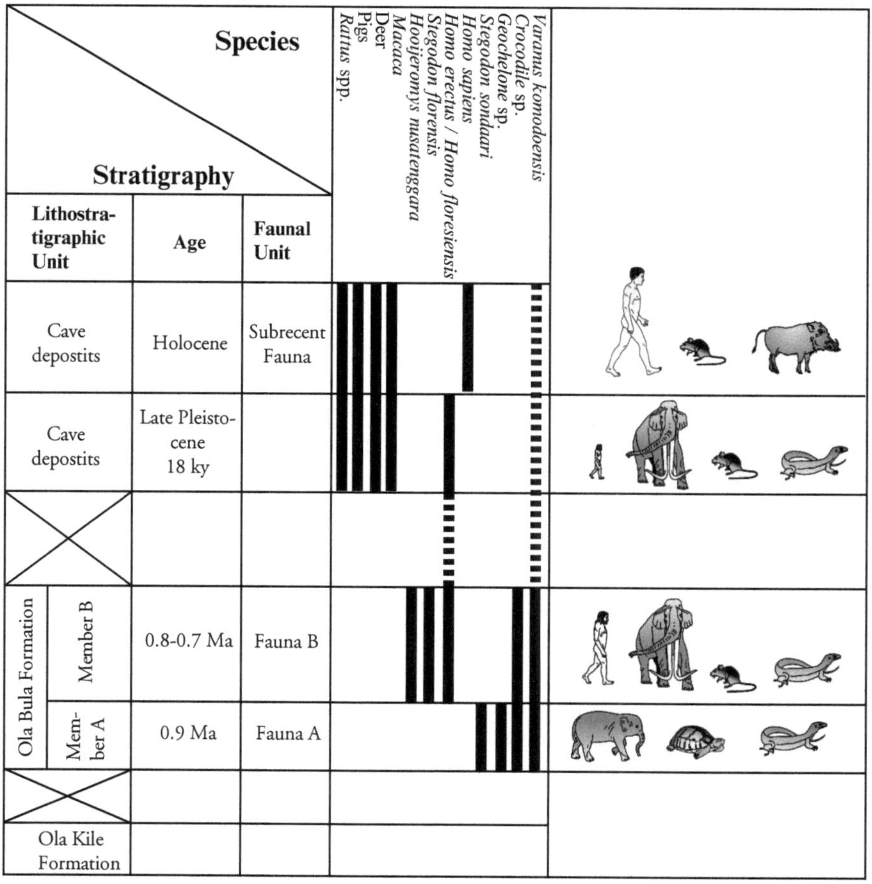


Fig. 5 Biostratigraphy and faunal turnovers on Flores during the Pleistocene (after van den Bergh et al., 2001).

Based on the association with *S. trigonocephalus florensis*, the artefacts, described as a number of pebble tools and retouched flakes mostly made of volcanic rock, were inferred to be middle or late Pleistocene and it was speculated that *H. erectus*? might have reached the Lesser Sunda islands.

The large to medium-sized *Stegodon* from the localities Ola Bula, Boa Leza, Mata Menge and Dhozo Dhalu is slightly more advanced than *Stegodon trigonocephalus* from Java (Hooijer 1953b, 1972b). The taxonomic status is, like the Cretan large elephant, under discussion. Hooijer (1957a) considered it to be a subspecies of the Javanese species (*Stegodon trigonocephalus florensis*), while it was considered to be a species on its own (*Stegodon florensis*) by van den Bergh (1999) and van den Bergh et al. (2001). Additionally, fossil remains of the giant rat *Hooijeromys nusatenggara* have been found at Mata Menge, Ola Bula, Dhozo Dhalu, and Boa Leza. Maringer and Verhoeven (1970) had discovered this murid earlier in the same area (Musser, 1981). Further, at Mata Menge a few teeth of a small crocodile were found, while at Dhozo Dhalu teeth of *Varanus komodoensis* occurred in association with the younger fauna. This still extant species seems to have been the only element from the older vertebrate fauna that did not become extinct.

In 1982 a new locality was discovered 2.5 km east of Mata Menge and 250 m southeast of Ola Bula, yielding fossils of a giant tortoise and a pygmy *Stegodon* (Sondaar, 1987). This locality, known as Tangi Talo, contained a distinct fauna that is stratigraphically below the Ola Bula excavation of Verhoeven (Sondaar et al., 1994; van den Bergh, 1999; van den Bergh et al., 2001). Based on faunal correlations, the Tangi Talo fauna was also inferred to be older than the artefact-bearing layer at Mata Menge. The fossil locality near Tangi Talo is the only one that yielded the giant tortoise-pygmy *Stegodon* fauna. The pygmy *Stegodon* from Tangi Talo represents a distinct species, *Stegodon sondaari*, differing from the pygmy stegodonts known from Java, Sulawesi and Timor, and the tortoise is on average much smaller than the fossil giant tortoises known from the other islands (van den Bergh, 1999; van den Bergh et al., 2001). According to Erick Setiabudi (personal communication), who currently studies the fossil giant tortoises from the various Indonesian islands, every island has its own species of *Colossochelys*, a genus also known from the Upper Siwaliks. At Tangi Talo remains of *Varanus komodoensis* were also recovered from the same layer. No artefacts have been found in the fossiliferous layer at Tangi Talo or in the tuffaceous interval in which this layer occurs.

The colonization of the island by humans coincides with a faunal turnover on the island. An endemic island fauna with giant tortoise and a pygmy *Stegodon* is replaced by an endemic island fauna with *Stegodon florensis* and *Hooijeromys nusatenggara*. This younger fauna is associated with artefacts at the localities Mata Menge and Boa Leza, indicating the co-occurrence with humans. Palaeomagnetic dating results suggest a late early Pleistocene age for the Tangi Talo fauna and early Middle Pleistocene for the Mata Menge fauna (Sondaar et al., 1994), indicating an age of *c*. 0.7 Ma for the artefact-bearing layer. Fission track ages of the same layer by Morwood et al. (1998) range between 0.88 ± 0.07 and 0.80 ± 0.07 Ma. More recent excavations carried out in 2004–2005 have yielded a much larger sample of stone tools from Mata Menge, which shows resemblance with the Late Pleistocene artefact assemblage from Liang Bua, which have been attributed to *H. floresiensis* recovered from the same deposits (Brumm et al., in press).

Brown et al. (2004) and Morwood et al. (2005) announced a new small-bodied hominin (*H. floresiensis*) from the cave Liang Bua. It was 1 m tall and had an endocranial volume of 380 $cm^3$. According to the authors the most likely explanation for its existance is long-term isolation, with subsequent dwarfing, of an ancestral *H. erectus*. Based on three alternative dating methods, the age was considered to be between >38 ka and *c.* 13 ka (Morwood et al., 2004). In the Liang Bua deposits *H. floresiensis* co-occurs with *Stegodon* and *Varanus komodoensis*. The ancestor of the Liang Bua *Stegodon* is *S. florensis*, and it can be classified as a distinct chrono-subspecies, *S. florensis insularis*, which is characterized by a 30% linear size reduction and more advanced molar ridge formula compared to its ancestor *S. florensis florensis* from the Middle Pleistocene Soa Basin localities (van den Bergh et al., in press).

The excavations in Liang Bua also yielded a large series of micro vertebrate fossils from the Late Pleistocene and Holocene, mostly murids, but also shrews (van den Hoek Ostende et al., 2006) and bats. Whereas currently only one species of giant rat, *Papagomys armandvillei*, survives on Flores, it was already apparent from excavations at another cave, Liang Toge, that several large murid species were present during the the Holocene. Hooijer (1957b) had described from the cave *Spelaeomys florensis* and *Papagomys verhoeveni*, as well as a subspecies of the recent *Papagomys*, *P. a. besar*. Musser (1981) subscribed Hooijer's conclusion that a second, smaller species of *Papagomys* was present at Liang Toge, but demonstrated that the holotype of *P. verhoeveni* was referable to *P. armandvillei*. A new type was designated and the smaller species was named *P. theodorverhoeveni*. The three giant murids from Liang Toge have all been found at Liang Bua. In addition, a third, larger species of *Papagomys* appears to be present in the Liang Bua fauna.

The Flores murid fauna also clearly shows that smaller species remain present next to the giants. Musser (1981) described middle-sized material from Liang Toge as *Komodomys rintjanus* and *Floresomys naso*, later changing the latter into *Paulamys naso*, as *Floresomys* proved to be occupied (Musser et al., 1986). The discovery of a recent (new or extant?) specimen of this species showed it to be in fact referable to the genus *Bunomys* (Kitchener et al., 1991a). Both *Komodomys rintjanus* and *Bunomys naso* are present in the extensive material from Liang Bua. Apart from these fossils a small-sized murid has been found, which had hitherto not been noted in literature. Part of this material is probably referable to *R. hainaldi*, an endemic rat from Flores with was discovered at the end of the last century (Kitchener et al., 1991b). It seems plausible that more than three taxa of small and middle-sized murids are present in the Liang Bua material, which is still awaiting taxonomical study.

### *4.2.3. Timor, Sumba, and Sangihe*

On Timor Verhoeven (1964) discovered the first remains of a dwarf *Stegodon* at Weaiwe. The posterior part of an M3 of this pygmy *Stegodon timorensis* was described and figured by Sartono (1969). More molar material from this Timor species was described and figured by Hooijer (1969), who also reported a large-sized *Stegodon* from the island, represented by a slightly worn DM3 from Sadilaun. Hooijer classified this fossil as *S. timorensis* subspecies D. More material of both the dwarf and the large-sized *Stegodon* was collected during a 1970 expedition in which Hooijer participated. Hooijer (1972b) described molar remains as well as postcranial elements obtained during this expedition. Sartono and Marino (1978) described additional material of the dwarf species from Timor.

Sartono (1979) announced the discovery of a pygmy *Stegodon* from Sumba. Unfortunately, the stratigraphic context is not clear.

The small island of Sangihe is located between the northern tip of Sulawesi and the Mindanao (the Philippines). In 1989, Dr. F. Aziz and Dr. Shibazaki collected some stegodont material from the island, including an upper tusk and some molar remains. All material originates from the Pintareng Formation on the southeast of Sangihe Island. The age of this formation is thought to be Pleistocene. The material was briefly described and figured by Aziz (1990) and attributed to *S*. sp. B cf. *trigonocephalus*.

### 4.3. The Philippines

From the islands of the Philippines only a few mammal fossils are found and described (de Vos and Bautista, 2003). Based on the size and morphology of the molars there is only one species of *Stegodon*, namely *Stegodon luzonensis* and a large species *Elephas*. *Stegodon luzonensis* is a little smaller than the continental form. Postcranial elements of the proboscideans show that there is a small and large proboscidean.

Other than that, there is a relatively small rhino, *Rhinoceros philippinensis* and material of a giant tortoise (de Vos and Bautista, 2003). This fauna composition indicates an endemic island fauna.

The stratigraphic position of the various fossils is not well known. Archaeological data (Fox and Paralta, 1974) suggest that the faunal remains are of Mid-Pleistocene age (following Von Koenigswald *in* Durkee and Pederson, 1961, p. 160) and that at least some of the tools found in the Philippines are coeval with the fossils (*in*: Wasson and Cochrane, 1979). Possibly this is the same succession as in Flores, that there first was a pygmy proboscidean and a giant tortoise, followed by a large proboscidean and artefacts.

## 5. Observations and Remarks

This overview of the results of the research on palaeo-isles in the latter half of the 20th century and beyond, allows us to make the following observations and remarks. Island evolution follows distinct patterns. A characteristic of island faunas is their unbalanced nature. They consist mainly of herbivores, like elephants, hippos, and cervids. Carnivores are absent, although both on Gargano and on Crete an otter has been found; presumably their semi-aquatic lifestyle enabled them to reach the islands, whereas other mammal predators could not. Perissodactyles, too, seem to be absent as a rule. Horses are absent in the Pleistocene faunas from Java, showing that even filter dispersal provides a barrier. Nevertheless, on Luzon a rhinoceros, was found, and fossil rhino's are also known from Japan. Possibly the gases in their gut provided enough floating capacity to reach the island via sweepstake dispersal.

The most common elements on the Quaternary islands are proboscideans, hippopotamuses, and cervids. Whether such a selectivity for certain families exists among smaller mammals is not clear. The island forms we know mainly belong to glirids (*Leithia*, *Stertomys*) and murids (*Mikrotia*, *Kritimys*, *Hooijeromys*), but it can easily be argued that these were common elements on the mainland at the time of colonization and that their arrival on islands, by, for example, rafting on driftwood, was simply a matter of chance. This also holds true for *Deinogalerix*, which is probably derived from *Parasorex*, a very common insectivore in the Late Miocene of Europe (van den Hoek Ostende, 2001b).

Apart from a preference for certain taxa, the telltale signs of an island fauna are dwarfism for the larger and gigantism among the smaller mammals. However, these phenomena are not necessarily clear-cut. Extreme dwarfism occurs, such as in the Sicilian *E. falconeri*. On the other hand, the *Elephas* found in the Middle and Late Pleistocene of Crete is only a little smaller than the mainland *E. antiquus*. In Wallacea and the Philippines, too, the Middle and Late Pleistocene proboscideans are usually somewhat smaller than the mainland forms, with the exception of the Late Pleistocene *S. florensis insularis*, which had an estimated body weight of between 350 and 900 kg. Other true pygmies occur on the Early Pleistocene islands of the region. The smallest pygmy *Stegodon* so far recorded, the Early Pleistocene *S. sondaari*, had an estimated body weight of between 200 and 500 kg (van den Bergh, 1999). That dwarfism among proboscideans is not restricted to the Mediterranean or Wallacea, is clear by the finds of pygmy mammoths on the Channel Islands of California (Agenbroad, 2003 and references therein).

Not all large mammals will reduce in size, nor will all small mammals obtain large proportions. The niches for small rodents are still available, as is clear from the large number of relatively unchanged rodents, insectivores and lagomorphs on Gargano. Also, on Flores, in Late Pleistocene sedimentary rocks in the Liang Bua cave, we find, apart from the giant murids, *Papagomys armandvillei*, *P. theodorverhoeveni*, *Papagomys* sp. and *Spelaeomys florensis* (Musser, 1981) at least three small to middle-sized rats. Another example of the absence of gigantism in smaller mammals is the Early Miocene fauna from Oschiri Road Cut on Sardinia (de Bruijn and Rümke, 1974). The largest glirid, *Glis major*, is of respectable dimensions, but certainly not a giant as *Leithia* or *Stertomys*. The Oschiri assemblage was identified as an insular fauna on the basis of its unbalanced composition and endemic forms as the mole *Nuragha*, but not on the presence of giants.

The specific characteristics of island faunas help us to determine whether or not an area was an island in the past. This is, for instance, clear from the Tertiary mammals of the Greek archipelago. Since they have a similar composition as the contemporaneous continental faunas, it is clear these islands were at the time still connected to the mainland. However, one has to bear in mind that if an island is sufficiently large, or if its time of isolation is limited, no island fauna will develop. This holds true, for example for the large islands of the Sunda Shelf. The only exception is the Early Pleistocene Satir fauna on Java, but probably large parts of present-day Java were still submerged when this island fauna developed (Bergh et al., 1996a). Whether or not an island fauna developed on Borneo, as indicated by *Stegolophodon latidens*, is not yet clear.

The islands of the Mediterranean and Wallacea do not only show striking similarities in the pattern of evolution, but also in the timing of certain events. Both on Crete and on, for example, Flores, a faunal turnover took place near the Early to Middle Pleistocene transition. This is also the period of a major faunal change on the continent in Eurasia, as the Villafranchian faunas gave way to the mammals of the Galerian. Since the turnover in insular faunas occurred on different sides of the globe, the explanation must be sought in a global cause. Probably this is related to the more extreme climatic fluctuations of the Middle Pleistocene. Glacial periods did not only cause major changes in the ecosystem on the continent, but also a considerable drop in sea level. Thus islands became more readily accessible and new faunal elements could reach the islands, either actively replacing existing faunas or simply filling up the empty space left after extinction. Increased competition between large-sized island herbivore species like proboscideans and intermediate-sized herbivores could be held responsible for the reduced degree of dwarfing seen in some of the island faunas (Raia and Meiri, 2006). With the niches of intermediate-sized herbivores already occupied, megaherbivores like proboscideans seem to dwarf to a lesser degree than when these intermediate-sized species would be missing. The total lack of predators may further enhance dwarfing of large herbivores to a minimal size with an optimal energetic balance (Palombo, 2007).

In the case of Flores this lead to the colonization of the island by *H. erectus* (Bergh et al., 1996b; Sondaar et al., 1996; Bergh, 1999), ultimately leading to the development of *H. floresiensis* (Brown et al., 2004), showing that Man too could be affected by the patterns of insular evolution. The presence of a large predator (*H. floresiensis*) could be the reason why *Stegodon florensis* did not became as small as *Stegodon sondaari*.

## Acknowledgements

The completion of this paper has been a long and bumpy road. Different versions of the manuscript were read and commented upon by various colleagues. We thank Andy Currant, Steve Donovan, Don MacFarlane, Fiona Fearnhead, Rienk de Jong, and Leo Kriegsman for their textual and conceptual comments. We are grateful to the editor Willem Renema for his patience and help in seeing this project through.

## References

Abbazzi, L., Benvenuti, M., Boschian, G., Dominci, S., Masini, F., Mezzabotta, C., Piccini, L., Rook, L., Valleri, G. and Torre, D., 1996, Revision of the Neogene and Pleistocene of the Gargano region (Apulia, Italy), the marine and continental successions and the mammal fauna assemblages in an area between Apricena and Poggio Imperiale (Foggia), *Memorie della Societa Geologica Italiana* **51** (1): 383–402.

Agenbroad, L.D., 2003, New absolute dates and comparisons for California's *Mammuthus exilis*, in: Reumer, J.W.F., de Vos, J., and Mol, D. (eds), *Advances in Mammoth Research. Proceedings of the Second International Mammoth Conference*, Rotterdam, May 16–20 1999, Deinsea **9**: 1–16.

d'Allesandro, A., Laviano, A., Ricchetti, G., and Sardella, A., 1979, Il Neogene del monte Gargano, *Bolletino della Società Paleontologia Italiana* **18**: 9–116.

Alvarez-Sierra, M.A., Daams, R. and van der Meulen, A.J., 1987, The mammals from the lower Miocene of Aliveri (Island of Evia, Greece); VII, The Eomyidae, *Proceedings Koninklijke Nederlandse Akademie van Wetenschappen*, Series B **90** (1): 47–56.

Athanassiou, A., 1996, Contribution to the study of the fossil mammals of Thessaly, PhD thesis, University of Athens (in Greek).

Aziz, F., 1990, Pleistocene mammal faunas of Sulawesi and their bearings to paleozoogeography, PhD thesis, Kyoto University.

Badoux, D.M., 1959, Fossil mammals from two deposits at Punung (Java), PhD. thesis, Utrecht University.

Bate, D.M.A., 1904, Further note on the remains of *Elephas Cypriotes* from a cave-deposit in Cyprus. *Philosophical Transactions of the Royal Society London B* **197**: 347–360.

Bate, D.M.A., 1905, Four and a half months in Crete in search of Pleistocene mammal remains. *Geological Magazine Series 5* **2**: 193–202.

Bate, D.M.A., 1907, On elephant remains from Crete, with description of *Elephas creticus*, sp. n. *Proceedings of the Zoological Society London* **1907**: 238–250.

Bate, D.M.A., 1909, Preliminary note on a new artiodactyle from Majorca, *Myotragus balearicus*, gen. et sp. nov. *Geological Magazine Series 5* **6**: 385–388.

Benda, L., Hiltermans, H., Kuss, S.E., and Symeonides, N.K., 1968, Der erste *Mastodon* Fund der insel Kreta, *Annales Géologiques des Pays Helléniques* **12**: 167–177.

Benda, L., Meulenkamp, J.E. and Zachariasse, W.J., 1974, Biostratigraphic correlations in the eastern Mediterranean Neogene: 1. Correlation between planktonic foraminiferal, uvigerinid, sporomorphal and mammal zonations of the Cretan and Italian Neogene, *Newsletter of Stratigraphy* **3**: 205–217.

Bernor, R., Solounias, N., Swisher, C., III, and Van Couvering, J.A., 1996, The correlation of three classical 'Pikermian' mammal faunas – Marageh, Samos, and Pikermi – with the European MN Unit System, in: Bernor, R., Fahlbusch, V., and Mittmann, H. (eds), *The Evolution of Western Eurasian Neogene Mammal Faunas*, Columbia University Press, New York, pp. 137–156.

Besenecker, H. and Symeonidis, N.K., 1974, Der erste Säugetierfund aus dem Neogen der *Griechischen Insel Psara (Ostägäis), Annales de Géologie des Pays Helleniques* **46**: 647–652.

Bover, P. and Alcover, J.A., 2003, Understanding Late Quaternary extinctions: the case of *Myotragus balearicus* (Bate, 1909), *Journal of Biogeography* **30** (5): 771–781.

Brown, P., Sutikna, T., Morwood, M.J., Soejono, R.P., Jatmiko, Wayhu Saptomo, E., and Rokus Awe Due, 2004, A new small-bodied hominin from the Late Pleistocene of Flores, Indonesia, *Nature* **431**: 1055–1061.

Brinkerink, H.P., 1996, Tortoises and turtles from the Pleistocene of Crete, in: Reese, D.E. (ed.), *Pleistocene and Holocene Fauna of Crete and Its First Settlers, Monographs in World Archaeology* **28**: 207–209.

Brumm, A., Aziz, F., van den Bergh, G.D., Morwood, M.J., Moore, M.W., Kurniawan, I., Hobbs, D.R., and Fullagar, R., 2006, Early stone technology on Flores and its implications for *Homo floresiensis*, *Nature*, **441**: 624–628.

Butler, M., 1980, The giant erinaceid insectivore, *Deinogalerix* Freudenthal, from the upper Miocene of Gargano, Italy, *Scripta Geologica* **57**: 1–72.

Capasso Barbato, L.C. and Petronio, C., 1986, *Cervus major* n. sp. of Bate Cave (Rethymnon, Crete), *Atti della Accademia Nazionale dei Lincei, Memorie Lincee, Scienze Fisische e Naturali, Serie VIII* **18** (2): 59–100.

de Bonis, L., Koufos, G.D., and Sen, S., 1997a, A giraffid from the Middle Miocene of the island Chios, Greece, *Palaeontology* **40** (1): 121 –123.

de Bonis, L., Koufos, G.D., and Sen, S., 1997b, The Sanitheres (Mammalia, Suoidea) from the middle Miocene the island Chios, Greece, *Revue de Paléobiologie* **16** (1): 259–270.

de Bonis, L., Koufos, G.D., and Sen, S., 1998. Ruminants (Bovidae and Tragulidae) from the Middle Miocene (MN 5) of the island Chios, Aegean sea (Greece), *Neues Jahrbuch für Geologie und Paläontologie, Abhandlungen* **210** (3): 399–420.

de Bruijn, H., Dawson, M.R., and Mein, P., 1970, Upper Pliocene Rodentia, Lagomorpha and Insectivora (Mammalia) from the island of Rhodes (Greece), I, II and III, *Proceedings Koninklijke Nederlandse Akademie van Wetenschappen B* **73** (5): 535–584.

de Bruijn, H. and Meulenkamp, J.E., 1972, Late Miocene rodents from the Pandanassa Formation (Prov. Rethymnon), Crete, Greece, *Proceedings Koninklijke Nederlandse Akademie van Wetenschappen B* **75** (1): 54–60.

de Bruijn, H. and Rümke, C.G., 1974, On a peculiar mammalian association from the Miocene of Oschiri (Sardinia). I and II, *Proceedings Koninklijke Nederlandse Akademie van Wetenschappen B* **77** (1): 44–79.

de Bruijn, H., Meulen, van der A.J., and Katsikatsos, G.C., 1980, The mammals from the lower Miocene of Aliveri (Island of Evia, Greece); Part 1, The Sciuridae, *Proceedings Koninklijke Nederlandse Akademie van Wetenschappen, B* **83** (3): 241–261.

de Bruijn, H. and Saraç, G., 1991, Early Miocene rodent faunas from the eastern Mediterranean area Part I. The genus *Eumyarion. Proceedings Koninklijke Nederlandse Akademie van Wetenschappen* **94** (1): 1–36.

de Giuli, C., Masini, F., and Torre, D., 1985, Effetto arcipelago: un esempio nelle fauna fossili del Gargano, *Bolletino della Società Paleontologia Italiana* **24**: 191–193.

de Giuli, C., Masini, F., Torre, D., and Boddi, V., 1986a, Endemism and biochronological reconstructions: the Gargano case history, *Bolletino della Società Paleontologia Italiana* **25**: 267–276.

de Giuli, C., Masini, F., Torre, D., and Valleri, G., 1986b, Mammal migration events in emerged areas of the Apulian platform during the Neogene, *Giornale di Geologia, Serie 3* **48** (1/2): 145–162.

de Giuli, C., Masini, F., Torre, D., and Valleri, G., 1987a, Paleogeography and mammal faunas in the Apulo-Dalmatic area, *Annals of the Hungarian Geological Institute* **70**: 471–476.

de Giuli, C., Masini, F., and Valleri, G., 1987b, Paleogeographic evolution of the Adriatic area since Oligocene to Pleistocene, *Rivista Italiana di Paleontologia e Stratigrafia* **93**: 109–126.

Daams, R. and Freudenthal, M., 1985, *Stertomys laticrestatus*, a new glirid (dormice, Rodentia) from the insular fauna of Gargano (Prov. of Foggia, Italy), *Scripta Geologica* **77**: 21–27.

Dermitzakis, M.D. and Sondaar, P.Y., 1978, The importance of fossil mammals in reconstruction paleogeography with special reference to the Pleistocene Aegean Archipelago, *Annales Géologiques des Pays Helléniques* **46**: 808–840.

Dexter, R.W., 1978, Some historical notes on Louis Agassiz's lectures on zoogeography. *Journal of Biogeography* **5**: 207–209.

Doukas, C.S., 1986, The mammals from the lower Miocene of Aliveri (Island of Evia, Greece); V, Insectivores, *Proceedings Koninklijke Nederlandse Akademie van Wetenschappen B* **89** (1): 15–38.

Doukas, C.S. and Athanassiou, A., 2003, Review of the Pliocene and Pleistocene Proboscidea (Mammalia) from Greece, in: Reumer, J.W.F., de Vos, J., and Mol, D. (eds), *Advances in Mammoth Research* (*Proceedings of the Second International Mammoth Conference*, Rotterdam, May 16–20 1999), *Deinsea* **9**: 97–110.

Drooger, C.W. and Meulenkamp, J.E., 1973, Stratigraphic contribution to geodynamics in the Mediterranean area; Crete as a case history, *Bulletin of the Geological Society of Greece* **10**: 193–200.

Dubois, E., 1889, Over de wenschelijkheid van een onderzoek naar diluviale fauna van Ned. Indië, in het bijzonder van Sumatra, *Natuurkundig Tijdschrift voor Nederlandsch-Indië* **48**: 148–165.

Durkee, E.F. and Pederson, S.L., 1961, Geology of Northern Luzon, Philippines, *American Association of Petrologists Geological Bulletin* **45**: 137–168.

Foster, J.B., 1964, Evolution of mammals on islands. *Nature* **202**: 234–235.

Fox, R.B. and Peralta, J.T., 1974, Preliminary report on the Palaeolithic archaeology of Cagayan Valley, Philippines and the Cabalwanian Industry, in: *Proceedings of the first Regional Seminar on South East Asian Prehistory and archseology*, National Museum of the Philippines, Manila, pp. 100–147.

Freudenthal, M., 1971, Neogene vertebrates from the Gargano Peninsula, Italy, *Scripta Geologica* **3**: 1–10.

Freudenthal, M., 1972, *Deinogalerix koenigswaldi* nov. gen., nov. spec., a giant insectivore from the Neogene of Italy, *Scripta Geologica* **14**: 1–19.

Freudenthal, M., 1976, Rodent stratigraphy of some Miocene fissure fillings in Gargano (prov. Foggia, Italy), *Scripta Geologica* **37**: 1–23.

Freudenthal, M., 1985, Cricetidae (Rodentia) from the Neogene of Gargano (Prov. of Foggia, Italy), *Scripta Geologica* **77**: 29–76.

Freudenthal, M., 2006, *Mikrotia* nomen novum for *Microtia* Freudenthal, 1976 (Mammalia, Rodentia), *Journal of Vertebrate Paleontology* **26** (3): 748.

Galobart, A., Pons-Moyà, J., Anton, M., and Maroto, J., 2003, Descripci6n del material de *Homotherium latidens* (Owen) de los yacimientos del Pleistocene inferior de Incarcal (Girona, NE de la Península Ibérica), *Paleontologia i Evolucío* **34**: 99–141.

van den Bergh, G.D., de Vos, J., Sondaar, P.Y., and Aziz, F., 1996a, Pleistocene zoogeographic evolution of Java (Indonesia) and glacio-eustatic sea-level fluctuations: a background for the presence of Homo. *Indo-pacific Prehistory Association Bulletin* (Chiang Mai Papers), **14**: 7–21.

van den Bergh, G.D., Mubroto, B., Sondaar, P.Y., and de Vos, J., 1996b, Did *Homo erectus* reach the island of Flores? *Indo-pacific Prehistory Association Bulletin* (Chiang Mai Papers), **14**: 27–36.

van den Bergh, G.D., 1999, The Late Neogene elephantoid-bearing faunas of Indonesia and their palaeozoogeographic implications; A study of the terrestrial faunal succession of Sulawesi, Flores and Java, including evidence for early hominid dispersal east of Wallace's line, *Scripta Geologica* **117**: 1–419.

van den Bergh, G.D., de Vos, J., and Sondaar P.Y., 2001, The Late Quaternary palaeogeography of mammal evolution in the Indonesian Archipelago, *Palaeogeography, Palaeoclimatology, Palaeoecology* **171**: 385–408.

van den Bergh, G.D., Rokus, A.D., Morwood, M.J., Sutikna, T., Jatmiko, P., and Wahyu Saptomo, E., in press, The youngest *Stegodon* remains in south-east Asia from the Late Pleistocene archaeological site Liang Bua, Flores, Indonesia. *Quaternary International*.

van den Hoek Ostende, L.W., 2001a, Insectivore faunas from the Lower Miocene of Anatolia – Part 8: stratigraphy, palaeoecology, palaeobiogeography, *Scripta Geologica* **122**: 101–122.

van den Hoek Ostende, L.W., 2001b, A revised generic classification of the Galericini (Insectivora, Mammalia) with some remarks on their palaeobiogeography and phylogeny, *Geobios* **34** (6): 681–695.

van den Hoek Ostende, L.W., van den Bergh, G., and Rokus Awe Due, 2006, First fossil insectivores from Flores, *Annales Geologiques de Pays Helleniques* **41**: 67–72.

Hooijer, D.A., 1947, On fossil and prehistoric remains of *Tapirus* from Java, Sumatra and China, *Zoologische Mededelingen* **27**: 253–299.

Hooijer, D.A., 1948a, Evolutie van zoogdieren in het Quartair van Z.O.-Azie, *Vakblad voor Biologen* **28** (7): 117–122.

Hooijer, D.A., 1948b, Pleistocene Vertebrates from Celebes. I. *Celebochoerus heekereni* nov. gen. nov. spec. *Proceedings of the Koninklijke Nederlandse Akademie van Wetenschappen* **51**: 1024–1032.

Hooijer, D.A., 1948c, Pleistocene vertebrates from Celebes. II. Testudo margae nov. spec. *Proceedings of the Koninklijke Nederlandse Akademie van Wetenschappen* **51**: 1169–1182.

Hooijer, D.A., 1949, Pleistocene vertebrates from Celebes IV. *Archidiskodon celebensis* nov. spec., *Zoologische Mededelingen* **30**: 205–226.

Hooijer, D.A., 1953a, Pleistocene vertebrates from Celebes V – lower molars of *Archidiskodon celebensis* Hooijer, *Zoologische Mededelingen* **31**: 311–318.

Hooijer, D.A., 1953b, Pleistocene vertebrates from Celebes VI – *Stegodon* sp., *Zoologische Mededelingen* **32**: 107–112.

Hooijer, D.A., 1953c, Pleistocene vertebrates from Celebes VII – milk molars and premolars of *Archidiskodon celebensis* Hooijer, *Zoologische Mededelingen* **32**: 221–230.

Hooijer, D.A., 1954a, Pleistocene vertebrates from Celebes XI – molars and a tusked mandible of *Archidiskodon celebensis* Hooijer, *Zoologische Mededelingen* **33**: 103–120.

Hooijer, D.A., 1954b, Pleistocene vertebrates from Celebes. X. Testudinata. *Proceedings of the Koninklijke Nederlandse Akademie van Wetenschappen* **57**: 486–489.

Hooijer, D.A., 1955, Fossil Proboscidea from the Malay Archipelago and India, *Zoologische Verhandelingen* **28**: 1–146.

Hooijer, D.A., 1957a, A *Stegodon* from Flores, *Treubia* **24**: 119–129.

Hooijer, D.A. 1957b, Three new giant prehistoric rats from Flores, Lesser Sunda Islands, *Zoologische Mededelingen* **35**: 299–314.

Hooijer, D.A., 1954, Pleistocene vertebrates from Celebes. VIII. Dentition and skeletton of *Celebochoerus heekereni*. Zoologische Verhandelingen, **24**: 3–46.

Hooijer, D.A., 1964, Pleistocene vertebrates from Celebes XII – Notes on pygmy Stegodonts, *Zoologische Mededelingen* **40**: 37–44.

Hooijer, D.A., 1969, The *Stegodon* from Timor, *Proceedings Koninklijke Nederlandse Akademie van Wetenschappen, Amsterdam, Series B* **72** (3): 201–210.

Hooijer, D.A., 1972a, Pleistocene vertebrates from Celebes. XIV. Additions to the *Archidiskodon – Celebochoerus* fauna, *Zoologische Mededelingen* **46**: 1–16.

Hooijer, D.A., 1972b, *Stegodon trigonocephalus florensis* Hooijer and *Stegodon timorensis* Sartono from the Pleistocene of Flores and Timor, *Proceedings Koninklijke Nederlandse Akademie van Wetenschappen, Series B* **75** (1): 12–33.

Kitchener, D.J., How, R.A., and Maharadatunkamsi, R.A., 1991a, *Paulamys* sp. cf. *P. naso* (Musser, 1981) (Rodentia: Muridae) from Flores Island, Nusa Tenggara, Indonesia – description from a modern specimen and a consideration of its phylogenetic affinities, *Records of the Western Australian Museum* **15** (1): 171–189.

Kitchener, D.J., How, R.A., and Maharadatunkamsi 1991b, A new species of *Rattus* from the mountains of West Flores, Indonesia, *Records of the Western Australian Museum* **15** (3): 611–626.

Klein Hofmeijer, G. and Bruijn, H. de, 1985, The mammals from the lower Miocene of Aliveri (Island of Evia, Greece); IV, The Spalacidae and Anomalomyidae, *Proceedings Koninklijke Nederlandse Akademie van Wetenschappen, Series B* **88** (2): 185–198.

Kondopulou, D., Bonis, L. de, Koufos, G.D., and Sen, S., 1993, Palaeomagnetic data and biostratigraphy of the middle Miocene vertebrate locality of Thymiana (Chios island, Greece), *Proceedings 2nd Congress Hellenic Geophysical Union*, Florina, Greece **2–3**: 626–635.

Kostopoulos, S.D., 1996, The Plio-Pleistocene artiodactyls of Macedonia (Greece); systematics, palaeoecology, biochronology, biostratigraphy, PhD Thesis, University of Thessaloniki (in Greek).

Koufos, G.D. and Kostopoulos, D.S., 1997, Biochronology and succession of the Plio-Pleistocene macromammalian localities of Greece, *Mémoires et Travaux EPHE, Inst. Montpellier* **21**: 619–634.

Koufos, G.D., de Bonis, L., and Sen, S., 1995, *Lophocyon paraskevaidisi*, a new viverrid (Carnivora, Mammalia) from the middle Miocene of Chios island (Aegean sea, Greece), Géobios **28**: 511–523.

Koufos, G.D., Kostopoulos, S.D., and Kaliadimou, K.K., 1991, Un nouveau gisement de mammifères dans le Villafranchien de Macédoine occidentale (Grèze), *Comptes Rendus de l'Académie des Sciences, Paris Série II* **313**: 831–836.

Koufos, G.D., Zouros, N., and Mourouzidou, O., 2003, *Prodeinotherium bavaricum* (Proboscidea, mammalia) from Lesvos island, Greece; the appearance of deinotheres in the Eastern Mediterranean, *Geobios* **36**: 305–315.

Kuss, S.E., 1965, Eine pleistozäne Säugetierfauna der Insel Kreta, *Berichte der Naturforschenden Gesellschaft zu Freiburg im Breisgau* **55** (2): 271–348.

Lehmann, U. and Tobien, H., 1995, Artiodaktyla-Fossilien (Mammalia) aus dem Miozän von Thymiana, C., *Annales de Géologie des Pays Helleniques* **60**: 403–414.

Leinders, J.J.M., 1984, Hoplitomerycidae fam. nov. (Ruminantia, Mammalia) from Neogene fissure fillings in Gargano (Italy), *Scripta Geologica* **70**: 1–68.

Lomolino, M.V., 2005, Body size evolution in insular vertebrates: generality of the island rule. *Journal of Biogeography* **32**: 1683–1699.

Lomolino, M.V., Riddle, B.R., and Brown, J.H., 2006, *Biogeography*, 3rd edition. Sindauer Associates, Sunderland, MA.

Lopez-Martinez, N., 1986, The mammals from the lower Miocene of Aliveri (Island of Evia, Greece); VI. The ochotonid lagomorph *Albertona balkanica* n.gen. n.sp. and its relationships, *Proceedings Koninklijke Nederlandse Akademie van Wetenschappen, Series B* **89** (2): 177–194.

Lydekker, R., 1885, Description of a tooth of *Mastodon latidens* Clift, from Borneo, *Proceedings Zoological Society*, London, pp. 777–779.

MacArthur, R.H. and Wilson, E.O., 1963, An equilibrium theory of insular zoogeography. *Evolution* **17**: 373–387.

MacArthur, R.H. and Wilson, E.O., 1967, *The Theory of Island Biogeography, Monographs in Population Biology*, Vol. 1, Princeton University Press, Princeton, NJ.

van der Made, J., 1996, Pre-Pleistocene land mammals from Crete, in: Reese, D.S. (ed.), *Pleistocene and Holocene Fauna of Crete and Its First Settlers, Monographs in World Archaeology* **28**: 69–79.

Maringer, J. and Verhoeven, T.H., 1970, Die Steinartefacte aus der *Stegodon*-Fossil-Schicht von Mengeruda auf Flores, Indonesien, *Anthropos* **65**: 229–247.

Mayhew, D.F., 1996, The extinct murids of Crete, in: Reese, D.S. (ed.), *Pleistocene and Holocene Fauna of Crete and Its First Settlers, Monographs in World Archaeology* **28**: 167–171.

Medway, L., 1979, The Niah excavations and an assessment of the impact of early man on ammals in Borneo, *Asian Perspectives* **20**: 51–69.

Meiri, S., Dayan, T., and Sinberloff, D., 2006, The generality of the island rule re-examined. *Biogeography* **33**: 1571–1577.

Melentis, J.K. and Tobien, H., 1968, Paläontologische Ausgrabungen auf der Insel Chios (Eine vorläufige Mitteilung), *Annales de Géologie des Pays Helleniques* **19**: 647–652.

van der Meulen, A.J. and de Bruijn, H., 1982, The mammals from the lower Miocene of Aliveri (Island of Evia, Greece); II, The Gliridae, *Proceedings Koninklijke Nederlandse Akademie van Wetenschappen, Series B* **85** (4): 485–524.

van der Meulen, A.J. and van Kolfschoten, T., 1986, Review of the Late Turolian to Early Biharian mammal faunas from Greece and Turkey, *Memoria Società Geologia Italiana* **31**: 201–211.

Meulenkamp, J.E., 1971, The Neogene in the Southern Aegean Area, in: Strid, A. (ed.), *Evolution in the Aegean, Opera Botanica* **30**: 5–12.

Mol, D., de Vos, J., van den Bergh, G.D., and Sondaar, P.Y., 1996, The taxonomy and ancestry of the fossil elephants of Crete: faunal turnover and a comparison with Proboscidean faunas of Indonesian islands, in: Reese, D.S. (ed.), *Pleistocene and Holocene Fauna of Crete and Its First Settlers, Monographs in World Archaeology* **28**: 81–98.

Morwood, M.J., Brown, P., Jatmiko, S.T., Wahyu, S.E., Westaway, K.E., Rokus, A.D., Roberts, R.G., Maeda, T., Wasisito, S., and Djubiantono, T., 2005, Further evidence for small-bodied hominins from the Late Pleistocene of Flores, Indonesia, *Nature* **437**: 1012–1017.

Morwood, M.J., O'Sullivan, P.B., Aziz, F., and Raza, A., 1998, Fission-track ages of stone tools and fossils on the east Indonesian island of Flores, *Nature* **392**: 173–176.

Morwood, M.J., Soejono, R.P., Roberts, R.G., Sutikna, T., Turney, C.S.M., Westaway, K.E., Rink, W.J., Zhao, J.-X, van den bergh, G.D., Rokus, A.D., Hobbs, D.R., Moore, M.W., Bird, M.I., and Fifield, L.K., 2004, Archaeology and age of a new hominin from Flores eastern Indonesia, *Nature* **431**: 1087–1091.

Musser, G.G., 1981, The giant rats of Flores and its relatives east of Borneo and Bali, *Bulletin of the American Museum of Natural History* **169** (2): 69–175.

Musser, G.G., van der Weerd, A., and Strasser, E., 1986, *Paulamys*, a replacement name for *Floresomys* Musser, 1981 (Muridae), and new material of that taxon from Flores, Indonesia, *American Museum Novitates*, **2850**: 1–10.

Palombo, M.R., in press, How can endemic proboscideans help us understand the "Island Rule"? A case study of Mediterranean islands. *Quaternary International.*

Paraskevaidis, I., 1940, Eine obermiocäne fauna von Chios, *Neues Jahrbuch für Mineralogie, Geologie und Paläontologie,* B **83**: 363–442.

Poulakakis, N., Mylonas, M., Lymberakis, P., and Fassoulas, C., 2002, Origin and taxonomy of the fossil elephants of the island of Crete (Greece): problems and perspectives, *Palaeogeography, Palaeoclimatology, Palaeoecology* **186** (1–2): 163–183.

Radulesco, C. and Samson, P., 1991, Traces d'activité humaine a la limite Pliocène/Pléistocène dans le Bassin Dacique (Roumainie), in: Bonifay, E. and Vandermeersch, V. (eds), *Les Premiers Européens, Paris, Éditions du C.T.H.S.*: 203–207.

Raia, P. and Meiri, S., 2006, The island rule in large mammals: paleontology meets ecology. *Evolution* **60** (8): 1731–1742.

Reese, D.S. (ed.), 1996, Pleistocene and Holocene fauna of Crete and its first Settlers, *Monographs in World Archaeology* **28**: 207–210.

Reumer, J.W.F. and Payne, S., 1986, Notes on the Soricidae (Insectivora, Mammalia) from Crete, II, The shrew remains from Minoan and Classical Kommos, *Bonner zoologische Beiträge* **37** (3): 173–182.

Reumer, J.W.F., 1996, Shrews (Soricidae) on islands, with special reference to *Crocidura zimmermanni* from Crete, in: Reese, D.S. (ed.), *Pleistocene and Holocene Fauna of Crete and Its First Settlers, Monographs in World Archaeology* **28**: 173–179.

Rothausen, K., 1977, Die mittelmiozänen Wirbeltierfundstellen südlich Thymiana (Insel Chios, Ägäis, Griechenland), 2 Teil: Geologie: Die Fundstellen und ihre Abfolge, *Annales de Géologie des Pays Helleniques* **28**: 495–515.

Sartono, S., 1969, *Stegodon timorensis*: a pygmy species from Timor (Indonesia), *Proceedings Koninklijke Nederlandse Akademie van Wetenschappen, Series B* **72**: 192–202.

Sartono, S. and Marino R., 1978, A mandibula and a maxilla of *Stegodon timorensis*. *Modern Quaternary Research in SE Asia*, 4, 41–50.

Sartono, S., 1979, The discovery of a pygmy *Stegodon* from Sumba, East Indonesia: an announcement, *Modern Quaternary Research in SE Asia* **5**: 57–63.

Schmidt-Kittler, N., 1983, The mammals from the lower Miocene of Aliveri (Island of Evia, Greece); III, On a new species of *Sivanasua* Pilgrim, 1931 (Feliformia, Carnivora) and the phylogenetic position of this genus, *Proceedings Koninklijke Nederlandse Akademie van Wetenschappen, Series B* **86** (3): 301–318.

Sen, S., 1977, *Megapedetes aegaeus* n. sp. (Pedetidae) et à propos d'autres "rongeurs africains" dans le Miocene d'Anatolie, *Geobios* **10** (6): 983–986.

Sen, S., Sondaar, P.Y., and Staesche, U., 1978, The biostratigraphical applications of the genus *Hipparion* with special references to the Turkish representatives, *Proceedings Koninklijke Nederlandse Akademie van Wetenschappen, Series B* **81**: 370–385.

Simpson, G.G., 1940, Mammals and land bridges. *Journal of the Washington Acadamy of Science* **30**: 137–163.

Sondaar, P.Y., 1976, The island sweepstakes. Why did pygmy elephants, dwarf deer and large mice populate the Mediterranean? *Natural History* **95** (9): 50–57.

Sondaar, P.Y., 1977, Insularity and its effect on mammal evolution, in: Hecht, M.N., Goody, P.C., and Hecht, B.M. (eds), *Major Patterns in Vertebrate Evolution*, Plenum, New York, pp. 671–707.

Sondaar, P.Y., 1987, Pleistocene man and extinctions of island endemics. *Memoirs du Societe Geologique de France, N.S.* **150**: 159–65.

Sondaar, P.Y. and Boekschoten, G.J., 1967, Quaternary mammals in the south Aegean island arc with notes on other mammals from the coastal regions of the Mediterranean, I/II, *Proceedings Koninklijke Nederlandse Akademie van Wetenschappen, Series B* **7**: 556–576.

Sondaar, P.Y., de Vos, J., and Dermitzakis, M.D., 1986, Late Cenozoic faunal evolution and Palaeogeography of the South Aegean Island arc, *Modern Geology* **10**: 249–259.

Sondaar, P.Y., Dermitzakis, M.D., and Vos, J. de, 1996, The Paleogeography and faunal Evolution of the Land Mammals of Crete, in: Reese, D.S. (ed.), *Pleistocene and Holocene Fauna of Crete and Its First Settlers, Monographs in World Archaeology* **28**: 61–67.

Sondaar, P.Y., van den Bergh, G.D., Mubroto, B., Aziz, F., de Vos J., and Batu Unkap L., 1994, Middle Pleistocene faunal turnover and colonization of Flores (Indonesia) by *Homo erectus*, *Comptes rendus de l'Academie des Sciences, Série II* **320**: 1255–1262.

Spaan, A., 1996, *Hippopotamus creutzburgi*: the case of the Cretan Hippopotamus, in: Reese, D.S. (ed.), *Pleistocene and Holocene Fauna of Crete and Its First Settlers, Monographs in World Archaeology* **28**: 99–110.

Spassov, N., 1998, A new late Villafranchian locality of vertebrate fauna, Slivnitsa (Bulgaria) and the carnivore dispersal events in Europe on the Pliocene/Pleistocene boundary, *Historia Naturalis Bulgarica* **9**: 101–113.

Storm, P., 2001, The evolution of humans in Australia from an environmental perspective, *Palaeogeography, Palaeoclimatology, Palaeoecology* **171**: 363–383.

Storm, P., Aziz, F., de Vos, J., Kosasih, D., Baskoro, S., Ngaliman and van den Hoek Ostende, L.W., 2005, Late Pleistocene *Homo sapiens* in a tropical rainforest fauna in East Java, *Journal of Human Evolution* **49**: 536–545.

Suzuki, M., Wikarno, B., Saefudin, I., and Itihara, M., 1985, Fission track ages of Pumice tuff, tuff layers and Javites of Hominid fossil bearing formations in Sangiran area, Central Java, in: Watanabe, N. and Kadar, D. (eds), *Quaternary Geology of the Hominid Fossil Bearing Formations in Java, Geological Research and Development Centre, Special Publication* **4**: 309–331.

Symeonides, N.K., Theodorou, G.E., and Giannopoulos, V.I., 2001, New data on *Elephas chaniensis* (Vamos cave, Chania, Crete), *Proceedings of the 1st International Congress "The World of Elephants"* Rome, 2001, pp. 510–513.

Thenius, E., 1956, Die Suiden und Tayassuiden des seirischen Tertiärs, *Sitzungsberichten Österreichische Akademie der Wissenschaften, Mathematisch-naturwissenschaflicht Klasse, Abteilung I* **165** (4): 337–382.

Theodorou, G., Spjeldnaes, N., Hanken, N.M., Lauritzen, S.E., Velitzelos, E., Athanassiou, A., and Roussiakis, S., 2000, Description and taphonomic investigations of Neogene Proboscidea from Rhodos, Greece, *Annales Géologiques des Pays Helléniques* **38** (C): 133–156.

Tobien, H., 1969, Wirbeltiergrabungen im Miozän der Insel Chios (Ägäis), *Praktika Akademie Athen* **43**: 151–157.

Tobien, H., 1980, A note on the skull and mandible of a new Choerolophodont Mastodont (Proboscidea, Mammalia from the Middle Miocene of Chios (Aegean Sea, Greece), in: Jacobs, L.L. (ed.), *Aspects of Vertebrate History*, Essays in Honor of E.H. Colbert, pp. 299–307.

Valleri, G., 1984, New data on planktonic foraminifera biostratigraphy from the Neogene of the Gargano Peninsula (Foggia, southern Italy), *Rivista Italiana di Paleontologia e Stratigrafia* **90** (3): 375–406.

Van Valen, L., 1973, Body size and the number of plants and animals. *Evolution* **27**: 27–35.

Verhoeven, Th., 1964, *Stegodon*-Fossilien auf der Insel Timor, *Anthropos* **59**: 634.

de Vos, J., 1983, The *Pongo* faunas from Java and Sumatra and their significance for biostratigraphical and paleo-ecological interpretations, *Proceedings Koninklijke Nederlandse Akademie van Wetenschappen, Amsterdam, Series B* **86** (4): 417–425.

de Vos, J., 1984a, The endemic Pleistocene deer of Crete, *Verhandeling der Koninklijke Nederlandse Akademie van Wetenschappen, Afdeling Natuurkunde, Eerste reeks* **31**: 1–100.

de Vos, J., 1984b, Reconsideration of Pleistocene cave faunas from South China and their relation to the faunas from Java, *Courier Forschungsinstitut Senckenberg* **69**: 259–266.

de Vos, J., 1996a, Taxonomy, Ancestry and speciation of the endemic Pleistocene deer of Crete compared with the taxonomy, ancestry and speciation of Darwin's Finches, in: Reese, D.S. (ed.), *Pleistocene and Holocene Fauna of Crete and Its First Settlers, Monographs in World Archaeology* **28**: 111–124.

de Vos, J., 1996b, Faunal turnovers in Java in relation to faunas of the continent, *Odontologie, Association for comparative Odontology* **1**: 32–36.

de Vos, J., 2000, Pleistocene deer fauna in Crete: its adaptive radiation and extinction, *Tropics* (special issue) **10** (1): 125–134.

de Vos, J., Aziz, F., Sondaar, P.Y., and van den Bergh, G.D., 1999, *Homo erectus* in S.E. Asia: time space and migration routes; a global model III. Migration routes and evolution, in: Gibert, J. Sánchez, F., Gibert, L., and Ribot, F. (eds), *The Hominids and Their Environment During the Lower and Middle Pleistocene of Eurasia, Proceedings of the International Conference of Human Paleontology*, Orce 1995, pp. 369–383.

de Vos, J. and Bautista, A., 2003, Preliminary notes on the vertebrate fossils from the Philippines, *Semantic and Systematics: Philippine Archaeology, Proceedings of the Society of Philippine Archaeologists* **1**: 42–62.

de Vos, J., Sartono, S., Hardja-Sasmita, H.S., and Sondaar, P.Y., 1982, The fauna from Trinil, type locality of *Homo erectus*; a reinterpretation, *Geologie en Mijnbouw*, **61**: 207–211.

de Vos, J. and Vu the Long, 2001, First settlements: relations between continental and Insular Southeast Asia, in: Sémah, F., Falguères, C., Grimaud-Hervé, D., and Sémah, A.-S. (eds), *Origine des peuplements et chronologie des Cultures paléolithiques dan le Sud-est Asiatique*, pp. 225–249.

de Vos, J., van der Made, J., Athanassiou, A., Lyras, G., Sondaar, P.Y., and Dermitzakis, M.D., 2002, Preliminary note on the Late Pliocene fauna from Vatera (Lesvos, Greece), *Proceedings of the 1st International Workshop "On Late Plio/Pleistocene extinction and evolution in the Paleartic. The Vatera site". Annales Géologiques des Pays Helléniques, première série* **29** A: 37–70.

Vu the Long, de Vos, J., and Ciochon, R., 1996, The fossil mammal fauna of the Lang Trang Caves, Vietnam, compared with southeast Asian fossil mammal faunas: the geographical implications, *Indo-Pacific Prehistory Association Bulletin* **14**: 101–109.

Wasson, R.J. and Cochrane, R.M., 1979, Geological and geomorphological perspectives on archaeological sites in the Cagayan valley, Northern Luzon, The Philippines, *Modern Quaternary Research SE Asia* **5**: 1–26.

van de Weerd, A., Reumer, J.W.F., and Vos, J. de, 1982, Pliocene mammals from the Apolakkia Formation (Rhodes, Greece*), Proceedings Koninklijke Nederlandse Akademie van Wetenschappen, B* **85** (1): 89–112.

Weesie, P.D.M., 1988, The Quaternary avifauna of Greece, *Palaeovertebrata* **18** (1): 1–94.

Westaway, K.E., Morwood, M.J., Roberts, R.G., Rokus, A.D., Zhao, J-X., Storm, P., Aziz., F., van den Bergh, G.D., Hadi, P., Jatmiko and de Vos, J., 2007, Age and biostratigraphic significance of the Punung rainforest fauna, East Java, Indonesia. *Proceedings of the National Academy of Science* (under review).

Willemsen, G.F., 1983, *Paralutra garganensis* sp. nov. (Mustelidae, Lutrinae), a new otter from the Miocene of Gargano, Italy, *Scripta Geologica* **72**: 1–9.

Willemsen, G.F., 1996, The Cretan Otter *Lutrogale cretensis*, in: Reese, D.S. (ed.), *Pleistocene and Holocene fauna of Crete and its first Settlers, Monographs in World Archaeology* **28**: 153–157.

Chapter 11

# Islands from a Snail's Perspective

E. GITTENBERGER
Nationaal Natuurhistorisch Museum. Naturalis, P.O. Box 9517, 2300 RA Leiden, The Netherlands, Gittenberger@naturalis.nnm.nl



**Abstract** Islands are broadly defined as inhabitable areas surrounded by a hostile environment, which makes island a relational term. The various kinds of islands from a snail's perspective are listed, with examples of species occurring in those places. Isolation may be brought about by surrounding water, ice, or a variety of other ecological factors. Palaeoislands have existed in the geological past but ended their existence afterwards, for example by a lowering of sea level. Since there are by definition no contact zones between islands, a species or subspecies concept based on reproductive isolation under natural circumstances cannot be applied there. It is concluded still that archipelagos constitute the most important reservoir of gastropod diversity, with a high degree of endemism. Some predictions of evolutionary theory about speciation and adaptation are tested in an island setting, with also palaeoislands taken into account. Allopatry may result in differentiation and speciation, but whether, when, and how this happens differs considerably among the taxa. The adaptation to a particular island habitat, for example a cave, may follow quite different routes, probably because of the random mutations that enable the process.



*W. Renema (ed.), Biogeography, Time, and Place: Distributions, Barriers, and Islands,* 347–363

# 1. Introduction

In everyday language, an island is a piece of land surrounded by water. From an ecological or a biogeographical point of view however, an island may more generally be considered an area with a particular, taxon-specific habitat, surrounded by a contrasting, hostile habitat, which acts as a barrier against dispersal or invasions. The term has a relational aspect, since a piece of land will hardly ever keep all kinds of organisms isolated and imprisoned. By various means boundaries can be crossed by particular taxa. Here we consider islands in relation to terrestrial snails.

# 2. Land Surrounded by Water

## 2.1. Islands

While stating that "Islands are prime reservoirs of land snail diversity", Solem (1984: 9) referred to regular islands, surrounded by water. Accepting that Solem (1984: 7) correctly estimated that there are at least 30,000–35,000 species of land snails worldwide, it may well be that about one third of the global diversity in land snail species is found on such islands. Several regions are still poorly known, as may be concluded on the basis of some recent monographs. However, this lack of knowledge concerns both mainland mountain ranges and islands, so that general conclusions about the distribution of molluscan diversity might not be much affected by this fact.

In his textbook on island biogeography, Whittaker (1998: 46) published species numbers with percentages of endemism for eight archipelagos, namely Hawaiian Islands (around 1,000 species, 99.9% endemic), Japan (492 species, 99%), Madagascar (380 species, 95%), New Caledonia (300 species, 99%), Madeira (237 species, 88%), Canary Islands (181 species, 77.9%), Mascarene Islands (145 species, 87.6%), and Rapa (>105 species? 100%). Somewhat different numbers can also be found in the basic literature, but that does not affect the general picture.

Cowie et al. (1995) listed nearly 800 described species for the Hawaiian Islands. The "possible 290 additional undescribed Hawaiian taxa" that were referred to by Solem (1984: 7) as "endodontoid", could not yet be included, which means that the total number might come close to 1,100.

Minato (1994) reported 149 species and subspecies of Clausiliidae from Japan. Together with a number of 711 taxa from other gastropod families, that were listed in an earlier review (Minato, 1988), this brings the total of Japanese terrestrial gastropod species and subspecies to 860.

During the last decade, Emberton (2002 and papers cited therein) has described hundreds of new species in a series of papers on the molluscan fauna of the island of Madagascar, so that the number of 380, mentioned by Whittaker (1998: 46), is far too low now.

Bank et al. (2002) published an impressive annotated checklist of the non-marine snails of the Macaronesian islands, i.e., the Azores, the Madeiras, and the Canaries. In that list 478 species are reported from those islands and over 30 new species or subspecies are additionally referred to. This brings the actual number of species there to over 500, with a level of endemism of nearly 80%. Bank et al. (2002) also make clear that the high diversity at the species level is partly caused by radiations in only a limited number of genera, which is not unusual in island faunas. Many species are not even restricted to a single island, but to only a part of an island.

Rosenberg (in literature, 2004) pointed out, referring to Espinosa and Ortea (1999), Rosenberg and Muratov (2005), Aguayo (1966) and unpublished work on Hispaniola by D. G. Robinson, that "the Greater Antilles alone have more than 2,500 endemic species, and the rest of the Caribbean islands must bring this total to at least 3,000". According to Climo (cf. Solem, 1984: 7) "there is a possible New Zealand native land snail fauna of more than 1,000 species". Rosenberg (in literature, 2004) estimated that Indonesia and the Philippines "could have another 3,000 species". In particular among the micro-snails, many species still await discovery here, as may be concluded on the basis of Vermeulen (1994), who found that in Borneo 40 of 73 species and subspecies of the genus *Opisthostoma* (Figs. 1–6) were undescribed.

Adding up the species numbers for the terrestrial snails occurring on the various islands and island groups, results in a number of about 10,000 species.

Regular islands are extremely diverse in their geological history as a terrestrial habitat, in size, altitude, diversity in soil types, climate, and remoteness. As a consequence, not all islands or archipelagos are equally rich in molluscan species numbers. For example, the islands of the Tristan–Gough archipelago, separated by over 2,400 km from the nearest land, support only two autochthonous genera of land snail, with a total of hardly more than ten species (Preece and Gittenberger, 2003). This may also serve to illustrate that without taking many aspects into account, comparisons between island faunas are hardly useful. What factors are relevant and how they interact are far from clear, not least because this may be taxon-dependent. For that reason, a study of less complex kinds of islands may be more promising from the perspective of evolutionary biology. This is certainly so when snails are used as model organisms.

## 2.2. Nunataks

A piece of land may be surrounded by frozen water, ice, or snow. Krajick (1998) published a beautifully illustrated account of such "nunataks" as flowering, green islands in a white sea of glaciers. Surveys of the molluscan fauna of recent nunataks could not be found in the literature. However, ancient nunataks figure prominently in phylogeographic studies on, for example, the Alpine terrestrial molluscan fauna.

In an early phylogeography paper, based on long-term field observations, Klemm (1973: 70) described the ongoing dispersal of *Cochlostoma henricae*

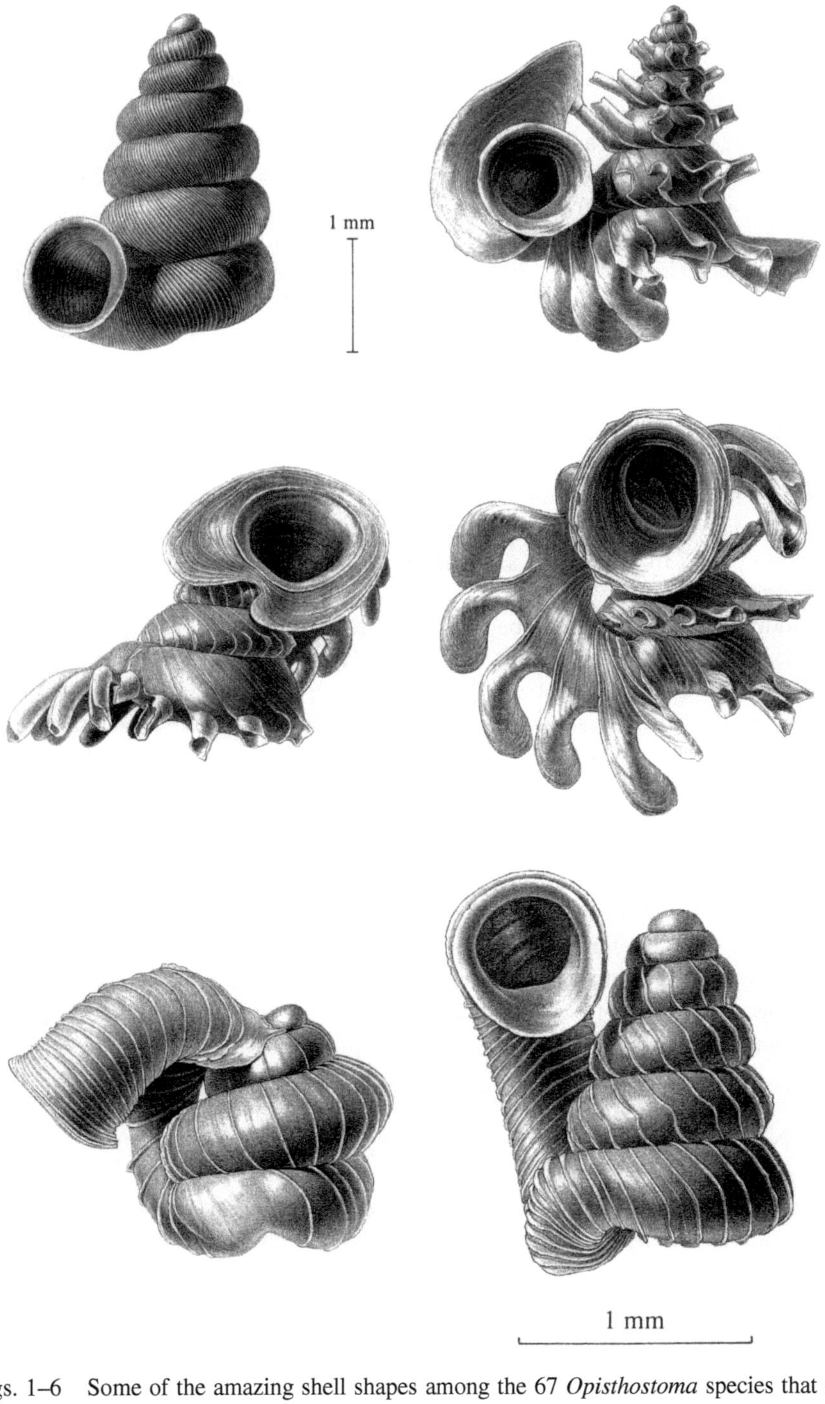


Figs. 1–6 Some of the amazing shell shapes among the 67 *Opisthostoma* species that are now known from Borneo, triggering questions about adaptive peaks and their uniqueness. Shell heights about 1–4 mm (after Vermeulen, 1994; Figs 61, 21a, 42a, 43a, 6a, and 50a).

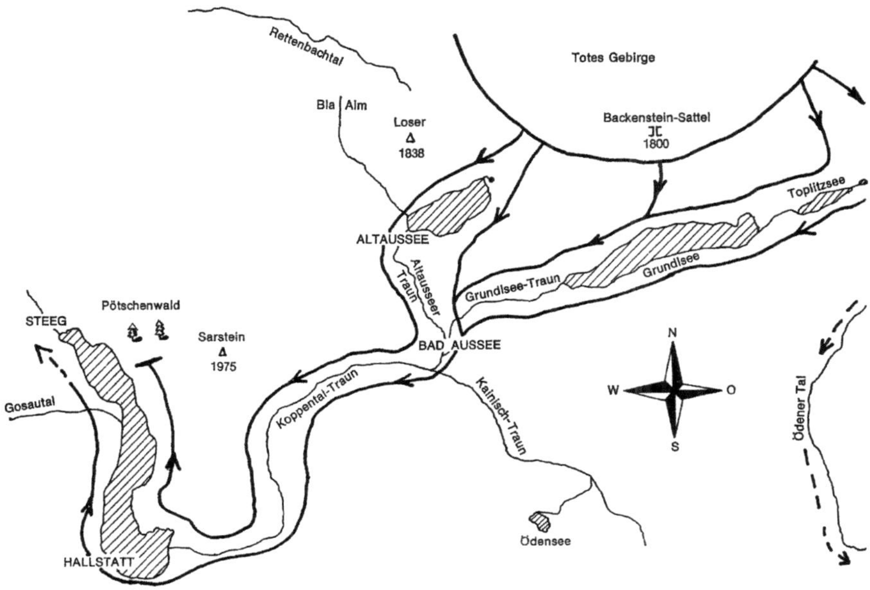


Fig. 7 Post-glacial range expansion of *Cochlostoma henricae*, in areas once covered by ice, south of its former refuge in the Totes Gebirge (after Klemm, 1973; Fig. 1).

(Strobel, 1851) along valleys starting at the Totes Gebirge in northern Austria (Fig. 7). He supposed that the species had survived the last glacial maximum (LGM) in that mountain range, which was never completely glaciated. Apparently *C. henricae* survived nowhere else in or near the northern Austrian Alps. As a consequence, this is a relatively simple case of dispersal from a known, island-like source area, quite different from what is known about the history of *Arianta arbustorum* (Linnaeus, 1758).

There are good reasons to suppose that *Arianta arbustorum*, differentiated taxonomically as *Arianta a. styriaca* (Kobelt, 1876), occurred with *Cochlostoma henricae* in the Totes Gebirge (and elsewhere) during the LGM. The complex evolutionary history of *A. arbustorum* was reconstructed on the basis of various techniques, including DNA analyses (Gittenberger et al., 2004). It turned out that this species survived on several nunataks, and in comparable but less completely isolated refugia, in both the Alps and the Pyrenees. Large populations also survived the LGM north of the Alps. These formed the main source of a postglacial recolonization of the Alps. Hybrid zones developed where invading *A. a. arbustorum* from the vast lowland refugia met the taxonomically differentiated snails, like *A. a. styriaca*, from the inner-alpine islands of refuge. This has determined the actual taxonomic diversity in the polytypic *A. arbustorum* and its distributional pattern of island-like subspecies in a sea of the widespread nominate form.

Bank and Gittenberger (2000) hypothesized that the distributional pattern of the alpine *Clausilia dubia alpicola* Clessin, 1878, in the Alps in Northeast Italy results from survival and differentiation during the LGM on former nunataks that are known to have existed in the area. *Clausilia d. alpicola* has an island-like range that is disjunctly distributed amidst *C. d. dubia* Draparnaud, 1805, which lives at lower altitudes. In contrast to the situation in *Arianta arbustorum*, sympatric occurrences or populations with morphologically intermediate specimens, indicative of a hybrid zone, are not known. Therefore, the taxonomic status of these taxa as either species or subspecies remains a matter of debate.

## 2.3. Palaeoislands

Snails are notoriously slow. As a consequence, they may still show ancient distributional patterns that have faded away by dispersal or other causes in most groups of organisms. It has been suggested that the configuration of some Aegean, Neogene palaeoislands can still be recognized in the modern distributional patterns of certain *Albinaria* taxa (Welter-Schultes and Williams, 1999; Welter-Schultes, 2000a, b). In the Miocene, from the early Tortonian (11–10 Ma), until the late Pliocene (3–2 Ma), there were several islands in the area of modern Crete (Fig. 8). These Cretan palaeoislands are supposed to have triggered allopatric speciation in *Albinaria*. In a stirring account, Welter-Schultes (2000a) tried to integrate geological data on the Neogene of central Crete and data on the eustatic sea-level changes in the Mediterranean Sea, with the distributional patterns shown by recent *Albinaria* species. Using this approach he attempted to provide a reliable palaeogeographical reconstruction of late Neogene, central Crete.

It is known that the Cretan *Albinaria* species can be roughly separated into a group that have limited ranges in areas that were not submerged during the Pliocene and a second group that occur also in regions that only emerged after the Pliocene (Schilthuizen et al., 1993; Welter-Schultes, 2000a). The polytypic *Albinaria hippolyti* (O. Boettger, 1878) exemplifies the former category (Schilthuizen et al., 1993: 139; Fig. 6). *Albinaria teres* (Olivier, 1801), widely distributed in eastern Crete with a westerly range only along the south coast,

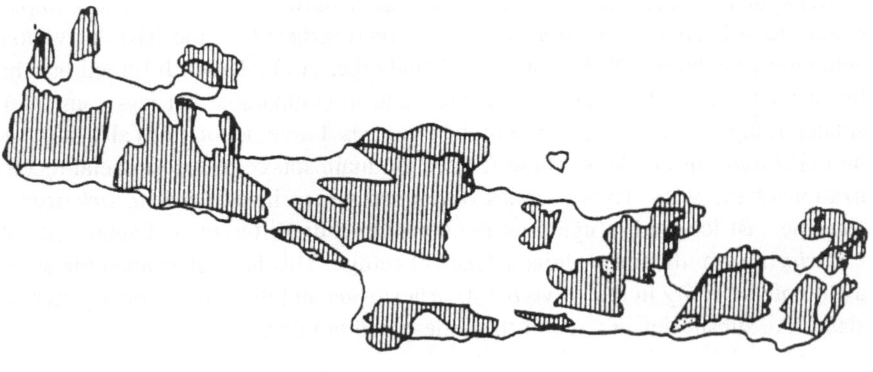

Fig. 8 Cretan Pliocene palaeoislands (after Vardinoyannis, 1994; Fig. 12) that might have been instrumental in allopatric speciation in terrestrial snails.

possibly resulting from westward dispersal (Welter-Schultes, 2000b; Fig. 64A), illustrates the alternative. More or less similar patterns occur in other genera of land snails, such as *Orculella* Gittenberger and Hausdorf, 2004 and *Mastus* (personal observation). *Orculella cretiminuta* (Gittenberger and Hausdorf, 2004), was able to spread all over modern Crete, filling the gaps between the palaeoislands, whereas other endemic Cretan *Orculella* species occupy much more restricted ranges, supposedly indicative of the palaeoislands of the ancient archipelago.

In a recent attempt to give the historical biogeography of the Cretan *Albinaria* taxa a firmer basis, Schilthuizen et al. (2004) used DNA sequences (ITS-1) to reconstruct the phylogeny of the *Albinaria hippolyti* subspecies and subsequently employed a molecular clock model to calculate the time of divergence of the various taxa. The results of this study do not support a scenario with major Miocene-Pliocene vicariance events triggering differentiation in *A. hippolyti*, but suggest a far more recent origin of the divergences, i.e., in the Pleistocene. This single result cannot be accepted as a decisive falsification of the compelling general hypothesis that palaeoislands have been important for the evolutionary history of snails in the Aegean. It underlines however, that the mutation rates of more DNA markers should be calibrated on the basis of more data. Other taxa should be investigated and compared.

## 3. Highlands Surrounded by Lowlands

In mountain ranges the summit areas act as ecological islands surrounded by lowland. Some strikingly dissimilar species, such as *Pyramidula rupestris* (Draparnaud, 1801), *Discus ruderatus* (W. Hartmann, 1821), or *Arianta arbustorum* (Linnaeus, 1758), seem to be largely independent of altitude in their occurrence (Jaeckel, 1962). Many species prefer intermediate heights or lowlands. A few gastropod species are restricted to higher altitudes, which results in a variety of distributional patterns. One extreme is exemplified by *Gittenbergia sororcula* (Benoit, 1859) (Fig. 9), a widespread minute gastropod, occurring at rather humid localities in rocky limestone habitats and screes, often hidden among moss. The species is restricted to high altitudes in mountains and has hardly ever been found below 1,000m altitude. As a consequence, its range is strongly fragmented (Fig. 10). Nevertheless, there is no differentiation in shell size, shape, or sculpture, between isolated populations, such as those from Mt. Psiloritis (Idhi Oros) in Crete, Greece, and those from the departments Drôme and Isère in France (Gittenberger, 1977a, 1993).

Species of Clausiliidae provide another extreme. For example, two neighbouring summits in the Parnon Oros in the eastern Peloponnese are inhabited by different, not congeneric, clausiliid species. *Idyla pelobsoleta* Gittenberger, 1993, and *Albinaria idyllica* (Gittenberger, 1987), occur on Mt. Kronion (i.e., Megala Tourlos), the highest elevation in the Parnon Mountains, whereas only *Albinaria grisea immensa* (O. Boettger, 1889) occurs in a similar habitat on a peak 10km northwest of Kosmas only 15km away.

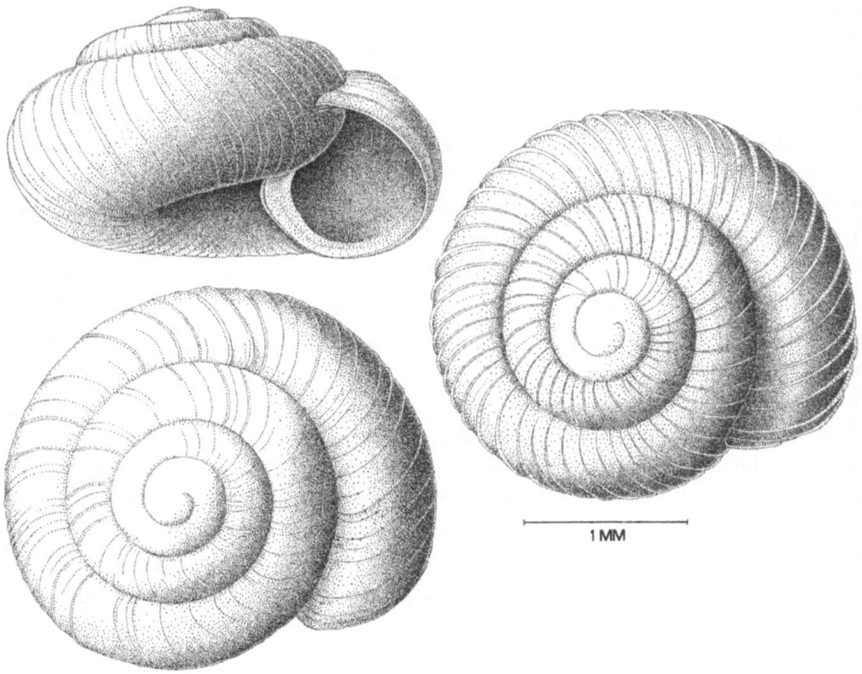


Fig. 9 *Gittenbergia sororcula* (Benoit, 1859), exemplifying a widespread, high-altitude species (see Fig. 10) without any obvious geographical variation in shell characters. The snails occur in moderately moist places, well hidden among litter, below rocks, etc.

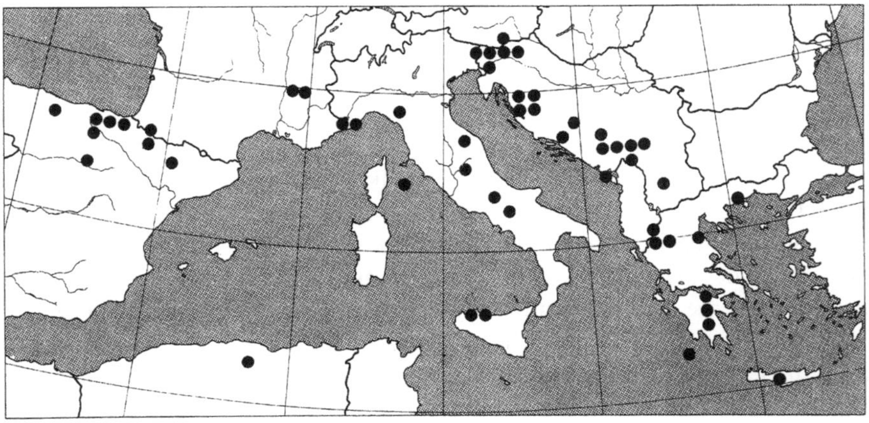

Fig. 10 UTM map of 50 $km^2$ with records of *Gittenbergia sororcula* (Benoit, 1859) (after Gittenberger, 1993, fig. 2), with UTM DK71, DK72, DK81, DK96, EK13, FG16, and KF53 (all in Greece) added after material in the molluscan collection of the Nationaal Natuurhistorisch Museum Naturalis, Leiden.

The so-called subspecies of *Idyla castalia* (Roth, 1856) show an intermediate pattern, restricted to separate mountain ranges. Here we have an unsolvable, island-dependent problem regarding the status of these taxa, since reproductive isolation or hybridization under natural conditions cannot be observed.

## 4. Stable Versus Varying Temperature, Humidity, and Light Conditions

Caves are not only dark, but are more generally relatively stable environments, surrounded by the outside world with its daily fluctuations in temperature, humidity, and light. For relatively small invertebrate animals like snails, caves are only artificially restricted to the places given that name by humans. Troglobionts may occur in narrow corridors and crevices that may considerably extend the cave habitat. Nevertheless, individual caves often have an island character, suggesting that long-range, subterranean connections between them are not common. In a series of caves that were studied in Montenegro (i.e., Crna Gora), for example, each cave turned out to have its own composition of snail species (Gittenberger, 1975). Some of these species seem to be restricted to a single cave, which implies that their range may be quite small. Very few species of snail inhabit caves and there is no category of common, catholic species found in all or nearly all caves. Only the cave entrance region may be a favourable habitat for a variety of species that may also be found elsewhere under similar conditions with limestone and a relatively high humidity.

In the "cave archipelago" of the southeast European karstic region, several endemic taxa seem to have a long, independent, evolutionary history in that they differ strongly from their sister taxa occurring outside caves. This unique phenomenon is exemplified by the aberrant, monotypic genus *Speleodentorcula* Gittenberger, 1985, which is known from only one cave in Euboea, Greece (Gittenberger, 1985). With a convincing cladistic analysis, using conchological and anatomical data, Hausdorf (1996) demonstrated that *Speleodentorcula* (Fig. 11) should be classified with the pupilloid family Argnidae Hudec, 1965. The strictly troglodytic pupilloid genera *Virpazaria* Gittenberger, 1969 (Fig. 12), and *Klemmia* Gittenberger, 1969 (Fig. 13), with five and two species, respectively, are another example of subterranean species that are easily recognizable in shell characters among the Valloniidae (Gittenberger, 1969). The most extreme example of obvious relic species in caves are the two southeast European species of *Pholeoteras* Sturany, 1904, the only representatives of the terrestrial Caenogastropoda, Cyclophoridae, in Europe, where the family disappeared from the fossil record after the Oligocene. The distribution of the *Pholeoteras* species is unique for a troglobiont, since those species are widespread, with extreme distributional gaps (Gittenberger and Bank, 2005). *Pholeoteras euthrix* Sturany, 1904 (Fig. 14), for example, is known from a single cave in Kerkyra (i.e., Corfu), Greece, and a group of caves about 350 km more to

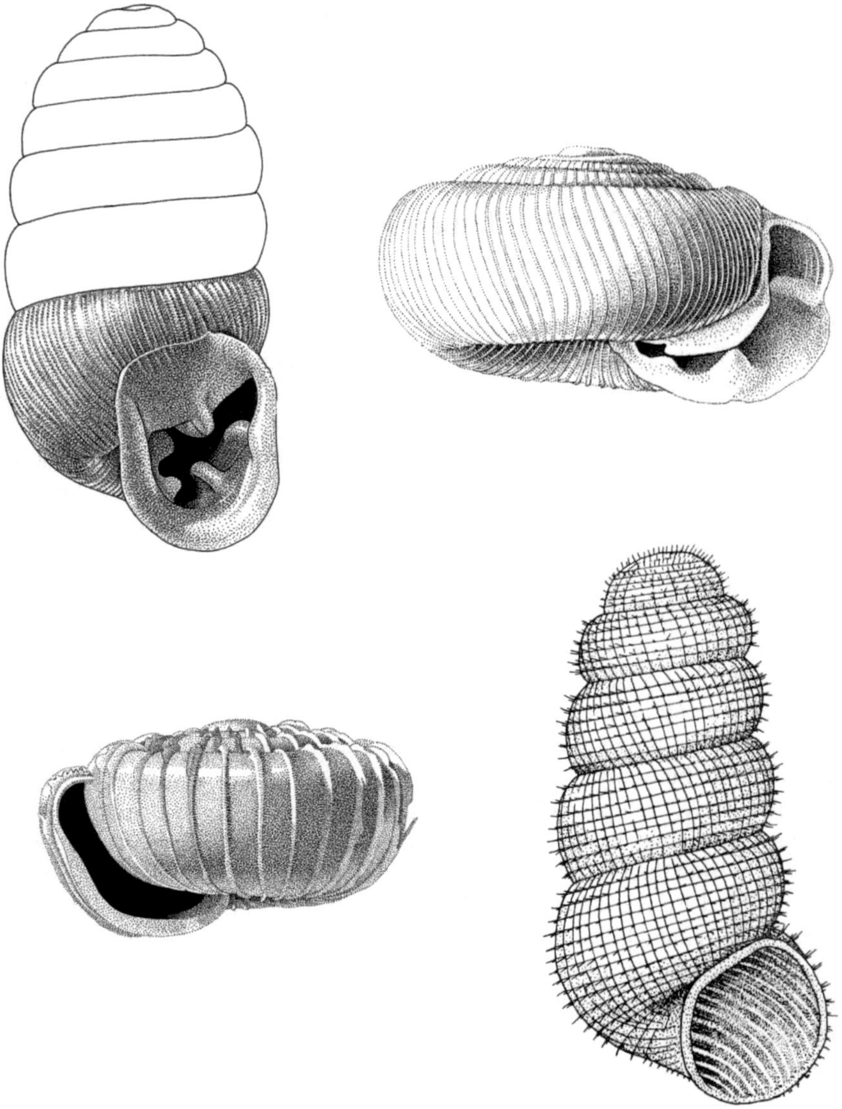

Figs. 11–14 Shells of southeast European snails from caves, exemplifying their heterogeneity in shell morphology. (11) *Speleodentorcula beroni* Gittenberger, 1985 (actual height 5.0 mm); (12) *Virpazaria (Aemiliella) ripkeni* Gittenberger, 1969 (actual width 4 mm); (13) *Klemmia magnicosta* Gittenberger, 1975 (actual width 3.4 mm); (14) *Pholeoteras euthrix* Sturany, 1944 (actual height about 2.5 mm; after Bole, 1975; fig. 7). Except for *Pholeoteras euthrix*, these species are known from their type localities only.

the north. In several other caves in Kerkyra that were thoroughly inspected, other cave snails were found (Gittenberger, 1977b). Among those is *Sciocochlea collasi* Sturany, 1904, a clausiliid species with a transparent, colourless shell, which is also known from only a single cave. *Sciocochlea* Sturany, 1904, with three species (Gittenberger, 2000), is another characteristic genus without any representatives outside caves. The various troglodytic snail species have very diverse shell shapes (Figs. 11–15), suggesting that there is not a single optimal adaptation in shell design related to their extreme habitat.

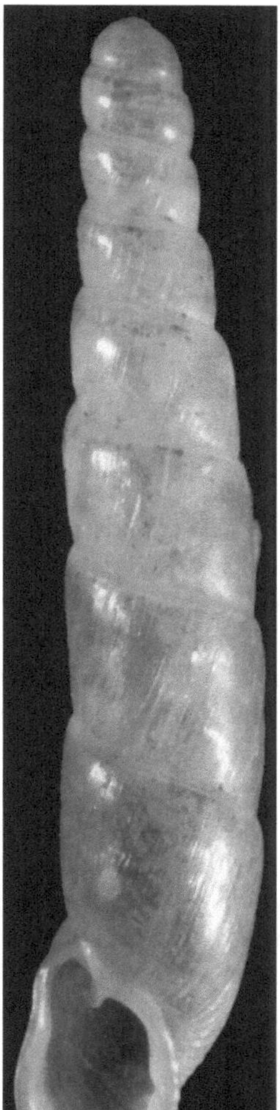

Fig. 15 The colourless, transparent shell of *Sciocochlea collasi* Sturany, 1904, from the cave Katsuri, southern slope of the Pantokrator, Kerkyra (i.e., Corfu); actual height 11.0 mm.

## 5. Calcareous Surrounded by Non-Calcareous Soils or Rocks

A calcareous environment is essential for many, though certainly not all snails. For calciphile snails, a limestone outcrop surrounded by non-calcareous soils is to be considered an island. Most probably, those snails will not actively cross areas without any limestone. Prudence is the mother of wisdom, here however, since malacologists usually prefer calcareous regions for their fieldwork, so that there are relatively few and more superficial observations from other areas, where species may still be represented in very low numbers of individuals (Schilthuizen et al., 1999).

When relatively large snails are involved, it is less likely that our views are biased. Ongoing research on Clausiliidae in Greece has resulted in many examples of distributional patterns that reflect the distributions of limestone rocks and certain gastropod taxa. From two isolated limestone cliffs in northern Greece, Macedonia, situated only around 10 km apart, Gittenberger (2002) described two subspecies of both *Montenegrina dennisi* Gittenberger, 2002, and *Macedonica pindica* Gittenberger, 2002, and *Montenegrina janinensis maasseni* Gittenberger, 2002, from only one of the limestone islands. Uit de Weerd et al. (2004: 55; Figs 4.2; 56, 4.3) published rough maps for several calciphile species, nicely illustrating the distribution of both calcareous soils and their endemic clausiliid snails.

## 6. Discussion

It has been argued repeatedly that islands can be seen as natural laboratories to be used for evolutionary studies (Whittaker, 1998). This being so, terrestrial snails are among the ideal model organisms. They are notoriously slow and can neither swim nor fly, so that they are strongly isolated in most kinds of islands. For these organisms with limited migratory capacities, there are more " 'islands" than for many other organisms. Gastropod shells can easily be collected for study without any consequences for the living animals. Evolutionary processes can be observed in relatively small areas (Cameron and Cook, 1992, 2001; Cameron et al., 1996).

Snails get easily locked up with their alleles. In that situation, gene-flow from elsewhere depends on passive dispersal, for example, by sea, attached to driftwood, or by aerial dispersal by strong winds or birds (Rees, 1965). On the basis of circumstantial evidence, incidental transport due to abiotic events or by organisms that are not affected by the same isolating factors has often been postulated more or less convincingly (Preece and Gittenberger, 2003).

Islands generate exemplary problems for the application of most species and subspecies concepts because of the criterion of reproductive isolation under natural conditions and difficulties in interpreting morphological differentiation. This may result in conspicuous differences in species and subspecies numbers cited in the literature, as for example, in the case of the radiation in Clausiliidae and Hygromiidae in the Maltese islands (Beckmann, 1992; Giusti et al., 1995).

In general, most systematists will classify forms that hybridize freely, producing fertile intermediates whenever they meet, as conspecific. Depending upon the distributional patterns, with or without relatively narrow hybrid zones, such forms may be considered subspecies. If no contact occurs between closely related but morphologically and genetically divergent populations on different islands, there is no objective method to decide whether subspecies or species status should be attributed to those taxa. Only in contact zones, can the amount of reproductive isolation and competitive interactions be estimated under natural conditions.

When subspecies are also counted, the biodiversity and endemism on islands become even more impressive. The data for the Japanese molluscan fauna make this clear. Whittaker (1998: 46) listed 492 Japanese species names, whereas Minato (1988, 1994), adding subspecies, counted 860 taxa.

An absolute or a relative sea-level lowering may unite smaller islands into larger ones. Secondary contacts may then initiate a phase of species selection with intrainsular, adaptive radiation on larger islands. Global warming may result in melting glaciers and changing nunatak configurations. Global cooling will push the alpine vegetational zones downwards, sometimes uniting formerly separate mountain summit populations. Erosion may change the extension and separation of limestone islands. All these phenomena in combination with palaeogeographical and palaeoclimatological data offer excellent opportunities for evolutionary biologists interested in a quasi experimental approach of long-term processes to compare the predictions of evolutionary theory with reality.

Islands figure prominently in biogeographical studies. They may have served as stepping stones, enabling dispersal as, for example, in two directions along the South Aegean island arc between Asia Minor and the Peloponnese. In that arc, the distributional patterns of snails show an obvious break between Kythira and Andikythira (Vardinoyannis, 1994; Gittenberger, unpublished data), suggesting that once the latter islet was connected to Crete and not or less recently to Kythira.

Evolutionary theory predicts that populations on different islands will diverge because of island-bound mutations, local adaptation, genetic bottle necks, or founder effects. In an archipelago this may eventually result in speciation according to the popular allopatry scenario, with often an initial phase of interinsular, non-adaptive radiation (Gittenberger, 1991, 2004). When similar habitats occur on different islands, niche specialization will not always be evident. From comparisons between the individual species in the land snail faunas of different islands we may learn that the dogmatic view that speciation simply results from adaptation is a naive oversimplification. There may be speciation in the absence of differential adaptation, in particular in cases of interinsular radiation. Intrainsular radiation may be truly adaptive, with several species of the clade in question occurring sympatrically (Gittenberger, 1991; Cameron et al., 1996).

To study gastropod biodiversity and the processes behind it, islands are fascinating places to go, providing wonderful drugs against the restraining effects of reductionism. Species that are subjected to similar environmental factors may differ considerably in their evolutionary pathway and resulting morphology. The popular metaphor of a landscape with adaptive peaks is misleading in that it suggests a far too static situation. At least, the peaks are not in stable positions. They shift while interacting with the potential of the climber, which is determined by its genes and their mutations. The metaphor of an adaptive landscape does not clarify the differences in shell shapes of the *Opisthostoma* species from Borneo (Figs 1–6).

What actually accounts for the difference between *Gittenbergia sororcula* on the one hand and several clausiliid species on the other hand remains unclear. It is understandable that evolutionary changes occur, but stasis as seen in at least the shell characters of *G. sororcula* (Fig. 9) and *Pholeoteras euthrix* (Fig. 11) remains far more difficult to explain. There is a tendency among evolutionary biologists to simply neglect such fundamental aspects, probably because an experimental approach in the laboratory is difficult to achieve. Islands of various kinds as natural laboratories may be helpful here.

Unfortunately island faunas in general and snails in particular, illustrate an evolutionary relevant phenomenon not yet mentioned here, i.e., mass extinctions (Cowie et al., 1995; Whittaker, 1998: 242). For several Pacific islands the story is already dramatic. About 75% of the nearly 800 species of land snails of the Hawaiian Islands, for example, are now thought to be extinct (Cowie et al., 1995). But as Bank et al. (2002) argue it is not yet too late for the Macaronesian terrestrial molluscan fauna, still "one of the most diverse in the world": "Needless to say, protection measures are necessary, if destruction of the Macaronesian molluscan fauna is to be avoided" (Bank et al., 2002: 89, 90).

## Acknowledgements

I would like to thank most cordially Dr. G. Rosenberg (Philadelphia) and an anonymous reviewer for helpful comments on the manuscript, as regards content and linguistics. Any remaining errors are still my own responsibility. I am also grateful to Dr. J.J. Vermeulen, who allowed me to reproduce his drawings of *Opisthostoma* sp.

## References

Aguayo, C.G., 1966, Una lista de los moluscos terrestres y fluviales de Puerto Rico, *Stahlia* **5**: 1–17.

Bank, R.A. and Gittenberger, E., 2000, On the polytypic and problematic *Clausilia dubia*: notes on its nomenclature and systematics (Gastropoda, Pulmonata, Clausiliidae), *Basteria* **64**: 15–27.

Bank, R.A., Groh, K., and Ripken T.E.J., 2002, Catalogue and bibliography of the non-marine Mollusca of Macaronesia, in: Falkner, M., Groh, K., and Speight, M.C.D. (eds), *Collectanea Malacologica, Festschrift für Gerhard Falkner*: 89–235. ConchBooks, Hackenheim, Germany.

Beckmann, K.-H., 1992, Catalogue and bibliography of the land- and freshwater molluscs of the Maltese islands, the Pelagi islands and the isle of Pantelleria, *Heldia* **2**, Sonderheft **2**: 1–60.

Bole, J., 1975, Anatomija in taksonomski polazaj vrste *Pholeoteras euthrix* Sturany 1904 (Gastropoda), *Razprave Dissertationes, Slovenska Akademija Znanosti in Umetnosti (IV, Historia Naturalis)* **18/2**: 33–46.

Cameron, R.A.D. and Cook, L.M., 1992, The development of diversity in the land snail fauna of the Madeiran archipelago, *Biological Journal of the Linnean Society* **46** (1–2): 105–114.

Cameron, R.A.D. and Cook, L.M., 2001, Madeiran snails: faunal differentiation on a small island, *Journal of Molluscan Studies* **67** (3): 257–267.

Cameron, R.A.D., Cook, L.M., and Hallows, J.D., 1996, Land snails on Porto Santo: adaptive and non-adaptive radiation, *Philosophical Transactions of the Royal Society of London* B **351**: 309–327.

Cowie, R.H., Evenhuis, N.L., and Christensen, C.C., 1995, Catalog of the native land and freshwater molluscs of the Hawaiian Islands. Backhuys Publishers, Leiden.

Emberton, K.C., 2002, *Parvedentulina* and edentate *Gulella* of Madagascar, *Archiv für Molluskenkunde* **131**: 67–165.

Espinosa, J. and Ortea, J., 1999, Moluscos terrestres del archipielago cubano, *Avicennia, Suplemento* **2**: i–iv and 1–137.

Gittenberger, E., 1969, Beiträge zur Kenntnis der Pupillacae. I, Die Spelaeodiscinae, *Zoologische Mededelingen*, Leiden **43**: 87–306.

Gittenberger, E., 1975, Cave snails found in Southern Crna Gora, *Glasnik Republickog Zavoda za zastitu Prirode i Prirodnjackog Muzeja u Titogradu* **8**: 21–37.

Gittenberger, E., 1977a, *Planogyra sororcula* (Benoit, 1857) (Pulmonata, Valloniidae), une espèce nouvelle pour la France, *Zoologische Mededelingen* **51**: 191–197.

Gittenberger, E., 1977b, Cave snails from Corfu, Greece, *Comunicacions del 6è. Simposium d'Espeleologia. Bioespeleologia,* Terrassa, 1977, pp. 47–53.

Gittenberger, E., 1985, Beiträge zur Kenntnis der Pupillacea. XI. *Speleodentorcula beroni* gen. and spec. nov. (Mollusca: Gastropoda: Orculidae) aus einer Höhle in Euboea, Griechenland, *Zoologische Mededelingen* **59**: 221–228.

Gittenberger, E., 1991, What about non-adaptive radiation? *Biological Journal of the Linnean Society of London* **43**: 263–272.

Gittenberger, E., 1993, On *Idyla pelobsoleta* spec. nov. (Mollusca: Gastropoda Pulmonata: Clausiliidae), and insularity in the molluscan fauna of the highest parts of the Parnon Oros, Peloponnisos, Greece, *Zoologische Mededelingen* **67**: 321–329.

Gittenberger, E., 2000, Serrulininae in Greece, there may be more (Gastropoda, Pulmonata, Clausiliidae), *Basteria* **64**: 81–87.

Gittenberger, E., 2002, Two limestone 'islands' in central northern Greece, six new clausiliid taxa, three kinds of microarmature (Gastropoda, Pulmonata, Clausiliidae), *Basteria* **66**: 129–138.

Gittenberger, E., 2004, Radiation and adaptation, evolutionary biology and semantics, *Organisms, Diversity & Evolution* **4**: 135–136.

Gittenberger, E. and Bank, R.A., 2005, The genus *Pholeoteras* Sturany, 1904, in Greece (Gastropoda, Caenogastropoda, Cyclophoridae), *Journal of Conchology* **38**: 713.

Gittenberger, E. and B. Hausdorf, 2004, The *Orculella* species of the South Aegean island arc, a neglected radiation (Gastropoda, Pulmonata, Orculidae), *Basteria* **68**: 93–124.

Gittenberger, E., Piel, W.H., and Groenenberg, D.S.J., 2004, The Pleistocene glaciations and the evolutionary history of the polytypic snail species *Arianta arbustorum* (Gastropoda, Pulmonata, Helicidae), *Molecular Phylogenetics and Evolution* **30**: 64–73.

Giusti, F., Manganelli, G., and Schembri, P.J., 1995, The non-marine molluscs of the Maltese Islands, *Monografie* **15**: 1–608.

Hausdorf, B., 1996, Die Orculidae Asiens (Gastropoda: Stylommatophora). *Archiv für Molluskenkunde* **125**: 1–86.

Jaeckel, S.G.A., 1962, 2. Ergänzungen und Berichtigungen zum rezenten und quartären Vorkommen der mitteleuropäischen Mollusken, in: Zilch, A. and Jaeckel, S.G.A. (eds), *Mollusken, Die Tierwelt Mitteleuropas* **2**(1, Ergänzung): 25–294.

Klemm, W., 1973, Die Verbreitung der rezenten Land-Gehäuse-Schnecken in Österreich, *Denkschriften der Österreichischen Akademie der Wissenschaften, mathematisch naturwissenschaftliche Klasse* **117**: 1–503.

Krajick, K., 1998, Nunataks. Icebound islands of life, *National Geographic* **194** (6): 60–71.

Minato, H., 1988, A systematic and bibliographic list of the Japanese land snails. Shirahama.

Minato, H., 1994, Taxonomy and distribution of the land snail family Clausiliidae (Gastropoda: Pulmonata) of Japan, *Venus* **2** (Suppl.): 1–212, Tables 1–6, pls 1–74.

Preece, R.C. and Gittenberger, E., 2003, Systematics, distribution and ecology of *Balea* (= *Tristania*) (Pulmonata: Clausiliidae) in the islands of the Tristan-Gough group, *Journal of Molluscan Studies* **69**: 329–348.

Rees, W.J., 1965, The aerial dispersal of Mollusca, *Proceedings of the Malacological Society of London* **36**: 269–282.

Rosenberg, G., and Muratov, I.V., 2005, Recent terrestrial mollusks of Jamaica. http://data.acnatsci.org/jamaica/landsnails.html.

Schilthuizen, M., Welter-Schultes, F.W. and Wiese, V., 1993, A revision of the polytypic Albinaria hippolyti (Boettger, 1878) from Crete (Gastropoda Pulmonata: Clausiliidae). *Zoologische Mededelingen, Leiden* **67**: 137–157.

Schilthuizen, M., Gutteling, E., Van Moorsel, C.H.M., Welter-Schultes, F.W., Haase, M., and Gittenberger, E., 2004, Phylogeography of the land snail *Albinaria hippolyti* (Pulmonata: Clausiliidae) from Crete, inferred from ITS-1 sequences, *Biological Journal of the Linnean Society* **83**: 317–326.

Schilthuizen, M., Vermeulen, J.J., Davison, G.W.H., and Gittenberger, E., 1999, Population structure in a snail species from isolated Malaysian limestone hills, inferred from ribosomal DNA sequences, *Malacologia* **41**: 283–296.

Solem, A., 1984, A world model of land snail diversity and abundance, in: Solem, A. and Van Bruggen, A.C. (eds), *World-Wide Snails*, E. J. Brill/Dr. W. Backhuys Publishers, Leiden, pp. 6–22.

Uit de Weerd, D.R., Gittenberger, E., and Piel, W.H., 2004, Widespread polyphyly among Alopiinae snail genera: when phylogeny mirrors biogeography more closely than morphology. *Molecular Phylogenetics and Evolution* **33**: 533–548.

Vardinoyannis, K., 1994, Biogeography of land snails in the south Aegean island arc. Ph.D. thesis, Athens.

Vermeulen, J.J., 1994, Notes on the non-marine molluscs of the island of Borneo 6. The genus *Opisthostoma* (Gastropoda Prosobranchia: Diplommatinidae), part 2, *Basteria* **58**: 75–191.

Welter-Schultes, F.W., 2000a [15.03], The paleogeography of late Neogene central Crete inferred from the sedimentary record combined with *Albinaria* land snail biogeography, *Palaeogeography, Palaeoclimatology, Palaeoecology* **157**: 27–44.

Welter-Schultes, F.W., 2000b [25.05], Approaching the genus *Albinaria* in Crete from an evolutionary point of view (Pulmonata: Clausiliidae), *Schriften zur Malakozoologie* **16**: 1–208.

Welter-Schultes, F.W. and Williams, M.R., 1999, History, island area and habitat availability determine land snail species richness of Aegean islands, *Journal of Biogeography* **26**: 239–249.

Whittaker, R.J., 1998, *Island Biogeography. Ecology, Evolution and Conservation*, Oxford University Press, Oxford, NY.

Chapter 12

# Morphological and Genetical Differentiation of Lizards (*Podarcis bocagei* and *P. hispanica*) in the Ria de Arosa Archipelago (Galicia, Spain) resulting from Vicariance and Occasional Dispersal

J.W. ARNTZEN[1] AND P. SÁ-SOUSA[2]
[1]Nationaal Natuurhistorisch Museum, Naturalis, P.O. Box 9517, 2300 RA Leiden, The Netherlands, arntzen@nnm.nl
[2]Conservation Biology Unit, Department of Biology, University of Evora, Pólo da Mitra, 7002-554 Évora, Portugal, psasousa@uevora.pt





*W. Renema (ed.), Biogeography, Time, and Place: Distributions, Barriers, and Islands,* 365–401

**Abstract** We studied morphological and molecular characters in *Podarcis bocagei* and *P. hispanica* lizards in the Ria de Arosa archipelago in coastal Galicia, Spain. Contrasting published information about insular distributions indicated that morphological species identification is problematic. Instead, we identified 145 lizards from 13 islands and several mainland populations by a panel of partially diagnostic nuclear protein loci and through the DNA sequencing of a stretch of 297 base pairs (bp) of the mitochondrial cytochrome-b gene. Correspondence between the molecular identifications was complete, with the exception of two lizards that carried *P. hispanica* mitochondrial DNA in a *P. bocagei* nuclear background. The combined results indicate past hybridization events and the oversea dispersal of a female lizard over a distance of about 500 m. Fourteen morphometric and 12 meristic characters were measured in 187 lizards from 15 islands and several mainland reference populations. Multivariate analysis revealed significant differences between sexes and between species. Two island populations for which no molecular data were available were identified as belonging to *P. bocagei* from morphology. Locally, *P. bocagei* and *P. hispanica* have an essentially parapatric distribution, with *P. bocagei* in the northeast and the islands of the inner Ria, and *P. hispanica* in the southwest and the islands of the outer Ria. Common patterns for island populations are an increase in absolute size, a decrease in three relative head width parameters, and a lower nuclear genetic heterozygosity than on the mainland. No correspondence was observed between morphological and molecular patterns of intra-specific differentiation. Also, linear distance between populations did not help to explain the results. Because the Ria de Arosa is a flooded river valley, the local bathymetry reflects the order in which the islands have become isolated from the mainland since the sea-level rise started *c.* 14,000 years ago. However, the temporal order of isolation was not associated with the pattern of population differentiation. The island populations of *P. hispanica* seem to have diverged over multiple genetic and morphological axes, with no obvious relationship to inter-population distance, coastline history, island ecology, or species habitat preferences. The results indicate that genetic drift is the main force driving in population divergence. Oversea dispersal, that is gene-flow, appears insufficient to counter the morphological differentiation between some island populations. Hypothesis on the biogeographical history of *P. bocagei* and *P. hispanica* in western Galicia are discussed.

## 1. Introduction

Ever since Darwin sailed the Galapagos Archipelago, island populations have been at the forefront of evolutionary research. Island populations are readily impacted by intrinsic factors, such as inbreeding and genetic drift, as well as by external factors, for example, aberrant weather conditions; tornados may wipe out a population overnight, vacating the habitat for new arrivals (Templeton, 1980; Spiller et al., 1998; Schoener et al., 2001). The island setting can be seen as a natural laboratory where evolutionary processes are intensified relative to the mainland. Moreover, the outcome of the experiment is more readily observed, appreciated, and interpreted than on the mainland. The main ingredients to the "natural experiment" are physical isolation in combination with limited oversea dispersal, small population size, the local niche, and differential competition through the presence or absence of similar and closely related forms. The geological and evolutionary processes that come into play are manifold and even simple settings may give rise to complex patterns of variation (e.g., Thorpe et al., 1996; Gübitz et al., 2000, 2005; Thorpe and Stenson, 2003).

Lizards have been a favourite group of organisms for island evolutionary studies. The prime reason is that lizards are widespread, potentially inhabiting the tiniest of islands and showing strong eco-morphological diversification. With few predators on islands, the focal species may be conveniently abundant. They are terrestrial, ground-dwelling, or arboreal organisms with little dispersal capacity, of the right size for easy handling and, with an often-colourful dress, appealing to laymen and professional alike.

Anole species of the Caribbean (genus *Anolis*) are a case in point. The anoles of the Greater Antilles have specialized morphologies adapted to a range of different ecologies. On a single island, related anole species will differ in the niche they occupy (e.g., trunk vs. twig vs. crown inhabiting), foraging and defensive behaviour, size, body proportions, and colouration. Williams (1972, 1983) designed a classification of "ecomorphs", with as many as six coexisting anole species distributed across the habitat from the ground to the canopy, in characteristic places. The association between specialized ecologies and matching morphologies is recurrent among the separate islands and parallel patterns of diversification are found in anoles from different islands, regardless of taxonomic affiliation (Losos, 1992). Most large islands have each of the six ecomorphs represented. Smaller islands have fewer species that are less extreme in size, and without some of the specialized niches and lifestyles. The ecomorph concept does not apply to lizards of the continent that have a lower abundance and form part of a more complex ecosystem than on the islands. While selective pressures on islands are mainly driven by interspecific competition, the main selective force on the mainland is predation (Irschick et al., 1997). However, from a lizard perspective, the islands of the Greater Antilles are almost like continents. To eventually unravel complicated patterns it may be best to start with the study of simpler systems as encountered on the smaller islands.

The Lesser Antilles has simple anole communities, with either one or two species. Most single species islands (20 out of 21) have intermediate-sized anoles, whereas most two-species islands (7 out of 8) are inhabited by one large- and one small-sized species (Roughgarden, 1995). This strong relationship between the number of species on an island and the size of its inhabitants is present in different anole groups, distributed over the northern and the southern arc of the Lesser Antilles (from Anguilla to Dominica and from Martinique to Bonaire, respectively). Both groups are monophyletic, suggesting that the observed relationship is the results of converging selection pressure and does not reflect common ancestry (Roughgarden, 1995; for molecular phylogenies see, e.g., Jackman et al., 1999; Schneider et al., 2001). As Losos et al. (1998) put it that adaptive change in similar environments can overcome historical contingencies to produce strikingly similar evolutionary outcomes.

Large or small, islands in the Caribbean have an ancient and heterogeneous origin (Dengo and Case 1990; Donovan and Jackson, 1994). The Caribbean is a geologically complex region that displays a variety of tectonic plate boundary interactions in which the origin of the Caribbean crust dates back to the Albian, 100 Ma (Jackson, 2002). The intricate geological history of the region complicates the reconstruction of the evolutionary biology of Caribbean *Anolis*. Younger and simpler settings are found in European *Podarcis* wall lizards. These lizards may also show a marked morphological variability in body size and chromatic patterns (e.g., giantism, melanism) among populations on smaller and more remote islands relative to the mainland (Klemmer, 1964; Clover, 1979; Galán, 1985; Böhme, 1986; Henle and Klaver, 1986; Lanza and Poggesi, 1986; Cirer, 1987; Cheylan, 1988). Most studies have shown limited amounts of genetic differentiation between populations (Gorman et al., 1975; Cirer and Guillaume 1986; Ramón et al., 1991; Capula, 1994a, b, 1996, 1997; Castilla et al., 1998a, b; Sá-Sousa et al., 2000; Capula and Ceccarelli, 2003; Poulakakis et al., 2003) with a trend for genetic distance to increase with bathymetry (reflecting the period of population isolation) and to decrease with island size (reflecting effective population size and the prominence of genetic drift) (Gorman et al., 1975; Cirer and Martínez-Rica, 1990; Ramón et al., 1991; Delaugerre and Cheylan, 1992; Vicente, 1999; Brehm et al., 2001; Podnar et al., 2004, 2005). We have chosen to study morphological evolution of lizards on islands in a simple and tractable setting, namely the Ria de Arosa in Galicia, Spain, differing from the Antilles along the temporal and spatial axes by orders of magnitude. As for these temporal and spatial dimensions, our study is between that of a strict natural setting, such as *Podarcis* lizards on single islands and archipelagos of Mediterranean Europe with a mostly Pleistocene history (Gorman et al., 1975; Capula, 1994a, 1996), and that of deliberate release and replicated field experiments (Thorpe and Malhotra, 1992; Losos et al., 1997; Losos and Spiller, 1999). The man-assisted introduction (Austin, 1999) of lizards can also be seen as falling into the latter category.

The islands of the Ria de Arosa have come into existence with the inundation of a river valley, resulting from the rise in sea level towards the end of the last glacial period, starting *c.* 14,000 years ago (Bard et al., 1996). Complicating factors are the local elevation of land (Pannekoek, 1966a) and climate fluctuations such as the "Little Ice Age" between the 15th and 19th century (Pethick, 1984; Dias et al., 2000). Prior to post-glacial drowning, the Rias of Galicia were vegetated with woodland (Rodríguez-Guitian et al., 1996). The depth between the islands is maximally 70 m. The two *Podarcis* species occurring in the northwestern corner of the Iberian Peninsula are *Podarcis bocagei* Seoane 1885, and "type 1" of *Podarcis hispanica* Steindachner 1870 (Sá-Sousa, 2000; Harris and Sá-Sousa, 2001; Galán, 2003a, b). *Podarcis bocagei* and *P. hispanica* are both generalist species that are broadly sympatric in southern Galicia and northern Portugal. In the preferred habitats both species can locally be abundant (Galán, 1986, 1999b). Despite a marked intraspecific morphological variability, mainland populations of the species can be identified by phenotype. Island populations differ from mainland ones in particular ways, e.g., in size, shape, and coloration pattern, rendering species identification problematic. The aims of the present paper are fourfold:

1. To develop molecular genetic marker techniques for the unambiguous identification of *P. bocagei* and *P. hispanica*
2. To document species distribution in the Ria de Arosa region
3. To search for evidence on dispersal events and interspecific hybridization
4. To document and analyse morphological differentiation of island populations relative to the mainland. As to this point, we demonstrate that morphological change in island populations of lizards can be rapid. Lack of congruence among populations from ecologically similar islands and genetic data indicate that the underlying process is genetic drift rather than natural selection

Finally, we present a hypothesis on the distributional history of lizards of the Ria de Arosa, with contrasting scenarios for *P. bocagei* and *P. hispanica*.

## 2. Material and Methods

### 2.1. Description of the Study Area

The Ria de Arosa is an inlet of the Atlantic Ocean at the northwestern coast of Spain. With a surface area of about 230 km$^2$, the Ria de Arosa is the largest, the most irregularly shaped and the most diversified among the four "Rias Bajas" of Galicia (Otto, 1975). Four larger islands (Sálvora, Arosa, Cortegada, and Toja Grande), a large number of smaller islands and rocks, and several secondary embayments produce a complex topography. The average width of the Ria de Arosa is 9 km. The longitudinal axis is directed from the northeast to the southwest and covers a distance of about 25 km, from Cortegada, situated in the mouth of the River Ulla, to Sálvora in the mouth of the Ria. The Ulla, Umia, and Beluso are the most important streams discharging into the Ria, but their influence is limited; the salinity of the water at the surface gradually increases from 32‰ in the northeast to 36‰ in the southwest of the Ria, in summer as well as in winter (Cadée, 1968; Otto, 1975). Wind direction is predominantly from the north, yielding northeast to southwest surface currents. The Ria de Arosa is divided into two parts, the outer Ria and the inner Ria. The border between them is formed by a chain of islands, including Jidoiro Arenoso, Jidoiro Pedregoso, and Rua. Most islands are situated within the 10 m isobath, except Sálvora and Rua, which are situated within the 20 m and 30 m isobaths, respectively (Fig. 1).

The islands can also be classified in terms of vegetation. Sandy islands with sparse vegetation are Jidoiro Arenoso and Vionta; rocky islands with bare rock

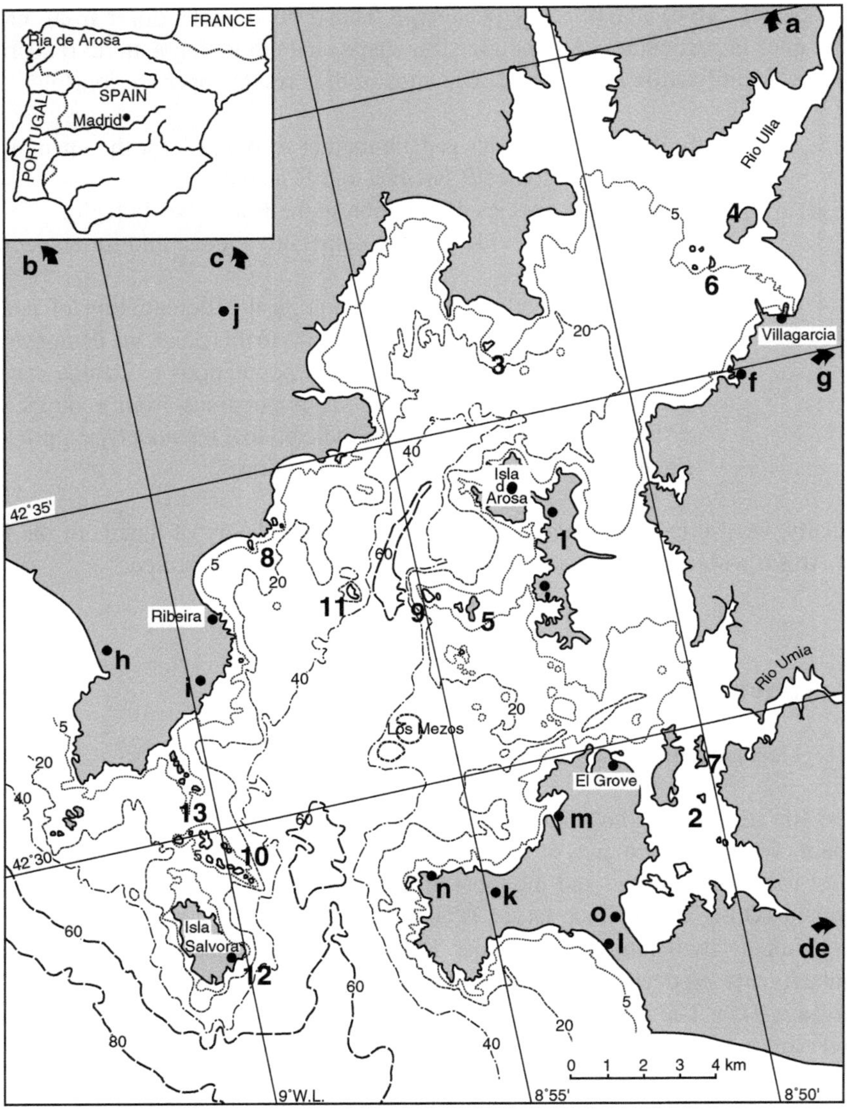


Fig. 1 Situation map and bathymetry of the Ria de Arosa, Galicia, Spain. (From Cadée, 1968.) Island and mainland localities where lizards (*Podarcis bocagei* and *P. hispanica*) were sampled are indicated by numbers (1–13) and letters (a–o), respectively. For locality names and details see Appendix I. The Island Toja Grande (unnumbered) is located in between El Grove and the mouth of the Rio Umia. Samples from the mainland were pooled according to their origin from north and south of the River Ulla, respectively.

surfaces and with dense vegetation mostly at the bases of boulders of granite rock are Coroso, Jidoiro Pedregoso, Noro, and Rua. Rocky islands covered with abundant vegetation are Beiro, Benencia, San Bartolomé (i.e., Malveira Grande), and Toja Pequeña. The larger islands (Arosa, Cortegada, and Sálvora) have mixed characteristics. The island Arosa (and also Toja Grande where we did not find lizards) are the only islands with a sizable human population. The island of Rua stands out because of the walls of the lighthouse and man-made dry stone walls bordering small pasture fields, providing ample shelter for a large lizard population. Whereas the larger island have wells that supply freshwater, on the smaller islands freshwater is only generated through the development of fog, as a consequence of the cooling of the lower layers of air by the relatively cold sea water, which increases the relative humidity. This generation of fog is important for the lizard populations in the Ria de Arosa, since in summer this is a principal source of freshwater. Smaller islands inhabited by lizards are Beiro, Coroso, and San Bartolomé. Classifications of the islands in terms of distances to one another and the mainland and potential patterns of dispersal through "stepping stones" are readily retrieved by inspection of Fig. 1. A formal classification in terms of the temporal order of isolation caused by a rising water level was made on the basis of bathymetric and geographical distance, on an ordinal scale as follows: *P. bocagei* locality numbers 14 and 15, 4, 7, 2, 3, 6, 1, and 5; *P. hispanica* localities 16 and 17, 8, 9, 13, 10, 12, and 11 (see Fig. 1). Nonn (1966) and Pannekoek (1966b, 1970) described the geomorphological setting of the Ria de Arosa and Galicia. For a palaeobotanical perspective of the area, see Ramil-Rego and Orellana (1996).

## 2.2. Sampling

*Podarcis bocagei* and *P. hispanica* lizards were collected in July and August 1962–1964 by M.S. Hoogmoed and W.J. Roosdorp (see Brongersman and Pannekoek, 1966) and in July 2002 by M.S. Hoogmoed, J.M. Oliveira and the senior author, from 13 islands and islets in the Ria and from 15 localities on the mainland surrounding the Ria de Arosa (Fig. 1). All material is stored as vouchers at the Nationaal Natuurhistorisch Museum Naturalis, Leiden, the Netherlands (Appendix I).

## 2.3. Morphological Data and Multivariate Statistical Analysis

We measured snout-vent length (SVL) and 13 other continuous morphometric variables with 0.02 mm precision digital callipers, and we counted 12 scalation characters using a fixed magnifying glass in 187 adult lizards. Bilateral variables were observed at the right side of the body in principal. For a further description of the variables, see Fig. 2 and Table 1.

The 14 measurements were ln-transformed to increase statistical normality of the data and to reduce possible effects of non-linear growth. The standardized residuals were calculated for the regression of ln < character > on ln < SVL> to

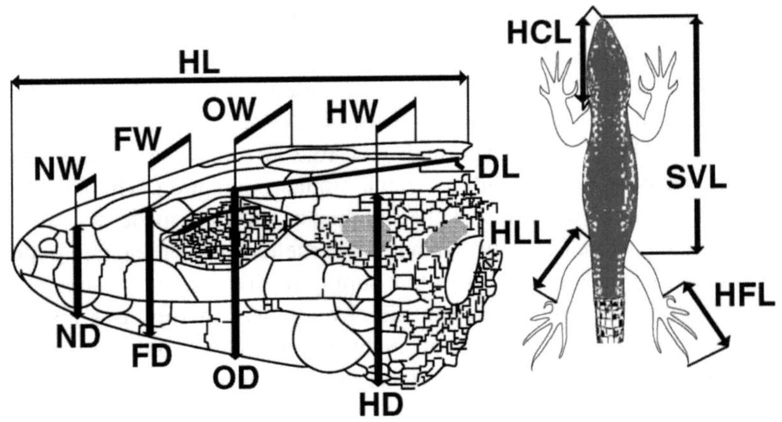


Fig. 2 Fourteen morphometric characters measured in lizards from the Ria de Arosa, Galicia, Spain. Characters are described in Table 1, alongside the descriptions and codes of 12 meristic characters.

*Table 1. Fourteen continuous morphometric variables (upper panel, see also Fig. 2) and 12 discrete meristic variables (lower panel) measured in 187 adult lizards* (Podarcis bocagei *and* P. hispanica) *from the Ria de Arosa, Galicia, Spain. The material is listed in Appendix I.*

| | |
|---|---|
| 1. | SVL – snout-vent length |
| 2. | HCL – head-chest length, measured from the tip of snout to the edge of the collar scales |
| 3. | HL – head length, measured from the tip of snout to rear edge of parietal scales |
| 4. | DL – diagonal length, measured from the outer edge of the suture of the second–third supraoculars and the rear edge of the occipital scale |
| 5. | HW – head width, maximum distance between parietal scale edges |
| 6. | OW – inter-orbital width, measured between the opposite sutures of the second–third supraocular plates |
| 7. | FW – frontal width, measured between the opposite sides of the snout near the edges of the first supraoculars plates (i.e., near the midline of the frontal scale) |
| 8. | NW – inter-nasal width, measured between the opposite sides of the snout near the midline of the internasal scale |
| 9. | HD – head depth – measured at the mid-position of parietal scale |
| 10. | OD – orbital depth, measured between the second and third supraocular sutures and the chin |
| 11. | FD – frontal depth, measured between the frontal scale and the chin |
| 12. | ND – nasal depth, measured between the internasal scale and the chin |
| 13. | HFL – hind foot length, measured from the tip of the fourth toe to the ankle |
| 14. | HLL – hind leg length, measured from the groin ring to the ankle |
| 1. | DOR – number of dorsal scales around the mid-body |
| 2. | GUL – number of gular scales counted along the throat midline, from the collar to the confluence of the maxillaries |
| 3. | COL – number of differentiated scales in the collar |
| 4. | VTR – number of inner ventral scales counted longitudinally, from the thorax to the lower belly |
| 5. | CAN – number of scales that circle the anal plate |
| 6. | TAI – number of caudal scales around the (dorsal) fifth whorl equalling second ventral tail whorl |
| 7. | L4F – number of lamellar scales beneath the fourth finger |
| 8. | L4T – number of lamellar scales beneath the fourth toe |
| 9. | FPO – number of femoral pores beneath the thighs |
| 10. | TPL – number of supratemporal scales, near the edge of the parietal plate |
| 11. | SOC – number of supraocular scales |
| 12. | GRA – number of supraciliary granules |

reduce the impact of overall size in the analysis. With two exceptions the meristic characters showed no statistical correlation with SVL under Bonferroni correction and we used standardized counts in the analyses. The 0.3% of missing data were replaced for values estimated by linear regression ($n$ = 5, morphometric data) or by the grand average value ($n$ = 9, meristic data). Initial data exploration by discriminant analysis revealed significant morphological differentiation between the four groups of *P. bocagei* males and females, and *P. hispanica* males and females. Because we are more interested in comparing species than comparing sexes, from this point onwards the morphological data were analysed for species in conjunction and for sexes separately. The computer software used in the (multivariate) statistical analysis was SPSS 12 (SPSS, 2003). The congruence between classifications derived from different data sets is expressed by Cohen's kappa. Matrix comparisons were made by the Mantel test in NTSYS 1.7 (Rohlf, 1992). In $t$-tests the number of degrees of freedom (df) was adjusted downwards in cases where equal variances could not be assumed, as indicated by the Levene's statistic. When multiple $t$-tests were used to compare average values, this was done as a data exploratory exercise and not to ascertain the true statistical significance of results. Moreover, the Bonferroni correction was not applied in this exercise and nominal significant results were only taken as a convenient cut-off point assisting our search for trends in the underlying data. The contrasting results revealed by this analysis indicated that the pooling of island samples is not warranted as this might suppress important information. Discriminant analysis on morphological data was therefore carried out with just populations from the islands Arosa, Jidoiro Arenoso, Jidoiro Pedregoso, Noro, and Rua, that have larger than average distance to the mainland (i. e., presumably subject to a longer period of isolation) and have comparatively large samples available.

## 2.4. Molecular Genetics and the Analysis of Sequence Data

Small amounts (10–20 μg) of stomach tissue from 145 recently collected lizards were incubated overnight at 55°C with Proteinase K, and a lysis-buffer containing edetic-acid and sodium dodecyl sulphate. DNA extractions were performed with the "DNeasy" extraction kit (QIAgen). DNA was separated from proteins and other components (contaminants and enzyme inhibitors) by selective binding to a silica membrane from which it was eluted with the elution buffer provided with the kit. DNA was PCR (polymerase chain reaction) amplified in 25 μL reactions in a MJ-research PTC-200 thermal cycler under the following conditions; 4% glycerol (Merck), 4 mM $MgCl_2$ (QIAgen), 0.2 mM dNTPs (Amersham), 2.5 U Taq polymerase (QIAgen) in manufacturer's PCR buffer, and 0.4 μM of H and L primer (Isogen). Oligonucleotide primers for cytochrome-b were H15149 and L14841 (Kocher et al., 1989). PCR profiles were a 2 min denaturing step at 95°C followed by 35 cycles of denaturing for 15 s at 95°C, primer annealing for 1 min at 60°C and elongation for 1 min at 72°C, with a final 10 min elongation step at 72°C. The products were cleaned with the QIAgen PCR purification kit according to manufacturer's instructions. Amplified products were analysed by comparison with the Eurogentec smartladder MW 1700–10 by electrophoresis (100 V, 30 min) in a 1% agarose gel and post-staining with ethidium bromide.

Cycle sequencing was performed on PCR products using the BigDye Terminator v.1.1 (Applied Biosystems) reaction premix for 30 cycles of 95°C for 30 s, 50°C for 1 min and 60°C for 4 min. Nucleotide sequences were determined with an ABI377 (Applied Biosystems). One amplification consistently failed. This sample alongside with six reference samples were instead amplified and sequenced for a part of the subunit 4, NADH dehydrogenase gene using the primers ND4 and Leu (Areválo et al., 1994) with the help of technical information provided by S.D. Busack (personal communication 2004). The fragment sizes obtained were from *c.* 350 bp of the cytrochrome-b gene and *c.* 710 bp comprising the subunit 4, NADH dehydrogenase gene.

Sequences were aligned by hand, assisted by the program Clustal W 1.7 (Thompson et al., 1994), and checked for unexpected results such as gaps and stop codons after the translation into amino acids with MacClade 4.0 (Maddison and Maddison, 2000). Gaps and stop codons were not inferred. Additional sequence information was taken from GenBank. We selected information representing *P. bocagei* (AF372087, AF372088, AF469424 and AF469426) and *P. hispanica* (AF372084, AF 469445, AF 469447, and AF 469449) from nine localities in northwestern Iberia (Harris and Sá-Sousa, 2002). Three sequences derived from *P. carbonelli* Pérez-Mellado, 1981, were added for comparison (AF372079-372081). PAUP* 4.0b10 (Swofford, 2002) was used to construct a UPGMA tree of all different haplotypes on the basis of pair-wise percent sequence divergences ("uncorrected p-distance", *d*) over a stretch from 297 bp of the cytochrome b-gene.

## 2.5. Enzyme Electrophoresis and Data Analysis

Blood and liver tissue were dissected from 145 recently collected and freshly sacrificed adult or subadult lizards. Erythrocytes were removed from the blood by brief centrifugation. The liver was homogenized in cold buffer (100 mM Tris, 1 mM EDTA, and 50 μM NADP, adjusted to pH 7.0 with HCl) and centrifuged. The aqueous supernatant was decanted and stored at −80°C for future electrophoresis on starch gels. From a pilot study in which 37 inferred protein loci were surveyed, eight enzymes were selected for detailed investigation following the criteria of genetic variability, diagnostic properties, and practicality, such as ease of handling, enzyme-staining properties, zymogram resolution and consistency of results in starch gel electrophoresis. Selected enzymes were aspartate transaminase (ATA-1, E. C. Number 2.6.1.1), malic enzyme (ME, E.C. 1.1.1.40) and mannose phosphate isomerase (MPI, E. C. 5.3.1.8), that were run on "type V" buffer, and lactate dehydrogenase (LDH-1, E. C. 1.1.1.27) and phosphogluconate dehydrogenase (6-PGD, E.C. 1.1.1.44) that were run on "type I' buffer. The buffers used are described in Shaw and Prasad (1970). Two esterases (E.C. 3.1.1.1) expressed in the liver (L.EST-2 and -3) and one esterase active in blood plasma (P.EST-2) showed good activity, separation and reproducibility on either system. Enzymes were stained following Shaw and Prasad (1970) and Arntzen and García-París (1995). Presumptive loci and alleles (electromorphs) were assigned numbers and letters, respectively, in sequence starting from the most anodally migrating forms. Allele frequencies were determined directly from the observed genotypes. To assess the degree of genetic variability shown by individual protein loci and for populations, we calculated heterozygosity under the assumption that each population and locus was in Hardy–Weinberg equilibrium. These values were compared with the observed values. GenePop 3.4 (Raymond and Rousset, 1995) was used to assess deviations from hypothesized Hardy–Weinberg equilibrium by exact probabilities following Fisher's method in a Markov-chain procedure. Hierarchical relationships of populations were analysed with BIOSYS (Swofford and Selander, 1981). We opted for the construction of a distance-Wagner tree on Rogers' genetic distance, and not UPGMA, reckoning with the possibility that genetic restructuring and drift and not mutation rate would mostly influence the enzyme genetic profiles.

## 2.6. Molecular Identification

Lizards were identified to species, either *Podarcis bocagei* or *P. hispanica*, by combining the results obtained by principal coordinate analysis (PCoA) on nuclear and cytoplasmic (mitochondrial DNA) genetic characters. Firstly, PCoA was performed on DNA sequences representing 145 individuals. Invariable nucleotide positions were omitted. Nucleotide positions with alternative character states and multistate positions were recoded by hand into binary characters. The subroutine SIMQUAL of the program NTSYS 1.7 (Rohlf, 1992) performed a comparison between all pairs of sequences and calculated a pairwise similarity coefficient, for which we chose the "simple matching" coefficient. The subroutine DCENTER was used to transform the similarity matrix into scalar product form, after which it was factored using the subroutine EIGEN. Secondly, for the enzyme data a binary data set was generated for all the same individuals. The presence (1) or absence (0) of each allele at each locus was defined as a separate character state. Character states were assumed to be independent, although in reality limited to a maximum of two scores of 1 per locus. Homozygotes were not distinguished from heterozygotes (i.e., they were represented by a single score of 1). The pairwise similarity coefficient we choose is that of Jaccard, because it ignores joint absences. Finally, analysis of molecular variance (AMOVA) was performed with the software Arlequin 2.1 (Schneider et al., 2000) to assess the degree to which genetic variability was partitioned between species, between populations and within populations.

## 2.7. Correlates of Population Differentiation

Mantel tests were used to investigate the relationship between genetic distance of populations (1, Rogers' genetic distance for the enzyme data and 2, uncorrected p-distance for the sequence data) and, alternatively, Euclidian distances over the first and second PCoA axis; morphological distance (Euclidian distance of population means over the first, second and third axis of a principal component analysis for 3, males and 4, females), 5, geographic distance between populations and 6, the temporal order in which the islands may have become isolated following a rising sea level, as described in the introduction. Geographic distances were straight-line distances between islands, with and without the mainland localities as pooled under numbers 14c, 15d, 16c, and 17d. Mantel tests were done for *P. bocagei* and *P. hispanica,* separately. This is because the potentially more powerful partial-Mantel test, with taxonomic affiliation as a second independent variable may be critically flawed (Raufaste and Rousset, 2002).

## 3. Results

### 3.1. Gross Morphological Differentiation

Discriminant analysis on the entire morphological data set in which lizards had been identified to gender by eye and to the species by molecular markers (see below) produced discriminant functions that explained 78.6% and 20.2% of the total variation, along the first and second axis, respectively. The first discriminant function effectively separates males and females. The morphological variables correlated to this axis are, in decreasing order of effect, HL, HD, OD, DL, HW, FD, HCL, VTR, ND, OW, HLL, HFL, NW, FW, and FPO. The second discriminant function effectively separates *P. bocagei* and *P. hispanica.* The morphological variables correlated to this axis are GUL and L4T. The classification table reveals no individuals misallocated for gender (Cohens' kappa = 1.0) and six individuals misallocated for taxonomic affiliation (Cohens' kappa = 0.94). The latter cases all refer to island lizards (three males and three females).

### 3.2. Genetic Variability at the Mitochondrial Level

The different sequences observed were interpreted as haplotypes. In UPGMA clustering, these fall in two distinct groups representing 39 individuals (the "b" group), and 106 individuals (the "h" group), respectively. Four *P. bocagei* and four *P. hispanica* haplotype sequences from GenBank were added as species references and three *P. carbonelli* sequences were added for comparison. *Podarcis bocagei* and *P. hispanica* haplotypes firmly group with or within the "b" and "h" groups, respectively. The most frequently observed haplotype in *P. bocagei* was equivalent to sequence AF469426 from Malpica in northern Galicia, Spain. The most frequently observed sequence in *P. hispanica* showed no difference to sequence AF372084 from Vila Real, Portugal, except for an undetermined character state at position 195 of the original GenBank sequence. The interspecific difference between the most commonly observed *P. bocagei* and *P. hispanica* haplotypes was $d = 8.4\%$ and the interspecific distance between either of these species and *P. carbonelli* was about 13.0%. The "b" and "h" groups of haplotypes are therefore taken to represent *P. bocagei* and *P. hispanica* mitochondrial genes, respectively. Six ND4 sequences could be placed in two groups corresponding to *P. bocagei* and *P. hispanica*, in line with cyt-b results (results not shown). One lizard from the southern mainland for which amplification of the cyt-b DNA consistently failed could, according to the ND4 sequence, be determined to carry *P. hispanica* mtDNA. The mitochondrial genotype of this lizard was, for convenience, coded as the common cytochrome-b haplotype for the species.

## 3.3. Genetic Variability at the Nuclear Level

We observed from two to seven alleles per selected enzyme locus (Table 2). Direct count heterozygozity ($H_{dc}$) was near zero for the locus *6-Pgd,* with the allele *6-Pgd*$^{b}$ fixed in one set of populations and the allele *6-Pgd*$^{c}$ nearly fixed in all other populations, indicating the species diagnostic properties of this locus and simultaneously suggested the absence of $F_1$ hybrids in our sample. $H_{dc}$ was intermediate ($H_{dc}$ = 0.12–0.17) for the loci *Ata-1*, *Ldh-1*, and *L.Est-2,* and high for *Mpi* ($H_{dc}$ = 0.27), *Me* ($H_{dc}$ = 0.31), *P.Est-2* ($H_{dc}$ = 0.42) and *L.Est-3* ($H_{dc}$ = 0.59). The latter locus was out of Hardy–Weinberg equilibrium in three populations, producing either an excess of heterozygotes (in population Noro and in mainland population 15) or a deficit of heterozygotes (in population Rua). Other loci were not significantly out of Hardy–Weinberg equilibrium in any population. For the loci selected in this study, the weighted average $H_e$ for mainland populations was 0.33 in *P. bocagei* and 0.38 in *P. hispanica*; the weighted average $H_e$ for island populations was 0.22 in *P. bocagei* and 0.27 in *P. hispanica*. Nested ANOVA analysis indicated no significant difference in level of heterozygosity between the species ($F_{s\,[1,\,2]} = 0.17$, $P > 0.05$) and a significant added variance component for mainland versus island origin of the populations ($F_{s\,[2,\,11]} = 7.51$, $P < 0.01$).

Distance-Wagner trees on the basis of Rogers' genetic distance between populations showed two groups of populations, corresponding to *P. bocagei* and *P. hispanica* (Fig. 3). Within *P. bocagei* populations from Benencia, Jidoiro Arenoso, and San Bartolomé are grouped together, as are the populations from the southern and northern mainland.

Within *P. hispanica*, populations from Rua and Jidoiro Pedrogoso are placed together as are those from Vionta, Sálvora, Noro and the northern mainland, as separate from populations from Coroso and the southern mainland. For the loci selected in this study, Rogers' genetic distance averages at $D_R = 0.21$ for *P. bocagei* populations, at $D_R = 0.25$ for *P. hispanica* populations and at $D_R = 0.57$ between both groups of populations. Principal coordinate analysis of the binary coded enzyme allelic data produces a first, second and third principal coordinate which account for 23.5%, 10.1%, and 6.6% of the total variation in the data set, respectively. Visual inspection of the scores from the first and second axis plotted

*Table 2.* *Allele frequencies for eight polymorphic gene loci surveyed over 15 lizard populations from the Ria de Arosa, Galicia, Spain. To the bottom: "direct count" heterozygosity ($H_{dc}$) and heterozygosity estimated under the assumption of Hardy–Weinberg equilibrium ($H_e$) with their standard errors (SE). For locality information see Appendix I.*

| *Species* | *Podarcis bocagei* | | | | | | | *Podarcis hispanica* | | | | | | | |
|---|---|---|---|---|---|---|---|---|---|---|---|---|---|---|---|
| Locality number | 1 | 3 | 5 | 6 | 7 | 14 | 15 | 8 | 9 | 10 | 11 | 12 | 13 | 16 | 17 |
| Sample size | 8 | 3 | 10 | 1 | 2 | 3 | 14 | 7 | 6 | 12 | 27 | 5 | 5 | 12 | 30 |
| Locus and allele | | | | | | | | | | | | | | | |
| *Ata-1* | | | | | | | | | | | | | | | |
| a | 0.25 | | 0.05 | | | | 0.07 | 0.71 | 1.00 | 0.54 | 1.00 | 0.70 | 0.60 | 0.71 | 0.33 |
| b | 0.75 | 1.00 | 0.95 | 1.00 | 1.00 | 1.00 | 0.93 | 0.29 | | 0.46 | | 0.30 | 0.40 | 0.29 | 0.67 |
| *L.Est-2* | | | | | | | | | | | | | | | |
| a | 1.00 | 1.00 | 1.00 | 1.00 | 1.00 | 1.00 | 0.68 | 0.29 | 0.25 | 0.04 | 0.28 | 0.10 | | 0.04 | 0.13 |
| b | | | | | | | 0.32 | 0.71 | 0.75 | 0.96 | 0.72 | 0.90 | 1.00 | 0.96 | 0.87 |
| *L.Est-3* | | | | | | | | | | | | | | | |
| a | 0.88 | 0.83 | 0.55 | 0.50 | 0.75 | 0.33 | 0.82 | 0.79 | 0.92 | 0.50 | 0.82 | 0.40 | 0.40 | 0.58 | 0.57 |
| b | | | | | | 0.17 | | | | | | | | | |
| c | | 0.17 | 0.25 | 0.50 | 0.25 | | 0.04 | 0.07 | | | | 0.10 | | | 0.05 |
| d | 0.13 | | 0.20 | | | 0.50 | 0.14 | 0.14 | 0.08 | 0.50 | 0.19 | 0.50 | 0.60 | 0.42 | 0.37 |
| e | | | | | | | | | | | | | | | 0.02 |
| *P.Est-2* | | | | | | | | | | | | | | | |
| a | | 0.50 | | | | 0.17 | | | | | | | | | |
| b | | | 0.20 | | | | | | | | | | | | |
| c | | | | | | | 0.04 | | | | | | | | |
| d | 0.25 | | | | 0.25 | 0.50 | 0.43 | 0.14 | 0.67 | 0.04 | 0.59 | | 0.40 | 0.25 | 0.25 |
| e | | | 0.35 | 1.00 | | | 0.07 | | | | | | | | 0.05 |
| f | 0.75 | 0.50 | 0.45 | | 0.75 | 0.33 | 0.43 | 0.86 | 0.33 | 0.96 | 0.41 | 0.80 | 0.60 | 0.75 | 0.63 |
| g | | | | | | | 0.04 | | | | | 0.20 | | | 0.07 |

| | | | | | | | | | | | | | | | |
|---|---|---|---|---|---|---|---|---|---|---|---|---|---|---|---|
| *Ldh-1* | | | | | | | | | | | | | | | |
| a | 1.00 | 1.00 | 0.95 | 1.00 | 1.00 | 0.83 | 0.96 | 0.64 | 1.00 | 0.38 | 0.28 | | | 0.17 | 0.52 |
| b | | | 0.05 | | | 0.17 | 0.04 | 0.36 | | 0.63 | 0.72 | 1.00 | 1.00 | 0.83 | 0.48 |
| *Me* | | | | | | | | | | | | | | | |
| a | 0.13 | 0.67 | 0.65 | 1.00 | 0.25 | 0.67 | 0.68 | 0.07 | | | 0.07 | 0.10 | | 0.50 | 0.22 |
| b | 0.88 | 0.33 | 0.35 | | 0.75 | 0.33 | 0.29 | 0.93 | 1.00 | 0.42 | 0.93 | 0.90 | 0.90 | 0.46 | 0.75 |
| c | | | | | | | 0.04 | | | 0.58 | | | 0.10 | 0.04 | 0.03 |
| *Mpi* | | | | | | | | | | | | | | | |
| a | | | | | 0.50 | 0.67 | 0.36 | | 0.42 | | 0.02 | | | | 0.22 |
| b | 0.75 | 1.00 | 1.00 | 1.00 | 0.50 | 0.33 | 0.64 | 1.00 | 0.50 | 0.83 | 0.26 | 0.90 | 1.00 | 0.96 | 0.55 |
| c | | | | | | | | | | 0.17 | 0.04 | | | | 0.02 |
| d | 0.25 | | | | | | | | 0.08 | | 0.69 | 0.10 | | 0.04 | 0.22 |
| *6-Pgd* | | | | | | | | | | | | | | | |
| a | | | | | | | | | | | | | | | 0.02 |
| b | 1.00 | 1.00 | 1.00 | 1.00 | 1.00 | 1.00 | 1.00 | | | | | | | | |
| c | | | | | | | | 1.00 | 1.00 | 1.00 | 1.00 | 1.00 | 1.00 | 1.00 | 0.98 |
| *Hdc* | 0.22 | 0.17 | 0.21 | 0.13 | 0.31 | 0.17 | 0.33 | 0.27 | 0.23 | 0.33 | 0.25 | 0.28 | 0.18 | 0.33 | 0.40 |
| SE | 0.07 | 0.09 | 0.09 | 0.13 | 0.13 | 0.09 | 0.08 | 0.08 | 0.10 | 0.12 | 0.07 | 0.11 | 0.10 | 0.10 | 0.08 |
| *He* | 0.21 | 0.18 | 0.25 | 0.13 | 0.27 | 0.36 | 0.32 | 0.27 | 0.21 | 0.31 | 0.28 | 0.26 | 0.23 | 0.29 | 0.42 |
| SE | 0.07 | 0.09 | 0.10 | 0.13 | 0.10 | 0.11 | 0.08 | 0.07 | 0.09 | 0.08 | 0.07 | 0.08 | 0.09 | 0.08 | 0.07 |

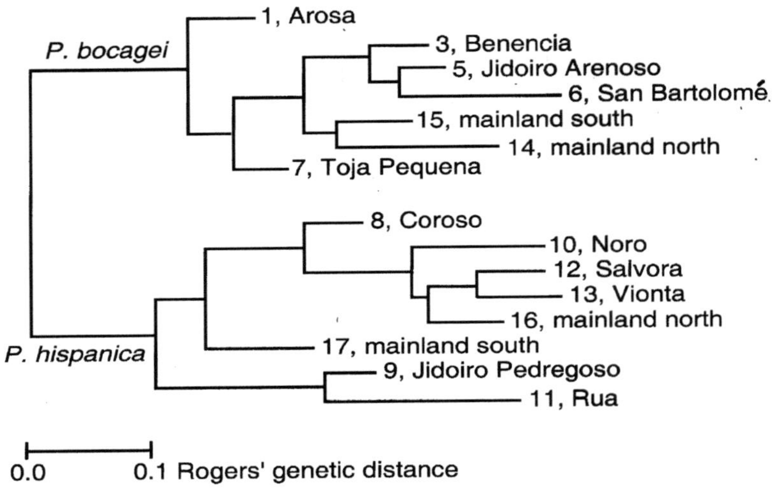


Fig. 3 Distance-Wagner tree of 11 island and four mainland populations of *Podarcis bocagei* and *P. hispanica* from the Ria de Arosa, Galicia, Spain. Rogers' genetic distance was used on allelic data over eight polymorphic enzyme loci (see Table 2). The tree is arbitrarily rooted at the midpoint of the longest path.

in ordination space reveals two groups with scores along the first axis smaller or larger than a certain value (Fig. 4), in which individuals from the respective groups belong to populations that were already tentatively identified as *P. bocagei* and *P. hispanica*. This grouping simultaneously produced maximum congruence between the classifications based upon mitochondrial and nuclear genetic markers. AMOVA on the enzyme data shows that 42.7% of the total variation is distributed between species, 11.6% between populations within species and 45.7% within populations.

## 3.4. Molecular and Morphological Identification of *P. bocagei* and *P. hispanica*

The association between species classification results by nuclear and cytoplasmic genetic markers is tight (Cohen's kappa 0.97). Disconcordance was observed for one male lizard from the island Jidoiro Arenoso (RMNH 35207) and for one male lizard from the southern mainland (RMNH 35322). Both these individuals carry *P. hispanica* mitochondrial DNA in a *P. bocagei* nuclear background. Thirty adult lizards from the mainland were identified through the expert opinion of the junior author. The results from this independent exercise entirely matched the results obtained from nuclear markers (Cohens' kappa = 1.00). A similar exercise for island lizards paralleled the nuclear genetic results (Cohens' kappa = 1.00), with the exception of lizards from Jidoiro

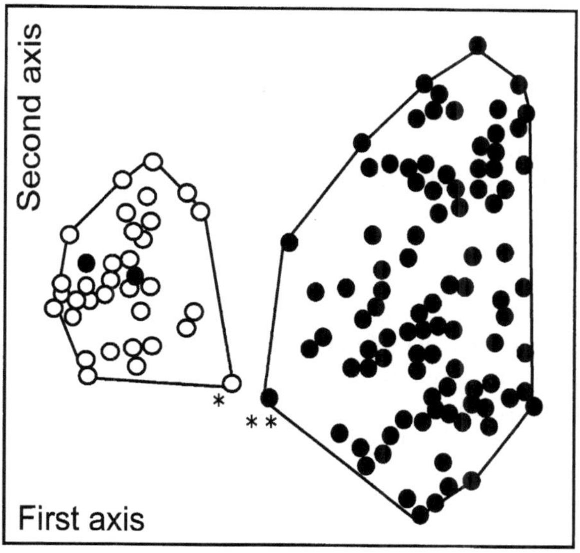


Fig. 4 Bivariate plot of the scores for 145 lizards from the Ria de Arosa, Galicia, Spain along the first and second axis of a Principal Coordinate Analysis on enzyme electrophoretic data. The convex polygons to the left and to the right enclose individuals carrying *Podarcis bocagei* (open dots) and *P. hispanica* (solid dots) mitochondrial DNA, respectively, with two exceptions; these two lizards carry *P. hispanica* mitochondrial DNA in a *P. bocagei* nuclear background. Individuals shown by asterisks are considered potential $F_2$-hybrids (details see text).

Arenoso, Jidoiro Pedregoso, and San Bartolomé that we were not sure about. No molecular data were available for the island populations of Beiro and Cortegada. Instead, four males and three females from either island were subjected to a discriminant analysis of morphological data, in conjunction with known *P. bocagei* (42 males, 32 females) and known *P. hispanica* (57 males, 41 females). Thirteen out of 14 lizards from these islands were unambiguously classified as *P. bocagei* ($P_{bocagei} > 0.99$).

## 3.5. Species Distribution and Tests for Interspecific Hybridization

Islands at the inner Ria are all inhabited by *P. bocagei* whereas islands at the outer Ria are all inhabited by *P. hispanica* (Fig. 1, Appendix I). Hence, a north to south line can be drawn running between Jidoiro Arenoso and Jidoiro Pedregoso delimiting the species, with *P. bocagei* to the east and *P. hispanica* to the west. This east–west separation is also found for the lizard populations on the mainland, with observed syntopy (i.e., sympatry at the scale of the lizard neighbourhood size) at the localities “c” and “d” (see Fig. 1). Interestingly, of the two lizards known to carry alien mtDNA, one is from the syntopic locality “d” (RMNH 35322), suggesting interspecific gene flow. The other case (RMNH 35207) is from the island Jidoiro Arenoso, in close proximity to Jidoiro Pedregoso that is occupied by *P. hispanica*, indicating the possibility of hybridization following dispersal. These two individuals were selected for further discriminant analysis together with one lizard identified as potential $F_2$ or backcross hybrids on the basis of an intermediate score at the first principal coordinate axis (RMNH 35327 marked with an asterisk in Fig. 4; note that the equally intermediate lizard marked with two asterisks has no morphological data available.) Discriminant functions with all lizards of the same sex included in the analysis support morphological intermediacy of RMNH 35322 ($P_{bocagei} = 0.27$) and do not support morphological intermediacy for RMNH 35327 ($P_{bocagei} = 1.0$) and RMNH 35207 ($P_{bocagei} = 0.94$). With just local (and smaller) reference samples (i.e., mainland males, mainland females, males Jidoiro Arenoso and males Jidoiro Pedregoso), the three lizards were allocated to *P. bocagei* with a probability of near unity. Lizard RMNH 11066 was initially identified as a male *P. bocagei* by its phenotype, but due to its outlier position in the overall species distribution it was also subjected to discriminant analysis. Morphologically the lizard qualifies as *P. hispanica* whether all males were used as reference samples ($P_{hispanica} = 0.98$) or just mainland males ($P_{hispanica} = 1.0$). We excluded this individual from the analysis and put it aside as a candidate for future molecular identification (see Appendix I). Cross-validation by the jackknife ("leave one out") procedure showed that classifications based upon the full data sets were more robust than the classifications with just the local samples as the reference. Under jackknifing, kappa values dropped from 0.92–0.95 to 0.72–0.77 for the full data and from unity to near zero for some of the local data sets.

### 3.6. Association Between Molecular Differentiation, Morphological Differentiation, and Bathymetry

Associations between estimators of population differentiation (nuclear – mitochondrial, nuclear – morphological, mitochondrial – morphological) were not significant in Mantel tests, for either species. Also, associations with geographical distance and bathymetry were all not significant ($P > 0.05$).

### 3.7. Morphological Character State Distributions

One size character (SVL), 13 continuous size characters adjusted for SVL and 12 meristic characters were explored by univariate statistical comparison (*t*-test) for each island population versus the corresponding *P. bocagei* or *P. hispanica* mainland reference population. With altogether 676 tests performed and an *a priori* accepted significance level of alpha = 0.05, the number of nominally significant results due to chance alone is about 34. In actual fact, 109 such cases were observed (43 in *P. bocagei* and 66 in *P. hispanica*), suggesting statistically relevant intraspecific morphological differentiation among island and mainland populations. Inspection of the results reveals the following trends (Tables 3 and 4). Island lizards are frequently larger than their mainland counterparts, more often so in *P. hispanica* than in *P. bocagei*. The relative size characters HW, OW, FW, and HLL are commonly smaller in island populations than on the mainland. Among the meristic characters, COL has higher and CAN has lower counts on several of the islands than on the mainland. The characters L4F and GRA show contrasting results within species, with either higher counts (Jidoiro Pedregoso and San Bartolomé) or lower counts (Coroso and Jidoiro Arenoso) on some islands than on the mainland.

Discriminant analysis on the basis of the characters listed in Table 3, for selected island populations (Aroso, Jidoiro Arenoso, Jidoiro Pedregoso, Noro, and Rua) and mainland lizard populations as a reference, confirmed these contrasting results. In males, the discriminant function scores over the first and second axis are overlapping between *P. hispanica* from Rua and *P. bocagei* from Arosa, Jidoiro Arenoso and the mainland, and distinct from mainland *P. hispanica*. *Podarcis hispanica* from Jidoiro Pedregoso and Noro have overlapping scores, distinct from the other populations (Fig. 5A). Along the third axis, *P. hispanica* from Jidoiro Pedregoso and Rua are different from the other populations, contributing to the further discrimination between *P. bocagei* and *P. hispanica*. The results for females are similar to those obtained for males, with less overlap between populations (Fig. 5). The correspondence between the patterns of morphological discrimination among the sexes is statistically significant (Mantel test, $r = 0.50$, $P < 0.05$ over the first and second axis; $r = 0.61$, $P < 0.01$ over the first, second and third axis). Morphometric and meristic characters, when analysed separately, yielded more or less similar patterns of population differentiation (Mantel test, $0.23 < r < 027$, $0.13 < P < 0.17$ for both sexes over two or three discriminant axes).

## 4. Discussion

We obtained a near perfect match between the distribution of nuclear and mitochondrial genetic markers. This allowed species identification without reference to morphology, colouration characteristics, or distribution. With molecular identifications available, the expert phenotypic identification of lizards turned out to be straightforward on the mainland, although some rare morphologically intermediate individuals may remain problematic. To the contrary, morphological species identification of island populations is problematic; in particular, for the lizard populations from Beiro, Noro, Sálvora, and Toja Pequeña (Mateo, 1997; Galán, 1999a, 2003a), and from Jidoiro Arenoso and Jidoiro Pedregoso (Galán, 1999a, 2003a, present study). Morphological discrimination was better for females than for males, a result that was also reported by Sá-Sousa and Harris (2002) and Sá-Sousa et al. (2002).

The locus *6-Pgd* was fully diagnostic in western Galicia (present study) as well as in northern Portugal (Pinho et al., 2003). Given the absence of heterozygous individuals at this locus, $F_1$ hybrids are not represented in our sample. What about further interspecific gene flow? Some individuals stand out as potential backcross hybrids

*Table 3.* *Univariate analysis of morphological characters in island versus mainland populations of* Podarcis bocagei *(left panel) and* P. hispanica *(right panel) from the Ria de Arosa, Galicia, Spain. Morphological characters are introduced in Fig. 2 and Table 1 and quantified in Table 4. Population numbers are as in Appendix I; m = males, f = females. Tests were pairwise comparisons by t-test. "Plus" and "minus" symbols indicate an increase respectively decrease of character state values of the island sample relative to the mainland reference sample ($P < 0.05$). Note that the results are not necessarily interpreted as statistically significant; they do just represent convenient cut-off points for trend analysis.*

| *Species* | Podarcis bocagei | | | | | | | | | | | | | | Podarcis hispanica | | | | | | | | | | | |
|---|---|---|---|---|---|---|---|---|---|---|---|---|---|---|---|---|---|---|---|---|---|---|---|---|---|---|
| Island number | 1 | | 2 | | 3 | | 4 | | 5 | | 6 | | 7 | | 8 | | 9 | | 10 | | 11 | | 12 | | 13 | |
| Sex | m | f | m | f | m | f | m | f | m | f | m | f | m | f | m | f | m | f | m | f | m | f | m | f | m | f |
| Sample size | 9 | 5 | 4 | 3 | 6 | 4 | 4 | 3 | 11 | 10 | 2 | 3 | 6 | 3 | 6 | 2 | 9 | 6 | 6 | 6 | 10 | 4 | 3 | 3 | 4 | 3 |
| Absolute size character | | | | | | | | | | | | | | | | | | | | | | | | | | |
| SVL | | | | | + | | | | | + | | | | | + | + | + | + | | + | + | + | | | + | + |
| Relative size character | | | | | | | | | | | | | | | | | | | | | | | | | | |
| HCL | | | | | | | | | | | | | | | | | | | | | | | | | | |
| HL | | | | + | | | | | | - | | | | + | - | | | | | | | | | | | |
| DL | | | | | | | | | | | | | | + | | | | | | | | | | | | |
| HW | | | | + | | | | | | - | | | | + | - | - | | + | - | | - | | - | | | |
| OW | | | | | | - | | - | | - | | - | | + | | - | | | - | | | | | | | |
| FW | | - | | - | | - | | | | | | | | | | | | | - | - | - | | - | | | |
| NW | | | | | | | | | | | | | | | | | | | | | | | | | | |
| HD | | | | | | | | | | | | | | | + | | | | + | | | | | | | |
| OD | | | | | | | | | | - | + | | | | | | | | + | | | | | | | |
| FD | | | + | | | | + | | | | | | + | | | | | | | | - | | | | + | |
| ND | | | | | | | | | | | | | | | | | | | | | | + | | | | |
| HFL | | | | | | | | | | | | | | | | + | | + | + | | | | | | | |
| HLL | - | - | | | - | | | | | - | | | | | - | - | | | | | - | | | | | |

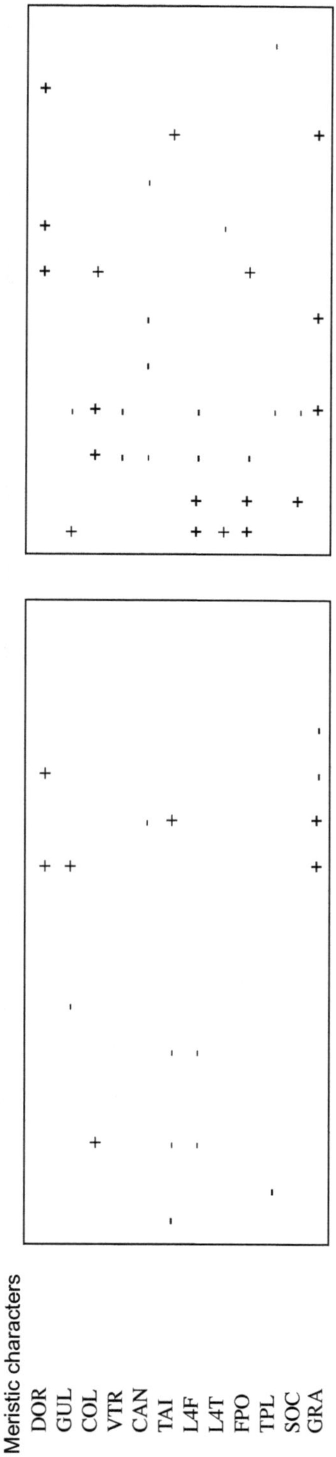
Meristic characters
DOR
GUL
COL
VTR
CAN
TAI
L4F
L4T
FPO
TPL
SOC
GRA

*Table 4. Morphological data for 26 external characters in seven islands plus pooled mainland population of* Podarcis bocagei *(n = 88) and six islands plus pooled mainland population of* P. hispanica *(n = 98) from the Ria de Arosa, Galicia, Spain. Data include sample size, mean and standard deviation (SD) for sexes separately. One lizard (RMNH 11066) remained unidentified.*

| *Podarcis bocagei* | Sex | N | | SVL | HCL | HL | DL | HW | OW | FW | NW | HD | OD | FD | ND | HLL | HFL |
|---|---|---|---|---|---|---|---|---|---|---|---|---|---|---|---|---|---|
| Locality | | | | | | | | | | | | | | | | | |
| 1, Arosa | males | 9 | mean | 55.44 | 19.66 | 13.58 | 7.30 | 6.48 | 5.48 | 4.44 | 3.39 | 6.22 | 5.38 | 4.74 | 3.77 | 14.60 | 14.14 |
| | | | SD | 4.26 | 1.21 | 0.95 | 0.59 | 0.60 | 0.37 | 0.34 | 0.37 | 0.60 | 0.58 | 0.39 | 0.19 | 1.16 | 0.84 |
| | females | 5 | mean | 54.31 | 17.23 | 11.38 | 5.82 | 5.56 | 4.90 | 3.89 | 2.91 | 4.93 | 4.46 | 4.01 | 3.19 | 11.90 | 11.70 |
| | | | SD | 2.53 | 0.84 | 0.37 | 0.26 | 0.30 | 0.23 | 0.23 | 0.19 | 0.35 | 0.37 | 0.28 | 0.31 | 0.76 | 0.28 |
| 2, Beiro | males | 4 | mean | 58.46 | 21.30 | 14.62 | 7.92 | 7.14 | 5.99 | 4.82 | 3.59 | 6.77 | 6.09 | 5.36 | 4.18 | 14.96 | 15.23 |
| | | | SD | 1.65 | 1.69 | 0.59 | 0.45 | 0.30 | 0.33 | 0.41 | 0.22 | 0.07 | 0.22 | 0.17 | 0.29 | 0.82 | 0.52 |
| | females | 3 | mean | 55.24 | 17.56 | 12.32 | 6.46 | 6.01 | 5.23 | 4.12 | 3.23 | 5.48 | 4.96 | 4.44 | 3.42 | 13.01 | 11.85 |
| | | | SD | 2.90 | 0.33 | 0.52 | 0.41 | 0.37 | 0.08 | 0.11 | 0.17 | 0.24 | 0.24 | 0.15 | 0.17 | 1.14 | 0.54 |
| 3, Benencia | males | 6 | mean | 65.41 | 22.95 | 16.02 | 8.30 | 7.65 | 6.23 | 5.34 | 4.07 | 7.52 | 6.54 | 5.61 | 4.47 | 17.15 | 16.88 |
| | | | SD | 3.29 | 1.51 | 0.94 | 0.64 | 0.62 | 0.45 | 0.31 | 0.34 | 0.46 | 0.45 | 0.48 | 0.32 | 1.17 | 1.24 |
| | females | 4 | mean | 60.46 | 18.66 | 12.68 | 6.50 | 6.19 | 5.23 | 4.36 | 3.40 | 5.87 | 5.12 | 4.62 | 3.85 | 13.83 | 13.27 |
| | | | SD | 3.31 | 0.53 | 0.20 | 0.24 | 0.14 | 0.15 | 0.14 | 0.10 | 0.31 | 0.27 | 0.14 | 0.10 | 0.87 | 0.33 |
| 4, Cortegada | males | 4 | mean | 59.72 | 21.01 | 14.13 | 7.62 | 7.25 | 5.75 | 5.25 | 3.79 | 7.03 | 6.24 | 5.65 | 4.26 | 16.30 | 15.76 |
| | | | SD | 1.40 | 0.57 | 0.13 | 0.28 | 0.32 | 0.25 | 0.42 | 0.07 | 0.29 | 0.25 | 0.12 | 0.28 | 0.50 | 0.37 |
| | females | 3 | mean | 55.41 | 17.66 | 11.78 | 6.37 | 5.77 | 4.82 | 4.20 | 2.98 | 5.26 | 4.71 | 4.28 | 3.46 | 12.82 | 12.78 |
| | | | SD | 2.56 | 0.70 | 0.15 | 0.28 | 0.13 | 0.10 | 0.13 | 0.20 | 0.10 | 0.02 | 0.19 | 0.28 | 1.25 | 0.59 |
| 5, Jidoiro Arenoso | males | 11 | mean | 61.17 | 21.45 | 14.70 | 7.79 | 7.18 | 6.14 | 5.11 | 3.78 | 6.95 | 6.06 | 5.23 | 4.03 | 16.03 | 15.23 |
| | | | SD | 5.17 | 1.58 | 1.05 | 0.59 | 0.43 | 0.40 | 0.47 | 0.27 | 0.56 | 0.49 | 0.36 | 0.31 | 1.84 | 1.28 |
| | females | 10 | mean | 61.75 | 18.40 | 12.41 | 6.49 | 6.01 | 5.31 | 4.47 | 3.34 | 5.71 | 5.10 | 4.56 | 3.60 | 13.70 | 12.92 |
| | | | SD | 5.66 | 1.90 | 0.93 | 0.44 | 0.53 | 0.30 | 0.29 | 0.24 | 0.45 | 0.45 | 0.44 | 0.32 | 1.28 | 0.88 |
| 6, San Bartolomé | males | 2 | mean | 57.31 | 20.13 | 14.30 | 7.67 | 6.99 | 5.94 | 5.39 | 4.01 | 7.30 | 6.57 | 5.67 | 4.57 | 15.86 | 15.25 |
| | | | SD | 9.55 | 5.30 | 3.38 | 2.26 | 2.62 | 1.99 | 1.68 | 1.29 | 2.55 | 1.99 | 2.05 | 1.58 | 2.60 | 2.60 |
| | females | 3 | mean | 59.40 | 18.86 | 12.60 | 6.56 | 6.22 | 5.10 | 4.51 | 3.27 | 5.68 | 4.74 | 4.59 | 3.56 | 14.71 | 13.65 |
| | | | SD | 1.81 | 0.45 | 0.23 | 0.31 | 0.18 | 0.17 | 0.37 | 0.17 | 0.47 | 0.32 | 0.28 | 0.08 | 1.05 | 0.45 |
| 7, Toja Pequena | males | 6 | mean | 58.61 | 20.46 | 14.48 | 7.68 | 7.08 | 5.76 | 5.10 | 3.66 | 6.80 | 6.04 | 5.45 | 4.02 | 15.85 | 14.07 |
| | | | SD | 4.98 | 1.81 | 1.24 | 0.73 | 0.70 | 0.57 | 0.38 | 0.34 | 0.65 | 0.53 | 0.51 | 0.32 | 1.61 | 1.34 |
| | females | 3 | mean | 50.15 | 15.82 | 11.05 | 6.12 | 5.49 | 4.89 | 3.96 | 2.91 | 4.86 | 4.32 | 3.87 | 3.01 | 12.34 | 11.53 |
| | | | SD | 4.36 | 1.52 | 0.53 | 0.24 | 0.33 | 0.23 | 0.40 | 0.22 | 0.34 | 0.37 | 0.34 | 0.25 | 1.48 | 1.07 |
| 14 and 15, mainland | males | 8 | mean | 56.80 | 19.82 | 13.59 | 7.43 | 6.77 | 5.68 | 4.74 | 3.54 | 6.35 | 5.60 | 4.94 | 3.79 | 15.65 | 14.25 |
| | | | SD | 7.20 | 2.44 | 1.33 | 0.71 | 0.88 | 0.55 | 0.48 | 0.45 | 0.86 | 0.77 | 0.65 | 0.44 | 1.89 | 1.11 |
| | females | 7 | mean | 54.82 | 17.05 | 11.43 | 5.99 | 5.75 | 5.03 | 4.29 | 3.08 | 5.11 | 4.65 | 4.13 | 3.26 | 13.31 | 12.50 |
| | | | SD | 5.39 | 1.21 | 0.98 | 0.69 | 0.56 | 0.41 | 0.38 | 0.29 | 0.68 | 0.57 | 0.47 | 0.32 | 1.59 | 0.97 |

| *Podarcis bocagei* | Sex | N | | DOR | GUL | COL | VTR | CAN | TAI | L4F | L4T | FPO | TPL | SOC | GRA |
|---|---|---|---|---|---|---|---|---|---|---|---|---|---|---|---|
| Locality | | | | | | | | | | | | | | | |
| 1, Arosa | males | 9 | mean | 58.0 | 26.8 | 10.2 | 26.9 | 6.6 | 36.3 | 15.9 | 23.0 | 18.0 | 5.6 | 5.8 | 9.0 |
| | | | SD | 3.6 | 0.8 | 1.2 | 1.5 | 0.7 | 1.2 | 0.6 | 1.2 | 1.7 | 1.0 | 0.8 | 1.8 |
| | females | 5 | mean | 56.5 | 25.3 | 10.3 | 30.3 | 6.0 | 33.3 | 17.0 | 22.3 | 16.0 | 5.3 | 5.8 | 8.5 |
| | | | SD | 2.2 | 1.6 | 1.5 | 0.9 | 0.0 | 1.6 | 2.5 | 1.9 | 0.7 | 0.8 | 0.5 | 0.9 |
| 2, Beiro | males | 4 | mean | 58.3 | 24.8 | 11.5 | 25.0 | 7.0 | 35.5 | 14.5 | 22.0 | 15.5 | 5.8 | 5.8 | 8.5 |
| | | | SD | 3.4 | 1.5 | 0.6 | 1.4 | 1.2 | 1.3 | 1.0 | 2.0 | 1.7 | 1.0 | 1.0 | 2.4 |
| | females | 3 | mean | 55.3 | 25.0 | 10.0 | 29.7 | 7.0 | 34.0 | 14.7 | 22.0 | 14.7 | 5.7 | 4.7 | 7.3 |
| | | | SD | 1.5 | 0.0 | 1.0 | 0.6 | 1.0 | 2.6 | 1.2 | 1.0 | 1.5 | 0.6 | 2.5 | 3.1 |
| 3, Benencia | males | 6 | mean | 57.0 | 25.3 | 9.5 | 24.8 | 7.3 | 35.7 | 13.8 | 22.4 | 16.2 | 5.7 | 5.7 | 11.3 |
| | | | SD | 3.0 | 1.6 | 1.0 | 2.1 | 0.8 | 2.4 | 1.2 | 1.1 | 1.0 | 1.2 | 0.5 | 3.3 |
| | females | 4 | mean | 56.0 | 24.0 | 9.7 | 29.7 | 6.7 | 34.7 | 14.0 | 20.3 | 14.7 | 6.0 | 7.0 | 9.7 |
| | | | SD | 1.0 | 0.8 | 0.6 | 1.0 | 0.8 | 5.4 | 0.5 | 1.0 | 0.5 | 1.3 | 0.5 | 1.5 |
| 4, Cortegada | males | 4 | mean | 61.3 | 26.8 | 10.5 | 26.0 | 7.3 | 35.8 | 15.3 | 23.7 | 18.8 | 5.5 | 5.5 | 8.0 |
| | | | SD | 5.0 | 1.3 | 1.9 | 0.8 | 0.5 | 2.9 | 0.5 | 0.6 | 2.5 | 0.6 | 0.6 | 1.6 |
| | females | 3 | mean | 56.0 | 25.3 | 10.0 | 29.7 | 6.0 | 36.3 | 15.3 | 21.0 | 16.3 | 6.7 | 5.7 | 9.0 |
| | | | SD | 3.6 | 0.6 | 1.0 | 1.5 | 0.0 | 4.0 | 1.2 | 4.4 | 1.5 | 1.2 | 0.6 | 2.0 |
| 5, Jidoiro Arenoso | males | 11 | mean | 62.4 | 28.7 | 10.9 | 27.0 | 6.5 | 40.6 | 15.9 | 22.7 | 18.2 | 6.1 | 6.2 | 10.9 |
| | | | SD | 3.8 | 1.0 | 0.5 | 1.3 | 0.8 | 3.0 | 0.7 | 1.5 | 1.0 | 1.0 | 1.1 | 1.0 |
| | females | 10 | mean | 57.5 | 26.3 | 10.2 | 30.7 | 5.8 | 38.3 | 15.1 | 22.5 | 15.8 | 6.3 | 6.1 | 10.4 |
| | | | SD | 3.2 | 1.8 | 1.1 | 1.1 | 0.4 | 2.8 | 1.1 | 1.5 | 1.5 | 0.8 | 0.6 | 1.9 |
| 6, San Bartolomé | males | 2 | mean | 64.0 | 27.0 | 9.5 | 26.0 | 8.0 | 40.5 | 14.0 | 23.0 | 16.5 | 5.5 | 5.5 | 5.5 |
| | | | SD | 0.0 | 2.8 | 0.7 | 0.0 | 0.0 | 0.7 | - | 2.8 | 0.7 | 0.7 | 0.7 | 0.7 |
| | females | 3 | mean | 54.0 | 27.0 | 9.3 | 28.7 | 6.3 | 35.0 | 15.0 | 23.5 | 16.7 | 5.7 | 6.0 | 5.7 |
| | | | SD | 1.0 | 2.0 | 0.6 | 0.6 | 0.6 | 3.0 | 1.0 | 3.5 | 1.5 | 0.6 | 1.0 | 0.6 |
| 7, Toja Pequena | males | 6 | mean | 60.3 | 24.8 | 9.8 | 25.3 | 7.2 | 40.2 | 15.8 | 22.8 | 17.7 | 6.0 | 6.0 | 7.5 |
| | | | SD | 1.5 | 2.3 | 0.8 | 0.8 | 1.3 | 1.6 | 1.0 | 0.4 | 1.4 | 1.1 | 0.9 | 1.4 |
| | females | 3 | mean | 57.7 | 24.3 | 10.0 | 28.3 | 6.3 | 34.7 | 15.3 | 23.7 | 15.0 | 6.0 | 5.7 | 6.7 |
| | | | SD | 2.3 | 2.1 | 0.0 | 0.6 | 0.6 | 1.5 | 0.6 | 1.2 | 1.0 | 0.0 | 0.6 | 2.1 |
| 14 and 15, mainland | males | 8 | mean | 58.75 | 26.0 | 10.3 | 26.4 | 7.1 | 39.4 | 15.8 | 22.1 | 17.1 | 5.8 | 5.5 | 8.9 |
| | | | SD | 2.9 | 1.4 | 1.2 | 1.1 | 0.8 | 3.1 | 0.9 | 3.3 | 1.8 | 0.7 | 0.8 | 2.2 |
| | females | 7 | mean | 54.9 | 26.3 | 10.3 | 29.4 | 6.6 | 35.4 | 14.9 | 21.1 | 15.7 | 6.6 | 6.3 | 8.3 |
| | | | SD | 2.8 | 1.5 | 0.8 | 2.1 | 0.8 | 2.2 | 1.1 | 2.0 | 1.8 | 0.8 | 1.0 | 1.5 |

(continued)

*Table 4.* *(continued)*

| *Podarcis hispanica* | Sex | N | | SVL | HCL | HL | DL | HW | OW | FW | NW | HD | OD | FD | ND | HLL | HFL |
|---|---|---|---|---|---|---|---|---|---|---|---|---|---|---|---|---|---|
| Locality | | | | | | | | | | | | | | | | | |
| 8, Coroso | males | 6 | mean | 65.07 | 22.88 | 14.99 | 8.14 | 7.25 | 6.06 | 5.14 | 3.85 | 6.31 | 5.74 | 5.03 | 4.18 | 14.98 | 16.09 |
| | | | SD | 5.07 | 1.91 | 1.33 | 0.75 | 0.54 | 0.40 | 0.43 | 0.39 | 0.51 | 0.50 | 0.54 | 0.48 | 1.19 | 0.83 |
| | females | 2 | mean | 66.06 | 20.02 | 12.90 | 6.77 | 5.91 | 5.22 | 4.42 | 3.36 | 5.29 | 4.61 | 4.12 | 3.46 | 12.59 | 15.27 |
| | | | SD | 7.67 | 1.38 | 1.23 | 0.59 | 0.52 | 0.36 | 0.22 | 0.11 | 0.54 | 0.62 | 0.59 | 0.51 | 0.42 | 0.32 |
| 9, Jidoiro Pedregoso | males | 9 | mean | 62.27 | 22.47 | 14.84 | 7.71 | 7.05 | 5.91 | 4.96 | 3.72 | 6.39 | 5.62 | 4.80 | 3.89 | 16.16 | 15.50 |
| | | | SD | 4.54 | 2.22 | 0.90 | 0.65 | 0.56 | 0.50 | 0.27 | 0.29 | 0.63 | 0.56 | 0.51 | 0.40 | 0.87 | 0.77 |
| | females | 6 | mean | 60.21 | 18.55 | 12.44 | 6.63 | 6.16 | 5.18 | 4.33 | 3.27 | 4.98 | 4.44 | 3.89 | 3.20 | 13.38 | 13.04 |
| | | | SD | 2.64 | 0.93 | 0.53 | 0.41 | 0.13 | 0.28 | 0.28 | 0.28 | 0.28 | 0.29 | 0.18 | 0.17 | 0.66 | 0.37 |
| 10, Noro | males | 6 | mean | 55.51 | 20.04 | 13.12 | 7.07 | 6.29 | 5.18 | 4.25 | 3.20 | 5.84 | 5.14 | 4.36 | 3.50 | 14.33 | 14.81 |
| | | | SD | 6.07 | 1.60 | 1.37 | 0.72 | 0.61 | 0.41 | 0.41 | 0.17 | 0.71 | 0.58 | 0.49 | 0.38 | 1.65 | 1.46 |
| | females | 6 | mean | 59.10 | 18.25 | 11.69 | 6.14 | 5.75 | 5.06 | 3.93 | 2.94 | 4.96 | 4.55 | 3.88 | 3.08 | 12.54 | 12.85 |
| | | | SD | 5.62 | 1.26 | 0.88 | 0.60 | 0.35 | 0.55 | 0.23 | 0.26 | 0.46 | 0.43 | 0.33 | 0.23 | 1.04 | 0.40 |
| 11, Rúa | males | 10 | mean | 64.99 | 22.87 | 15.07 | 8.11 | 7.36 | 5.99 | 4.83 | 3.73 | 6.77 | 5.84 | 4.89 | 3.89 | 15.96 | 15.41 |
| | | | SD | 6.00 | 2.12 | 1.09 | 0.69 | 0.63 | 0.58 | 0.30 | 0.40 | 0.68 | 0.55 | 0.51 | 0.29 | 1.67 | 1.09 |
| | females | 4 | mean | 60.22 | 19.00 | 12.58 | 6.68 | 6.00 | 5.15 | 4.34 | 3.22 | 5.31 | 4.74 | 4.10 | 3.47 | 12.47 | 12.76 |
| | | | SD | 4.19 | 0.87 | 0.54 | 0.24 | 0.26 | 0.15 | 0.15 | 0.24 | 0.27 | 0.24 | 0.19 | 0.14 | 0.98 | 0.92 |
| 12, Salvora | males | 3 | mean | 56.76 | 19.88 | 13.61 | 7.09 | 6.35 | 5.26 | 4.56 | 3.36 | 5.80 | 5.02 | 4.39 | 3.51 | 15.19 | 14.64 |
| | | | SD | 3.27 | 1.55 | 0.82 | 0.30 | 0.31 | 0.31 | 0.24 | 0.24 | 0.62 | 0.40 | 0.23 | 0.11 | 0.56 | 0.41 |
| | females | 3 | mean | 55.14 | 17.66 | 11.02 | 6.02 | 5.49 | 4.93 | 4.10 | 2.99 | 4.45 | 4.11 | 3.67 | 2.99 | 12.83 | 12.32 |
| | | | SD | 9.12 | 1.82 | 1.42 | 1.01 | 0.56 | 0.42 | 0.33 | 0.28 | 0.38 | 0.35 | 0.25 | 0.12 | 1.22 | 0.80 |
| 13, Vionta | males | 4 | mean | 61.47 | 21.49 | 14.75 | 8.00 | 7.21 | 5.79 | 4.98 | 3.55 | 6.67 | 5.90 | 5.14 | 3.87 | 17.72 | 14.54 |
| | | | SD | 6.97 | 2.31 | 1.55 | 0.79 | 1.02 | 0.66 | 0.58 | 0.29 | 0.76 | 0.61 | 0.48 | 0.33 | 1.93 | 1.46 |
| | females | 3 | mean | 58.82 | 18.37 | 11.67 | 6.50 | 5.72 | 4.85 | 4.31 | 3.04 | 4.90 | 4.42 | 4.09 | 3.31 | 13.95 | 12.03 |
| | | | SD | 5.70 | 0.43 | 0.66 | 0.34 | 0.54 | 0.51 | 0.20 | 0.16 | 0.78 | 0.72 | 0.66 | 0.34 | 1.23 | 0.65 |
| 16 and 17, mainland | males | 19 | mean | 53.07 | 18.72 | 12.66 | 6.73 | 6.24 | 5.22 | 4.46 | 3.19 | 5.15 | 4.60 | 4.06 | 3.23 | 14.41 | 13.13 |
| | | | SD | 3.79 | 0.86 | 0.70 | 0.39 | 0.34 | 0.26 | 0.22 | 0.33 | 0.39 | 0.35 | 0.33 | 0.28 | 1.25 | 1.03 |
| | females | 17 | mean | 49.21 | 15.08 | 10.00 | 5.31 | 4.90 | 4.38 | 3.71 | 2.59 | 3.92 | 3.55 | 3.25 | 2.57 | 10.93 | 10.42 |
| | | | SD | 3.28 | 0.72 | 0.52 | 0.37 | 0.26 | 0.24 | 0.27 | 0.19 | 0.47 | 0.36 | 0.37 | 0.18 | 0.99 | 0.77 |
| *Incertae sedis* | | | | | | | | | | | | | | | | | |
| 15f (RMNH 11066) | male | 1 | mean | 58.22 | 18.47 | 13.83 | 7.45 | 6.85 | 5.54 | 5.01 | 3.59 | 5.69 | 5.00 | 4.54 | 3.88 | 14.92 | 12.91 |

| *Podarcis hispanica* | Sex | N | | DOR | GUL | COL | VTR | CAN | TAI | L4F | L4T | FPO | TPL | SOC | GRA |
|---|---|---|---|---|---|---|---|---|---|---|---|---|---|---|---|
| 8, Coroso | males | 6 | mean | 57.2 | 30.7 | 10.3 | 26.3 | 7.0 | 35.3 | 17.8 | 25.3 | 18.0 | 5.7 | 5.3 | 9.7 |
| | | | SD | 1.6 | 2.3 | 0.5 | 0.5 | 1.1 | 2.9 | 0.8 | 2.5 | 1.5 | 1.8 | 0.5 | 1.6 |
| | females | 2 | mean | 54.0 | 29.0 | 10.5 | 30.0 | 7.0 | 31.0 | 18.0 | 25.0 | 17.5 | 6.5 | 6.0 | 10.0 |
| | | | SD | 1.4 | 1.4 | 0.7 | 2.8 | 1.4 | 0.0 | 1.4 | 1.4 | 2.1 | 0.7 | 0.0 | 1.4 |
| 9, Jidoiro Pedregoso | males | 9 | mean | 56.3 | 26.8 | 10.9 | 25.6 | 6.3 | 37.3 | 14.6 | 23.3 | 14.4 | 6.1 | 5.3 | 9.6 |
| | | | SD | 1.3 | 1.3 | 0.6 | 1.3 | 0.7 | 2.8 | 1.6 | 1.7 | 1.5 | 1.2 | 1.0 | 1.3 |
| | females | 6 | mean | 55.4 | 24.0 | 11.2 | 28.4 | 6.3 | 33.0 | 14.0 | 22.8 | 14.3 | 5.3 | 5.2 | 10.3 |
| | | | SD | 1.7 | 1.1 | 0.4 | 1.5 | 0.5 | 2.7 | 1.4 | 1.0 | 1.4 | 1.0 | 0.8 | 1.2 |
| 10, Noro | males | 6 | mean | 54.3 | 27.5 | 10.0 | 26.7 | 5.7 | 37.2 | 15.7 | 23.8 | 15.7 | 6.0 | 5.8 | 8.3 |
| | | | SD | 2.2 | 2.7 | 0.9 | 0.5 | 0.5 | 2.6 | 0.8 | 1.3 | 1.6 | 1.3 | 0.8 | 1.5 |
| | females | 6 | mean | 52.8 | 27.5 | 10.8 | 30.3 | 6.2 | 33.2 | 15.7 | 22.8 | 13.8 | 6.8 | 5.3 | 10.5 |
| | | | SD | 2.9 | 2.1 | 1.2 | 1.4 | 0.4 | 1.8 | 1.0 | 1.3 | 1.2 | 1.2 | 1.0 | 1.2 |
| 11, Rúa | males | 10 | mean | 62.9 | 28.1 | 11.8 | 25.9 | 7.5 | 38.3 | 16.2 | 22.6 | 17.5 | 5.8 | 5.8 | 8.4 |
| | | | SD | 2.8 | 2.7 | 0.8 | 1.7 | 0.8 | 3.2 | 1.0 | 1.6 | 1.0 | 0.9 | 0.9 | 1.3 |
| | females | 4 | mean | 61.8 | 26.8 | 11.5 | 29.5 | 6.3 | 34.0 | 14.5 | 22.0 | 15.8 | 6.3 | 5.3 | 8.0 |
| | | | SD | 1.7 | 1.0 | 1.3 | 0.6 | 0.5 | 2.7 | 1.7 | 1.4 | 1.3 | 1.0 | 0.5 | 0.8 |
| 12, Salvora | males | 3 | mean | 57.3 | 25.8 | 10.0 | 27.5 | 6.3 | 38.3 | 15.0 | 22.2 | 15.7 | 6.7 | 6.0 | 11.2 |
| | | | SD | 2.1 | 1.0 | 0.0 | 0.9 | 1.5 | 1.5 | 0.0 | 1.0 | 1.2 | 0.6 | 1.0 | 1.3 |
| | females | 3 | mean | 54.7 | 26.3 | 9.0 | 31.3 | 7.3 | 37.3 | 15.3 | 23.3 | 14.7 | 5.3 | 6.7 | 10.7 |
| | | | SD | 1.2 | 1.5 | 1.0 | 0.6 | 1.2 | 1.5 | 1.2 | 0.6 | 0.6 | 0.6 | 1.2 | 0.6 |
| 13, Vionta | males | 4 | mean | 51.8 | 27.8 | 10.0 | 27.0 | 7.5 | 34.3 | 15.3 | 22.0 | 16.3 | 5.5 | 5.5 | 8.3 |
| | | | SD | 1.5 | 2.1 | 0.8 | 2.0 | 0.6 | 2.4 | 1.2 | 2.0 | 1.0 | 1.0 | 1.0 | 2.2 |
| | females | 3 | mean | 51.0 | 27.0 | 9.7 | 30.7 | 7.3 | 34.0 | 15.5 | 22.7 | 14.3 | 6.0 | 6.0 | 7.3 |
| | | | SD | 3.6 | 2.0 | 1.2 | 1.5 | 1.2 | 1.0 | 0.7 | 2.3 | 0.6 | 0.0 | 1.0 | 1.5 |
| 16 and 17, mainland | males | 19 | mean | 55.4 | 28.2 | 9.5 | 26.7 | 7.6 | 36.7 | 15.9 | 23.7 | 15.8 | 6.5 | 5.8 | 9.4 |
| | | | SD | 3.3 | 2.1 | 1.1 | 1.0 | 0.8 | 2.3 | 0.9 | 1.9 | 1.5 | 1.2 | 0.6 | 2.4 |
| | females | 17 | mean | 54.6 | 26.0 | 10.4 | 30.2 | 6.9 | 33.1 | 15.6 | 23.8 | 14.5 | 6.8 | 5.7 | 8.4 |
| | | | SD | 3.3 | 2.5 | 1.1 | 1.6 | 0.8 | 2.6 | 0.7 | 1.2 | 1.6 | 1.4 | 0.5 | 1.2 |
| *Incertae sedis* | | | | | | | | | | | | | | | |
| 15f (RMNH 11066) | male | 1 | mean | 60 | 27 | 9 | 28 | 8 | 41 | - | - | 15 | 7 | 5 | 10 |

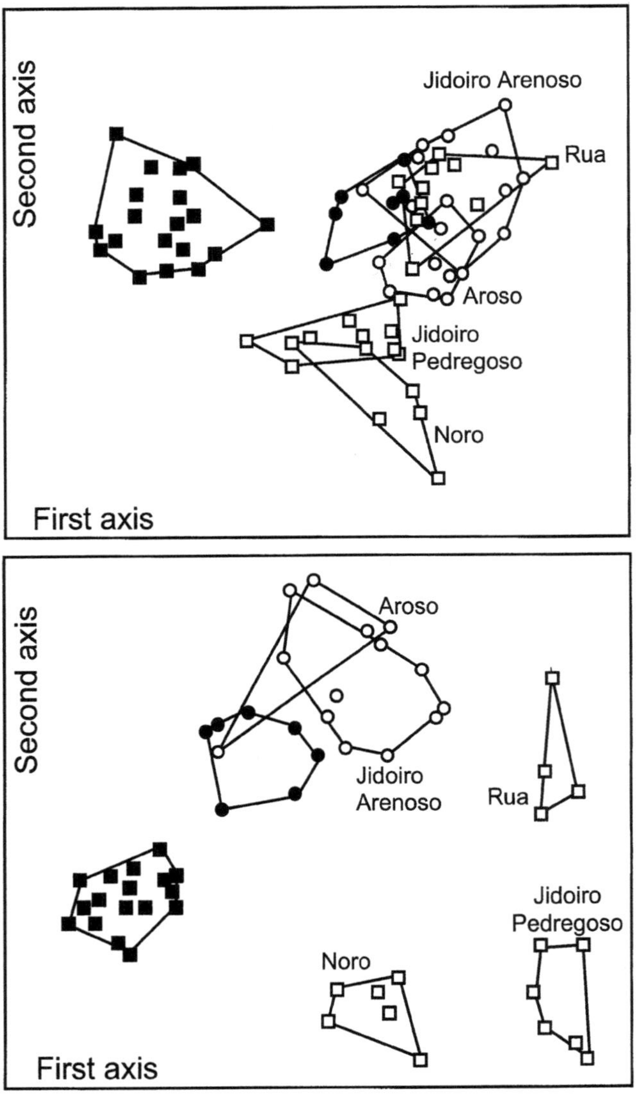


Fig. 5 Bivariate plot of the scores along the first and second discriminant function axis for *Podarcis bocagei* and *P. hispanica* from the mainland and selected island localities. The species are represented by round and square symbols, respectively; mainland and island localities are indicated by closed and open symbols, respectively. The five island localities that were included have a large distance to the mainland and fair sample sizes. Top panel: males; bottom panel: females.

on account of a more or less intermediate position along the species diagnostic first axis in a principal component analysis of nuclear genetic data (Fig. 4), and the presence of alien (*P. hispanica*) mtDNA within *P. bocagei*, as observed in the syntopic population of the southern mainland and on Jidoiro Arenoso. However, neither an intermediate nuclear genotype (i), nor the presence of alien mtDNA (ii), was coupled to an intermediate morphology. The results are consistent with interspecific hybridization being rare, even in syntopic populations. Instead of signalling hybridization, the alien mtDNA could represent an ancestral polymorphism and incomplete mtDNA lineage sorting. This explanation we consider unlikely given the 8.4% sequence divergence, corresponding to a long period over which the *P. bocagei* and *P. hispanica* lineages have been independent (cf. Bromham, 2002).

It must be kept in mind that the genetic systems studied display different modes of propagation and provide different, yet complementary information. Nuclear genes are subject to segregation and recombination. Under hybridization and subsequent introgression, marker genes may become swamped by the genetic background of the opposite species. Conversely, the mitochondrial genome behaves as a single, non-recombining, uniparentally (i.e., maternally) inherited unit (for a review see Piganeau et al., 2004). Upon continued back-crossing through the female lineage individuals carrying the mtDNA-molecule may become rare in the population, but the molecule and marker propensities remain intact.

The observed distribution of lizards (*P. bocagei* and *P. hispanica*) in the Ria de Arosa is essentially parapatric, on the mainland as well as over the archipelago. Syntopic populations were found in two mainland localities (localities "c" and "d"). The borderline between the species runs in north to south direction through both these points and across the archipelago, east of Jidoiro Arenoso and Toja Grande, and west of Benencia, Arosa, Jidoiro Pedregoso, Toja Pequeña, and Beiro (Fig. 1). Interestingly, this border is also the place were two genetically mixed individuals were found (locality "c" and Jidoiro Arenoso, respectively). Syntopy of *P. bocagei* and *P. hispanica* on the islands has not been observed (Galán, 2003a and present study) and we consider it an unlikely condition. Island syntopy would invoke interspecific competition on a small, restricted surface. Species population sizes would be smaller than they currently are, with chances for stochastic extinction increasing accordingly. However, the observation of alien mtDNA on Jidoiro Arenoso suggests that *P. bocagei* and *P. hispanica* have met in the past. The most parsimonious scenario to explain this observation is dispersal over about 500 m of one female *P. hispanica* from Jidoiro Pedregoso into Jidoiro Arenoso. Because the individual carrying the alien mtDNA is not a $F_1$ hybrid the genealogical distance is at least two generations. It must be noted that the detection of oversea dispersal in the Ria de Arosa is limited to the firmly parapatric parts of the species border (i.e., Jidoiro Arenoso versus Jidoiro Pedregoso; Toja Grande versus Beiro and Toja Pequeña) and the statistical power provided by the system is low. More highly variable genetic systems such as microsatellites (Pinho et al., 2004), capable of detecting intraspecific population differentiation, would allow us to gain better insight into the structured distribution of lizards over the archipelago. This would include cases in which islands are recolonized after stochastic population extinction (Foufopoulos and Ives, 1999). Islands currently vacant include Isla Ratas, Centolleiros and Turis (visited in the 1960's) and Malveira Chica and Isla Briñas (visited in 2002).

Absolute size as represented by SVL tends to be larger on islands than in mainland populations of the Ria de Arosa and more strongly so in *P. hispanica* than in *P. bocagei*. Increase in absolute size is commonly observed in onshore island populations of lizards, including the congeneric *Podarcis atrata*, *P. carbonelli*, *P. muralis*, *P. pityusensis* (Klemmer, 1964; Cheylan, 1988; Cirer and Martínez-Rica, 1990; Castilla et al., 1998a, b; Sá-Sousa et al., 2000). A possible correlate to large body size is the feeding on vegetable matter rather than animal prey (van Damme, 1999); examples include *Gallotia* and *Mabuya* lizards on the Atlantic archipelagos (Carranza et al., 2001). In Mediterranean *Podarcis* deviations from an insectivore diet (nectar, pollen, and fruits) appears to be opportunistic rather than structural (Pérez-Mellado and Corti, 1993; Pérez-Mellado and Traveset, 1999). However, large size could, in organisms with indeterminate growth such as lizards, also reflect old age, for example, associated with a reduced level of predation in island relative to mainland populations (Cirer and Martínez-Rica, 1990) but up to the present a relationship with predation or competition could not be demonstrated (Cheylan, 1988; Cirer and Martínez-Rica, 1990; Delaugerre and Cheylan, 1992; Vicente, 1999). To properly distinguish between the hypotheses would require a combined dietary and population demographic analysis.

Several relative size characters, in particular those describing head width (HW, OW, and FW) and hind limb length (HLL), show the tendency to be smaller on islands than on the mainland in both *P. bocagei* and *P. hispanica*. Populations of *Anolis* lizards that were experimentally introduced onto small islands showed a marked and extremely rapid change in hind limb length. The introduced populations had significantly diverged from the founder population and the changes appeared related to characteristics of the vegetation (Losos et al., 1997). On islands with large trees and bushes with thick branches, two species had by evolving larger relative hind limb length, whereas populations from islands with thinner vegetation had developed smaller limbs (Losos et al., 2001). The degree to which morphological change is genetically based or reflects environmentally driven phenotypic plasticity is currently under debate (Losos et al., 1998, 2000; Thorpe et al., 2005). In the Ria de Arosa, morphological differentiation of island populations (e.g., from Aroso) versus mainland populations is not pronounced in *P. bocagei*. On the other hand, island populations of *P. hispanica*, e.g., from Jidoiro Pedregoso and Noro and Rua, are morphologically rather different from one another and from the mainland. These populations appear to have gone different evolutionary avenues, even though they all inhabit similar, mostly rocky habitats. Either morphological change is mainly under genetic drift and not governed by phenotypic plasticity or natural selection (Cheylan, 1988; Vicente, 1999) or we fail to appreciate environmental differences between the localities. Field observations indicate that island population sizes are mostly small. Concomitantly small effective population sizes are in line with the reduced level of heterozygosity that we observed at nuclear genes, also suggesting genetic drift as the main operating evolutionary mechanism.

How and when did the *Podarcis* lizards colonize the various islands of the Ria de Arosa? Two hypotheses are that (i) the lizards were present on the proto-islands when the Ria was formed, and (ii) the islands were reached by overseas dispersal, either from neighbouring islands or from the mainland. These hypotheses are contrasting but not mutually exclusive, considering, for example, the recolonization of an island where the original island population has gone extinct. Support for the first hypothesis comes from the observed distribution of *P. bocagei* and *P. hispanica* that is consistently parapatric over islands and the mainland. This scenario would suggest that the currently observed pattern has been more or less stationary ever since the Ria de Arosa came into existence. Support for the second hypothesis comes from the mixed genotype observed in Jidoiro Arenoso, the origin of which we explained by the oversea dispersal of a female lizard from the neighbouring Jidoiro Pedregoso. Oversea dispersal has also been inferred in Caribbean *Anolis* (Calsbeek and Smith, 2003) and in *Podarcis* lizards over the Strait of Gibraltar (Harris et al., 2002).

A key observation to the understanding of the biogeographical history of the Ria de Arosa lizards is the presence of *P. hispanica* on the sandy island Vionta where environmental conditions would have predicted the occurrence of *P. bocagei*. On Vionta, *P. hispanica* lizards occupy the area with a few large stones and not the entire island as – we think – *P. bocagei* would have done (Galán, 2003a; personal observations). Other large islands that have no *P. bocagei* despite the presence of seemingly suitable habitat (but up to seven other reptile species, Galán, 2003a, b) are the greater islands of Ons (428 ha) and Cíes (433 ha) in the outer part of the neighbouring Ria de Vigo. Two hypotheses explaining the absence of *P. bocagei* on those islands are that (i) *P. bocagei* is a long time resident of western Galicia but could not reach vacant habitats lying across the parapatric species border, and (ii) *P. bocagei* is a recent arrival that only reached western Galicia when the more remote islands of the archipelago had already formed. A third hypothesis, in which *P. hispanica* is a late arrival (i.e., more recent than *P. bocagei*) can be directly dismissed because *P. hispanica* is present on the more remote islands and absent from the less remote ones. Assessing the plausibility of the first "spatial exclusion" hypothesis requires knowledge on the ecological relationships of *P. bocagei* and *P. hispanica* and the conditions under which sympatry or parapatry would prevail. The second "recent arrival" hypothesis might be tested through the reconstruction of the comparative phylogeography of *P. bocagei* and *P. hispanica*. The reconstruction of a glacial refugium in western Galicia exclusive for *P. bocagei* would be incompatible with this hypothesis.

The two *Podarcis* species involved exhibit differential ecologies on the mainland. *Podarcis bocagei* is a ground-dwelling lizard that prefers agricultural areas, paths, and clearings in woodland plots, meadows, and sandy areas, whereas *P. hispanica* is a thermophilous rock-dwelling species that also colonizes stone structures made by humans, such as dry stone walls, ruins, and quarries (Galán, 1986, 2002, 2003b; Sá-Sousa and Pérez-Mellado, 2002). In western Galicia, *P. hispanica* predominates or is the sole species present in the exposed areas of the Rias Bajas, along the caps and peninsulas, including, Corrubedo, O Grove, Punta del Couso, and Silleiro, at the slopes of the Castrove and Los Paramos mountains and the Sierra de Barbanza. *Podarcis bocagei* is often present in the river valleys and floodplains that drain the area east of the Ria de Arosa (Fig. 1; Galán, 2003a). Even if habitat segregation produces a parapatric distribution, such as appears to be the case on mainland Ria de Arosa, at a larger scale the species have broadly overlapping ranges, with scattered cases of syntopy (Galán, 2002, 2003a, b). On the whole habitat partitioning – not distributional displacement – is the major factor causing spatial segregation of *P. bocagei* and *P. hispanica*. This indeed is the common pattern in lizards and in lacertid lizards such as *Podarcis* in particular (Arnold, 1987; Pianka 1993; Mayer and Beyerlein, 1999). We therefore doubt that the parapatry of *P. bocagei* and *P. hispanica* observed in the Ria de Arosa is more than a chance effect. To settle this issue a more thorough survey is required, focussing on habitats favouring the presence of the species yet undisclosed. If the two species distribution in the Ria de Arosa is mixed and mosaic and spatial exclusion does not apply, then the absence of *P. bocagei* on Vionta and other islands is, by default, best explained by the its late arrival in the area. Positive support for the late arrival hypothesis is provided by a phylogeographic analysis of *P. bocagei* on the basis of enzyme, microsatelite and mitochondrial genetic data that suggests that the last glacial refugium of the species was located in northern Portugal and that the current distribution in western Galicia represents a recent range expansion (Pinho et al., 2002; and Pinho, personal communication 2005). The most likely epoch for *P. bocagei* to have colonized the eastern part of the Ria de Arosa is the "Little Ice Age" event (cf. Pethick, 1984; Dias et al., 2000), which involved a temporary drop in the sea level with which islands within the current about 10 m isobath became temporarily reconnected to the mainland.

## Acknowledgements

We thank S. Busack (Raleigh, USA), M.S. Hoogmoed (Belém, Brazil) and C. Pinho (Porto, Portugal), who contributed with unpublished information on DNA-primers, wildlife and enzymes, respectively. R. Glas (Leiden, The Netherlands) and L. de Groot (Amsterdam, The Netherlands) prepared the DNA-sequences, and W. van Ginkel (Amsterdam) assisted with enzyme electrophoresis. N. Baptista (Évora, Portugal) assisted with morphological databases. Collection was done under license nr. 16/2002 supplied by the "Servicio de Conservación da Natureza" (V. Piorno, Pontevedra, Spain) with assistance of M.S. Hoogmoed and J.M. Oliveira (Coimbra, Portugal).

## References

Areválo, E., Davis, S.K., and Sites J.W., 1994, Mitochondrial-DNA sequence divergence and phylogenetic-relationships among 8 chromosome races of the *Sceloporus grammicus* complex (Phrynosomatidae) in central Mexico, *Systematic Biology* **43**: 387–418.

Arnold, E.N., 1987, Resource partition among lacertid lizards in southern Europe, *Journal of Zoology B, London* **1**: 739–782.

Arntzen, J.W. and García-París, M., 1995, Morphological and allozyme studies on midwife toads (genus *Alytes*), including the description of two new taxa from Spain, *Contributions to Zoology* **65**: 5–34.

Austin, C.C., 1999, Lizards took express train to Polynesia, *Nature* **397**: 113–114.

Bard, E., Hamelin, B., Arnold, M., Montaggioni, L., Cabioch, G., Faure, G., and Rougerie, F., 1996, Deglacial sea-level record from Tahiti corals and the timing of global meltwater discharge, *Nature* **382**: 241–244.

Böhme, W., 1986, *Podarcis wagleriana* (Gistel 1868) – Sizilianische Mauereidechse, in: Böhme, W. (ed.), *Handbuch der Amphibien und Reptilien Europas (vol. 2/II - Sauria III - Podarcis)*, Aula-Verlag, Wiesbaden, Germany, pp. 377–387.

Brehm, A., Khadem, M., Jesus, J., Andrade, P., and Vicente, L., 2001, Lack of congruence between morphometric evolution and genetic differentiation suggests a recent dispersal and local habitat adaptation of the Madeiran lizard *Lacerta dugesii*, *Genetics Selection and Evolution* **33**: 671–685.

Bromham, L., 2002, Molecular clocks in reptiles: life history influences rate of molecular evolution, *Molecular Biology and Evolution* **19**: 302–309.

Brongersma, L.D. and Pannekoek, A.J., 1966, Investigations in and around the Ria de Arosa, north-west Spain, 1962–1964, *Leidse Geologische Mededelingen* **37**: 1–5.

Cadée, G.C., 1968, Molluscan biocoenoses and thanatocoenoses in the Ria de Arosa, Galicia, Spain, *Zoologische Verhandelingen* **95**: 1–121.

Calsbeek, R. and Smith, T. B., 2003, Ocean currents mediate evolution in island lizards, *Nature* **426**: 552–555.

Capula, M. and Ceccarelli, A., 2003, Distribution of genetic variation and taxonomy of insular and mainland populations of the Italian wall lizard, *Podarcis sicula*, *Amphibia-Reptilia* **24**: 483–495.

Capula, M., 1994a, Genetic variation and differentiation in the lizard, *Podarcis wagleriana* (Reptilia, Lacertidae), *Biological Journal of the Linnean Society* **52**: 177–196.

Capula, M., 1994b, Evolutionary relationships of *Podarcis* lizards from Sicily and the Maltese islands, *Journal Zoological of Systematics and Evolutionary Research* **32**: 180–192.

Capula, M., 1996, Evolutionary genetics of the insular lacertid lizard *Podarcis tiliguerta*: genetic structure and population heterogeneity in a geographically fragmented species, *Heredity* **77**: 518–529.

Capula, M., 1997, High genetic variability in insular populations of the lacertid *Podarcis muralis*, *Biochemical Systematics and Ecology* **25**: 411–417.

Carranza, S., Arnold, E.N., Mateo, J.A., and López-Jurado, L.F., 2001, Parallel gigantism and complex colonization patterns in the Cape Verde scincid lizards *Mabuya* and *Macroscincus* (Reptilia: Scincidae) revealed by mitochondrial DNA sequences, *Proceedings of the Royal Society of London, B* **268**: 1595–1603.

Castilla, A.M., Fernández-Pedrosa, V., Harris, D.J., González, A., Latorre, A., and Moya, A., 1998a, Mitochondrial DNA divergence suggests that *Podarcis hispanica atrata* (Squamata: Lacertidae) from the Columbretes islands merits specific distinction, *Copeia* **1998**: 1037–1040.

Castilla, A.M., Fernandez-Pedrosa, V., Backeljau, T., Gonzalez, A., Latorre, A., and Moya, A., 1998b, Conservation genetics of insular *Podarcis* lizards using partial cytochrome b sequences, *Molecular Ecology* **7**: 1407–1411.

Cheylan, M., 1988, Variabilité phénotipique du lézard des murailles *Podarcis muralis* sur les îles de la côte provençale, France, *Revue d'Ecologie, Terre et Vie* **43**: 287–321.

Cirer, A.M. and Guillaume, C.-P., 1986, Electrophoretic analysis of the Pityusic Islands lizard, in: Rocek, Z. (ed.), *Studies in Herpetology*, Charles University, Prague, pp. 201–206.

Cirer, A.M. and Martínez-Rica, J.P., 1990, The polymorphism of *Podarcis pityusensis* and its adaptative evolution in Mediterranean Isles, *Herpetological Journal* **1**: 465–473.

Cirer, A.M., 1987, Aplicación de técnicas estadísticas multivariantes a las poblaciones del lacértido *Podarcis pityusensis* (Boscà, 1883), *Revista Española de Herpetología* **2**: 145–163.

Clover, R.C., 1979, Phenetic relationships among populations of *Podarcis sicula* and *P. melisellensis* (Sauria, Lacertidae) from islands in the Adriatic Sea, *Systematic Zoology* **28**: 284–298.

Damme, R. van, 1999, Evolution of herbivory in lacertid lizards: effects of insularity and body size, *Journal Herpetology* **33**: 663–674.

Delaugerre, M. and Cheylan, M., 1992, *Atlas de répartition des batraciens et reptiles de Corse*. Parc Naturel Regional de Corse, École Pratique des Hautes Études de Montpellier, Pamplona, Spain.

Dengo G. and Case J.E., (eds), 1990, *The Geology of North America, Volume H, The Caribbean Region*, Geological Society of America, Boulder, CO.

Dias, J.M.A., Boski, T., Rodrigues, A., and Magalhães, F., 2000, Coast line evolution in Portugal since the Last Glacial Maximum until present – a synthesis, *Marine Geology* **170**: 177–186.

Donovan, S.K. and Jackson T.A. (eds), 1994, *Caribbean Geology. An Introduction*, University of the West Indies Publishers' Association, Kingston, Jamaica.

Foufopoulos, J. and Ives, A.R., 1999, Reptile extinctions on land-bridge islands: life-history attributes and vulnerability to extinction, *American Naturalist* **153**: 1–25.

Galán, P., 1985, Morfología de las *Podarcis bocagei* (Seoane, 1884) (Sauria, Lacertidae) de dos islas atlanticas ibéricas: Sisargas y Berlengas, *Trabajos Compostelanos de Biología* **12**: 121–139.

Galán, P., 1986, Morfología y distribución del genero *Podarcis*, Wagler 1830 (Sauria, Lacertidae) en el noroeste de la Península Ibérica, *Revista Española de Herpetología* **1**: 87–142.

Galán, P., 1999a, Contribución al conocimiento de la herpetofauna de las islas de Galicia: inventario faunístico y notas sobre la ecología y morfología de algunas poblaciones, *Chioglossa* **1**: 147–163.

Galán, P., 1999b, Demography and population dynamics of the lacertid lizard *Podarcis bocagei*, *Journal of Zoology, London* **249**: 203–218.

Galán, P., 2002, *Podarcis bocagei*, in: Pleguezuelos, J. M., Márquez, R., and Lizana, M. (eds), *Atlas y Libro Rojo de los Anfíbios y Reptiles de España*, Dirección General de la Naturaleza, Ministerio de Medio Ambiente, Asociación Herpetológica Española, Madrid, pp. 239–241.

Galán, P., 2003a, *Anfibios y reptiles del parque nacional de las islas atlánticas de Galicia: faunística, biología y conservación*. Ministerio de Medio Ambiente, Madrid.

Galán, P., 2003b, Lagartija de Bocage – *Podarcis bocagei*, in: Carrascal, L. M. and Salvador, A. (eds), *Enciclopedia Virtual de los Vertebrados Españoles*, Museo Nacional de Ciencias Naturales, Madrid. http://www.vertebradosibericos.org/

Gorman, G.C., Soule, M., Yang, S.Y., and Nevo, E., 1975, Evolutionary genetics of insular Adriatic lizards, *Evolution* **29**: 52–71.

Gübitz, T., Thorpe, R.S., and Malhotra, A., 2000, Phylogeography and natural selection in the Tenerife gecko *Tarentola delalandii*: testing historical and adaptive hypotheses, *Molecular Ecology* **9**: 1213–1221.

Gübitz, T., Thorpe, R.S., and Malhotra, A., 2005, The dynamics of genetic and morphological variation on volcanic islands, *Proceedings of the Royal Society, B* **272**: 751–757.

Harris, D.J. and Sá-Sousa, P., 2001, Species distinction and relationships of the western Iberian *Podarcis* lizards (Reptilia, Lacertidae) based on morphology and mitochondrial DNA sequences, *Herpetological Journal* **11**: 129–136.

Harris, D.J. and Sá-Sousa, P., 2002, Molecular phylogenetics of Iberian wall lizards (*Podarcis*): is *Podarcis hispanica* a species complex? *Molecular Phylogenetics and Evolution* **23**: 75–81.

Harris, D.J., Carranza, S., Arnold, E.N., Pinho, C., and Ferrand, N., 2002, Complex biogeographical distribution of genetic variation within *Podarcis* wall lizards across the Strait of Gibraltar, *Journal of Biogeography* **29**: 1257–1262.

Henle, K. and Klaver, J.J., 1986, *Podarcis sicula* (Rafinesque-Schalmtz, 1810) Ruineneidechse, in: Böhme, W. (ed.), *Handbuch der Amphibien und Reptilien Europas (vol. 2/II - Sauria III - Podarcis)*, Aula-Verlag, Wiesbaden, Germany, pp. 254–342.

Irschick, D.J., Vitt, L.J., Zani, P.A., and Losos, J.B., 1997, A comparison of evolutionary radiations in mainland and Caribbean Anolis lizards, *Ecology* **78**: 2191–2203.

Jackman, T.R., Larson, A., Queiroz, K. de, and Losos, J.B., 1999, Phylogenetic relationships and tempo of early diversification in *Anolis* lizards, *Systematic Biology* **48**: 254–285.

Jackson, T.A. (ed.), 2002, *Caribbean Geology into the Third Millennium – Transactions of the Fifteenth Caribbean Geological Conference*, University of the West Indies Press, Kingston, Jamaica.

Klemmer, K., 1964, Die westlichen Randformen der Mauereidechse *Lacerta muralis* (Reptilia, Lacertidae), *Senckenbergiana Biologica* **45**: 491–499.

Kocher, T.D., Thomas, W.K., Meyer, A., Edwards, S.V., Paäbo, S., Villablanca, F.X., and Wilson, A.C., 1989, Dynamics of mitochondrial evolution in animals: amplification and sequencing with conserved primers, *Proceedings National Academy Sciences, USA* **86**: 6196–6200.

Lanza, B. and Poggesi, M., 1986, Storia naturale selle isole satelliti della Corsica, *L'Universo, Firenze* **66**: 1–198.

Losos, J.B. and Spiller, D.A., 1999, Differential colonization success and asymmetrical interactions between two lizard species, *Ecology* **80**: 252–258.

Losos, J., 1992, The evolution of convergent structures in Caribbean *Anolis* communities, *Systematic Biology* **41**: 403–420.

Losos, J.B., Creer, D.A., Glossip, D., Goellner, R., Hampton, A., Roberts, G., Hasskell, N., Taylor, P., and Ettling, J. 2000, Evolutionary implications of phenotypic plasticity in the hindlimb of the lizard *Anolis sagrei*, *Evolution* **54**: 301–305.

Losos, J.B., Jackman, T.R., Larson, A., Queiroz, K. de, and Rodriguez-Schettino, L., 1998, Contingency and determinism in replicated adaptive radiations of island lizards, *Science* **279**: 2115–2118.

Losos, J.B., Schoener, T.W., Warheit, K.I., and Creer, D., 2001, Experimental studies of adaptive radiation in Bahamian *Anolis* lizards, *Genetica* **112**: 399–415.

Losos, J.B., Warheit, K.I., and Schoener, T.W., 1997, Adaptive differentiation following experimental island colonization in *Anolis* lizards, *Nature* **387**: 70–73.

Maddison D.R. and Maddison, W.P., 2000, *MacClade 4.0. Computer Program*, Sinauer Associates, Sunderland, MA.

Mateo, J.A., 1997, Las islas e islotes del litoral ibérico, in: Pleguezuelos, J. M. (eds), *Distribución y Biogeografía de los Anfibios y Reptiles en España y Portugal. Monografías de Herpetología, 3*, Editorial Universidad de Granada and Asociación Herpetológica Española, Granada, pp. 343–350.

Mayer, W. and Beyerlein, P., 1999, Ecological niche segeregation of seven sympatric lacertid lizards in the Peloponnese highlands, *Natura Croatica* **8**: 339–344.

Nonn, H., 1966, Les régions cotières de la Galice (Espagne): étude géomorphologique, *Publications de la Faculté des Lettres de l'Université de Strasbourg* **3**: 1–591.

Otto, L., 1975, Oceanography of the Ría de Arosa (N.W. Spain), *Mededelingen en Verhandelingen, Koninklijk Nederlands Meteorologisch Instituut* **96**: 1–211.

Pannekoek, A.J., 1966a, The Ria problem. The role of antecedence, deep weathering, and Pleisocene slope-wash in the formation of the west-Galician Rias, *Tijdschrift Koninklijk Nederlands Aardrijkskundig Genootschap* **83**: 289–297.

Pannekoek, A.J., 1966b, The geomorphology of the surroundings of the Ria de Arosa (NW Spain), *Leidse Geologische Mededelingen* **37**: 7–32.

Pannekoek, A.J., 1970, Additional geomorphological data on the ria area of western Galicia (Spain), *Leidse Geologische Mededelingen* **37**: 185–194.

Pérez-Mellado, V. and Corti, C., 1993, Dietary adaptations and herbivory in lacertids lizards of the genus *Podarcis* from Western Mediterranean islands (Reptilia: Sauria), *Bonner Zoologische Beiträge* **44**: 193–220.

Pérez-Mellado, V. and Traveset, A., 1999, Relationships between plants and Mediterranean lizards, *Natura Croatica* **8**: 275–285.

Pethick, J., 1984, *An introduction to coastal geomorphology*. Edward Arnold, London.

Pianka, E.R., 1993, The many dimensions of a lizard's ecological niche, in: Valakos, E. D., Böhme, W., Pérez-Mellado, V., and Maragou, P. (eds), *Lacertids of the Mediterranean Region*, Hellenic Zoological Society, Athens, Bonn, Alicante, pp. 121–154.

Piganeau, G., Gardner, M., and Eyre-Walker, A., 2004, A broad survey of recombination in animal mitochondria, *Molecular Biology Evolution* **21**: 2319–2325.

Pinho, C., Harris, D.J., and Ferrand, N., 2002, Post-glacial expansion of *Podarcis bocagei*: evidence from protein electrophoretic data. *Abstracts of the 7th Portuguese-Spanish Congress of Herpetology*, Portuguese Herpetological Society, Évora, p. 147.

Pinho, C., Harris, D.J., and Ferrand, N., 2003, Genetic polymorphism of 11 allozyme loci in populations of wall lizards (*Podarcis* sp.) from the Iberian Peninsula and North Africa, *Biochemical Genetics* **41**: 343–359.

Pinho, C., Sequeira, F., Godinho, R., Harris, D.J., and Ferrand, N., 2004, Isolation and characterization of nine microsatellite loci in *Podarcis bocagei* (Squamata: Lacertidae), *Molecular Ecology Notes* **4**: 286–288.

Podnar, M., Mayer, W., and Tvrtkovic, N., 2004, Mitochondrial phylogeography of the Dalmatian wall lizard, *Podarcis melisellensis* (Lacertidae), *Organisms, Diversity and Evolution* **4**: 307–317.

Podnar, M., Mayer, W., and Tvrtkovic, N., 2005, Phylogeography of the Italian wall lizard, *Podarcis sicula* as revealed by mitochondrial DNA sequences, *Molecular Ecology* **14**: 575–588.

Poulakakis, N., Lymberakis, P., Antoniou, A., Chalkia, D., Zouros, E., Mylonas, M., and Valakos, E., 2003, Molecular phylogeny and biogeography of the wall-lizard *Podarcis erhardii* (Squamata: Lacertidae), *Molecular Phylogenetics and Evolution* **28**: 38–46.

Ramil-Rego, P., and Gomez Orellana, L., 1996, Dinamica climatica y biogeografica del area litoral-sublitoral de Galicia durante el Pleistoceno superior-Holocene, in: Ramil-Rego, P., Fernández-Rodríguez, C., and Rodríguez Guitian, M., (eds), *Biogeografia Pleistocena - Holocena de la Península Ibérica*, Consellería de Cultura, Xunta de Galicia, Santiago de Compostela, Spain pp. 43–71.

Ramón, M.M., Castro, J.A., and Arranz, M.J., 1991, A study of the Balearic lizard *Podarcis lilfordi* based on morphological characters and enzyme polymorphisms, *Evolución Biológica* **5**: 75–92.

Raufaste, N. and Rousset, F., 2002, Are partial Mantel tests adequate? *Evolution* **56**: 1871–1873.

Raymond, M. and Rousset, F., 1995, GenePop (version 1.2) - population genetics software for exact tests and ecumenicism, *Journal of Heredity* **86**: 248–249.

Rodríguez-Guitian, M., Ramil-Rego, P., Muñoz-Sobrino, C., and Gómez-Orellana, L., 1996, *Consideraciones sobre la migracion holocena de Fagus a traves de la 'Via Pirenaico Cantabrica'*, in: Ramil-Rego, P., Fernández-Rodríguez, C., and Rodríguez Guitian, M., (eds), *Biogeografia Pleistocena - Holocena de la Península Ibérica*. Consellería de Cultura, Xunta de Galicia, Santiago de Compostela, Spain, pp. 98–114.

Rohlf, F.J., 1992, *NTSYS-1.70. Numerical Taxonomy and Multivariate Analysis System. Computer Program*, Exeter, NY.

Roughgarden, J., 1995, *Anolis Lizards of the Caribbean: Ecology, Evolution, and Plate Tectonics*, Oxford Series in Ecology and Evolution, Oxford University Press, NY.

Sá-Sousa, P. and Harris, D.J., 2002, *Podarcis carbonelli* Pérez-Mellado, 1981 is a distinct species, *Amphibia-Reptilia* **23**: 459–468.

Sá-Sousa, P. and Pérez-Mellado, V., 2002, *Podarcis hispanica*, in: Pleguezuelos, J. M., Márquez, R., and Lizana, M., (eds), *Atlas y Libro Rojo de los Anfíbios y Reptiles de España*, Dirección General de la Naturaleza, Ministerio de Medio Ambiente, Asociación Herpetológica Española, Madrid, pp. 244–245.

Sá-Sousa, P., 2000, A predictive distribution model for the Iberian wall lizard (*Podarcis hispanicus*) in Portugal, *Herpetological Journal* **10**: 1–11.

Sá-Sousa, P., Almeida, A.P., Rosa, H.D., Vicente, L.A., and Crespo, E.G., 2000, Genetic and morphological relationships of the Berlenga wall lizard (*Podarcis bocagei berlengensis*: Lacertidae), *Journal of Zoological Systematics and Evolutionary Research* **38**: 95–102.

Sá-Sousa, P., Vicente, L., and Crespo, E.G., 2002, Morphological variability of *Podarcis hispanica* (Sauria: Lacertidae) in Portugal, *Amphibia-Reptilia* **23**: 55–69.

Schneider, C.J., Losos, J.B., and Queiroz, K. de, 2001, Evolutionary relationships of the *Anolis bimaculatus* group from the northern Lesser Antilles, *Journal of Herpetology* **35**: 1–12.

Schneider, S., Roessli, D., and Excoffier, L., 2000, *Arlequin 2.0. A Software for Population Genetics Data Analysis*, University of Geneva, Geneva, Switzerland.

Schoener, T.W., Spiller, D.A., and Losos, J.B., 2001, Natural restoration of the species-area relation for a lizards after a hurricane, *Science* **294**: 1525–1528.

Shaw, C.R. and Prasad, R., 1970, Starch gel electrophoresis of enzymes - a compilation of recipes, *Biochemical Genetics* **4**: 297–320.

Spiller, D.A., Losos, J.B., and Schoener, T.W., 1998, Impact of a catastrophic hurricane on island populations, *Science* **281**: 695–697.

SPSS, 2003, *Statistical Package for the Social Sciences 12. Computer program.* SPSS,Chigago.

Swofford, D.L. and Selander, R.B., 1981, Biosys-1: a FORTRAN program for the comprehensive analysis of electrophoretic data in population genetics and systematics, *Journal of Heredity* **72**: 281–283.

Swofford, D.L., 2002, *PAUP*. Computer Program*, Sinauer Associates, Sunderland, MA.

Templeton, A.R., 1980, The theory of speciation via the founder principle, *Genetics* **94**: 1011–1038.

Thompson, J.D., Higgins, D.G., and Gibson, T.J., 1994, CLUSTAL W: improving the sensitivity of progressive multiple sequence alignment through sequence weighting, positions-specific gap penalties and weight matrix choice, *Nucleic Acids Research* **22**: 4673–4680.

Thorpe, R.S. and Malhotra, A., 1992, Are *Anolis* lizards evolving? *Nature* **355**: 505–506.

Thorpe, R.S. and Stenson, A.G., 2003, Phylogeny, paraphyly and ecological adaptation of the colour and pattern in the *Anolis roquet* complex on Martinique, *Molecular Ecology* **12**: 117–132.

Thorpe, R.S., Black, H., and Malhotra, A., 1996, Matrix correspondence tests on the DNA phylogeny of the Tenerife lacertid elucidate both historical causes and morphological adaptation, *Systematic Biology* **45**: 335–343.

Thorpe, R.S., Reardon, J.T., and Malhotra, A., 2005, Common garden and natural selection experiments support ecotypic differentiation in the Dominican anole (*Anolis oculatus*), *American Naturalist* **165**: 495–504.

Vicente, L., 1999, Evolutionary strategies in insular environments, *Natura Croatica* **8**: 301–323.

Williams, E.E., 1972, The origin of faunas. Evolution of lizard congeners in a complex island fauna: a trial analysis, *Evolutionary Biology* **6**: 47–89.

Williams, E.E., 1983, Ecomorphs, faunas, island size, and diverse end points in island radiations of *Anolis*, in: Huey, R. B., Pianka, E. R., and Schoener, T. W., (eds), *Lizard Ecology: Studies of a Model Organism*, Harvard University Press, Cambridge, MA, pp. 326–370.

# Appendix 1.

Lizards figuring in the present study, investigated for cytoplasmic (c), nuclear (n), and morphological (m) characteristics. Island and mainland localities are shown in Fig. 1 by numbers (1–13) and letters (a–o), respectively. Geographical coordinates were taken in the field with a GPS. All material is deposited at the Nationaal Natuurhistorisch Museum Naturalis, Leiden, The Netherlands, the museum code of which is "RMNH".

## *Podarcis bocagei*

Islands – 1. Arosa (42°32('449 N, 8°52′528 W), males RMNH 11076–9 (m), 35198–9 (c, n, m), 35201–2 (c, n, m), females 11074–5 (m), 35200 (c, n, m), 35203 (c, n, m), Arosa, Mirador, Con do Forno (42°33′945 N, 8°52′784 W), male 35204 (c, n, m), Arosa, East end of village (42°33′448 N, 8°52′089 W), female 35205 (c, n, m);

2. Beiro, males 11161–4 (m), females 11165–7 (m);
3. Benencia (42°36′107 N, 8°52′619 W), males 11200–3 (m), 35292–3 (c, n, m), females 11204–6 (m), 35294 (c, n, m);
4. Cortegada, males 11147 (m), 12936–7 (m), 12940 (m), females 11148 (m), 12938–9 (m);
5. Jidoiro Arenoso (42°32′445 N, 8°54′117 W), males 11114–5 (m), 11120 (m), 11122 (m), 35207–11 (c, n, m), 35249–50 (c, n, m), females 11116–8 (m), 11121 (m), 11169–71 (m), 35206 (c, n, m), 35212 (c, n, m), 35248 (c, n, m);
6. San Bartolomé (Malveira Grande) (42°36′692 N, 8°47′921 W), males 11152 (m), 35295 (c, n, m), females 11149–51 (m);
7. Toja Pequeña (42°29′462 N, 8°50′061 W), males 11153–6 (m), 35261–2 (c, n, m) and females 11157–9 (m).

14. Mainland north – (a) Bridge over Rio Ulla, near Puentecesures, males 11071–2 (m);
    (b) Playa de Louro, N. of Ria de Muros y Noia (42°45′417 N, 9°06′269 W), male 35214 (c, n, m);
    (c) Mirador de la Curotá near windmill plant of Serra de Barbanza(42°38′936 N, 8°57′948 W), male 35307 (c, n) and female 35308 (c, n).
15. Mainland south – (d) Between Mostero da la Armenteira and Mostero del Poio (42°27′654 N, 8°41′925 W), males 35328 (c, n), 35336–7 (c, n, m), females 35327 (c, n, m), 35330–1 (c, n, m);
    (e) Mostero da la Armenteira, NE of Sanxenxo (42°27′862 N, 8°44′484 W), males 35321–2 (c, n, m), female 35323–5 (c, n, m), 35340–1 (c, n), juvenile 35326 (c, n);
    (f) Punta Preguntoiro, female 11068 (m);
    (g) quarry near Caldas de Reyes, male 11494 (m).

## *Podarcis hispanica*

Islands – 8. Coroso (42°33′950 N, 8°58′172 W), males RMNH 11207 (m), 35285–9 (c, n, m), females 35290–1 (c, n, m);

9. Jidoiro Pedregoso (42°32′784 N, 8°54′918 W), males 11105–9 (m), 11112 (m), 35217 (c, n, m), 35219–20 (c, n, m), females 11110–1 (m), 11113 (m), 35215–6 (c, n, m), 35218 (c, n, m);
10. Noro (42°29′301 N, 8°59′863 W), males 35269–74 (c, n, m), females 35263 (c, n, m), 35264–8 (c, n, m);
11. Rua (42°32′978 N, 8°56′391 W), males 35221 (c, n, m), 35223–4 (c, n, m), 35225 (c, n), 35226 (c, n, m), 35227 (c, n), 35229–30 (c, n, m), 35232–3 (c, n), 35235–9 (c, n), 35240–1 (c, n, m), 35242–3 (c, n), 35244–5 (c, n, m), females 35222 (c, n, m), 35228 (c, n), 35231 (c, n, m), 35234 (c, n), 35246–7 (c, n, m);
12. Sálvora (42°28′103 N, 9°00′140 W), males 11495 (m), 35280–1 (c, n, m), females 35282–4 (c, n, m);
13. Vionta (42°29′968 N, 9°00′265 W), males 11199 (m), 35275–7 (c, n, m), females 11198 (m) and 35278–9 (c, n, m).
16. Mainland north – (c) Mirador de la Curotá near windmill plant of Serra de Barbanza (42°38′936 N, 8°57′948 W), female 35342 (c, n, m);
    - (h) Between Vilar and Corrubedo, males 11043 (m), 11045–7 (m), females 11042 (m);
    - (i) Castineiras, slope of Mount Los Paramos, female 11139 (m);
    - (j) Mirador de la Curotá (42°37′399 N, 8°57′659 W), males 35296–8 (c, n, m), 35300 (c, n), 35302–4 (c, n), females 35299 (c, n, m), 35301 (c, n), 35305 (c, n) and 35306 (c, n, m).
17. Mainland south – (d) Between Mostero da la Armenteira and Mostero del Poio (42°27′654 N, 8°41′925 W), males 35329 (c, n, m), 35332 (c, n, m), 35333–4 (c, n), 35338 (c, n), females 35335 (c, n), 35339 (c, n, m);
    - (k) Mirador, Con de la Siradella, Peninsula del Grove (42°28′256 N, 8°54′854 W), male 35213 (c, n);
    - (l) Playa de Lanzada, male 11062 (m), female 11064–5 (m),
    - (m) Punta Moreiras, Peninsula del Grove, (42°29′114 N, 8°53′421 W), female 35309 (c, n);
    - (n) Punta San Vincente, Peninsula del Grove, males 11052–3 (m), 11055–6 (m), females 11057–60 (m);
    - (o) quarry in Coto de Caza San Martin near Ardia (42°27′316 N, 8°52′466 W), males 35251 (c, n, m), 35252 (c, n), 35253 (c, n, m), 35254 (c, n), 35311–3 (c, n), 35315–7 (c, n, m), 35318–9 (c, n), females 35255–6 (c, n), 35257–9 (c, n, m), 35260 (c, n), 35310 (c, n, m), 35314 (c, n) and 35320 (c, n, m).

## *Incertae sedis*, i.e., taxonomic affiliation uncertain

15. Mainland south – (f) Punta Preguntoiro, male 11066 (m).

# Index

MIX
Papier aus verantwortungsvollen Quellen
Paper from responsible sources
FSC® C105338

If you have any concerns about our products,
you can contact us on
**ProductSafety@springernature.com**

In case Publisher is established outside the EU,
the EU authorized representative is:
**Springer Nature Customer Service Center GmbH**
**Europaplatz 3, 69115 Heidelberg, Germany**

Printed by Libri Plureos GmbH
in Hamburg, Germany